Studies in Classification, Data Analysis, and Knowledge Organization

Springer
Berlin
Heidelberg
New York
Barcelona
Hong Kong
London
Milan
Paris
Tokyo

Titles in the Series

H.-H. Bock and P. Ihm (Eds.)
Classification, Data Analysis, and Knowledge Organization. 1991 (out of print)

M. Schader (Ed.)
Analyzing and Modeling Data and Knowledge. 1992

O. Opitz, B. Lausen, and R. Klar (Eds.)
Information and Classification. 1993 (out of print)

H.-H. Bock, W. Lenski, and M. M. Richter (Eds.)
Information Systems and Data Analysis. 1994 (out of print)

E. Diday, Y. Lechevallier, M. Schader, P. Bertrand, and B. Burtschy (Eds.)
New Approaches in Classification and Data Analysis. 1994 (out of print)

W. Gaul and D. Pfeifer (Eds.)
From Data to Knowledge. 1995

H.-H. Bock and W. Polasek (Eds.)
Data Analysis and Information Systems. 1996

E. Diday, Y. Lechevallier and O. Opitz (Eds.)
Ordinal and Symbolic Data Analysis. 1996

R. Klar and O. Opitz (Eds.)
Classification and Knowledge Organization. 1997

C. Hayashi, N. Ohsumi, K. Yajima, Y. Tanaka, H.-H. Bock, and Y. Baba (Eds.)
Data Science, Classification, and Related Methods. 1998

I. Balderjahn, R. Mathar, and M. Schader (Eds.)
Classification, Data Analysis, and Data Highways. 1998

A. Rizzi, M. Vichi, and H.-H. Bock (Eds.)
Advances in Data Science and Classification. 1998

M. Vichi and O. Opitz (Eds.)
Classification and Data Analysis. 1999

W. Gaul and H. Locarek-Junge (Eds.)
Classification in the Information Age. 1999

H.-H. Bock and E. Diday (Eds.)
Analysis of Symbolic Data. 2000

H. A. L. Kiers, J.-P. Rasson, P. J. F. Groenen, and M. Schader (Eds.)
Data Analysis, Classification, and Related Methods. 2000

W. Gaul, O. Opitz and M. Schader (Eds.)
Data Analysis. 2000

R. Decker and W. Gaul
Classification and Information Processing at the Turn of the Millenium. 2000

S. Borra, R. Rocci, M. Vichi, and M. Schader (Eds.)
Advances in Classification and Data Analysis. 2001

W. Gaul and G. Ritter (Eds.)
Classification, Automation, and New Media. 2002

Krzysztof Jajuga · Andrzej Sokołowski
Hans-Hermann Bock (Eds.)

Classification, Clustering, and Data Analysis

Recent Advances and Applications

With 84 Figures and 65 Tables

Springer

Prof. Krzysztof Jajuga
Wroclaw University
of Economics
ul. Komandorska 118/120
53-345 Wroclaw
Poland
jajuga@manager.ae.wroc.pl

Prof. Hans-Hermann Bock
Technical University of Aachen
Institute of Statistics
Wuellnerstrasse 3
52056 Aachen
Germany
bock@stochastik.rwth-aachen.de

Prof. Andrzej Sokołowski
Department of Statistics
Cracow University
of Economics
ul. Rakowicka 27
31-510 Cracow
Poland
sokolows@ae.krakow.pl

ISSN 1431-8814
ISBN 3-540-43691-X Springer-Verlag Berlin Heidelberg New York

Cataloging-in-Publication Data applied for
Die Deutsche Bibliothek – CIP-Einheitsaufnahme
Classification, clustering and data analysis: recent advances and applications / Krzysztof Jajuga ... (ed.). – Berlin; Heidelberg; New York; Barcelona; Hong Kong; London; Milan; Paris; Tokyo: Springer, 2002
(Studies in classification, data analysis, and knowledge organization)
ISBN 3-540-43691-X

Springer-Verlag Berlin Heidelberg New York
a member of BertelsmannSpringer Science+Business Media GmbH

http://www.springer.de

Printed in Germany

Softcover-Design: Erich Kirchner, Heidelberg

SPIN 10880070 43/2202-5 4 3 2 1 0 – Printed on acid-free paper

Preface

The present volume contains a selection of papers presented at the Eighth Conference of the International Federation of Classification Societies (IFCS) which was held in Cracow, Poland, July 16-19, 2002. All originally submitted papers were subject to a reviewing process by two independent referees, a procedure which resulted in the selection of the 53 articles presented in this volume.

These articles relate to theoretical investigations as well as to practical applications and cover a wide range of topics in the broad domain of classification, data analysis and related methods. If we try to classify the wealth of problems, methods and approaches into some representative (partially overlapping) groups, we find in particular the following areas:

- Clustering
- Cluster validation
- Discrimination
- Multivariate data analysis
- Statistical methods
- Symbolic data analysis
- Consensus trees and phylogeny
- Regression trees
- Neural networks and genetic algorithms
- Applications in economics, medicine, biology, and psychology.

Given the international orientation of IFCS conferences and the leading role of IFCS in the scientific world of classification, clustering and data analysis, this volume collects a representative selection of current research and modern applications in this field and serves as an up-to-date information source for statisticians, data analysts, data mining specialists and computer scientists.

This is well in the mainstream of the activities of the International Federation of Classification Societies which considers as one of its main purposes the dissemination of technical and scientific information concerning data analysis, classification, related methods and their applications. Note that the IFCS comprizes, as its members, the following twelve regional or national classification societies:

- British Classification Society (BCS)
- Associaao Portuguesa de Classificação e Análise de Dados (CLAD)
- Classification Society of North America (CSNA)
- Gesellschaft für Klassifikation (GfKl)
- Japanese Classification Society (JCS)
- Korean Classification Society (KCS)

- Société Francophone de Classification (SFC)
- Società Italiana di Statistica (SIS)
- Sekcja Klasyfikacji i Analizy Danych PTS (SKAD)
- Vereniging voor Ordinatie en Classificatie (VOC)
- Irish Pattern Recognition and Classification Society (IPRCS)
- Central American and Carribean Society of Classification and Data Analysis (CASCCDA)

Previous conferences of IFCS took place in Aachen (1987), Charlottesville (1989), Edinburgh (1991), Paris (1993), Kobe (1996), Rome (1998), and Namur (2000).

The editors of this volume would like express their gratitude towards the authors of the papers contained in this volume for their valuable contributions. In particular, they thank the reviewers for their professional, careful and often time-consuming work when reviewing the submitted papers. Our special thanks and gratitude go to Dr. Katarzyna Kuziak of Wrocław University of Economics for her outstanding work and great help in the process of reviewing and preparing final version of the book. We express our gratitude to Dr. Andrzej Bąk of Wrocław University of Economics for his professional contribution in the process of the formatting and technical editing of this volume. Thanks also to Dr. Martina Bihn from Springer Verlag (Heidelberg) for the smooth and timely production and dispatch of these Proceedings.

Aachen, Wrocław, and Cracow
July 2002

Hans-Hermann Bock
Krzysztof Jajuga
Andrzej Sokołowski

Contents

Part I

Clustering and Discrimination

Clustering

Some Thoughts about Classification

Frank Hampel

Seminar for Statistics of ETH, Swiss Federal Institute of Technology, Leonhardstrasse 27, LEO D2, CH-8092 Zurich, Switzerland

Abstract. The paper contains some general remarks on the high art of data analysis, some philosophical thoughts about classification, a partial review of outliers and robustness from the point of view of applications, including a discussion of the problem of model choice, and a review of several aspects of robust estimation of covariance matrices, including the pragmatic choice of a weight function based on empirical and theoretical evidence. Several sections contain new (or at least original) ideas: There are some proposals for incorporating robustness into Bayesian practice and theory, including weighted log likelihoods and Bayes' theorem for weighted data. Some small ideas refer to artificial classification in a continuum, to a "robust" (Prohorov-type) metric for high-dimensional data, and to the use of multiple minimum spanning trees. A promising but difficult research idea for clustering on the real line, based on a new smoothing method, concludes the paper.

1 Introduction

When I was asked to give a keynote lecture at this grand conference, I first hesitated and thought it might be some kind of misunderstanding. After all, I had never properly done research in classification or clustering, my research in robustness theory (which might and should be a general background tool at this conference) lies decades back, I did not follow up the literature, and my (intensive) practical work in data analysis is long past. On the other hand, my present work on the foundations of statistics can clarify many concepts and controversies and can reconcile frequentists and Bayesians on a higher level, but apart from the concepts (and a few simple examples) it is not yet applicable.

But then I thought that many researchers, both in theory and applications, are so engulfed in what they are presently doing, and busy reading the most modern literature of their respective narrow fields, that it might occasionally be a good idea for them to hear a talk "from the outside", which gives them a bird eye's view of statistics, which reminds them of the roots of some of the things they are doing (in the hope of often still badly needed clarification), and which exposes them to a number of fresh ideas, some of which may become a useful tool or may lead to fruitful further research.

In this spirit, I am starting with some general philosophical thoughts about classification, and then I recall some basic ideas, tools and results in robustness theory, which are still often unknown, distortedly known, misunderstood or ignored, but which should be in the tool box of every practical statistician, including every pragmatic Bayesian. An outline of how the results of robustness theory might be incorporated into a more canonical Bayesian theory (which may mean an extension of canonical Bayesian theory) follows, including the use of weighted likelihoods and Bayes' theorem for weighted data.

I apologize if some of my proposals, here and later in the paper, are already (unknown to me) somewhere in the literature; if they are not new, they are at least original, and I think it is better an idea is published twice than it is not published at all.

After mentioning the future potential of the use of upper and lower (instead of only ordinary) probabilities in statistics, I return to details of robust covariance matrices and associated weight functions. Several small subsections on more or less tentative ideas follow. The section on subdividing a continuum may be mathematically simple, but logically interesting (leading to a nontransitive definition of "potentially equal"). The robust metric introduced is of potential use anywhere in multivariate statistics where a metric occurs; its practical usefulness has to be tried out. The idea of superimposing several minimum spanning trees for interactive exploratory data analysis has its roots in some practical experiences, but again it has to be tried out in order to learn about its practical value. The last proposal is more an extensive research program: to use the existing beginnings of a "truly smooth" (neither biased nor locally wiggly), but very complicated smoothing method (an improved counterpart of splines, in a way) for developing tests for clusters, modes, and mixture components for data on the real line.

Before we go into the details of the next sections, we should remember the high standards of good data analysis (which are missing in mathematical theories). For example, in regression (and elsewhere), we should be "fitting equations to data" (Daniel and Wood 1980), and not "fitting data to equations" (or preconceived models), as is frequently done. Daniel (cf. also the highly recommendable book by Daniel 1976 on ANOVA) shows that in order to be an excellent data analyst, it suffices to know very little mathematics, but to have a lot of common sense, including a deep intuitive understanding of simple mathematical formulae, a down-to-earth interpretation of the data, and a clear and thorough understanding of the practical problem to be solved.

This does not mean that no high-power mathematics is needed in statistics; for example, the very basic problem of the violation of the independence assumption of supposedly i.i.d. data (even measurement data in the hard sciences), of which first-class statisticians such as Newcomb, K. Pearson, "Student" and Jeffreys were clearly aware, could only be modeled and attacked successfully after Kolmogorov invented abstractly, and Mandelbrot introduced into parts of statistics, the mathematically highly demanding and "pathological" increment processes of self-similar processes; but the consequences for practical statistics (cf. Hampel et al. 1986, Ch. 8.1; Kuensch et al. 1993) are in full agreement with what C. Daniel and undoubtedly other first-rate practitioners knew qualitatively already from experience. (This topic of lack of independence is not discussed here further.)

Unfortunately, what is very often missing (even in theoretical discussions), is the interpretation of mathematical results in the light of the problems of practical data analysis (see Hampel 1998a). Often there is a deep gap, and a lot of misuse of the results of mathematical statistics. What should also be more emphasized is the iterative going back and forth between the statistical analysis and the background knowledge from the subject matter science; it is often nonsense to require that a purely statistical analysis be complete in itself.

An early source for openminded and thorough data analysis (not blinded by any mathematical theory), which should be compulsory reading for every data analyst, is "Student" (1927). There are also many excellent data analyses in Jeffreys (1939). For more recent examples of creative and innovative data analyses (besides C. Daniel), see the writings by J.W. Tukey, F. Mosteller, G.E.P. Box and others. There are also some condensed but deep data analyses in Hampel (1987), including germs for new methods in design of experiment. An interesting example of modern thinking about data analysis can be found in Davies (1995).

2 Some philosophical thoughts

Classification, the separation and naming of appearances, is one of the most basic cultural activities of humanity; it is a fundament for our science and civilization. Thus, astronomy started with naming the stars and constellations and separating out the planets (in the old sense). Modern geology began with the statistical separation (by the number of surviving fossils) and naming (from Eocene to Pleistocene) of consecutive geological layers by Lyell. Modern biology began with Linné's system of nomenclature, which basically is still used today.

Names of human persons, gods (or God), spirits, and animals often had a mythical importance and power attached to them (we still find remnants of this). Nowadays, names are more and more replaced (at least in part) by numbers (such as passport number, social security number), a trend enhanced by the computer. But even the simple step from "one" to "two" needs the separation of the world into (at least) two entities.

In a deep philosophical sense, the world is basically One. This attitude is explored in detail in the old Indian philosophy of Advaita Vedanta, and is also found in mystics of probably all religions, including the Christian one (for example in Meister Eckhart). But as they all emphasize, this "Oneness" cannot appropriately be described by our language, only circumscribed and approximated from the outside, as it were, because language means words, and words automatically mean separation. "Every word a lie." - However, outside rare mystical experiences, we all live on the other side of the "veil of Maya", in a dualistic world with differences and with language, which reflects these differences. It is interesting to note that in the Bible one of the first tasks God gives to Adam (still in Paradise) is to name and thereby separate all the animals and birds. This can be seen as the beginning of classification.

In modern research on classification, especially in cluster analysis, it may be useful to remember occasionally that, in a sense, all classification is more or less arbitrary, that its boundaries are fuzzy (even those between life and death, as modern medicine had to acknowledge), and that we probably should ask more often inhowmuch (not: whether or not) a given classification scheme matches our observations. For example, in biology the only halfways reasonably definable classification concept (limiting arbitrariness) is the species concept; but even on this level, there is a permanent merging and splitting going on (as I am noting for birds and orchids). And what about a species like the Common Ringed Plover (Charadrius hiaticula), a small shorebird, which breeds from Baffin Land and Greenland through northern Europe, Asia and North America, gradually changing in what is called a Cline, until it meets its relatives again on Baffin Land, but now it looks and behaves like a different species (Ch. semipalmatus)? (If one wants to cling to the species concept, a split has to be made at an arbitrary line - for convenience, the Bering Sea was chosen.)

In many situations, there are "natural" groups, separated from each other (apart, perhaps, from a few "hybrids"), to be described by the classification scheme. Even there, much may depend on the degree of refinement of the differentiation, and the overall structure may be more like a complicated tree, with branches at different "heights" (as in the dendrograms of hierarchical clustering). It is also possible that another than the "best-fitting" structure is required by outside reasons, most often a linear sequence for listing for convenience "canonically", for example, all birds of the world, although every specialist is aware of a better fitting tree structure (which also is changing over time), with a few small "loose" (hard to localize) branches.

In other situations, there is a continuum to be separated artificially, for convenience, into several classes. The continuum may be on a qualitative scale (e.g., hardness) or described by 1 (e.g., brightness of stars), 2 (e.g., off-the-peg clothing) or more quantitative variables; and the values of these variables may be determined more or less accurately. For example, many interpretations of a biplot or correspondence analysis by sociologists, psychologists etc. would fit in here. There are also diseases measured on a quantitative scale (e.g., blood particle counts); and while usually a "norm" region is interpreted as meaning "healthy" and all other values as meaning "sick", one might ask whether for some purposes it might not be better to define 3 or more classes, such as "clearly healthy", "clearly sick", and some region of increasing suspicion in between. (It is interesting to note that on a hard technical and empirical level, methods for dealing with outliers which include a smooth transition zone have proven to be clearly superior to all methods which use "hard", sudden "rejection of outliers", cf. Hampel 1985).

In more difficult situations, nothing or not much is known about the type of structure, which is to be inferred solely or mainly from the data. For example, for a long time it was not known whether double stars and star clusters (such as the Pleiades) also belong together physically (some do, some do partially).

Probably many clusters as found in cluster analysis can be described as a scatter around a zero-dimensional point; but some clusters may be scatters around a linear or curved one-dimensional structure, or around a higher-dimensional structure. It can be seen as a higher level task of cluster analysis to find not only points belonging together, but also the dimension and structure describing the clusters. (Some techniques needed may be based upon factor analysis. It should also be noted that solutions need not be unique: a zero-dimensional cluster with a highly elongated scatter ellipsoid may look the same as a linear one-dimensional cluster, and so on; and the dimension might even vary within the cluster.)

3 A brief review of outliers and gross errors

A problem with which all practical statisticians, be they frequentists or Bayesians, have to deal, is that of outliers and gross errors. Outliers - or more generally, outlying substructures - are data which are "far away" from the bulk of the data, or more precisely, which do not fit the pattern suggested by the majority of the data. "Outlier" is an ill-defined concept, with only artificial borders within a fuzzy transitional zone; nevertheless it is a useful concept, as long as one keeps its very limitations in mind. (As in other situations with classification, there are clear cases of class A, there are clear cases of class B, and there are other cases which do not fit in clearly.)

Outliers are not the same as gross errors. Gross errors are cases where "something went wrong"; for example, equipment failure, transmission errors, false categories in a structured design, or mistaken measurement of a member of a different population. The percentage of gross errors differs of course with the quality of the data; but it may still be surprising to some that even for routine data (i.e. data not taken with very special care) in the exact sciences, 1-10% gross errors appear to be the rule rather than the exception (cf., e.g., Hampel et al. 1986, Ch. 1.4). For medical data, about 10% and even about 25% gross errors have been cited. And even after careful checking, to the satisfaction of the authors, 15 data sets in the behavioral sciences still contained around 1% and up to 4% (and only in one case zero) gross errors (Rosenthal 1978). Thus, it is clear that we have to live with them. And since a single, sufficiently distant gross error can completely spoil a normal theory (or least squares) analysis, we also have to cope with them.

Outliers may be gross errors; but they may also come from a genuinely longtailed distribution (e.g., the effects on one of the four levels of the famous hierarchical random effects model in Bennett 1954 are best modeled by a Cauchy among all t-distributions); or they may come from a different distribution and (though formally still a "gross error") may give a valuable - if not the most valuable - unsuspected new information.

Clearly, known gross errors should be given zero weight in the analysis of the bulk of the data; but they should be looked at separately and not be discarded immediately, in order to see what can be learned from them. If an outlier is known or assumed to be a proper observation, it is a common mistake to give it full weight in, for example, a normal theory analysis. This is a costly mistake because the outlier clearly shows that the model of the normal distribution is wrong, and that the true model is longer-tailed; and the best (e.g., maximum likelihood and Bayes) methods for longtailed distributions give very low influence (cf. the influence function in, e.g., Hampel et al. 1986) and hence very low weight to outlying points. (The only case where a full weight is justified is a truly nonparametric situation (e.g., when a mean or total is required no matter what the distribution), with its associated severe difficulties.) Thus, fortunately, the treatment of outliers of different types in connection with the bulk of the data is qualitatively, and can be made exactly, the same in most situations; it is only their interpretation which differs.

4 The question of model choice

If, as it occurs frequently in classification and cluster analysis, a model of multivariate normality is tentatively entertained and there appear to be outliers which are not gross errors, many scientists, both Bayesians and frequentists, are inclined to change the model, by adding more parameters and including longtailed distributions. I do not think this is the best idea. It prevents the worst — and in that sense it is defensible — but it generally leads to rather mediocre procedures, as is shown by a number of 3-parameter and adaptive procedures in the Princeton Monte Carlo study of location estimates (Andrews et al. 1972), including those named after Hogg, L.J. Savage, Takeuchi, Jaeckel, and others. Such models cannot claim either to be the exact "true model"; they are more complicated, mathematically less nice and harder to interpret; they either lose efficiency by switching between simple models, or they try to estimate ill-determined parameters and thus are in danger of doing overfitting (which may be a partial explanation for their surprisingly mediocre performance); and they contradict one of the deepest principles of experienced data analysis (C. Daniel, in his Berkeley lectures 1968): use (and first search for) the simplest model reasonably possible, even if it is "significantly wrong"(!), because it is more useful, more reliable and better generalizable (C.D.'s term) than a more complicated one. (If a complicated model is not too much ad hoc and can be interpreted, or if it is even called for by the background knowledge of the problem, and if simultaneously the sample size is large enough, then such a complicated model may be the simplest one reasonably possible.)

Thus, in general I'd suggest to keep the simple model of normality (or whatever the most basic model is, such as normality after a simple transformation(!), or exponentiality), but to treat it explicitly as only an approximation to reality. This is done in robustness theory, the stability theory of statistical procedures (cf. Huber 1981 and Hampel et al. 1986, especially Ch. 1; for a recent overview, including also the problem of violation of the independence assumption, which is not treated in this talk, cf. also Hampel 2002). The theory allows at the same time for outliers, gross errors, longtailed (and shorttailed) distributions, small visible distortions of the model distribution and the imperceptible deviations from the ideal model distribution which still can have large effects on the efficiency (cf., e.g., Tukey 1960). There has been a rich technology built up, with concepts such as M-estimators (or estimating equations; slight but powerful generalizations of maximum likelihood estimators), influence curve or influence function IF (describing the local effect of each data point or potential observation on whatever is being estimated; closely related to the jackknife and various forms of sensitivity curves, to perturbation theory and sensitivity analysis, and, more mathematically, to derivatives of functionals and - highly useful - Taylor expansions), and breakdown point BP (a global robustness concept, essentially the smallest fraction of arbitrary gross errors or outliers which can make a procedure totally unreliable). On this general basis (which holds for every "reasonable" parametric model!), a specific technology was developed, and checked by Monte Carlo studies, for how to deal with approximately normal data with outliers in practice, leading to 2- and 3-part redescending M-estimators, Tukey's biweight, and other estimators of the same type (cf. Andrews et al. 1972, Hampel 1985, Hampel 1997). (Huber-estimators and related ones are only a first, though important and necessary step; in the same way, in which Huber-estimators sacrifice a few percent efficiency under the strict normal distribution for being arbitrarily much better than least squares under more realistic alternatives, good redescenders sacrifice a few percent under Huber's least favorable distribution, for being at least 10-20% more efficient than Huber-estimators under more realistic distributions which include outliers. But Huber was right in warning against the careless and thoughtless use of redescenders, without sufficient understanding about the problem of multiple solutions.)

Bayesians seem to have problems with the idea of an approximate model, which does not fit into their traditional paradigm. Some suggestions for them are given below (Sec. 5).

Some decades ago, many statisticians switched to nonparametric procedures "because normality is not true"; now, for the same reason, some are switching to models with upper and lower probabilities . Either method may be perfectly appropriate in a given situation, but they are not, or at least not fully justified by the reason given. There is no need to "pour out the baby with the bath water" and abandon the model (e.g. normality) entirely.

We can try to summarize a fairly simple treatment of data from an approximate model (which is still sufficiently general for most purposes):

(i) give central ("clearly good") observations weight one, and distant outliers (which in classification and cluster analysis may already belong to another population) weight zero;
(ii) decrease the weight continuously, and not too quickly, in the "zone of doubt" from one to zero.

The quantitative details are not so important, but it is essential that the weights decrease rather smoothly and not too quickly, especially in the range where they are still high. Otherwise inefficiency and instability may result. (Cf. Hampel 1997.)

For some more details on which redescending methods to use in practice, see also Hampel (1980).

It should be stressed that not only "clear outliers", but also "doubtful outliers" should be looked at separately (and interpreted, if possible); in connection with other knowledge, they may give valuable information (cf. the example in Hampel 1987, Sec. 3, p. 101ff).

With the downweighting, we are not doing a "test whether the value is an outlier". If one really wants to know how "unlikely" a value is, it is not appropriate to use the normal distribution as basis; both Bayesians and frequentists would have to estimate the longtailedness of the true distribution of the "good" data (which would lead to very different results); but most good scientists will prefer practical expertise to a formal rejection rule (for the interpretation of outliers. For assessing the structure of the bulk of the data, "doubtful outliers" should still be kept with their partial weight).

If for nonstatistical reasons (e.g. because of simplicity) a "classification" (separation) of the data into only two "classes" ("good" and "bad") is required (although this is highly artificial), then rejection rules based on the median deviation or MAD (the median of the absolute deviations from the sample median) work best, and even they lose at least about 10-20% efficiency unnecessarily, compared with good "smooth redescenders". Subjective rejection (if well done) also prevents the worst, but it has been shown to lose also about 10-20% efficiency unnecessarily (Relles and Rogers 1977, Hampel et al. 1986, Ch. 1.4). It has often been said that the whole methodology of rejection of outliers is faulty in principle; but in addition most rejection rules proposed in the literature (cf., e.g., Barnett and Lewis 1994) are only mediocre to bad. Even the frequently used largest Studentized residual ("Grubbs's rule") is only mediocre, with a breakdown point around 10%, and works only for fairly high-quality data. See Hampel 1985 for details. Not even the interquartile range can fully replace the MAD (cf. Andrews et al. 1972, Ch.7E3).

A few words might be added on the question of efficiency. Many statisticians have objected to robust procedures for the normal model because robust procedures "lose efficiency". But this argument is nonsensical. These statisticians lose not a few, but dozens percent efficiency with normal theory methods, without even noticing it, the misleading and false suggestions or claims by Cox and Hinkley (1968), Stigler (1977) and the Gauss-Markov theorem (to cite a few sources) notwithstanding (cf., e.g., Hampel 1998a). On the other hand, a few dozen percent efficiency loss may often be bearable in practice, as long as the analysis is good otherwise (with respect to systematic errors, model choice, etc). It may also be stressed that the arithmetic mean is "much less nonrobust" than the empirical variance (for which the efficiency is in fact not rarely close to zero), so the mean without extreme outliers may well be used with rather modest efficiency standards. (With high standards, if trying to have just a few percent avoidable efficiency loss, one needs the utmost of robustness technology.) In the topics of this conference, simple and flexible tools seem often more important than the last refinements, hence details about how to "optimize" robust procedures will at most be alluded to.

5 Some suggestions for Bayesians

5.1 General remarks

Bayesians seem to have problems with robustness, especially with robustness against deviations from the parametric model (there is now quite a bit of work concerning certain aspects of robustness against changes of the prior distribution). The most common way out in practice still seems to be the replacement of the original parametric model, such as normality, by another, more complicated ad hoc model. These models are, strictly speaking, as unrealistic as the original model; if (as is frequently the case) they are chosen with good intuition, they do work for a full neighborhood of the original model, but this can only be proven by robustness theory. For an old numerical data set, see Dempster (1975) and the subsequent discussion, where the results (apart from the formalism) are virtually identical with those of "classical" robustness theory.

It may also be noted that a sequence of Bayes estimators (indexed by n, for a fixed prior) can often also be viewed as a sequence of functionals, a different one for each n. In order to apply asymptotic theory for a Bayes estimator for a given n, one ought to use the functional for that fixed n (which the sequence of Bayes estimators "crosses" on its smooth but nondirect way to infinity), go to infinity with it (and NOT with the Bayes estimators) and extrapolate back from infinity onto this fixed n. This allows also to transfer concepts such as influence function and breakdown point onto Bayes estimators for each given n in a meaningful and informative way.

5.2 Robustified likelihood function

Some Bayesians may want to cling to their original model and to an unmodified likelihood function, yet be somewhat robust. For them I offer the following tentative suggestion. All they have to do is to replace the most extreme observations by pseudo-observations, which behave like data from the ideal model and do not contain dangerous outliers. It is well known that transition to ranks loses amazingly little information, and it is also known that the estimator derived from the Fisher-Yates-Terry-Hoeffding-van der Waerden test or normal scores test is robust and asymptotically fully efficient under the model of normality, although already in small neighborhoods of the normal, it is empirically surpassed by the Hodges-Lehmann-estimator derived from the Wilcoxon test, as predicted by its lower breakdown point (Hampel 1983). The basic idea is now to determine iteratively a central region with unchanged data, and replace the most extreme ones by something related to the normal scores.

More precisely, let (under the ideal model) for $i \leq n$ the X_i be *i.i.d.* $\sim N(\theta, 1)$, and assume the X_i are already ordered. Let t be a trial value for $\hat{\theta}$ (starting with the median), then keep all X_i with $|X_i - t| \leq c$ (with $c = 2$ or perhaps $= 1.5$). Let X_k be the largest $X \leq t + c$; let $\Phi(c) = d$, where Φ is the cumulative standard normal distribution; then, for all $j > k$, replace X_j by something like $t + \Phi^{-1}(aj + b)$ with $a = (1-d)/(n-k)$ and $b = 1-(n+1)(1-d)/(n-k)$ (for variants, see below). That is, replace the upper tail by an "ideal" normal sample. Do analogously for the lower tail. Call the new pseudo-observations, depending on t (including the central ones) Y_i. Compute the new $t = \Sigma Y_i/n$ and iterate until convergence. This means, solve the implicit equation $\Sigma(Y_i(t) - t) = 0$.

The new pseudosample would contain about the same information as the original one if that came from an exact normal distribution, but it does not contain dangerous outliers anymore. Its likelihood can be plugged into Bayes' theorem in the usual way.

Some variants for the argument of the inverse cumulative normal are the following. One may try to put some more "natural" variability into the tail pseudovalues; this can be achieved by replacing j by $j + \epsilon$, where the ϵ's are independent uniform on $[-1/2, 1/2]$. One may also try to get a smoother transition between $X_k (= Y_k)$ and Y_{k+1}, for example by putting $Y_{k+1} = X_{k+1}$ and starting the (perhaps approximately) linear scale in the argument of Φ^{-1} from there.

A related approach would be the following. Start with a trial value t and an $\epsilon > 0$, such as $\epsilon = 1/(n+1)$, and replace each $X_i - t$ by $\Phi^{-1}((1-\epsilon)\Phi(X_i - t))$. Compute the mean as the correction of the old t and iterate until convergence. Instead of Φ, it is even easier to use the cumulative logistic distribution. This procedure can also be restricted to the tails of the sample, as above; it contains a bit more information from the original sample and is also somewhat robust.

If t is a good robust starting value, the first step of the iterations (in all these cases) may already suffice (cf. the experiences in Andrews et al. (1972) and Bickel's (1975) proofs of asymptotic equivalence).

5.3 Bayes' theorem for weighted data

One of the most useful and important descriptions and tools in robust statistics is downweighting of outlying and nearly outlying points. The weights are to be determined, e.g., by the influence function of the M-estimator used, and thus are determined randomly by the whole sample. (M-estimators are basic for being able to reject outliers smoothly.) In general, the weights change with the parameter component being estimated, and the class of multivariate estimates obtained with "random weighting" (one weight only for all parameters in each data point, as in "iteratively reweighted least squares") is not the same as that obtained by "random transformations" (e.g., iterative Huberizing), cf. Hampel (1978). For multiple regression (with fixed x-variables), they are the same, but not in multivariate analysis. The class of all M-estimators is even more general. Nevertheless, for reasons of simplicity, we may often restrict ourselves to estimators based on "random weighting", as has been implicitly done in some sections above.

Since the weights are random weights, frequentists have to worry about their variability (which caused some complications in the seventies when such estimators were first discussed). I believe Bayesians (and adherents of the likelihood school) have a great advantage here, because they believe in the strong likelihood principle. For them, it does not matter how the weights were obtained; all they have to accept is that somehow these weights measure the internal consistency of a data point with the rest of the sample (independently of any prior belief), and then they can treat them as fixed weights.

Since the log likelihood ratios $\log(l_x(\theta_1)/l_x(\theta_2))$ measure the evidence of one parameter value over another one, and since the evidence can be weakened, even down to zero, by multiplication with the weight attached to it, it appears very natural to define the joint likelihood of a sample of i.i.d. observations, weighted by weights w_i $(0 \leq w_i \leq 1)$, as $\Pi l_{x_i}^{w_i}(\theta)$, and the log likelihood as $\Sigma w_i \log l_{x_i}(\theta)$. ($\Sigma w_i$ can be taken as the effective sample size.) With apriori-distribution $\alpha(\theta)$, we thus obtain Bayes' theorem for weighted data:

$$\alpha(\theta|x_1, \ldots, x_n) = \alpha(\theta) \cdot \Pi l_{x_i}^{w_i}(\theta) \ / \ [\int \alpha(\theta')\Pi l_{x_i}^{w_i}(\theta')d\theta']$$

This formula, together with just some reasonable weights derived from robustness considerations, allows Bayesians to make inference which is robust against deviations of the distribution of the data from the assumed model.

In most cases, the w_i are $\equiv 1$ for the "good" data - which have to exist for defining a model structure. However, sometimes we may discount the whole data batch for outside reasons, for example, if we doubt its overall quality, and then we may multiply all weights by a factor < 1, so that we obtain $\max w_i < 1$.

It is clear that one needs "enough" data for a purely internal "consistency check" (at least 3, preferably many more, in the case of the normal with unknown mean and variance). However, if also the prior is used for checking the reasonableness of the data, even a single observation can be downweighted. This gives an automatic solution to the problem of obvious discrepancy between prior and the data (more or less keeping the prior), although in some cases we would rather trust the data than the prior. (We could still treat the data as a more or less internally consistent group of "outliers" compared with our prior belief, which describe a "new" population of data sources.)

Bayesians who are not quite so pragmatic as to accept weights from some "outside" source such as common sense or robustness theory, may try to develop a purely Bayesian internal check for consistency of the data. (Perhaps this has been done already.) However, if they are sufficiently pragmatic to still care a little bit about frequentist properties, they should try to develop this check in alignment with the empirical and theoretical results of robustness theory.

5.4 Upper and lower probabilities

An extension of Bayesian theory which may help to model the state of our knowledge more flexibly and appropriately (and which can also precisely model the "state of total ignorance" with respect to a parameter), is the incorporation of upper and lower probabilities. This has been done by professed Bayesians such as Dempster (1967, 1968) and Good (cf. e.g., Good 1983 and the literature therein); it has been worked out by Shafer (1976), Smets and others to the so-called Dempster-Shafer belief function theory; and it is also in the background of the work by Berger (1984) and others on robustness against changes of the apriori-distribution. Upper and lower probabilities are more appropriate to describe uncertain and ambiguous knowledge and hence for statistical inference, while Bayesian proper probabilities are in general only suitable for decisions (if this distinction is being made). A unifying statistical theory which combines the aleatory probabilities of the Neyman-Pearson theory with the epistemic probabilities of the Bayesians, using (new) frequentist epistemic probabilities as a bridge which contain Fisher's fiducial probabilities in a corrected form, is outlined in Hampel (1998b, 2000). This theory is not yet suitable for general practical use, but it clarifies already many concepts, up to almost complete symmetry between the (extended) frequentist and the Bayesian approach.

I do not know how much has already been done, but I can imagine the (sensible) introduction of upper and lower probabilities into classification and cluster analysis to become very fruitful.

6 Some remarks on robust covariance matrices

6.1 General remarks

A basic and central tool in much of multivariate analysis are covariance matrices. It is clear that they can suffer much from outliers and gross errors (remember that an empirical variance is even "much more nonrobust" than an arithmetic mean); and unfortunately, their robustification encounters difficulties because of the "curse of dimension" ("too many directions in a high-dimensional space"). Robust M-estimators of location and scatter are an elegant and valuable tool for low dimensions; but it came as a shock to the "robustniks" community when Maronna (1976) (and later others) proved that under weak conditions the breakdown point for all equivariant M-estimators and in fact for all "smooth" and "reasonable" equivariant estimators is $\leq 1/p$, where p is the dimension of the data. (Hence, starting with dimension around 10, the reliability of these methods is too low.)

As a reaction, "high breakdown point" estimators, such as the Stahel-Donoho estimator (Stahel 1981a,b; Donoho 1982), the minimum covariance determinant (MCD) estimator and others were developed for covariance matrices (and similarly for multiple regression, the "least median of squares" estimator going in fact back to Hampel 1975), which theoretically keep a breakdown point 1/2 for all dimensions, but which have to be approximated by a cumbersome computer search "in all directions" of a high-dimensional space.

While I think these methods are a legitimate and even fascinating topic for theoretical research, I do not think they should be recommended for general routine practical use. (There are always problems where even the most crazy looking methods invented in mathematical theory may be most appropriate. Thus, I recall a scientist who had the problem of discovering a main effect analysis of variance structure in his data, which contained outliers, without knowing the experimental conditions (that is, the levels of the effects); but this is precisely the problem addressed by the sequence of estimators beyond the "shordth" (Hampel 1975, p. 380), the "shordth" being also the basis for the "least median deviation" or "least median of squares" estimator, which is also briefly discussed for general nonlinear models in that paper.)

In particular, the condition of equivariance has been overly stressed (for mathematical convenience?). Even though an affine equivariant "ideal" model is often appropriate, this is not true for the gross errors, which tend to occur quite often in single coordinates, thus breaking this structure. It may often be more appropriate first to downweight and eliminate the worst outliers in single coordinates, and then to "robustify" the rest in an affine equivariant way (thus still achieving "approximate" affine equivariance). Because of the "curse of dimension", it may well be that no simple routine method would be fully appropriate under all circumstances; but in any case, I would think that a box of flexible and easily understandable tools combined with interactive analysis and interpretation of the data might be quite useful in real practice.

6.2 Some aspects of weight functions

Given a sample of p-dimensional vectors X_i, we may start by first "robustifying" in single coordinates. In each coordinate, the median med is the most robust (in many ways) estimate of location, and the median deviation or median absolute deviation MAD (Hampel 1968, 1974; Andrews et al. 1972) is the most robust estimate of scale. Using them, we can center and scale the whole sample (by linear transformations) to robust location zero and robust dispersion diagonal one. Given this, we can already downweight and reject clear outliers in single coordinates, using the weights ($w(x) = \psi(x)/x$) corresponding to one of the "smoothly redescending M-estimators" (defined by $\int \psi((x-T)/MAD)dF(x) = 0$), such as 25A or biweight (cf. Section 4). - By the way, if deemed desirable, med and MAD can also be replaced by more efficient though similarly robust estimators.

A problem is to get good robust estimates of the correlations. There are a number of proposals for the correlation of each pair of variables. One possibility are rank correlations, such as Spearman's or Kendall's rank correlation. Their (moderately good) robustness properties, including their (somewhat intricate) breakdown properties, were first explored by Grize (1978). The most extreme correlation in some ways, the quadrant correlation, does often seem to lose too much information; but a less extreme class obtained by "smooth limiting" (cf. Huber 1981, Ch. 8.3), transforming the standardized data by a monotone ψ-function and taking the (biased) correlations of the transformed values, may be better applicable and leads to a positive semidefinite covariance matrix. Otherwise, one of the nicest proposals in this context (highly robust starting points) is probably the application of the idea in G-nanadesikan (1977, Ch. 5.2, Formula (70); there are many more proposals in this chapter): let Y_{ki} and Y_{kj} be the standardized *i-th* and *j-th* coordinates of the X_k; then define $r_{ij} = ((MAD_k\{Y_{ki} + Y_{kj}\})^2 - (MAD_k\{Y_{ki} - Y_{kj}\})^2)/((MAD_k\{Y_{ki} + Y_{kj}\})^2 + (MAD_k\{Y_{ki} - Y_{kj}\})^2)$. The main disadvantage is that if these pairwise highly robust correlations are put together, the resulting covariance matrix is (in general) not positive semidefinite.

There are other possibilities (cf. also the proposals in Hampel 1975; one claim there is wrong, cf. Hampel et al. 1986, p. 431). But for proceeding pragmatically and simply, we may tentatively just multiply all the weights ($= \psi(x)/x$, with ψ defining, e.g., 25A, cf. Andrews et al. 1972) of the single coordinates together to give a fixed weight for each data point (eliminating all clear outliers in single coordinates). Then we may iteratively compute weighted covariance matrices, with these fixed weights multiplied with iterative weights derived from the "robust Mahalanobis distances" in each step. The final multiplied weights can be used in an ordinary Bayesian analysis of the weighted sample (cf. Section 5.3 above), as if the weights had been given apriori. Frequentists still should at least approximately acknowledge the effects of the weights being only estimated, e.g. by using the ψ-function formulas of M-estimators.

If the first weight function is $\equiv 1$ in a large central region, the resulting procedure is even approximately affine equivariant, if all data are "good". The first weight function can be seen as giving a partial "precleaning" of the data.

For the choice of the weight function, compare also the general remarks towards the end of Section 4. If a quick descent is required for outside reasons, e.g. in order to separate nearby clusters, then the form of the tanh-estimator should be used or approximated, the "gross-error sensitivity" can and probably should be decreased, and the advantage of a lower "rejection point" is to be weighted against the danger of local instability (for concepts and arguments, cf. e.g. Hampel et al. 1986 and Hampel 1997).

Even more specifically: If r is the standardized distance of a point from the center, or the Mahalanobis distance given by a weighted covariance matrix during the iterations, we may put $w(r) \equiv 1$ for $r \leq c$ and $w(r) \equiv 0$ for $r > z$, with $z > c$ (preferably $z > 3c$). (Some default values might be $c = 2$ or 2.5 and $z = 8$ for one-dimensional data, but there is a whole range of smooth redescenders, e.g. from 25A to 12A, see Andrews et al. 1972.) In between c and z, the most primitive idea would be to go down linearly (which, by the way, is not the same as the linear descent of ψ in HMD or in the third part of the three-part redescenders). However, it is better first to go down more slowly, as in a quadratic with derivative 0 at c (and only later linearly). A popular form of weight function is Tukey's biweight (Beaton and Tukey 1974); however, it redescends in one region almost as quickly as it ascends, and this could cause problems with very special data sets (Hampel et al. 1986, p. 408f).

The "curse of dimension" shows itself when we consider the choices of c and z for higher dimensions p. If c is constant with respect to p, then a smaller and smaller percentage of data will receive weight 1. If c (and z) is increasing with p (e.g. such that this percentage stays constant), then in each direction it becomes harder and harder to downweight and reject data points; but this is to be expected, because there are "so many" directions in which data and hence also outliers could be.

Covariance matrices can be used to characterize classes of clusters. It appears natural to allow data points to have positive weights in two (or more) classes or clusters. (This happens, somewhat similarly, also in "fuzzy clustering", cf., e.g., the FANNY method in Kaufman and Rousseeuw 1990.) This is often almost a necessity in a first round (for safety and efficiency reasons), and as long as only inference is considered, it may just describe the fact that the point fits into two (or more) populations. But when a decision is required (even if it be artificial), then each point may be put into the population where it has the largest weight. Depending on circumstances, we may even use rather quickly redescending weight functions in a later round (up to "hard rejection"), in order to separate the populations better.

The foregoing discussion may be naive in some ways, in trying to simplify the highly complex situation with robust covariance matrices to some essentials (in fact, some problems with the speed of convergence have already been noted), and there is room for many improvements and refinements. But that the basic ideas actually do work very nicely, has been shown by Hennig (1998, 2001, Hennig and Christlieb 2002) in his work on fixed point clusters, starting iterative weighting more than once from small homogeneous subsets, where he gives both theoretical proofs of properties and successful empirical results of the procedures.

7 Artificial classes in a continuum

7.1 The problem in the one-dimensional case

Assume measurements or counts Y are being made in a finite interval on R (or Z), and that they can be modeled as an ideal value or parameter X (unknown) plus a random term ϵ, whose distribution is known (or can be estimated). Assume we want to subdivide the interval of X-values into finitely many classes for convenience (for example, in order to distinguish and name different degrees of severeness of a disease, when the severeness can be approximately measured), but because of the random fluctuation ϵ, there will be misclassifications when Y is used for the unknown X. How do we get a maximum number of classes (and hence maximum information from the classification) while keeping the probabilities of misclassification small in a suitable way?

First, it is clear that for X near a boundary of two classes, the probability of misclassification is at least near 50%, at least for misclassification into a neighboring class, if not also other classes. All we can reasonably require is that the probability of misclassification into a nonneighboring class is small, say, at most 2α (e.g., = 10% or 5%) for all nonneighboring classes. But then we can solve the problem by successive computation of the boundaries.

If we have a location or shift model (with restricted parameter range), we can start with a whole class for the parameter on the lower boundary of the range of X, take the upper α-point of the distribution of Y under this parameter as the next class boundary point, the upper α-point under this next point as the following one, and so on. If we have a general continuous one-parameter model, we have to check each time whether the previous point is below the lower α-point of the distribution under the present boundary value, otherwise we have to shift the latter upwards until this condition is fulfilled. (The reason for this slight complication is obviously that the variance or distribution of ϵ may change with the parameter.)

In a discrete case, say, the binomial distribution, we try to classify the possible discrete values in such a way that 2 observations in nonneighboring classes most likely do not come from the same parameter value. This means, we try to find out how many parameter groups and observation groups we can "half-"distinguish (in the sense that only misclassifications into neighboring groups are allowed to be likely). It will be interesting to compute how many classes are possible, for example, for $n = 20$ compared with $n = 10$ in the binomial case.

7.2 The two-dimensional continuous case

The two-dimensional case can occur, for example, when the results of a PCA, biplot or a correspondence analysis based on a small sample (with appreciable random variability of the points) are to be classified into as many "half"-distinguishable groups as possible.

In the two-dimensional case, we have the additional problem (and flexibility) of the shape of the classes. Assume we have a circular distribution of Y with constant error distribution. A naive idea would be to divide the plane up into squares of a suitable size, by intersecting two orthogonal sets of equidistant parallels. But then each square would have 8 neighboring squares with which it could easily be confused. A better idea might be to shift every second row by half a square-length; then each square has only 6 neighbors (though with a smaller distance of the nearest non-neighbor). Still better appears to be a filling of the plane with regular hexagons, because presumably each hexagon (if of equal size as a square) is better isolated (farther away) from its non-neighbors, reducing the error rate, or equivalently the hexagons can be made smaller (to be checked by computations).

The basic considerations (not the mathematical details) of this section may be of some value for the recent tendencies of strong subdivision of the Larus argentatus/Larus cachinnans complex (Herring gull complex) in ornithology. I do not think a geographical subdivision in which nonneighbors cannot be distinguished anymore (with reasonable certainty) would make much sense.

8 Some ideas concerning clustering

8.1 A robust metric

The following idea can be applied anywhere in multivariate analysis where a metric is used and it is feared that gross errors might occur in single coordinates. Instead of searching for the outliers separately, one simply "almost ignores" them by using a metric which is hardly affected by a few outliers.

The idea is inspired by the Prohorov distance between two probability distributions (Prohorov 1956), which has found a nice interpretation in robustness theory. We consider the same basic idea, but in a different manner. Let X_i and X_j be two p-dimensional vectors with coordinates X_{ik} and X_{jk}.

Starting with the basic metric $d(X_i, X_j) = \max_k |X_{ik} - X_{jk}|$, we define the "robustified metric" $d^*(X_i, X_j) = \inf\{\epsilon : |X_{ik} - X_{jk}| \leq \epsilon$ except for a fraction ϵ of coordinates $\}$ or $d^*(X_i, X_j) = \inf\{\epsilon : \#k$ with $|X_{ik} - X_{jk}| > \epsilon$ is $\leq \epsilon p\}$. This implies that data vectors with $1, 2, \ldots$ clear outliers in single coordinates have at least distance $1/p$, $2/p, \ldots$ (also depending on the distances of the other coordinates), but not arbitrarily close to ∞; for pairs of vectors with $d < 1/p$ (without outliers), d and d^* agree. The incongruence of the two ϵ's in the definition, one describing a distance and the other describing a fraction of outliers, is also an advantage: we can scale the basic metric arbitrarily and thus decide in what distance we want to treat a pair of coordinates as containing an outlier, by putting that distance $= 1/p$ by multiplication of the basic metric with a constant. (By the way, this rescaling is also sometimes incorporated into the definition of the Prohorov distance.)

This robust metric also partly answers a remark by W.J.J. Rey (2001, orally), who correctly noted that the naive use of the breakdown point for whole data vectors in high-dimensional multivariate statistics is inappropriate, because for large p almost every data vector will contain some outlying coordinates. The problem was already noted, together with other problems, in Tukey's talk 1970 on robust regression in the Princeton robustness seminar (cf. Hampel 1997, p. 137).

8.2 Triple minimum spanning trees

Minimum spanning trees (MST) are a fast and valuable tool for getting a first hold of the (possibly complicated) structure of a data set in high dimensions. (There can and should be first many considerations concerning the choice of the metric, but we skip these here.) The MST describes a simple network covering all points and giving some information on closest neighbors of a point. But my impression is that it does not give enough information about the neighborhood of any point. In order to get a first feeling for a local structure, we need some more neighboring points, but not too many, otherwise we would get lost again.

Therefore I suggest to try the tentative idea of an overlay of a few, preferably perhaps 3, MST. The basic MST is very fast to compute, so the others should not take much longer. The idea for the second MST is to determine an MST, but leaving out all connections already in the first tree. Similarly, the third MST must not contain any connections used in the first two MST.

The hope is that these trees together give a first indication of local clustering, local dimension, structural peculiarities, and so on, while still allowing to "look at the data" (with their tree connections). They might perhaps serve as a tool in exploratory analysis of high-dimensional data sets.

8.3 One-dimensional clusters (and modes) via a new smoothing method

Assume a (moderately large) number of points on the real line are given. Are they scattered "randomly", or do they form certain clusters? (As an example, take the observation dates of a migrating bird species - with perhaps several populations - on the time axis. Are they distributed "uniformly" (or unimodally) over the whole migration period, or are there several distinct modes which should be reproducible with future observations, and which might belong to different populations or subspecies?) A related problem is that of discovering several modes (or even "hidden modes", from different subpopulations) in a histogram from possibly heterogeneous data.

There are many, many smoothing methods for one-dimensional data. However, to my limited knowledge, they all may contain local "wiggles" which are not justified by the data, or they may contain a strong bias, deviating more globally from the data structure. In trying to formalize Tukey's ideal of a "freehand smooth", which avoids either fault, I developed the ingredients of a new smoothing methodology which minimizes the number of points of inflection in the curve and, in principle, all its derivatives. A "necessary" point of inflection describes a true feature of the data. The ingredients were (partly) put together in the thesis by Mächler (1989; 1995a and b) with a lot of sophistication, and resulting in a first, preliminary, but working computer program.

The program gives a solution essentially for each number (and approximate location) of points of inflection (and hence modes and antimodes in the derivative) given. A further, probably very demanding continuation of this work would be to develop tests (presumably by cumbersome simulations) for the necessity of a point of inflection (to test whether the "smooth" with it, or a pair of them, is significantly better than the "smooth" without). These tests could then be applied to the empirical cumulative distribution function of the observed points (which does not suffer from the disadvantages of the histogram), in order to find out which "apparent" clusters appear to be real.

Acknowledgments: I am grateful to C. Hennig, M. Mächler, G. Shafer and W. Stahel for several valuable discussions and comments.

References

ANDREWS, D.F., BICKEL, P.J., HAMPEL, F.R., HUBER, P.J., ROGERS, W.H., and TUKEY, J.W. (1972): *Robust Estimates of Location; Survey and Advances.* Princeton University Press, Princeton, N.J.

BARNETT, V., and LEWIS, T. (1994): *Outliers in Statistical Data.* Wiley, New York. Earlier editions: 1978, 1984.

BEATON, A.B., and TUKEY, J.W. (1974): The Fitting of Power Series, Meaning Polynomials, Illustrated on Band-Spectroscopic Data. *Technometrics, 16,* 2, 147–185, with Discussion –192.

BENNETT, C.A. 1954: Effect of measurement error in chemical process control. *Industrial Quality Control*, *11*, 17–20.

BERGER, J.O. (1984): The robust Bayesian viewpoint. In: J.B. Kadane (Ed.): *Robustness of Bayesian Analyses*. Elsevier Science, Amsterdam.

BICKEL, P.J. (1975): One-step Huber estimates in the linear model. *J. Amer. Statist. Ass.*, *70*, 428–434.

COX, D.R., and HINKLEY, D.V. (1968): A note on the efficiency of least-squares estimates. *J. R. Statist. Soc. B* , *30*, 284–289.

DANIEL, C. (1976): *Applications of Statistics to Industrial Experimentation*. Wiley, New York.

DANIEL, C., and WOOD, F.S. (1980): *Fitting Equations to Data*. Wiley, New York. Second edition.

DAVIES, P.L. (1995): Data Features. *Statistica Nederlandica*, *49*, 185–245.

DEMPSTER, A.P. (1967): Upper and lower probabilities induced by a multivalued mapping. *Ann. Math. Statist.*, *38*, 325–339.

DEMPSTER, A.P. (1968): A generalization of Bayesian inference. *J. Roy. Statist. Soc.*, *B 30*, 205–245.

DEMPSTER, A.P. (1975): A subjectivist look at robustness. *Bull. Internat. Statist. Inst.*, *46*, Book 1, 349–374.

DONOHO, D.L. (1982): *Breakdown properties of multivariate location estimators*. Ph.D. qualifying paper, Department of Statistics, Harvard University, Cambridge, Mass.

GNANADESIKAN, R. (1977): *Methods for Statistical Data Analysis of Multivariate Observations*. Wiley, New York.

GOOD, I.J. (1983): *Good Thinking; The Foundations of Probability and Its Applications*. University of Minnesota Press, Minneapolis.

GRIZE, Y.L. (1978): *Robustheitseigenschaften von Korrelationsschätzungen*. Diplomarbeit, Seminar für Statistik, ETH Zürich.

HAMPEL, F. (1968): *Contributions to the theory of robust estimation*. Ph.D. thesis, University of California, Berkeley.

HAMPEL, F. (1974): The influence curve and its role in robust estimation. *J. Amer. Statist. Assoc.*, *69*, 383–393.

HAMPEL, F. (1975): Beyond location parameters: Robust concepts and methods (with discussion). *Bull. Internat. Statist. Inst.*, *46*, Book 1, 375–391.

HAMPEL, F. (1978): Optimally bounding the gross-error-sensitivity and the influence of position in factor space. *Invited paper ASA/IMS Meeting. Amer. Statist. Assoc.* Proc. Statistical Computing Section, ASA, Washington, D.C., 59–64.

HAMPEL, F. (1980): Robuste Schätzungen: Ein anwendungsorientierter Überblick. *Biometrical J. 22*, 3–21.

HAMPEL, F. (1983): The robustness of some nonparametric procedures. In: P.J. Bickel, K.A. Doksum and J.L Hodges Jr. (Eds.): *A Festschrift for Erich L. Lehmann*. Wadsworth, Belmont, California, 209-238.

HAMPEL, F. (1985): The breakdown points of the mean combined with some rejection rules. *Technometrics*, *27*, 95–107.

HAMPEL, F. (1987): Design, modelling, and analysis of some biological data sets. In: C.L. Mallows (Ed.): *Design, Data, and Analysis, by some friends of Cuthbert Daniel*. Wiley, New York, 93–128.

HAMPEL, F. (1997): Some additional notes on the "Princeton Robustness Year". In: D.R. Brillinger, L.T. Fernholz and S. Morgenthaler (Eds.): *The Practice of Data Analysis: Essays in Honor of John W. Tukey.* Princeton University Press, Princeton, 133–153.

HAMPEL, F. (1998a): Is statistics too difficult? *Canad. J. Statist.*, *26*, 3, 497–513.

HAMPEL, F. (1998b): On the foundations of statistics: A frequentist approach. In: Manuela Souto de Miranda and Isabel Pereira (Eds.): *Estatística: a diversidade na unidade.* Edições Salamandra, Lda., Lisboa, Portugal, 77–97.

HAMPEL, F. (2000): An outline of a unifying statistical theory. Gert de Cooman, Terrence L. Fine and Teddy Seidenfield (Eds.): *ISIPTA'01 Proceedings of the Second International Symposium on Imprecise Probabilities and their Applications.* Cornell University, June 26-29, 2001. Shaker Publishing BV, Maastricht, Netherlands (2000), 205–212.

HAMPEL, F. (2002): Robust Inference. In: Abdel H. El-Shaarawi and Walter W. Piegorsch (Eds.): *Encyclopedia of Environmetrics*, *3*, 1865–1885.

HAMPEL, F.R., Ronchetti, E.M., Rousseeuw, P.J., and Stahel, W.A. (1986): *Robust Statistics: The Approach Based on Influence Functions.* Wiley, New York.

HENNIG, C. (1998) Clustering and outlier identification: Fixed Point Clusters. In: A. Rizzi, M. Vichi, and H.-H. Bock (Eds.): *Advances in Data Science and Classification.* Springer, Berlin, 37–42.

HENNIG, C. (2001) Clusters, Outliers, and Regression: Fixed Point Clusters. *J. Multivariate Anal.* Submitted.

HENNIG, C., and CHRISTLIEB N. (2002): Validating visual clusters in large data sets: Fixed point clusters of spectral features. *Computational Statistics and Data Analysis*, to appear.

HUBER, P. (1981): *Robust Statistics.* Wiley, New York.

JEFFREYS, H. (1939): *Theory of Probability.* Clarendon Press, Oxford. Later editions: 1948, 1961, 1983.

KÜNSCH, H.R., BERAN, J., and HAMPEL F.R. (1993): Contrasts under long-range correlations. *Ann. Statist.*, *21* 2, 943–964.

KAUFMAN, L., and ROUSSEEUW, P.J. (1990): *Finding Groups in Data: An Introduction to Cluster Analysis.* Wiley, New York.

MÄCHLER, M.B. (1989): *Parametric' Smoothing Quality in Nonparametric Regression: Shape Control by Penalizing Inflection Points.* Ph. D. thesis, no 8920, ETH Zurich, Switzerland.

MÄCHLER, M.B. (1995a): Estimating Distributions with a Fixed Number of Modes. In: H. Rieder (Ed.): *Robust Statistics, Data Analysis, and Computer Intensive Methods – Workshop in honor of Peter J. Huber, on his 60th birthday.* Springer, Berlin, Lecture Notes in Statistics, Volume 109, 267–276.

MÄCHLER, M.B. (1995b): Variational Solution of Penalized Likelihood Problems and Smooth Curve Estimation. *The Annals of Statistics.* *23*, 1496–1517.

MARONNA, R.A. (1976): Robust M-estimators of location and scatter. *Ann. Statist.*, *4*, 51–67.

PROHOROV, Y.V. (1956): Convergence of random processes and limit theorems in probability theory. *Theor. Prob. Appl.*, *1*, 157–214.

RELLES, D.A., and ROGERS, W.H. (1977): Statisticians are fairly robust estimators of location. *J. Amer. Statist. Assoc.*, *72*, 107–111.

ROSENTHAL, R. (1978): How often are our numbers wrong? *American Psychologist*, *33*, 11, 1005–1008.

SHAFER, G. (1976): *A Mathematical Theory of Evidence.* Princeton University Press, Princeton, N. J.

STAHEL, W. (1981a): *Robust estimation: Infinitesimal optimality and covariance matrix estimators (in German)* Ph. D. thesis, no 6881, ETH Zurich, Switzerland.

STAHEL, W. (1981b): *Breakdown of covariance estimators.* Research Report 31, ETH Zurich, Switzerland.

STIGLER, S.M. (1977): Do robust estimators work on real data? *Ann. Statist.*, *6*, 1055–1098.

"STUDENT" (1927): Errors of routine analysis. *Biometrika*, *19*, 151–164.

TUKEY, J.W. (1960): A survey of sampling from contaminated distributions. In: I. Olkin, S.G. Ghurye, W. Hoeffding, W.G. Madow, and H.B. Mann (Eds.): *Contributions to Probability and Statistics.* Stanford University Press, 448–485.

Partial Defuzzification of Fuzzy Clusters

Slavka Bodjanova

Department of Mathematics, Texas A&M University-Kingsville,
Kingsville, TX 78363, U.S.A.

Abstract. Two methods of partial defuzzification of fuzzy clusters in a fuzzy k-partition are proposed. The first method is based on sharpening of fuzzy clusters with respect to the threshold $1/k$. Sharpening is accomplished by application of the generalized operator of contrast intensification of fuzzy sets to $k-1$ fuzzy clusters. The second method utilizes the idea of strong sharpening. In this approach the very small membership grades in each fuzzy cluster are reduced to zero. It is explained how these methods can lead to the well known approximation of a fuzzy partition by its crisp maximum membership partition.

1 Introduction

Let $X = \{x_1, \ldots, x_n\}$ be a given set of objects. Denote by $F(X)$ the set of all fuzzy sets defined on X. Let X be partitioned into k classes, $k \geq 2$. The degenerate fuzzy partition space associated with X was defined by Bezdek (1981) as follows:

$$P_{fko} = \{U \in V_{kn} : u_{ij} \in [0,1] : \sum_i u_{ij} = 1 \text{ for all } j; \sum_j u_{ij} \geq 0 \text{ for all } i\}, \tag{1}$$

where V_{kn} is the usual vector space of real $k \times n$ matrices. Let $U \in P_{fko}$. The least certain (the most fuzzy) partitioning of object $x_j \in X$ into k clusters of U is obtained when $u_{ij} = 1/k$ for all i. On the other hand, the closer the coefficients u_{ij} are to 0 or 1, the more certain (less fuzzy) is the partitioning of x_j in U. We will measure the amount of fuzziness of $U \in P_{fko}$ by partition entropy $H(U)$ introduced by Bezdek (1981) as follows:

$$H(U) = \frac{1}{n} \sum_i \sum_j h(u_{ij}), \tag{2}$$

where $h(u_{ij}) = u_{ij} \log_a(u_{ij})$, $a \in (0, \infty)$ and $h(0) = 0$. We will use $a = 1/k$. Then $0 \leq H(U) \leq 1$ for all $U \in P_{fko}$.

In applications, $U \in P_{fko}$ is often approximated by its maximum membership partition $MM(U)$ derived as follows: for all j, if $u_{rj} = \max_i\{u_{ij}\}$ then $MM(u_{rj}) = 1$ and $MM(u_{ij}) = 0$ for all $i \neq r$. However, if $u_{rj} = u_{sj} = \max_i\{u_{ij}\}$ and $r \neq s$ then $MM(U)$ is not unique. Also, if the number of clusters is large (e.g., $k \geq 4$) then $u_{rj} = \max_i\{u_{ij}\}$ might be too small (e.g.,

$u_{rj} = 1/k \leq 0.25$) and therefore it should not be approximated by 1.
We propose an approximation of $U \in P_{fko}$ by a partially defuzzified partition. A natural way of partial defuzzification of U is to increase its large membership grades and to decrease those which are small. In this paper we consider sharpening of membership grades with respect to $1/k$. The threshold $1/k$ separates "large" and "small" membership grades. Jang et al. (1997) introduced the operator of contrast intensification of fuzzy sets which enables one to sharpen fuzzy sets with respect to $1/2$. This operator was generalized in Bodjanova (2001) to the operator of α-contrast intensification, where the threshold α can be any number from the interval $(0, 1)$. In Section 2 we study how the operator of α-contrast intensification for $\alpha = 1/k$ can be used in partial defuzzification of $U \in P_{fko}$. The method of contrast intensification, or CI method, will be proposed. It is obvious that the "ultimate" sharpening of a fuzzy partition should be a crisp partition. However, the CI method never changes a fuzzy partition to crisp. Therefore, in Section 3 we propose defuzzification based on strong sharpening of membership grades. In this approach the very small membership grades will be sharpened down to zero. Because the threshold $\delta \in (0, 1/k)$ which separates "very small" and "small" coefficients needs to be chosen a priori, we call this method the delta membership, or DM method.

2 Method of contrast intensification

Let $U \in P_{fko}$. By sharpening of membership grades u_{ij} we will obtain a fuzzy partition $U^s \in P_{fko}$ such that $H(U^s) \leq H(U)$. Sharpening proceeds as follows:
Step 1: If $1/k < u_{ij} < 1$ then $u_{ij} < u^s_{ij} < 1$.
Step 2: If $0 < u_{ij} < 1/k$ then $0 < u^s_{ij} < u_{ij}$.
Step 3: If $u_{ij} \in \{0, 1/k, 1\}$ then $u^s_{ij} = u_{ij}$.

In Jang et al. (1997) the operator of contrast intensification, INT, was introduced for sharpening of fuzzy sets. This operator, when applied to a fuzzy set $A \in F(X)$, creates the fuzzy set $INT(A) \in F(X)$ defined for all $x \in X$ by the membership function

$$INT(A)(x) = \begin{cases} 2[A(x)]^2 & \text{if } A(x) \leq 0.5, \\ \neg 2[\neg A(x)]^2 & \text{if } A(x) \geq 0.5, \end{cases} \tag{3}$$

where $\neg A(x)$ denotes the complement of $A(x)$ given by $\neg A(x) = 1 - A(x)$. $INT(A)$ satisfies the following properties:
P1. If $1/2 < A(x) < 1$ then $A(x) < INT(A)(x) < 1$.
P2. If $0 < A(x) < 1/2$ then $0 < INT(A)(x) < A(x)$.
P3. If $A(x) \in \{0, 1/2, 1\}$ then $A(x) = INT(A)(x)$.
Therefore, operator INT can be used for sharpening of membership grades u_{ij} of fuzzy partition $U \in P_{fko}$ when $k = 2$. It can easily be checked that $INT(U) = [INT(u_{ij})] \in P_{f2o}$ such that $H(INT(U)) \leq H(U)$. We will prove

only that $\sum_i INT(u_{ij}) = 1$.
Let, e.g., $u_{1j} \leq 1/2$. Then $\sum_i INT(u_{ij}) = INT(u_{1j}) + INT(u_{2j}) = 2u_{1j}^2 + [1 - 2(1 - u_{2j})^2] = 2u_{1j}^2 + [1 - 2u_{1j}^2] = 1$.

Operator INT was generalized in Bodjanova (2001) to the operator of α-contrast intensification, INT_α, for $\alpha \in (0, 1)$.

$$INT_\alpha(A)(x) = \begin{cases} \frac{1}{\alpha}[A(x)]^2 & \text{if } A(x) \leq \alpha, \\ \neg_\alpha(\frac{1}{\alpha}[\neg_\alpha A(x)]^2) & \text{if } A(x) \geq \alpha, \end{cases} \tag{4}$$

where $\neg_\alpha$ is α-fuzzy complement given by $\neg_\alpha A(x) = \frac{1-A(x)}{1+\lambda A(x)}$, and $\lambda = \frac{1-2\alpha}{\alpha^2}$. Obviously, $\alpha < A(x) < INT_\alpha(A)(x) < 1$ and $0 < INT_\alpha(A)(x) < A(x) < \alpha$ and $INT_\alpha(A)(x) = A(x)$ if $A(x) \in \{0, \alpha, 1\}$. Therefore, operator INT_α, for $\alpha = 1/k$ can be used for sharpening of membership grades u_{ij} of fuzzy partition $U \in P_{fko}$. However, $INT_{1/k}(U)$ is not necessarily a fuzzy k-partition. We propose the method of contrast intensification (CI method), which will change $U \in P_{fko}$ into $CI(U) \in P_{fko}$ such that $H(CI(U)) \leq H(U)$.

Algorithm 1
Input: Let $U \in P_{fko}, I = \{1, \ldots, k\}$. Assign $CI(U) := U$. Then put $j := 1$.
Step 1: Find $I_{1j} = \{u_{ij} | u_{ij} \in (1/k, 1), i \in I\}$. If card $I_{1j} = 1$ then go to Step 2. Else find $I_{2j} = \{u_{ij} | u_{ij} \in (0, 1/k), i \in I\}$. If card $I_{2j} = 1$ then go to Step 3. Else go to Step 4.
Step 2: Let $I_{1j} = \{u_{rj}\}, r \in I$. Then for all $i \in I - \{r\}, CM(u_{ij}) = INT_{1/k}(u_{ij}) = ku_{ij}^2$ and $CM(u_{rj}) = 1 - \sum_{i \neq r} CM(u_{ij})$. Go to Step 4.
Step 3: Let $I_{2j} = \{u_{sj}\}, s \in I$. Then for all $i \in I - \{s\}, CM(u_{ij}) = INT_{1/k}(u_{ij}) = \neg_{1/k}(k[\neg_{1/k} u_{ij}]^2)$. If $\sum_{i \neq s} CM(u_{ij}) < 1$ then $CM(u_{sj}) = 1 - \sum_{i \neq s} CM(u_{ij})$. Otherwise, for all $i \in I, CM(u_{ij}) := u_{ij}$.
Step 4: If $j = n$ then stop. Else $j := j + 1$ and go to Step 1.

Example 1 Let $U \in P_{f4o}$ be a partition of $X = \{x_1, \ldots, x_7\}$ given by matrix

$$U = \begin{pmatrix} 0.20 & 0.10 & 0.00 & 0.40 & 0.10 & 0.25 & 0.52 \\ 0.30 & 0.20 & 0.50 & 0.30 & 0.70 & 0.30 & 0.10 \\ 0.25 & 0.20 & 0.30 & 0.10 & 0.20 & 0.30 & 0.22 \\ 0.25 & 0.50 & 0.20 & 0.20 & 0.00 & 0.15 & 0.16 \end{pmatrix}.$$

Then

$$CI(U) = \begin{pmatrix} 0.16 & 0.04 & 0.00 & 0.40 & 0.04 & 0.25 & 0.67 \\ 0.34 & 0.16 & 0.50 & 0.30 & 0.80 & 0.35 & 0.04 \\ 0.25 & 0.16 & 0.30 & 0.10 & 0.16 & 0.35 & 0.19 \\ 0.25 & 0.64 & 0.20 & 0.20 & 0.00 & 0.05 & 0.10 \end{pmatrix}.$$

Fuzzy partition $CI(U)$ is sharper than U. We can check that $H(CI(U)) = 0.721$ and $H(U) = 0.855$. Hence CI method reduced the fuzziness of U.

Because $CI(U)$ is a fuzzy partition, it can be further defuzzified by the CI method. Let $CI(CI(U)) = CI^2(U), CI(CI^2(U)) = CI^3(U), \ldots,$
$CI(CI^{n-1}(U)) = CI^n(U)$. We will show that if we have a fuzzy partition $U \in P_{fko}$ where all $u_{ij} \neq 1/k$ and each column of matrix U has the obvious maximum (only one large membership grade in the column) then for every small $\epsilon > 0$ we can find number n of repetitions of the CI method on U such that $\max_{ij} |CI^n(u_{ij}) - MM(u_{ij})| < \epsilon$.

Proposition 1 *Let $U \in P_{fko}$ such that all $u_{ij} \neq 1/k$ and for each $j \in \{1, \ldots, n\}$, card $I_{1j} = \{u_{ij} | u_{ij} \in (1/k, 1), i \in \{1, \ldots, k\}\} = 1$. Then $\lim_{n\to\infty} CI^n(U) = [\lim_{n\to\infty} CI^n(u_{ij})] = MM(U)$.*

Proof: For each $j, I_{1j} = \{u_{rj}\}, r \in \{1, \ldots, k\}$. Therefore $MM(u_{rj}) = 1$ and $MM(u_{ij}) = 0$ for all $i \neq r$. According to the CI method, for all $i \neq r, CI(u_{ij}) = k(u_{ij}^2), CI^2(u_{ij}) = k(ku_{ij}^2)^2 = k^3u_{ij}^4, CI^3(u_{ij}) = k(k^3u_{ij}^4)^2 = k^7u_{ij}^8, \ldots, CI^n(u_{ij}) = k^{(2^n-1)}u_{ij}^{2^n} = (1/k)(ku_{ij})^{2^n}$. Because for all $i \neq r, u_{ij} < 1/k$, we obtain that $ku_{ij} < 1$. Therefore $\lim_{n\to\infty}(1/k)(ku_{ij})^{2^n} = 0 = MM(u_{ij})$. Then $CI^n(u_{rj}) = 1 - \sum_{i\neq r} CI^n(u_{ij})$, and
$\lim_{n\to\infty} CI^n(u_{rj}) = 1 - \sum_{i\neq r} \lim_{n\to\infty} CI^n(u_{ij}) = 1 - 0 = 1 = MM(u_{rj})$.

3 Method of delta membership

The method of delta membership (DM method) transforms a partition $U \in P_{fko}$ into partition $DM(U) \in P_{fko}$ such that $H(DM(U)) \leq H(U)$. Transformation is based on strong sharpening of membership grades u_{ij} with respect to $1/k$. In this type of sharpening, large membership grades will be sharpened up to 1 (1 included) and small coefficients will be sharpened down to 0 (0 included). An a priori chosen threshold $\delta \in (0, 1/k)$ identifies very small coefficients ($u_{ij} \leq \delta$). These coefficients will be reduced to zero. The most uncertain membership grade, $u_{ij} = 1/k$, will remain unchanged because it is neither large nor small with respect to the threshold $1/k$.

Algorithm 2
Input: Let $U \in P_{fko}, \delta \in (0, 1/k)$. Assign $DM(U) := U$. Then put $j := 1$.
Step 1: Find $I_j = \{u_{ij} | u_{ij} \leq \delta, i \in \{1, \ldots, k\}\}$. If $I_j = \emptyset$ then go to Step 3.
Step 2: Let $r_j =$ card I_j and $S_j = \sum_{u_{ij}\in I_j} u_{ij}$. Compute $\beta_j = \frac{S_j}{r_j}$. If $u_{ij} \notin I_j$ then $DM(u_{ij}) = \frac{1}{k} + \frac{1}{1-k\beta_j}(u_{ij} - \frac{1}{k})$. Otherwise $DM(u_{ij}) = 0$.
Step 3: If $j = n$ then stop. Else $j := j + 1$ and go to Step 1.

Note: It may happen that $\sum_j u_{ij} > 0$ but $\sum_j DM(u_{ij}) = 0$.

Proposition 2 *Let $U \in P_{fko}$ and $\delta \in (0, 1/k)$. Then $DM(U)$ satisfies the following properties:*
P1. For all $j, \sum_i DM(u_{ij}) = 1$.

P2. If $u_{ij} \in (0, \delta]$ *then* $DM(u_{ij}) = 0$.
P3. If $u_{ij} \in (\delta, 1/k)$ *then* $0 < DM(u_{ij}) < u_{ij}$.
P4. If $u_{ij} \in \{0, 1/k, 1\}$ *then* $DM(u_{ij}) = u_{ij}$.
P5. If $u_{ij} \in (1/k, 1)$ *then* $u_{ij} < DM(u_{ij}) \leq 1$.
Therefore $DM(U)$ *belongs to* P_{fko} *and it is sharper (less fuzzy) than* U.

Proof: Properties P2 and P4 are obvious.
Proof of P1: $\sum_i DM(u_{ij}) = \sum_{u_{ij} \in I_j} DM(u_{ij}) + \sum_{u_{ij} \notin I_j} DM(u_{ij}) = 0 + \sum_{u_{ij} \notin I_j} \left(\frac{1}{k} + \frac{1}{1-k\beta_j}(u_{ij} - \frac{1}{k}) \right) = \frac{k-r_j}{k} + \frac{1}{1-k\beta_j} \sum_{u_{ij} \notin I_j} u_{ij} - \frac{k-r_j}{(1-k\beta_j)k} = \frac{k-r_j}{k} + \frac{1}{1-k\beta_j}(1 - S_j) - \frac{k-r_j}{(1-k\beta_j)k} = \frac{k-r_j}{k} + \frac{1}{1-k\beta_j}(1 - r_j\beta_j) - \frac{k-r_j}{(1-k\beta_j)k} = 1.$
Proof of P3: Let $\delta < u_{ij} < 1/k$. Because $\beta_j = \frac{S_j}{r_j} \leq \delta < \frac{1}{k}$, we have that $k\beta_j < 1$ and $\frac{1}{1-k\beta_j} > 1$. Then $DM(u_{ij}) - \frac{1}{k} = \frac{1}{1-k\beta_j}(u_{ij} - \frac{1}{k}) < u_{ij} - \frac{1}{k}$, which means that $DM(u_{ij}) < u_{ij}$. We also need to show that $0 < DM(u_{ij})$. Because $\beta_j \leq \delta < u_{ij} < 1/k$, we obtain that $|u_{ij} - \frac{1}{k}| < \frac{1}{k} - \beta_j = \frac{1}{k}(1 - k\beta_j)$, and therefore $\frac{1}{1-k\beta_j}|u_{ij} - \frac{1}{k}| < \frac{1}{k}$. Hence $DM(u_{ij}) = \frac{1}{k} + \frac{1}{1-k\beta_j}(u_{ij} - \frac{1}{k}) = \frac{1}{k} - \frac{1}{1-k\beta_j}|u_{ij} - \frac{1}{k}| > 0$.
Proof of P5: If $1/k < u_{ij} < 1$ then $DM(u_{ij}) - \frac{1}{k} = \frac{1}{1-k\beta_j}(u_{ij} - \frac{1}{k}) > u_{ij} - \frac{1}{k}$, which means that $DM(u_{ij}) > u_{ij}$. Because from P1 we know that $\sum_i DM(u_{ij}) = 1$ for all j, we must have $DM(u_{ij}) \leq 1$.

Example 2 Let U be the fuzzy partition from Example 1. Let $\delta = 0.20$. Then

$$DM(U) = \begin{pmatrix} 0 & 0 & 0 & \frac{5}{8} & 0 & \frac{1}{4} & \frac{13}{16} \\ \frac{1}{2} & 0 & \frac{2}{3} & \frac{3}{8} & 1 & \frac{3}{8} & 0 \\ \frac{1}{4} & 0 & \frac{1}{3} & 0 & 0 & \frac{3}{8} & \frac{3}{16} \\ \frac{1}{4} & 1 & 0 & 0 & 0 & 0 & 0 \end{pmatrix}.$$

Fuzzy partition $DM(U)$ is sharper than U. We can check that $H(DM(U)) = 0.402$. Because $H(U) = 0.855$, we conclude that the DM method substantially reduced the fuzziness of U.

Proposition 3 *Let* $U \in P_{fko}$ *such that all* $u_{ij} \neq 1/k$ *and for each* $j \in \{1, \ldots, n\}$, *card* $I_{1j} = \{u_{ij} | u_{ij} \in (1/k, 1), i \in \{1, \ldots, k\}\} = 1$. *Let* $\delta = \max_{i,j}\{u_{ij} | u_{ij} < 1/k\}$. *Then* $DM(U) = MM(U)$.

Proof: For each j, let $I_j = \{u_{rj}\}, r \in \{1, \ldots, k\}$. Then $MM(u_{rj}) = 1$, and $MM(u_{ij}) = 0$ for all $i \neq r$. Obviously, $\beta_j = \frac{1-u_{rj}}{k-1}$. Then $DM(u_{rj}) = \frac{1}{k} + \frac{1}{1-k\beta_j}(u_{rj} - \frac{1}{k}) = \frac{1}{k} + \frac{k-1}{k-1-k+ku_{rj}} \frac{ku_{rj}-1}{k} = \frac{1}{k} + \frac{k-1}{ku_{rj}-1} \frac{ku_{rj}-1}{k} = 1$. Therefore $DM(u_{rj}) = MM(u_{rj})$. For all $u_{ij}, i \neq r, u_{ij} \leq \delta$. Hence $DM(u_{ij}) = 0 = MM(u_{ij})$.

4 Conclusion

The main reason for defuzzification of fuzzy partitions is to simplify interpretation of the results of classification. Sharpening of membership functions of fuzzy clusters may also improve handling of fuzzy partitions in further applications. However, fuzziness is one of the most important characterizations of fuzzy partitions and its reduction to zero, when using approximation by the maximum membership method, may not always be appropriate. The only information which one can get from $MM(U)$ about the original fuzzy partition U is the location of maximum membership grade for each object $x_j \in X$, provided that the maximum is unique. Our proposed methods of partial defuzzification give researchers more flexibility in reduction of fuzziness. A researcher may regulate intensity of defuzzification by specifying number n of repetitions of the CI method or by specifying parameter δ in the DM method. Results of both methods provide information about locations of all large membership grades and all small membership grades in the original fuzzy partition. They also leave the most uncertain element $u_{ij} = 1/k$ unchanged.

From the description of the CI method it is clear that only the partitioning of object $x_j \in X$ which exhibits obvious maximal or obvious minimal membership grade in $U \in P_{fko}$ may be defuzzified by Algorithm 1. Perhaps a partitioning of $x_j \in X$ with no obvious maximum or minimum is "too fuzzy" and should not be replaced by its sharper version. This problem will be studied in our future work. In the DM method the notion of "obvious minimum" was extended to all very small membership grades, i.e., grades $u_{ij} \leq \delta$, where δ is chosen a priori. Again, only the partitioning of object $x_j \in X$ which has at least one very small membership grade in U may be defuzzified by Algorithm 2.

The main difference between the two methods is that the CI method does not sharpen the membership coefficients from the open interval $(0, 1)$ to zero or one while the DM method does. Also the DM method may decrease the number of nonempty fuzzy clusters (by transforming a weak fuzzy cluster to the empty set) while the CI method keeps the original number of clusters unchanged. The method of contrast intensification as well as the method of delta membership can be further modified. For example, the threshold $\alpha = \frac{1}{k}$ for sharpening might be chosen separately for each column of matrix $U \in P_{fko}$. We suggest $\alpha_j = \frac{1}{k-z_j}$, where z_j is the number of zero membership grades in the jth column of U. Some further modifications are under investigation.

References

BEZDEK, J.C. (1987): *Pattern Recognition with Fuzzy Objective Function Algorithms.* Plenum Press, New York.

BODJANOVA, S. (2001): Operators of Contrast Modification of Fuzzy Sets. In: *Proceedings of Joint 9th IFSA World Congress and 20th NAFIPS International Conference*, IEEE, Vancouver, 1478–1483.

JANG, J.-S.R, SUN, C.T. and MIZUTANI, E. (1997) *Neuro-Fuzzy and Soft Computing. A Computational Approach to Learning and Machine Intelligence.* Prentice Hall, Upper Saddle River.

A New Clustering Approach, Based on the Estimation of the Probability Density Function, for Gene Expression Data

Noël Bonnet[1,2], Michel Herbin[2], Jérôme Cutrona[1,2], and Jean-Marie Zahm[1]

[1] Inserm Unit 514 (UMRS, IFR53)
45 rue Cognacq Jay, 51092 Reims cedex, France
[2] LERI, IUT Léonard de Vinci, University of Reims
Rue des Crayères, BP 1035, 51687 Reims cedex, France

Abstract. Many techniques have already been suggested for handling and analyzing the large and high-dimensional data sets produced by newly developed gene expression experiments. These techniques include supervised classification and unsupervised agglomerative or hierarchical clustering techniques. Here, we present an alternative approach that does not make assumption on the shape, size and volumes of the clusters. The technique is based on the estimation of the probability density function (pdf). Once the pdf is estimated, with the Parzen technique (with the right amount of smoothing), the parameter space is partitioned according to methods inherited from image processing, namely the skeleton by influence zones and the watershed. We show some advantages of this suggested approach.

1 Introduction

Techniques are rapidly developing that allow to monitor the expression of thousands of genes in parallel. These gene expressions can be recorded as a function of time (along the cell cycle for instance), as a function of different tissues (normal and different malignant ones for instance) or as a function of different actions performed on the set of genes. As for many experimental techniques, this group of techniques produces huge amounts of data, which require in turn more and more sophisticated data analysis techniques.

During the past three or four years, many techniques have been (re)–developed for the automatic or semi–automatic handling and analysis of gene expression data sets. Emphasis has been put on clustering techniques, with the aim of finding classes of genes that behave similarly, i.e. that constitute co–regulated families. Briefly, several groups of automatic classification techniques have been investigated:

- some supervised approaches: support vector machines (SVM) (Brown et al. (2000)); significance analysis of micro–arrays (SAM) and its generalization as cluster scoring (Tibshirani et al. (2001))
- mostly, unsupervised approaches (clustering):

  - hierarchical clustering: standard average linkage agglomerative clustering (Weinshtein and Meyers (1997); Eisen et al. (1998); Sherf et al. (2000)); FITCH (Wen et al. (1998)); self–organizing tree algorithm (SOTA), a hierarchically growing neural network of the Kohonen type (Herrero et al. (2001)); model–based divisive hierarchical clustering (Alon et al. (1999))
  - partitioning clustering: K–means (KM) (Tavazoie et al. (1999)), Fuzzy C–means (FCM) (Mjolsness et al. (1999)); self–organizing mapping (SOM) (Tamayo et al. (1999)); algorithms based on the random graph theory: cluster affinity search technique (CAST) (Ben–Dor et al. (1999)), highly connected sub graph (HCS) (Hartuv et al. (1999)); model–based clustering (Yeung et al. (2001))
- some statistical methods, such as Principal Components Analysis and Singular Value Decomposition (SVD) have also been presented as clustering techniques (Wall et al. (2001)), but we do not consider them as such (see below).

Taking into account these different proposals, one may wonder why it should be useful to suggest another one. In fact, we think that no clustering method can be efficient in all the situations that can be encountered in practice. All the methods cited above are known to have deficiencies in specific situations. Briefly speaking, agglomerative hierarchical techniques are known to be sensitive to noise, the K–means and FCM techniques assume hyper–spherical clusters of equal size (when the Euclidean distance is used) and hyper–elliptical clusters when the Mahalanobis distance is used; model–based clustering almost always assume that Gaussian clusters are present.

Recently, we have developed a clustering technique that does not make such assumptions on the shape and size of clusters. We do not claim that it will systematically outperform all the techniques already in use. But we think it possesses some advantages in many situations and that it could be useful to have it also in the toolbox.

2 Presentation of the technique

Preliminary remark: In many papers devoted to the analysis of gene expression data, we think that some confusion is present concerning dimensionality reduction and the classification itself. In our philosophy (see for instance Bonnet (2000)), these two activities have to be clearly differentiated. Our method implies that dimensionality reduction is performed before attempting to perform clustering. It is so for theoretical reasons (the sparsity of high dimensional spaces and the curse of dimensionality problem) as well as for practical reasons. It is always useful to *see* how the data set looks like, rather than performing automatic classification blindly.

2.1 Dimensionality reduction

We start from data in a very high dimensional space: the dimensionality may be several tenths (the number of features) when the objects studied are the genes, or several thousands (the number of genes) when the objects to classify are the experimental conditions, i.e. the features of the genes. As stated above, trying to work directly in these high–dimensional spaces is very risky and should not be recommended, from our point of view.

Many techniques have been developed for dimensionality reduction. Briefly, we will mention: feature selection or feature reduction, i.e. building a new reduced set of features from the original large set. This can be done through linear methods (Principal Components Analysis, Karhunen–Loeve analysis, etc) or through non linear methods (Sammon's mapping, Multidimensional scaling, Self–Organizing mapping, Auto–associative neural networks, etc). For us, none of these techniques is a clustering technique. They only help to reduce the dimensionality of the data set in order to visualize it and to prepare the classification step. Sometimes, linear techniques suffice to reduce the dimensionality to two or three without loosing much of the information. In this favorable situation, two– or three–dimensional scatter diagrams (two– or three–dimensional histograms) can be built and the distribution of objects into several clouds of points can be visualized and interpreted. But often, performing such a dimensionality reduction without loosing too much information is impossible. We can then try to apply non linear methods, which are able to better reduce the dimensionality, at the price of a greater distortion of the data set. From our experience of dimensionality reduction in other fields of application (Bonnet (1998), Guerrero et al. (2000)), we suggest to perform a preliminary reduction with linear methods, to something like ten components. Then, non linear methods can be used to further reduce the dimension of the data set to two or three.

2.2 Clustering

Our clustering philosophy is that a cluster is represented by a local peak of the probability density function (pdf). So, two (or more) clusters can be recognized by the existence of a valley between the local peaks, or modes, of the pdf. The data set represents a sampling of the pdf. So, the pdf can be estimated from the data set or, better, its mapping onto a low dimensional space.

One of the most often used method for estimating the pdf from the data points is the Parzen–Rosenblatt method, and we suggest to use it. Of course, the size of the (Gaussian or Epaneknikov) kernel is very important in this context, since a small kernel will lead to many modes and a large kernel to a few modes. Due to the lack of space, we do not consider this problem here, and refer to Herbin et al. (2001) and references therein.

Once the pdf is estimated, the number of classes is equal to the number of modes of the pdf. Note that these modes play the role of class centers in the KM of FCM techniques, but without any constraint on the shape, size and volume of clusters. The next step consists in defining the boundaries of the classes in the parameter space. For this, we suggested to use techniques already in use for image processing, especially mathematical morphology techniques. Our first suggestion (Herbin et al. (1996)) was to use the skeleton by influence zones (SKIZ) to iteratively thresholded versions of the estimated pdf. Our updated suggestion (Bonnet et al. (1997), Bonnet (2000)) was to use the watersheds instead, which does not necessitate any thresholding.

At the end of the procedure, the parameter space is labeled into as many classes as the estimated pdf has modes. The last step is trivial: the objects to classify are labeled according to the coordinates of their mapping onto the reduced parameter space and the corresponding label.

3 Illustration of the procedure

We have applied the suggested procedure to several gene expression data sets publicly available. Here, we must limit ourselves to only one example for illustration of the method. The data set is the one published by Alon et al. (1999), which concerns tumor and normal colon epithelial cells probed by oligonucleotide arrays. The data set is composed of 2,000 active genes (out of 6,000) and the expression of each of these genes was recorded for 62 experimental conditions: 22 normal tissues and 40 tumor tissues. This data set was previously analyzed by Alon et al. using a model–based method which assumes that the data set is a mixture of two Gaussian probability density functions, separates them and then separates each of the two clusters into two new sub–clusters, and finally organizes the genes into a binary tree. This data set was also analyzed by Ben–Dor et al. (1999) for tissue classification.

The results we have obtained with our technique are now described and illustrated. As a dimensionality reduction approach, we have used Correspondence Analysis, a variant of Multivariate Statistical Analysis which is based on the chi–squares metric (Lebart et al. (1984), Fellenberg et al. (2001)). The mapping of the 62 tissues onto the subspace spanned by the first two factorial axes (representing 35% of the total variance) is displayed in figure 1a. The mapping of the 2,000 genes on the same subspace is displayed in figure 1b. Of course, we can also work with higher–dimensional subspaces, subspaces spanned by three or four factorial axes or subspaces obtained after non linear mapping, thus taking into account more than 35% of the variance. But for simplifying the illustration, we will continue the description of the procedure with these two–dimensional maps.

The next step consists in estimating the probability density functions. Figures 1c and 1d display the results of these estimations for tissues and

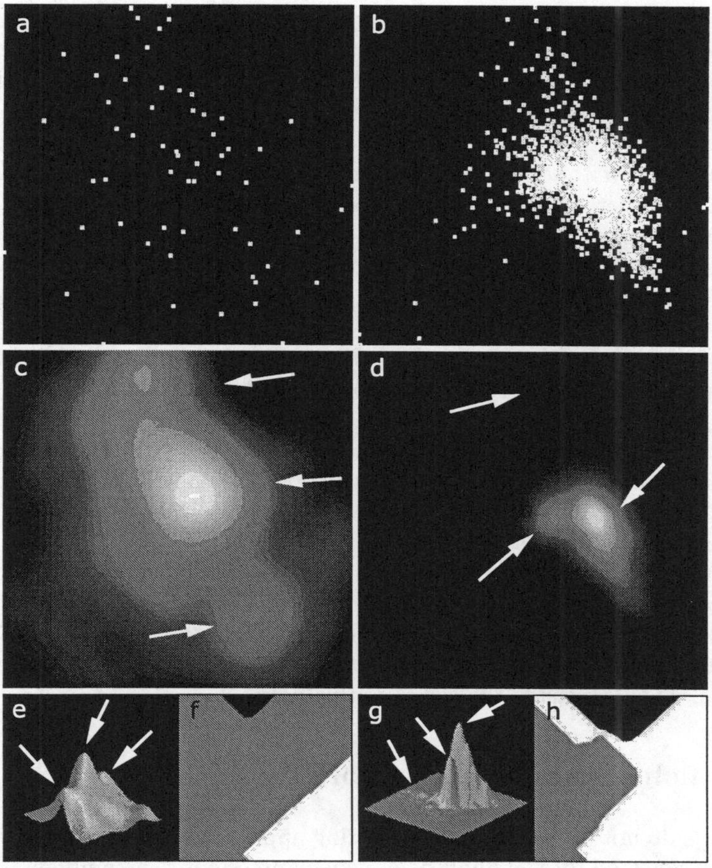

Fig. 1. Two–way clustering of a gene expression data set (60 tissues, 2000 genes). The left column concerns tissues and the right column concerns genes. a,b) Scatter-plots built with the first two factorial components, after Correspondence Analysis. c,d) Probability density functions (pdf) obtained with the Parzen technique. e,f) Estimated pdf visualized in pseudo–3D. g,h) Boundaries between the modes of the pdf, obtained with the watershed technique. Note that these boundaries differ from those obtained with the K–means technique, for instance. This is due to the fact that the watershed technique takes into account the shape of the clusters, and not only their centers.

genes, respectively, using Gaussian kernels. The way to choose the kernel size in order to obtain reliable estimates is described in (Herbin et al. (2001)). Here, we can see that the data set cannot be separated into two classes as expected. Instead, we found three classes: one containing normal tissues, one containing tumor tissues and one class containing a mixture of normal and tumor tissues. For the genes, we found also three classes, although the smaller class could also be aggregated to the largest one. As can be seen, the size and volume of the two main clusters are very different, which prevent any clustering by methods making assumptions on the shape and size of clusters. The estimated probability functions are displayed in pseudo–3D in figures 1e,g.

Then, we have to segment the parameter spaces, i.e. to find the boundaries between the different classes. Here, we applied the watersheds procedure. The results are displayed in figures 1f and 1h. One can notice that, contrary to the K–means procedure, the partitioning differs from a Voronoï partition, since the complete pdf is taken into account, and not only the centers of the classes.

The last step consists in labeling the objects, according to their position in the parameter space and the label associated to this position. In this specific case, we found that 7 tissues were classified in the first class and 15 tissues in the third class. 40 tissues were classified in the mixed class. When three factorial axes are used instead of two, it becomes possible to find two clusters and the percentage of good classification increases to 82% (intermediate results not shown).

We also found three stable clusters of genes: one cluster with 1648 genes, one with 320 genes and one with 32 genes. Of course, no ground truth is available for the genes and only a careful analysis of the results could indicate whether these results are more significant than those obtained by alternative methods. This is outside the scope of this note.

4 Conclusion and future work

Again, we do not claim that our clustering approach outperforms all the ones already suggested for the analysis of gene expressions. Certainly, we can find situations where it is less appropriate. But we claim that it offers a number of advantages:

- Our method is a partitioning technique, but it can easily be transformed into a hierarchical technique, just by changing the size of the kernel used for estimating the pdf. But in this case, the agglomeration is still performed on a global basis, and not on a local basis.
- It does not make any assumption concerning the shape, the size and the volume of clusters. It can even handle non convex cluster shapes.
- The need to perform dimensionality reduction as a pre–processing step may be seen as a handicap. But in fact, we think that trying to perform clustering directly in a one hundred– or ten thousands–dimensional space is a very risky procedure. In addition, the mapping onto a low–dimensional space offers many advantages in terms of visualization and qualitative interpretation of the data set.

Future work in the direction indicated here could include:

- the comparison of the method suggested here, which makes explicit use of the estimated pdf, with other clustering techniques which make implicit use of the pdf, such as the mean–shift procedure for instance (Cheng (1995), Comaniciu and Meer (2002)).
- the extension of the crisp clustering procedure described here to fuzzy variants of this approach that we have already performed in other contexts (Bonnet and Cutrona (2001), Cutrona et al. (2002)). This would allow, for instance, to effectively cluster genes (or experimental conditions) only when one of their grades of membership is high and put the others in a reject class.

References

ALON, U., BARKAI, N., NOTTERMAN, D.A., GISH, K., YBARRA, S., MACK, D., and LEVINE, A.J. (1999): Broad patterns of gene expression revealed by clustering analysis of tumor and normal colon tissues probed by oligonucleotide arrays. *Proc. Natl. Acad. Sci. USA*, *96*, 6745–6750.

BEN–DOR, A., SHAMIR, R., and YAKHINI, Z. (1999): Clustering gene expression patterns. *Journal of Computational Biology*, *6*, 281-297.

BONNET, N. (1998): Multivariate statistical methods for the analysis of microscope image series. *Journal of Microscopy*, *190*, 2-18.

BONNET, N. (2000): Artificial intelligence and pattern recognition techniques in microscope image processing and analysis. *Advances in Imaging and Electron Physics*, *114*, 1-77.

BONNET, N., HERBIN, M., and VAUTROT, P. (1997): Une méthode de classification non supervisée ne faisant pas d'hypothèse sur la forme des classes: application à la segmentation d'images multivariables. *Cinquièmes Rencontres de la Société Francophone de Classification.* Lyon. Proceedings pp 151-154.

BONNET, N., and CUTRONA, J. (2001): Improvement of unsupervised multi–component image segmentation through fuzzy relaxation. *IASTED International Conference on Visualization, Imaging and Image Processing (VI-IP'2001)* Marbella (Spain). Acta Press: 477-482.

BROWN, M., GRUNDY, W., LIN, D., CRISTIANINI, N., SUGNET, C., FUREY, T., ARES, M., and HAUSSLER, D. (2000): Knowledge–based analysis of microarray gene expression data by using support vector machines. *Proc. Nat. Acad. Sci. USA*, *97*, 262-267.

CHENG, Y. (1995): Mean shift, mode seeking and clustering. *IEEE Transactions on Pattern Analysis and Machine Intelligence*, *17*, 790-799.

COMANICIU, D., and MEER, P. (2002): Mean shift: a robust approach toward feature space analysis. *IEEE Transactions on Pattern Analysis and Machine Intelligence.* In press.

CUTRONA, J., BONNET, N., and HERBIN, M. (2002): A new fuzzy clustering technique based on pdf estimation. *Information Processing and Management of Uncertainty (IPMU'2002).* Submitted.

EISEN, M.B., SPELLMAN, P.T., BROWN, P.O., and BOTSTEIN, D. (1998): Cluster analysis and display of genome–wide expression patterns. *Proc. Natl. Acad. Sci. USA*, *95*, 14863-14868.

FELLENBERG, K., HAUSER, N.C., BRORS, B., NEUTZNER, A., HOHEISEL, J.D., and VINGRON, M. (2001): Correspondence analysis applied to microarray data. *Proc. Nat. Acad. Sci. USA*, *98*, 10780-10786.

GUERRERO, A., BONNET, N., MARCO, S., and CARRASCOSA, J. (2000): Comparative study of methods for the automatic classification of macromolecular image sets: preliminary investigations with realistic simulations. *Proc. SPIE - Applications of Artificial Neural Networks in Image Processing V*, *3962*, 92-103.

HARTUV, E., SCHMITT, A., LANGE, J., MEIER–EWERT, S., LEHRACH, H., and SHAMIR, R. (1999): An algorithm for clustering cDNAs for gene expression. *Third Int. Conf. on Computational Molecular Biology (RECOMB'99).* ACM Press, pp. 188-197.

HERBIN, M., BONNET, N., and VAUTROT, P. (1996): A clustering method based on the estimation of the probability density function and on the skeleton by influence zones. *Pattern Recognition Letters*, *22*, 1557-1568.

HERBIN, M., BONNET, N., and VAUTROT, P. (2001): Estimation of the number of clusters and influence zones. *Pattern Recognition Letters*, *17*, 1141-1150.

HERRERO, J., VALENCIA, A., and DOPAZO, J. (2001): A hierarchical unsupervised growing neural network for clustering gene expression patterns. *Bioinformatics*, *17*, 126-136.

LEBART, L., MORINEAU, A., and WARWICK, K.M. (1984): *Multivariate Descriptive Statistical Analysis.* Wiley & Sons, New York.

MJOLSNESS, E., NO, R.C., and WOLD, B. (1999): Multi–parent clustering algorithms for large scale gene expression analysis. *Technical report JPL-ICTR-99-5.*

SHERF U. et al. (2000): A gene expression database for the molecular pharmacology of cancer. *Nature Genetics*, *24*, 236-244.

TAMAYO, P., SLONIM, D., MESIROV, J., ZHU, Q., KITAREEWAN, S., D-MITROWSKY, E., LANDER, E., and GOLUB, T. (1999): Interpreting patterns of gene expression with self–organizing maps: methods and application to hematopoietic differentiation. *Proc. Nat. Acad. Sci. USA*, *96*, 2907-2912.

TAVAZOIE, S., HUGHES, J.D., CAMPBELL, M.J., CHO, R.J. and CHURCH, G.M. (1999): Systematic determination of genetic network architecture. *Nature Genetics*, *22*, 281-285.

TIBSHIRANI, R., HASTIE T., NARASIMHAN, B., EISEN, M, SHERLOCK, G., BROWN, P., and BOTSTEIN, D. (2001): Exploratory screening of genes and clusters from microarray experiments. *Internal report University of Stanford at http://www-stat.stanford.edu.*

WALL, M.E., DYCK, P.A., and BRETTIN, T.S. (2001): SVDMAN–singular value decomposition analysis of microarray data. *Bioinformatics*, *17*, 566-568.

WEINSHTEIN, J.N. et al. (1997): An information–intensive approach to the molecular pharmacology of cancer. *Science*, *275*, 343-349.

WEN, X., FUHRMAN, S., MICHAELS, G.S., CARR, D.B., SMITH, S., BARKER, J.L., and SOMOGYI, R. (1998): Large–scale temporal gene expression mapping of central nervous system development. *Proc. Natl. Acad. Sci. USA*, *95*, 334-339.

YEUNG, K.Y., FRALEY, C., MURUA, A., RAFTERY, A.E., and RUZZO, W.L. (2001): Model–based clustering and data transformations for gene expression data. *Bioinformatics*, *17*, 977-987.

Two-mode Partitioning: Review of Methods and Application of Tabu Search

William Castillo[1] and Javier Trejos[1,2]

[1] School of Mathematics, University of Costa Rica,
2060 San José, Costa Rica

[2] Department of Electric Engineering,
Metropolitan Autonomous University at Iztapalapa,
Av. Michoacán y La Purísima s/n, Col. Vicentina
México D.F. CP 09340, Mexico

Abstract. As the special contribution of this paper we deal with an application of tabu search, to the problem of two-mode clustering for minimization of the two-mode variance criterion. States are two-mode partitions, neighborhoods are defined by transfers of a single row or column-mode object into a new class, and the tabu list contains the values of the criterion. We compare the results obtained with those of other methods, such as alternating exchanges, *k*-means, simulated annealing and a fuzzy-set approach.

1 Introduction

Two mode partitioning seeks a partition of a two-mode matrix, such that clusters of rows and clusters of columns are found and there is some connection between these clusters, which represents the intensity in the relation between both modes. Several authors have proposed methods for performing this task, such as Govaert (1983) and Baier et al. (1997) who propose a method of the *k*-means type, Gaul and Schader (1996) use an alternating exchanges method, Trejos and Castillo (2000) use the simulated annealing technique, and Castillo et al. (2001) follow a fuzzy-set approach. All the above methods seek a two-mode partitioning and their aim is to minimize a variance-type criterion. There are some other approaches, such as permutation based by Hartigan (1974), and hierarchical two-mode methods, for instance the one proposed by Eckes and Orlik (1993); a recurrence formula of the Lance & William type is given in Castillo and Trejos (2000) for this method. Two-mode partitioning is very useful in some applied fields, such as marketing, where contingency tables that cross brands and some characteristics are analyzed (see Gaul and Schader (1996) or Baier et al. (1997)).

In this paper we use the optimization heuristic known as tabu search for two-mode partitioning. The article is organized as follows. Section 2 introduces some necessary notation and concepts. Section 3 describes some known methods that will be used later for comparisons. Section 4 recalls the main facts on tabu search and describes the method for performing two-mode partitioning with tabu search. Section 5 gives some useful updating formulas. Finally, section 6 shows some results and concludes the paper.

2 Notation

Let $X = (x_{ij})_{n\times m}$ be a two-mode matrix that crosses modes I and J such that $|I| = n$, $|J| = m$ and $I \cap J = \emptyset$. The entries in X are positive and reflect the degree of intensity in the association between the row and column modes, in such a way that the higher the value of x_{ij} is, the higher is the association between i and j; X could be, for instance, a contingency table. If $P = \{A_k | k = 1, \dots, K\}$ is a partition of mode I and $Q = \{B_l | l = 1, \dots, L\}$ a partition of mode J, a two-mode class of $I \times J$ has the form $A_k \times B_l$ and we say that (P, Q) is a **two-mode partition** of $I \times J$ in K, L classes. Partitions P and Q can be characterized by their indicator matrices, which will also be noted P and Q, the context being enough to distinguish which of the concepts is used.

Given the two-mode matrix X and fixing K and L, our problem is to find partitions $P = (p_{ik})_{n\times K}$ and $Q = (q_{jl})_{m\times L}$, and an association matrix $W = (w_{kl})_{K\times L}$ such that

$$\tilde{Z}(c, P, Q, W) = \sum_{i\in I}\sum_{j\in J}(x_{ij} - c - \hat{x}_{ij})^2$$

is minimum, where $\hat{x}_{ij} = \sum_k \sum_l p_{ik} q_{jl} w_{kl}$. We can see that in this model, called aditive model, x_{ij} is represented by $\hat{x}_{ij} + c$, and hence there is an error term ϵ_{ij} which we will assume satisfies $\sum_{i,j} \epsilon_{ij} = 0$. We suppose that the constant c models the "common" part of the entries in X, that is $c = \overline{x} = \frac{1}{pq}\sum_{i,j} x_{ij}$; then, we can assume that $c = 0$ if the x_{ij} are centered ($x_{ij} := x_{ij} - \overline{x}$).

Hence, the criterion to be minimized is:

$$Z = Z(P, Q, W) = \sum_{i\in I}\sum_{j\in J}\left(x_{ij} - \sum_{k=1}^{K}\sum_{l=1}^{L} p_{ik} q_{jl} w_{kl}\right)^2,$$

which can be written in the form (Gaul & Schader (1996)):

$$Z = Z(P, Q, W) = \sum_{k=1}^{K}\sum_{l=1}^{L}\sum_{i=1}^{n}\sum_{j=1}^{m} p_{ik} q_{jl} (x_{ij} - w_{kl})^2. \tag{1}$$

Given P and Q, the matrix W that minimizes Z is obtained for (see Gaul & Schader (1996)):

$$w_{kl} = \frac{1}{a_k b_l}\sum_{i\in A_k}\sum_{j\in B_l} x_{ij}, \tag{2}$$

where $a_k = |A_k|$ and $b_l = |B_l|$ are the cardinalities of the classes. We remark that W can be interpreted as a matrix of centroids or barycenters of the two-mode classes and it represents the degree of association between classes of both modes.

In what follows, we will write $Z(P, Q)$ instead of $Z(P, Q, W)$ and we suppose that, given P and Q, W is calculated with (2).

If the objects are not equally weighted, then the cardinalities in the expression of W in (2) should be replaced by the sum of weights of elements in the corresponding class; in this article we suppose that objects are equally weighted since the presentation of the methods below does not lose generality in this case.

The *Variance Accounted For* of the model is usually reported in the comparative results, which is defined as

$$\mathrm{VAF} = 1 - \frac{\sum_{i,j}(x_{ij} - \hat{x}_{ij})^2}{\sum_{i,j}(x_{ij} - \bar{x})^2}. \tag{3}$$

It is clear that maximizing $\mathrm{VAF} \in [0, 1]$ is equivalent to minimizing Z.

3 Two-mode partitioning methods

Most of the methods for performing two-mode partitioning are based on transfers of objects. We present a short description of four of these methods. For more details on the methods, we refer the interested reader to the references.

We note $A_k \xrightarrow{i} A_{k'}$ the transfer of the row mode i from class A_k to class $A_{k'}$, and analogously $B_l \xrightarrow{j} B_{l'}$ the transfer of column object j from class B_l to class $B_{l'}$. All the following algorithms begin with an initial two-mode partition (P, Q), which can be generated at random or it can be the result of some previous knowledge on the data (for instance, by the use of two-mode hierarchical clustering). We note ΔZ the difference in the criterion (1) between the two-mode partitions before and after a transfer. We suppose that the numbers of classes K and L are given and fixed throughout the algorithms.

Alternanting Exchanges (Gaul and Schader (1996))

1. Initialize P, Q; compute W according to (2).
2. Alternate the following steps until Z does not improve:
   (a) Make $A_k \xrightarrow{i} A_{k'}$ for i and k' chosen at random; if $\Delta Z < 0$ accept the transfer, redefine P and compute W using (2).
   (b) Make $B_l \xrightarrow{j} B_{l'}$ for j and l' chosen at random; if $\Delta Z < 0$ accept the transfer, redefine Q and compute W using (2).

k-means (Govaert (1983), Baier et al. (1997))

1. Initialize P^0, Q^0; compute W^0 according to (2); let $t := 0$.

2. Repeat the following steps until Z does not improve:
   (a) Given $W := W^t$ and $Q := Q^t$, define $\tilde{P}$ by

$$\tilde{A}_k := \{i \in I | \sum_l \sum_{j \in B_l} (x_{ij} - w_{kl})^2 \to \min_{k'} \sum_l \sum_{j \in B_l} (x_{ij} - w_{k'l})^2\}.$$

   (b) Given $W := W^t$ and $P := P^t$, define $\tilde{Q}$ by

$$\tilde{B}_l := \{j \in J | \sum_k \sum_{i \in A_k} (x_{ij} - w_{kl})^2 \to \min_{l'} \sum_k \sum_{i \in A_k} (x_{ij} - w_{kl'})^2\}.$$

   (c) Calculate W^t according to (2).
   (d) Let $t := t + 1$, $P^t := \tilde{P}$, $Q^t := \tilde{Q}$.

Remark 1: as usual in k-means methods, in case of equality in the minimization that define $\tilde{A}_k$ or $\tilde{B}_l$, the class with smaller index is chosen.
Remark 2: it can be proven that the sequence $Z_t := Z(P^t, Q^t)$ is decreasing (Castillo and Trejos (2000)).

Simulated Annealing (Trejos and Castillo (2000))

The user chooses parameters $\chi_0 \in [0.80, 0.95], \lambda \in I\!N, \gamma \in [0.8, 0.99]$.

1. Initialize P, Q; calculate W according to (2).
2. Estimate the initial temperature c_0 using the initial acceptance rate χ_0.
3. For $t := 0$ until λ do:
   (a) choose at random a mode (I or J) with uniform probability 1/2
   (b) make $A_k \xrightarrow{i} A_{k'}$ or $B_l \xrightarrow{j} B_{l'}$ in the following way:
      - choose at random i or j (according to the chosen mode);
      - choose at random k' or l', different from the current class index of i or j, noted k or l, depending on the corresponding case;
      - calculate ΔZ.
   (c) Accept the transfer if $\Delta Z < 0$, otherwise accept it with probability $\exp(-\Delta Z / c_t)$.
4. Make $c_{t+1} := \gamma c_t$ and return to step 3, until $c_t \approx 0$.

Fuzzy Approach (Castillo et al. (2001))

In this method, we handle fuzzy two-mode partitions (P, Q) where $p_{ik}, q_{jl} \in [0, 1]$ instead of $p_{ik}, q_{jl} \in \{0, 1\}$. Let $s > 1$ be a real number. The criterion to be minimized is

$$Z_s(P, Q, W) = \sum_{k=1}^{K} \sum_{l=1}^{L} \sum_{i=1}^{n} \sum_{j=1}^{m} p_{ik}^s q_{jl}^s (x_{ij} - w_{kl})^2, \tag{4}$$

and it is proved that, for fixed s, P and Q the best matrix W is given by

$$w_{kl} = \frac{\sum_{i=1}^{n} \sum_{j=1}^{m} p_{ik}^s q_{jl}^s x_{ij}}{\sum_{i=1}^{n} \sum_{j=1}^{m} p_{ik}^s q_{jl}^s}. \tag{5}$$

Define $d_{ik} = \sum_{j=1}^{m} \sum_{l=1}^{L} q_{jl}^s (x_{ij} - w_{kl})^2$ with $i \in I$ and $k \in \{1,\dots,K\}$ and $e_{jl} = \sum_{i=1}^{n} \sum_{k=1}^{K} p_{ik}^s (x_{ij} - w_{kl})^2$ with $j \in J$ and $l \in \{1,\dots,L\}$. By constrained derivation using the Lagrangian, for fixed Q and W, we obtain

$$p_{ik} = d_{ik}^{\frac{-1}{s-1}} / \sum_{k'=1}^{K} d_{ik'}^{\frac{-1}{s-1}} \tag{6}$$

and for fixed P and W, we obtain

$$q_{jl} = e_{jl}^{\frac{-1}{s-1}} / \sum_{l'=1}^{L} e_{jl'}^{\frac{-1}{s-1}}. \tag{7}$$

There are different ways to calculate p_{ik} and q_{jl} in the case of singularities (that is, if $d_{ik} = 0$ or $e_{jl} = 0$ for some i or j; see Castillo et al. (2001) for full details). For decreasing values of s (converging to 1), we want to find a two-mode fuzzy partition (P, Q) which approximates a 'crisp' partition, as is proved in Castillo et al. (2001).

1. Define initial value of $s > 1$, $s \approx 1$. Give initial fuzzy partitions P^0 and Q^0 (generated at random, for example). Compute W^0 according to (5).
2. Repeat Steps (a), (b) and (c) until $\frac{Z_s(P^t,Q^t,W^t)}{Z_s(P^{t-1},Q^{t-1},W^{t-1})} < \tau$, where $\tau \approx 0$, $\tau > 0$; or a maximum number of iterations is attained.
   For $t = 0, 1, \dots$ do:
   (a) Use Q^t and W^t for computing $P^{t+1} = (p_{ik}^{t+1})$, according to (6).
   (b) Use P^{t+1} and W^t for computing $Q^{t+1} = (q_{jl}^{t+1})$, according to (7).
   (c) Use P^{t+1} and Q^{t+1} for computing $W^{t+1} = (w_{kl}^{t+1})$, according to (5).

4 Algorithm for two-mode partitioning with tabu search

Generally speaking, tabu search (TS) can be described as an iterative heuristic for the minimization of a function defined on a discrete set (see Glover (1989) and Glover (1990)). In our case, the discrete set is the set of two-mode partitions (P, Q) of $I \times J$ in K, L classes and the function is $Z(P, Q)$.

TS starts from an initial solution and tries to find the global minimum by moving from one solution to another one, based on the concept of neighborhood, which is a new solution generated from the current one by a 'simple modification' or 'move'. In our case, this move will be the transfer of one single object:

$$A_{k'} \xrightarrow{i} A_{k''} \quad \text{or} \quad B_{l'} \xrightarrow{j} B_{l''}$$

such that $A_{k'}$ or $B_{l'}$ have more than one element. In this way, the neighborhood $N(P,Q)$ is a set of two-mode partitions of $I \times J$ in K, L classes directly defined from (P,Q) by a move.

At each iteration of the process, we generate the neighborhood $N(P,Q)$ and we move from (P,Q) to the best solution $(P,Q)^{\text{best}} \in N(P,Q)$ whether or not $Z((P,Q)^{\text{best}})$ is less than $Z(P,Q)$. Since $N(P,Q)$ may be too large for an exhaustive exploration (its size is $n(K-1)+m(L-1)$), we explore a subset $\tilde{N}(P,Q) \subseteq N(P,Q)$ instead of $N(P,Q)$.

TS handles a tabu list to prevent cycling. This tabu list is a queue and has length $|T| = t$ (the value of t is provided by the user), and it stores information about the t most recent moves in order to forbid coming back to recently visited solutions. In our implementation, the tabu list contains the values of Z of the t preceding iterations.

There could be another codification of states in the tabu list. For instance, the list could contain the class indicators of the objects that were transfered. In that case, an aspiration criterion could be applied, that is, the tabu status of the best partition in the neighborhood can be drop out if it is the best partition ever visited during the algorithm.

The **algorithm** is as follows:

1. Choose the maximum number of iterations *maxnum* and the size *sizeneigh* of the subset $\tilde{N}(P,Q)$.
   Initiate $e = (P,Q)$ at random. Calculate W according to 2.
2. Start the tabu list with $T := \{Z(P,Q)\}$. Let $e_{opt} := e$.
3. Repeat *maxnum* times:

   (a) Generate the set $\tilde{N}(e)$ by repeating *sizeneigh* times:
      - Choose the mode at random with uniform probability 0.5.
      - Choose an object i or j uniformly at random from I or J with probability $\frac{1}{n}$ or $\frac{1}{m}$, respectively. If the current class of the object has only one element, discard it and make a new choice.
      - Choose at random a class from P or Q, different from the current class of the object chosen in the preceding step, with probability $\frac{1}{K-1}$ or $\frac{1}{L-1}$ respectively, and then transfer the object to the chosen class.

   (b) Calculate $Z(e_{asp}) = \min\left\{Z(P',Q') \,\middle|\, (P',Q') \in \tilde{N}(e)\right\}$.

   (c) If $Z(e_{asp}) < Z(e_{opt})$ then $e^* = e_{asp}$ and $e_{opt} = e^*$. Otherwise let e^* be such that $Z(e^*) = \min\left\{Z(P',Q') \,\middle|\, (P',Q') \in \tilde{N}(e), Z(P',Q') \notin T\right\}$.

   (d) $e := e^*$.

   (e) Update T: if T has already length t its update consists on removing the older element which belongs to T and add $Z(e^*)$ to T. Otherwise you only have to add $Z(e^*)$ to the tabu list.

5 Updating formulas

An efficient implementation of the above TS method for two-mode partitioning, requires a simplified way to calculate the change in the criterion Z since it is used in every iteration for every neighbor in $\tilde{N}(P,Q)$. For this, we have the following updating formulas.

- Let $\widetilde{w}_{kl}$ denote the value of w_{kl} after the transfer $A_{k'} \xrightarrow{i} A_{k''}$. Then,

$$\widetilde{w}_{kl} = \begin{cases} w_{kl} & \text{if } k \neq k'', k' \\ \frac{a_{k'}}{a_{k'}-1} w_{k'l} - \frac{1}{(a_{k'}-1)b_l} \sum_{j\in B_l} x_{ij} & \text{if } k = k' \\ \frac{a_{k''}}{a_{k''}+1} w_{k''l} + \frac{1}{(a_{k''}+1)b_l} \sum_{j\in B_l} x_{ij} & \text{if } k = k'' \end{cases}$$

On the other hand, let ΔZ denote the difference in the criterion Z after the transfer minus the criterion before it. It is straightforward to see that

$$\Delta Z = \sum_l b_l \left[a_{k''} \left(w_{k''l} - \widetilde{w}_{k''l} \right)^2 + a_{k'} \left(v_{k'l} - \widetilde{w}_{k'l} \right)^2 \right] + \left(\widetilde{w}_{k'l} - \widetilde{w}_{k''l} \right) \left[-b_l \left(\widetilde{w}_{k'l} + \widetilde{w}_{k''l} \right) + 2 \sum_{j\in B_l} x_{ij} \right]$$

- In a similar way we can deduce formulas in the case $B_{l'} \xrightarrow{j} B_{l''}$:

$$\widetilde{w}_{kl} = \begin{cases} w_{kl} & \text{if } l \neq l'', l' \\ \frac{b_{l'}}{b_{l'}-1} w_{kl'} - \frac{1}{(b_{l'}-1)a_k} \sum_{i\in A_k} x_{ij} & \text{if } l = l' \\ \frac{b_{l''}}{b_{l''}+1} w_{kl''l} + \frac{1}{(b_{l''}+1)a_k} \sum_{i\in A_k} x_{ij} & \text{if } l = l'' \end{cases}$$

and

$$\Delta Z = \sum_k a_k \left[b_{l'} \left(w_{kl'} - \widetilde{w}_{kl'} \right)^2 + b_{l''} \left(w_{kl''} - \widetilde{w}_{kl''} \right)^2 \right] + \left(\widetilde{w}_{kl'} - \widetilde{w}_{kl''} \right) \left[-a_k \left(\widetilde{w}_{kl'} + \widetilde{w}_{kl''} \right) + 2 \sum_{i\in A_k} x_{ij} \right].$$

6 Some results

To test the proposed method we used the 'Sanitary data set' a (82×25) matrix which crosses 82 customers which qualify, in a $1-11$ scale, their buying intention of a product according to 5 attributs (price, recycling,...), each one with five levels (see Baier et al. (1997)). We applied the following methods: Alternating Exchanges (AE), k-means, Simulated Annealing, the fuzzy approach and the Tabu Search method proposed in this paper. For each algorithm we report the *Variance Accounted For* VAF (3), as indicated in Table 3.

Values of s in the fuzzy approach at convergence were: 1.005, 1.01, 1.01, 1.008 and 1.01 respectively for $K = L = 2, 3, 4, 5, 6$ classes, and parameters of SA were $\lambda = 1070$, 1605, 2140, 2675, 3210, 2996 and $c_0 = 100$, 120, 120, 120, 150, 150, respectively. All executions of SA used $\gamma = 0.85$. For TS,

$K = L$	AE	k-means	SA	Fuzzy	TS
2	0.2289	0.2289	0.2289	0.2289	0.2289
3	0.3234	0.3234	0.3234	0.3234	0.3234
4	0.3729	0.3729	0.3804^+	0.3703^-	0.3804^+
5	0.4293	0.4293	0.4293	0.4293	0.4293
6	0.4635	0.4579^-	0.4651	0.4641	0.4651^+
7	0.4987	0.4958^-	0.4990^+	—	0.4978

Table 1. Values of VAF for the sanitary data set using Alternating Exchanges (AE), k-means, Simulated Annealing (SA), the fuzzy approach and Tabu Search (TS).

sample neighborhood size were 70, 150, 150, 250, 400, 400 respectively for $K = L = 2, 3, 4, 5, 6, 7$ classes and length of the tabu list were 20, 80, 100, 100, 50, 100 respectively.

As can be seen, all methods lead to the same results for 2, 3 and 5 classes, and the SA and TS are better for 4 classes. SA is better for 7 classes. In case of 6 classes the best value found by TS is 0.465078 and by SS is 0.465057. The corresponding partitions were $(P_6^{ts}, Q_6^{ts}) \neq (P_6^{ss}, Q_6^{ss})$. That is why TS is slightly better than SS. The fuzzy approach is the worst one for 4 classes.

Further comparisons should be performed. In particular, we will simulate some data tables by the product PWQ', given appropriate matrices P, Q and W. This kind of experiment could give us an idea of the quality of the methods. For this, we should vary some of the parameters, such as the size of the matrices, the number of classes, the cardinalities of the classes, etc. Also, the experiment should give us some insight about the parameters of the methods, since in some cases these parameters are hard to tune.

References

BAIER, D., GAUL, W., and SCHADER, M. (1997): Two-mode overlapping clustering with applications to simultaneous benefit segmentation and market structuring. in: R. Klar, and O. Opitz (Eds.): *Classification and Knowledge Organization.* Springer, Heidelberg, 557–566.

CASTILLO, W., and TREJOS, J. (2000): Recurrence properties in two-mode hierarchical clustering. In: W. Gaul, and R. Decker (Eds.): *Classification and Information at the Turn of the Millenium.* Springer, Heidelberg, 68–73.

CASTILLO, W., GROENEN, P.J.F, and TREJOS, J. (2001): Optimization of a Fuzzy Criterion for Partitioning Two-Mode Data. Submitted to *Annals of Operations Research.*

ECKES, T., and ORLIK, P. (1993): An error variance approach to two-mode hierarchical clustering, *Journal of Classification, 10,* 51–74.

GAUL, W., and SCHADER, M. (1996): A new algorithm for two-mode clustering. In: H.-H. Bock, and W. Polasek (Eds.): *Data Analysis and Information Systems,* Springer, Heidelberg, 15–23.

GLOVER, F. (1989): Tabu search - Part I. *ORSA J. Comput.*, *1*, 190–206.

GLOVER, F. (1990): Tabu search - Part II. *ORSA J. Comput.*, *2*, 4–32.

GOVAERT, G. (1983): *Classification Croisée.* Thèse d'Etat, Université de Paris VI.

HARTIGAN, J. (1974): *Clustering Algorithms.* John Wiley & Sons, New York.

TREJOS, J., and CASTILLO, W. (2000): Simulated annealing optimization for two-mode partitioning. In: W. Gaul, and R. Decker (Eds.): *Classification and Information at the Turn of the Millenium*, Springer, Heidelberg, 135–142.

TREJOS, J., MURILLO, A., and PIZA, E. (1998): Global stochastic optimization for partitioning. In: A. Rizzi, M. Vichi, and H.-H. Bock (Eds.): *Advances in Data Science and Classification.* Springer, Heidelberg, 185–190.

Dynamical Clustering of Interval Data: Optimization of an Adequacy Criterion Based on Hausdorff Distance

Marie Chavent[1] and Yves Lechevallier[2]

[1] Mathématiques Appliquées de Bordeaux, UMR 5466 CNRS, Université Bordeaux 1 - 351, Cours de la libération, 33405 Talence Cedex, France
[2] INRIA- Institut National de Recherche en Informatique et en Automatique, Domaine de Voluceau- Rocquencourt B.P. 105, 78153 Le Chesnay Cedex, France

Abstract. In order to extend the dynamical clustering algorithm to interval data sets, we define the prototype of a cluster by optimization of a classical adequacy criterion based on Hausdorff distance. Once this class prototype properly defined we give a simple and converging algorithm for this new type of interval data.

1 Introduction

The main aim of this article is to define a dynamical clustering algorithm for data tables where each cell contains an interval of real values (Table 1 for instance). This type of data is a particular case of a symbolic data table where each cell can be an interval, a set of categories or a frequency distribution (Diday (1988), Bock and Diday (2000)).

	Pulse Rate	Systolic pressure	Diastolic pressure
1	[60,72]	[90,130]	[70,90]
2	[70,112]	[110,142]	[80,108]
3	[54,72]	[90,100]	[50,70]
4	[70,100]	[130,160]	[80,110]
5	[63,75]	[60,100]	[140,150]
6	[44,68]	[90,100]	[50,70]

Table 1. A data table for $n = 6$ patients and $p = 3$ interval variables

Dynamical clustering algorithms (Diday (1971), Diday and Simon (1976)) are iterative two steps relocation algorithms involving at each iteration the identification of a prototype (or center) for each cluster by optimizing an adequacy criterion. The k-means algorithm with class prototypes updated after all objects have been considered for relocation, is a particular case of dynamical clustering with adequacy criterion equal to variance criterion such that class prototypes equal to cluster centers of gravity (MacQueen (1967), Späth (1980)).

In dynamical clustering, the optimization problem is the following. Let Ω be a set of n objects indexed by $i = 1, ..., n$ and described by p quantitative variables. Then each object i is described by a vector $x_i \in \Re^p$. The problem is to find the partition $P = (C_1, ..., C_K)$ of Ω in K clusters and the system $Y = (y_1, ..., y_K)$ of class prototypes, optimum with respect to a partitioning criterion $g(P, Y)$. Two classical partitioning criteria are:

$$g(P, Y) = \sum_{k=1}^{K} \sum_{i \in C_k} d^2(x_i, y_k) \tag{1}$$

where $d(x, y) = ||x - y||_2$ is the L_2 distance, and:

$$g(P, Y) = \sum_{k=1}^{K} \sum_{i \in C_k} d(x_i, y_k) \tag{2}$$

where $d(x, y) = ||x - y||_1$ is the L_1 distance.

More precisely, the dynamical clustering algorithm converges and the partitioning criterion decreases at each iteration if the class prototypes are properly defined at each 'representation' step. Indeed, the problem is to find the prototype y of each cluster $C \subset \{1, ..., n\}$ which minimizes an adequacy criterion $f(y)$ measuring the "dissimilarity" between the prototype y and the cluster C. The two adequacy criteria corresponding to the partitioning criteria (1) and (2) are respectively:

$$f(y) = \sum_{i \in C} d^2(x_i, y) = \sum_{i \in C} \sum_{j=1}^{p} (x_i^j - y^j)^2 \tag{3}$$

and:

$$f(y) = \sum_{i \in C} d(x_i, y) = \sum_{i \in C} \sum_{j=1}^{p} |x_i^j - y^j| \tag{4}$$

The coordinates of the class prototype y minimizing criterion (3) are:

$$y^j = mean\{x_i^j \mid i \in C\} \tag{5}$$

and the coordinates of the class prototype y minimizing criterion (4) are:

$$y^j = median\{x_i^j \mid i \in C\} \tag{6}$$

In this latter case, the solution y^j is not always unique. If there is an interval of solutions, we usually choose y^j as the midpoint of this interval.

In this paper, we define the prototype y of a cluster C in the particular case of p-dimensional interval data. Each object i is now described on each variable j by an interval

$$x_i^j = [a_i^j, b_i^j] \in I = \{[a,b] \mid a, b \in \Re \ , \ a \leq b\}$$

and the coordinates of the class prototype y are also intervals of I noted $y^j = [\alpha^j, \beta^j]$. In other words, the vector x_i representing an object i and the class prototype y are vectors of intervals, i.e., (hyper-)rectangles in the euclidean space $\Re^p$.
The distance d between two vectors of intervals x_i and $x_{i'}$ will be based on the Hausdorff distance between two sets. This distance is given in section 2. Then we focus on the optimization problem for class prototypes and on its solution in section 3. Once the new class prototypes properly defined, a dynamical clustering algorithm of interval data is presented in section 4.

2 A distance measure between two vectors of intervals

There are several methods for measuring dissimilarities between interval data or more generally between symbolic objects (Chapters 8 and 11.2.2 of Bock and Diday (2000), De Carvalho (1998), Ichino and Yaguchi (1994)).

From our point of view, it is a natural approach to use Hausdorff distance, initially defined to compare two sets, to compare two intervals.

The Hausdorff distance d_H between two sets $A, B \in \Re^p$ is (Aubin, (1994)):

$$d_H(A, B) = \max(h(A, B), h(B, A)) \tag{7}$$

with

$$h(A, B) = \sup_{a \in A} \inf_{b \in B} ||b - a|| \tag{8}$$

By using L_2 norm in (8), the Hausdorff distance d_H between two intervals $A_1 = [a_1, b_1]$ and $A_2 = [a_2, b_2]$ is:

$$d_H(A_1, A_2) = \max(|a_1 - a_2|, |b_1 - b_2|) \tag{9}$$

In this paper, the distance d between two vectors of intervals

$$x_i = ([a_i^1, b_i^1], ..., [a_i^p, b_i^p])$$

and

$$x_{i'} = ([a_{i'}^1, b_{i'}^1], ..., [a_{i'}^p, b_{i'}^p])$$

representing two objects i and i' is defined as the sum for $j = 1, ..., p$ of the Haussdorf distance (9) between the two intervals $[a_i^j, b_i^j]$ and $[a_{i'}^j, b_{i'}^j]$.

Finally, the distance d is defined by:

$$d(x_i, x_{i'}) = \sum_{j=1}^{p} d_H(x_i^j, x_{i'}^j) = \sum_{j=1}^{p} \max(|a_i^j - a_{i'}^j|, |b_i^j - b_{i'}^j|) \tag{10}$$

In the particular case of intervals reduced to single points, this distance is the well-known L_1 distance between two points of $\Re^p$.

3 The optimization problem for class prototype

As presented in the introduction, the prototype y of a cluster C is defined in dynamical clustering by optimizing an adequacy criterion f measuring the "dissimilarity" between the prototype and the cluster. Here, we search the vector of intervals y noted:

$$y = (y^1, ..., y^p) = ([\alpha^1, \beta^1], ..., [\alpha^p, \beta^p])$$

which minimizes the following adequacy criterion:

$$f(y) = \sum_{i \in C} d(x_i, y) = \sum_{i \in C} \sum_{j=1}^{p} d_H(x_i^j, y^j) \tag{11}$$

where d is the distance between two vectors of intervals given in (10).

The criterion (11) can also be written:

$$f(y) = \sum_{j=1}^{p} \overbrace{\sum_{i \in C} d_H(x_i^j, y^j)}^{\tilde{f}(y^j)} \tag{12}$$

and the problem is now to find for $j = 1, ..., p$ the interval $y^j = [\alpha^j, \beta^j]$ which minimizes:

$$\tilde{f}(y^j) = \sum_{i \in C} d_H(x_i^j, y^j) = \sum_{i \in C} \max(|\alpha^j - a_i^j|, |\beta^j - b_i^j|) \tag{13}$$

We will see how to solve this minimization problem by transforming it into two well-known L_1 norm problems.

Let m_i^j be the midpoint of an interval $x_i^j = [a_i^j, b_i^j]$ and l_i^j be an half of its length:

$$m_i^j = \frac{a_i^j + b_i^j}{2}$$

$$l_i^j = \frac{b_i^j - a_i^j}{2}$$

and let μ^j and λ^j be respectively the midpoint and the half-length of the interval $y^j = [\alpha^j, \beta^j]$. According to the following property defined for x and y in $\Re$:

$$\max(|x - y|, |x + y|) = |x| + |y| \tag{14}$$

the function (13) can be written:

$$\begin{aligned}\tilde{f}(y^j) &= \sum_{i \in C} \max(|(\mu^j - \lambda^j) - (m_i^j - l_i^j)|, |(\mu^j + \lambda^j) - (m_i^j + l_i^j)|) \\ &= \sum_{i \in C} \max(|(\mu^j - m_i^j) - (\lambda^j - l_i^j)|, |(\mu^j - m_i^j) + (\lambda^j - l_i^j)|) \\ &= \sum_{i \in C} (|\mu^j - m_i^j| + |\lambda^j - l_i^j|) = \sum_{i \in C} (|\mu^j - m_i^j| + \sum_{i \in C} |\lambda^j - l_i^j|) \end{aligned} \tag{15}$$

This yields two well-known minimization problems in L_1 norm: Find $\mu^j \in \Re$ which minimizes:

$$\sum_{i \in C} |\mu^j - m_i^j| \tag{16}$$

and find $\lambda^j \in \Re$ which minimizes:

$$\sum_{i \in C} |\lambda^j - l_i^j| \tag{17}$$

The solutions $\hat{\mu^j}$ and $\hat{\lambda^j}$ are respectively the median of $\{m_i^j, i \in C\}$, the midpoints of the intervals $x_i^j = [a_i^j, b_i^j]$, $i \in C$, and the median of the set $\{l_i^j, i \in C\}$ of their half-lengths. Finally, the solution $\hat{y^j} = [\hat{\alpha^j}, \hat{\beta^j}]$ is the interval $[\hat{\mu^j} - \hat{\lambda^j}, \hat{\mu^j} + \hat{\lambda^j}]$.

4 The dynamical clustering algorithm

Iterative algorithms or dynamical clustering methods for symbolic data have already been proposed in Bock (2001), De Carvalho et al. (2001) and Verde et al. (2000). Here, we consider the problem of clustering a set $\Omega = \{1, ..., i, ..., n\}$ of n objects into K disjoint clusters $C_1, ..., C_K$ in the particular case of objects described on each variable j by an interval $x_i^j = [a_i^j, b_i^j]$ of $\Re$.

The dynamical clustering algorithm search for the partition $P = (C_1, ..., C_K)$ of Ω and the system $Y = (y_1, ..., y_K)$ of class prototypes which are optimum with respect to the following partitioning criterion based on the distance d defined in (10):

$$g(P,Y) = \sum_{k=1}^{K} \sum_{i \in C_k} d(x_i, y_k)$$
$$= \sum_{k=1}^{K} \sum_{i \in C_k} \sum_{j=1}^{p} \max(|a_i^j - \alpha_k^j|, |b_i^j - \beta_k^j|) \quad (18)$$

This algorithm proceeds like classical dynamical clustering by iteratively repeating an 'allocation' step and a 'representation' step.

4.1 The 'representation step'

During the 'representation' step, the algorithm computes for each cluster C_k the prototype y_k which minimizes the adequacy criterion given in (11). We have defined in section 3 the 'optimal' prototype y_k for this criterion. It is described on each variable j by the interval $[\alpha_k^j, \beta_k^j] = [\mu_k^j - \lambda_k^j, \mu_k^j + \lambda_k^j]$ where:

$$\mu_k^j = median\{m_i^j \mid i \in C_k\} \quad (19)$$

is the median of the midpoints of the intervals $[a_i^j, b_i^j]$ with $i \in C_k$ and

$$\lambda_k^j = median\{l_i^j \mid i \in C_k\} \quad (20)$$

is the median of their half-lenghts.

4.2 The 'allocation' step

During the 'allocation step', the algorithm performes a new partition by reassigning each object i to the closest class prototype y_{k*} where:

$$k* = arg \min_{k=1,\ldots,K} d(x_i, y_k)$$

and d is defined in (10).

4.3 The algorithm

Finally the algorithm is the following:

(a) Initialization
Choose a partition $(C_1, \ldots, C_K)$ of the data set Ω or choose K distinct objects $y_1, ..., y_K$ among Ω and assign each object i to the closest prototype y_{k*} ($k* = arg\min_{k=l,\ldots,K} \sum_{j=1}^{p} max(|a_i^j - \alpha_k^j|, |b_i^j - \beta_k^j|)$ to contruct the initial partition $(C_1, \ldots, C_K)$.

(b) 'Representation' step
For k in 1 to K compute the prototype $y_k = (y_k^1, ..., y_k^p)$ with $y_k^j = [\alpha_k^j, \beta_k^j] = [\mu_k^j - \lambda_k^j, \mu_k^j + \lambda_k^j]$ and:

$$\mu_k^j = median\{m_i^j \mid i \in C_k\}$$
$$\lambda_k^j = median\{l_i^j \mid i \in C_k\}$$

(c) 'Allocation' step
$test \leftarrow 0$
For i in 1 to n do
define the cluster C_{k*} such that

$$k* = arg \min_{k=l,...,K} \sum_{j=1}^{p} max(|a_i^j - \alpha_k^j|, |b_i^j - \beta_k^j|)$$

if $i \in C_k$ and $k* \neq k$
$test \leftarrow 1$
$C_{k*} \leftarrow C_{k*} \cup \{i\}$
$C_k \leftarrow C_k \backslash \{i\}$

(d) If $test = 0$ END, else go to (b)

5 Conclusion

We have proposed a dynamical clustering algorithm for interval data sets. The convergence of the algorithm and the decrease of the partitioning criterion at each iteration, is due to the optimization of the adequacy criterion (11) at each 'representation' step. The implementation of this algorithm is simple and the computational complexity is in $nlog(n)$.

References

AUBIN, J.P. (1994): *Initiation à l'analyse appliquée*, Masson.

BOCK H.H. (2001): Clustering algorithms and Kohonen maps for symbolic data. *Proc. ICNCB*, Osaka, 203–215.

BOCK H.H. and DIDAY, E. (eds.) (2000): *Analysis of symbolic data. Exploratory methods for extracting statistical information from complex data.* Springer Verlag, Heidelberg.

DE CARVALHO, F.A.T. (1998): Extension based proximities coefficients between boolean symbolic objects. In: C. Hayashi et al. (eds): *Data Science, Classification and Related Methods*, Springer Verlag, 370–378.

DE CARVALHO, F.A.T, DE SOUZA, R.M., VERDE, R. (2001): Symbolic classifier based on modal symbolic descriptions. *Proc. CLADAG2001*, Univ. de Palermo.

DIDAY, E. (1971): La méthode des nuées dynamiques. *Rev. Stat. Appliquées* , *XXX (2)*, 19–34.

DIDAY, E. (1988): The symbolic approach in clustering and related methods of data analysis: The basic choice. In: H.H. Bock (ed.): *Classification and related methods of data anlysis. Proc. IFCS-87*, North Holland, Amsterdam, 673-684.

DIDAY, E., and SIMON, J.C. (1976): Clustering analysis. In: K.S. Fu (ed.): *Digital Pattern Clasification.* Springer Verlag, 47–94.

ICHINO, M. and YAGUCHI, H. (1994): Generalized Minkowski metrics for mixed feature type data analysis. *IEEE Transactions on Systems, Man and Cybernetics, 24 (4)*, 698–708.

MACQUEEN, J. (1967): Some methods for classification and analysis of multivariate observations. In: L.M. LeCam et al. (eds.): *Proc. 5th Berkeley Symp. on Math. Stat. Proba.*, University of California Press, Los Angeles, vol 1, 281–297.

SPÄTH, H. (1980): *Cluster analysis algorithms*, Horwood Publishers/Wiley, New York.

VERDE, R., DE CARVALHO, F.A.T., LECHEVALLIER, Y. (2000): A dynamical clustering algorithm for multi-nominal data. In: H.A.L. Kiers et al. (eds.): *Data Analysis, Classification and Related methods.* Springer Verlag, 387–394.

Removing Separation Conditions in a 1 against 3-Components Gaussian Mixture Problem

Bernard Garel and Franck Goussanou

Statistics and Probability, ENSEEIHT, 2 rue Camichel, 31071 Toulouse cédex 7, France, garel@len7.enseeiht.fr

Abstract. In this paper we address the problem of testing homogeneity against a three components Gaussian mixture in the univariate case. As alternative we consider a contamination model. The likelihood ratio test statistic (LRTS) is derived up to an $o_p(1)$, and two separation conditions are removed. An example with real data is discussed.

1 Main result

In the last ten years, there has been a developing interest in mixture models as indicated by, for instance, a monograph by Lindsay(1995) and a book by McLachlan and Peel (2000). A paper by Bock (1996) shows the importance of these probability models in partitioning cluster analysis. Among the recent trends, we find interest in the following two problems:
(i) the study of the LRTS when the parameter space is not a compact set;
(ii) the problem of removing separation conditions on the parameters of the mixture components, imposed in order to restore identifiability.

In 1985, Hartigan proved that a statistic close to the LRTS for stating homogeneity against a Gaussian mixture of the means converges towards infinity in probability when n tends to infinity if the range of the unknown mean is unbounded. Bickel and Chernoff (1993) revisited this problem and showed that if the parameter set is unbounded, Hartigan's statistic approaches infinity with order $\log(\log n)$.

The problem (ii) is related to Ghosh and Sen's work (1985) on a two-components mixture. They assumed a separation condition between θ_1, the mean of the first component, and θ_2, the mean of the second component: $|\theta_2 - \theta_1| \geq \varepsilon_0 > 0$, where ε_0 is fixed. Note that in order to derive the LRTS it is necessary that the range of the means is a bounded interval.

Various authors have shown that, in several cases, it is possible to remove such a condition, for instance Garel (2001) and Goussanou (2001).

In this paper we address the problem of testing homogeneity against a three-components Gaussian mixture for a univariate random variable X. Let us denote by

$$f(x, \psi) = p_1\phi(x - \theta_1) + p_2\phi(x - \theta_2) + p_3\phi(x - \theta_3), \; x \in R$$

with

$$\phi(x) = \frac{1}{\sqrt{2\pi}} \exp(-\frac{x^2}{2}),$$

a three-components Gaussian mixture density with

$$p_1 + p_2 + p_3 = 1, \ \psi = (p_1, p_2, \theta_1, \theta_2, \theta_3), \text{ and } \theta_1, \theta_2, \theta_3 \in \Theta$$

where $\Theta \subseteq R$ is a real subset. In order to test

$$H_0 : \qquad f(\cdot, \psi) = \phi(\cdot - \theta_3), \ \theta_3 \in \Theta$$

against

$$H_1 : \qquad f(\cdot, \psi) = p_1\phi(\cdot - \theta_1) + p_2\phi(\cdot - \theta_2) + p_3\phi(\cdot - \theta_3)$$

$$p_1, p_2, p_3 \in]0, 1[, \quad \theta_1 \neq \theta_2 \neq \theta_3, \ \theta_1 \neq \theta_3$$

with independent random variables $X_1, ..., X_n$ with density f, Goussanou proposed to use the LRTS given by

$$\lambda_n = 2 \left\{ \sup_{\psi \in H_0 \cup H_1} L_n (X_1, ..., X_n, \psi) - \sup_{\psi \in H_0} L_n (X_1, ..., X_n, \psi) \right\}, \tag{1}$$

where $L_n (x_1, ..., x_n, \psi)$ denotes the log likelihood function

$$L_n (x_1, ..., x_n, \psi) = \sum_{i=1}^{n} \log f (x_i, \psi) .$$

His results require the following separation conditions:

$$|\theta_1 - \theta_2| \geq c_0 > 0, \ |\theta_1 - \theta_3| \geq c_1 > 0 \text{ and } |\theta_2 - \theta_3| \geq c_2 > 0. \tag{2}$$

The first inequality in (2) only means that we really have a three-components mixture, if there is a mixture. We would like to remove the two last separation conditions. It is possible to achieve this purpose if we consider a contamination model. In this case we test

$$H_0^* : \qquad f(\cdot) = \phi(\cdot)$$

against

$$H_1^* : \qquad f(\cdot, \psi) = (1 - p_1 - p_2)\phi(\cdot) + p_1\phi(\cdot - \theta_1) + p_2\phi(\cdot - \theta_2) \tag{3}$$

$$p_1, p_2 \in]0, 1[, \quad \theta_1 \neq 0, \ \theta_2 \neq 0, \quad |\theta_1 - \theta_2| \geq c_0 > 0, \ \theta_1, \theta_2 \in [-a, a]$$

where $a > 0$. This amounts to assuming θ_3 to be known and, without loss of generality, equal to 0.

An example of such a model can be found in clinical chemistry. If we are considering the number of white blood cells of a patient, we know that for healthy individuals the number of white blood cells is normally distributed. In case of specific illnesses the number of white blood cells increases. In a few other cases (after a chemotherapy, for instance) this number can decrease drastically. With no prior knowledge of a patient the observed distribution of white blood cells will give a contamination Gaussian model with three components (see section 3).

The general theory of Ghosh and Sen (1985) would impose $|\theta_1| \geq c_1 > 0$ and $|\theta_2| \geq c_2 > 0$. Allowing that θ_1 or θ_2 may approach 0 means that we are able to remove these separation conditions. Let us consider the parameter sets

$$\begin{aligned} \Theta &= [-a, 0[\cup]0, a] = [-a, a] \setminus \{0\}, \\ \mathcal{E} &= \{(\theta_1, \theta_2) \in \Theta \times \Theta \text{ such that } |\theta_1 - \theta_2| \geq c_0\}, \text{ and} \\ \mathcal{E}_1 &= \{\theta_1 \text{ such that } (\theta_1, \theta_2) \in \mathcal{E}\} = \{\theta_2 \text{ such that } (\theta_1, \theta_2) \in \mathcal{E}\}, \end{aligned}$$

the real value

$$D = (e^{\theta_1^2} - 1)(e^{\theta_2^2} - 1) - (e^{\theta_1 \theta_2} - 1)^2$$

and, finally, the random variable

$$\begin{aligned} T_n^2(\theta_1, \theta_2) = D^{-1} \Bigg\{ & (e^{\theta_2^2} - 1) \left[\frac{1}{\sqrt{n}} \sum_{i=1}^{n} \left(e^{(\theta_1 X_i - \frac{\theta_1^2}{2})} - 1 \right) \right]^2 \\ & + (e^{\theta_1^2} - 1) \left[\frac{1}{\sqrt{n}} \sum_{i=1}^{n} \left(e^{(\theta_2 X_i - \frac{\theta_2^2}{2})} - 1 \right) \right]^2 \\ & - 2(e^{\theta_1 \theta_2} - 1) \left[\frac{1}{\sqrt{n}} \sum_{i=1}^{n} \left(e^{(\theta_1 X_i - \frac{\theta_1^2}{2})} - 1 \right) \right] \\ & \cdot \left[\frac{1}{\sqrt{n}} \sum_{i=1}^{n} \left(e^{(\theta_2 X_i - \frac{\theta_2^2}{2})} - 1 \right) \right] \Bigg\} \end{aligned} \tag{4}$$

which yields an asymptotic approximation to the LRTS. We can prove:

Theorem 1 *Under the condition $|\theta_1 - \theta_2| \geq c_0 > 0$, the LRTS for testing H_0^* against H_1^* has for $n \to \infty$ the following asymptotic expansion:*

$$\lambda_n = \sup_{(\theta_1, \theta_2) \in \mathcal{E}} \left[T_n^2(\theta_1, \theta_2) \right] 1\!\!1_{\left\{ \sum_{i=1}^{n} \left(e^{(\theta_1 X_i - \frac{\theta_1^2}{2})} - 1 \right) \geq 0; \sum_{i=1}^{n} \left(e^{(\theta_2 X_i - \frac{\theta_2^2}{2})} - 1 \right) \geq 0 \right\}} + o_p(1) \tag{5}$$

where $o_p(1)$ is a quantity which tends to 0 in probability uniformly with respect to $(\theta_1, \theta_2) \in \mathcal{E}$ under H_0.

In order to prove Theorem 1 we need the following lemmas. Proofs are given in the last section.

Lemma 1 *Define, for $i = 1, ..., n$, $Y_i(\theta_1, \theta_2)$ by*

$$Y_i(\theta_1,\theta_2) = \begin{pmatrix} Y_{i1}(\theta_1) \\ Y_{i2}(\theta_2) \end{pmatrix} \quad \text{where } Y_{ij}(\theta) = \begin{cases} \frac{e^{X_i\theta - \frac{\theta^2}{2}} - 1}{\theta} & \text{if } \theta \neq 0 \\ X_i & \text{if } \theta = 0 \end{cases}.$$

Under H_0 the random field $\left\{ \frac{1}{\sqrt{n}} \sum_{i=1}^{n} Y_i(\theta_1,\theta_2) \mid (\theta_1,\theta_2) \in [-a,a]^2 \right\}$ is tight, i.e. the family of distributions of $n^{-1/2} \sum_{i=1}^n Y_i(\theta_1,\theta_2)$ is relatively compact; see Billingsley (1968). This entails that $\sup_{(\theta_1,\theta_2)\in[-a,a]^2} n^{-1/2} \sum_{i=1}^n Y_i(\theta_1,\theta_2)$ is $O_p(1)$.

Lemma 2 *Under H_0 a uniform strong law of large numbers holds for the variables*

$$\overline{Y_{nk}(\theta_k)Y_{nl}(\theta_l)} = n^{-1} \sum_{i=1}^{n} Y_{ik}(\theta_k) Y_{il}(\theta_l), \ 1 \leq k,l \leq 2$$

in the sense that for $k, l = 1, 2$ the values

$$\sup_{\theta_k,\theta_l \in [-a,a]} \left| n^{-1} \sum_{i=1}^{n} Y_{ik}(\theta_k)Y_{il}(\theta_l) - EY_{1k}(\theta_k)Y_{1l}(\theta_l) \right|$$

converge to 0 almost surely.

Lemma 3 *For $1 \leq k,l \leq 2$ consider the random variables:*

$$g_{kl}(X_1, p_1, p_2, \theta_1, \theta_2) = \frac{h_{kl}}{d_{kl}},$$

with

$$h_{kl} = \left(e^{-\frac{1}{2}(X_1-\theta_k)^2} - e^{-\frac{1}{2}X_1{}^2}\right)\left(e^{-\frac{1}{2}(X_1-\theta_l)^2} - e^{-\frac{1}{2}X_1{}^2}\right)$$

$$d_{kl} = \theta_k\theta_l \left\{ (1-p_1-p_2)\, e^{-\frac{1}{2}X_1{}^2} + p_1 e^{-\frac{1}{2}(X_1-\theta_1)^2} + p_2 e^{-\frac{1}{2}(X_1-\theta_2)^2} \right\}^2$$

with the convenient limits when θ_1 or θ_2 equals 0. Let us denote by $\|.\|$ the Euclidian norm. Then under H_0, for $1 \leq k,l \leq 2$ the following equalities hold:

$$\lim_{\delta \longrightarrow 0} E \left(\sup_{\|(p_1,p_2)\| < \delta, \theta_1,\theta_2 \in [-a,a]} |Y_{1k}(\theta_k)Y_{1l}(\theta_l) - g_{kl}(X_1,p_1,p_2,\theta_1,\theta_2)| \right) = 0. \tag{6}$$

$$\lim_{\delta\longrightarrow 0} E\left(\sup_{\|(p_1,\theta_2)\|<\delta, p_2\in[0,1],\ \theta_1\in[-a,a]} |Y_{1k}(\theta_k)Y_{1l}(\theta_l) - g_{kl}(X_1,p_1,p_2,\theta_1,\theta_2)|\right) = 0. \tag{7}$$

$$\lim_{\delta\longrightarrow 0} E\left(\sup_{\|(p_2,\theta_1)\|<\delta, p_1\in[0,1],\ \theta_2\in[-a,a]} |Y_{1k}(\theta_k)Y_{1l}(\theta_l) - g_{kl}(X_1,p_1,p_2,\theta_1,\theta_2)|\right) = 0. \tag{8}$$

$$\lim_{\delta\longrightarrow 0} E\left(\sup_{\|(p_1\theta_1,p_2\theta_2)\|<\delta, (p_1,p_2,\theta_1,\theta_2)\in[0,1]^2\times\overline{\mathcal{E}_1}^2} |Y_{1k}(\theta_k)Y_{1l}(\theta_l) - g_{kl}(X_1,p_1,p_2,\theta_1,\theta_2)|\right) = \tag{9}$$

2 Towards applications

In order to use the LRT for H_0^* against H_1^* with λ_n or its approximation (5) from Theorem 1 we need to calculate percentile points. This seems a very difficult task. A possible approach is the following one: let us define

$$Z_n(\theta_2) = \frac{1}{\sqrt{n}}\sum_{i=1}^{n}\frac{e^{(\theta_2 X_i - \frac{\theta_2^2}{2})} - 1}{\sqrt{e^{\theta_2^2} - 1}},$$

$$\mathcal{T}_n(\theta_1,\theta_2) = \frac{(e^{\theta_2^2} - 1)\frac{1}{\sqrt{n}}\sum_{i=1}^{n}[e^{(\theta_1 X_i - \frac{\theta_1^2}{2})} - 1] - (e^{\theta_1\theta_2} - 1)\frac{1}{\sqrt{n}}\sum_{i=1}^{n}[e^{(\theta_2 X_i - \frac{\theta_2^2}{2})} - 1]}{\sqrt{(e^{\theta_2^2} - 1)\Big((e^{\theta_1^2} - 1)(e^{\theta_2^2} - 1) - (e^{\theta_1\theta_2} - 1)^2\Big)}}.$$

Then we can derive from (4) and Theorem 1 that

$$\lambda_n = \sup_{(\theta_1,\theta_2)\in\mathcal{E}}\left[Z_n^2(\theta_2) + \mathcal{T}_n^2(\theta_1,\theta_2)\right]\mathbb{1}_{\{Z_n(\theta_2)\geq 0; \mathcal{T}_n(\theta_1,\theta_2)\geq 0\}} + o_p(1).$$

So we have the following simple majorization

$$\lambda_n \leq \sup_{\theta_2\in\mathcal{E}_1}[Z_n(\theta_2)\vee 0]^2 + \sup_{(\theta_1,\theta_2)\in\mathcal{E}}[\mathcal{T}_n(\theta_1,\theta_2)\vee 0]^2$$

where $\vee$ denotes the supremum. In order to investigate the behaviour of Z_n and $\mathcal{T}_n^2$ we can use the following results:

Lemma 4 *In* $C[0,a]$ *(continuous path processes on* $[0,a]$*), the process*

$$\tilde{Z}_n(\theta) = \begin{cases} Z_n(\theta) & \text{if} \quad \theta \neq 0 \\ n^{-1/2} \sum_{i=1}^{n} X_i & \text{if} \quad \theta = 0 \end{cases}$$

converges weakly to a centered Gaussian process with unit variance. The same result holds in $C[-a,0]$ *for*

$$\breve{Z}_n(\theta) = \begin{cases} Z_n(\theta) & \text{if} \quad \theta \neq 0 \\ -n^{-1/2} \sum_{i=1}^{n} X_i & \text{if} \quad \theta = 0 \end{cases}.$$

In $C([-a,0] \times [0,a])$, $C([0,a] \times [-a,0])$, $C([-a,0] \times [-a,0])$ *with* $\theta_1 \leq \theta_2$ *and* $C([0,a] \times [0,a])$ *with* $\theta_1 \geq \theta_2$ *the random field*

$$\widetilde{\mathcal{T}_n}(\theta_1,\theta_2) = \begin{cases} \mathcal{T}_n(\theta_1,\theta_2) & \text{if} \quad (\theta_1,\theta_2) \neq (0,0) \\ \frac{1}{\sqrt{n}} \sum_{i=1}^{n} \frac{X_i^2-1}{\sqrt{2}} & \text{if} \quad (\theta_1,\theta_2) = (0,0) \end{cases}$$

converges weakly to a centered Gaussian random field with unit variance. The same result holds in $C([-a,0] \times [-a,0])$ *with* $\theta_2 \leq \theta_1$ *and* $C([0,a] \times [0,a])$ *with* $\theta_1 \leq \theta_2$ *for*

$$\widehat{\mathcal{T}_n}(\theta_1,\theta_2) = \begin{cases} \mathcal{T}_n(\theta_1,\theta_2) & \text{if} \quad (\theta_1,\theta_2) \neq (0,0) \\ -\frac{1}{\sqrt{n}} \sum_{i=1}^{n} \frac{X_i^2-1}{\sqrt{2}} & \text{if} \quad (\theta_1,\theta_2) = (0,0). \end{cases}$$

In order to get percentile points for $\sup_{\theta_2 \in \mathcal{E}} \left[Z_n(\theta_2) \vee 0\right]^2$ we can use the bound of Davies (1977) and table 1, case 3.1 of Garel (2001), completed by Goussanou (2001). For $\sup_{(\theta_1,\theta_2)\in\mathcal{E}} \left[\mathcal{T}_n(\theta_1,\theta_2) \vee 0\right]^2$ Slepian's inequality can be used in order to get a bound.

The 316 following data come from automatic blood tests by the Cell-DYN 4000 automaton of the Hemathology Laboratory of ULB Erasme, Brussels (December 2001). Here we only give the histogram. In order to use (4) and (5) the original observations have been centered with respect to the mean of the individuals in good health 7.25 and standardized by 3.25. Then we assume that $\theta_1, \theta_2 \in [-3,3]$ and $|\theta_1 - \theta_2| \geq 0.5$. Then we get a very large value of the statistic and we reject the hypothesis of one component.

3 Proofs

Proof of Lemma 1

We have to show that each component of the random field is tight. This can be proved in analogy to Garel (2001), page 343.

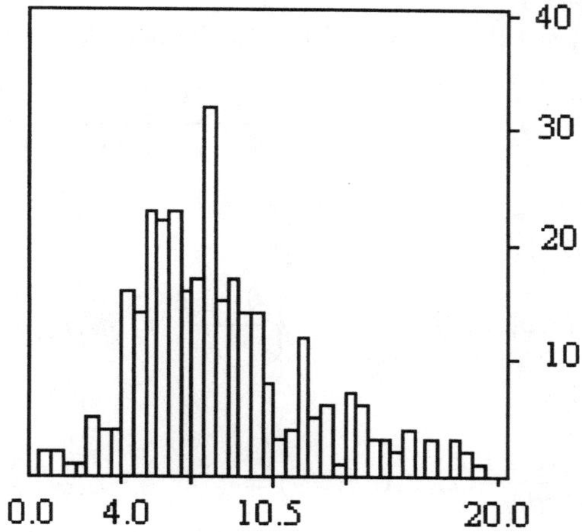

Fig. 1. White blood cells histogram of 316 patients.

Proof of Lemma 2

The same approach as Garel (2001) can be used.

Proof of Lemma 3

For $(\theta_1, \theta_2) \in \mathcal{E}$, that is to say under $|\theta_1 - \theta_2| \geq c_0$, we have

$$\begin{aligned}
\|(p_1\theta_1, p_2\theta_2)\| < \delta \iff & p_1^2\theta_1^2 + \theta_2^2\theta_2^2 < \delta^2 \\
\implies & |p_1\theta_1| < \delta \text{ and } |p_2\theta_2| < \delta \\
\implies & \left(|p_1| < \delta^{1/2} \text{ or } |\theta_1| < \delta^{1/2}\right) \text{ and } \left(|p_2| < \delta^{1/2} \text{ or } |\theta_2 < \delta^{1/2}\right) \\
\implies & \left(\|(p_1, p_2)\| < \sqrt{2\delta}\right) \text{ or } \left(\|(p_1, \theta_2)\| < \sqrt{2\delta}\right) \\
& \text{or } \left(\|(p_2, \theta_1)\| < \sqrt{2\delta}\right).
\end{aligned}$$

Let us define the following sets

$$\begin{aligned}
A_1(\delta) &= \left\{(p_1, p_2, \theta_1, \theta_2) \in [0,1]^2 \times \overline{\mathcal{E}}_1^2 \text{ such that } \|(p_1\theta_1, p_2\theta_2)\| < \delta\right\} \\
A_2(\delta) &= \left\{(p_1, p_2, \theta_1, \theta_2) \in [0,1]^2 \times \overline{\mathcal{E}}_1^2 \text{ such that } \|(p_1, p_2)\| < \sqrt{2}\delta^{\frac{1}{2}}\right\} \\
A_3(\delta) &= \left\{(p_1, p_2, \theta_1, \theta_2) \in [0,1]^2 \times \overline{\mathcal{E}}_1^2 \text{ such that } \|(p_1, \theta_2)\| < \sqrt{2}\delta^{\frac{1}{2}}\right\} \\
A_4(\delta) &= \left\{(p_1, p_2, \theta_1, \theta_2) \in [0,1]^2 \times \overline{\mathcal{E}}_1^2 \text{ such that } \|(p_2, \theta_1)\| < \sqrt{2}\delta^{\frac{1}{2}}\right\}.
\end{aligned}$$

The preceding inequalities show that $A_1 \subset A_2 \cup A_3 \cup A_4$. Equalities (2) to (8) can be proved using the same approach as Garel (2001), p. 344. Then (9) follows from (2) to (8).

Proof of Theorem 1

First we obtain the derivatives of the log-density. For $l = 1, 2$ we have:

$$\frac{\partial \log f}{\partial p_l}(x_i, p_1, p_2, \theta_1, \theta_2) = \frac{e^{-\frac{1}{2}(x_i-\theta_l)^2} - e^{\frac{1}{2}x_i^2}}{(1-p_1-p_2)e^{\frac{1}{2}x_i^2} + p_1 e^{-\frac{1}{2}(x_i-\theta_1)^2} + p_2 e^{-\frac{1}{2}(x_i-\theta_2)^2}}$$

and

$$\frac{\partial^2 \log f}{\partial p_l^2}(x_i, p_1, p_2, \theta_1, \theta_2) = -\left\{\frac{\partial \log f}{\partial p_l}(x_i, p_1, p_2, \theta_1, \theta_2)\right\}^2$$

which yields

$$\frac{\partial \log f}{\partial p_l}(x_i, 0, 0, \theta_1, \theta_2) = e^{-\frac{1}{2}x_i\theta_l - \frac{\theta_l^2}{2}} - 1$$

and

$$\frac{\partial^2 \log f}{\partial p_l^2}(x_i, 0, 0, \theta_1, \theta_2) = -\left(e^{-\frac{1}{2}x_i\theta_l - \frac{\theta_l^2}{2}} - 1\right)^2.$$

A Taylor expansion of the log likelihood function with respect to p_1 and p_2 yields :

$$L_n(x_1, ..., x_n, p_1, p_2, \theta_1, \theta_2) = L_n(x_1, ..., x_n, 0) + (p_1, p_2)\begin{pmatrix} \sum_{i=1}^n \left(e^{x_i\theta_1 - \frac{\theta_1^2}{2}} - 1\right) \\ \sum_{i=1}^n \left(e^{x_i\theta_2 - \frac{\theta_2^2}{2}} - 1\right)\end{pmatrix}$$

$$-\frac{1}{2}(p_1, p_2)\begin{pmatrix} \sum_{i=1}^n \theta_1^2 g_{11}(x_i, \alpha p_1, \alpha p_2, \theta_1, \theta_2) & \sum_{i=1}^n \theta_1\theta_2 g_{12}(x_i, \alpha p_1, \alpha p_2, \theta_1, \theta_2) \\ \sum_{i=1}^n \theta_1\theta_2 g_{12}(x_i, \alpha p_1, \alpha p_2, \theta_1, \theta_2) & \sum_{i=1}^n \theta_2^2 g_{22}(x_i, \alpha p_1, \alpha p_2, \theta_1, \theta_2)\end{pmatrix}\begin{pmatrix} p_1 \\ p_2 \end{pmatrix}$$

where $\alpha \in]0, 1[$ depends on p_1, p_2 and $(x_1, \ldots, x_n)$. We define a modified log likelihood function by :

$$\tilde{L}_n(x_1, ..., x_n, p_1, p_2, \theta_1, \theta_2)$$

$$= L_n(x_1, ..., x_n, 0) + (p_1, p_2)\begin{pmatrix} \sum_{i=1}^n e^{x_i\theta_1 - \frac{\theta_1^2}{2}} - 1 \\ \sum_{i=1}^n e^{x_i\theta_2 - \frac{\theta_2^2}{2}} - 1 \end{pmatrix} - \frac{1}{2}(p_1, p_2)\Lambda\begin{pmatrix} p_1 \\ p_2 \end{pmatrix}$$

with

$$\Lambda = \begin{pmatrix} \sum_{i=1}^{n}\left(e^{x_i\theta_1-\frac{\theta_1^2}{2}}-1\right)^2 & \sum_{i=1}^{n}\left(e^{x_i\theta_1-\frac{\theta_1^2}{2}}-1\right)\left(e^{x_i\theta_2-\frac{\theta_2^2}{2}}-1\right) \\ \sum_{i=1}^{n}\left(e^{x_i\theta_1-\frac{\theta_1^2}{2}}-1\right)\left(e^{x_i\theta_2-\frac{\theta_2^2}{2}}-1\right) & \sum_{i=1}^{n}\left(e^{x_i\theta_2-\frac{\theta_2^2}{2}}-1\right)^2 \end{pmatrix}$$

which can be written, for every $(\theta_1,\theta_2)\neq(0,0)$, in the form:

$$\tilde{L}_n\left(x_1,...,x_n,p,\theta_1,\theta_2\right) = L_n\left(x_1,...,x_n,0\right) + \left(n^{\frac{1}{2}}p_1\theta_1, n^{\frac{1}{2}}p_2\theta_2\right)\begin{pmatrix} \frac{1}{\sqrt{n}}\sum_{i=1}^{n} y_{i1}(\theta_1) \\ \frac{1}{\sqrt{n}}\sum_{i=1}^{n} y_{i2}(\theta_2)\end{pmatrix}$$

$$-\frac{1}{2}\left(n^{\frac{1}{2}}p_1\theta_1, n^{\frac{1}{2}}p_2\theta_2\right)\begin{pmatrix} \frac{1}{n}\sum_{i=1}^{n} y_{i1}^2(\theta_1) & \frac{1}{n}\sum_{i=1}^{n} y_{i1}(\theta_1)y_{i2}(\theta_2) \\ \frac{1}{n}\sum_{i=1}^{n} y_{i1}(\theta_1)y_{i2}(\theta_2) & \frac{1}{n}\sum_{i=1}^{n} y_{i2}^2(\theta_2)\end{pmatrix}\begin{pmatrix} n^{\frac{1}{2}}p_1\theta_1 \\ n^{\frac{1}{2}}p_2\theta_2\end{pmatrix}.$$

This modified log likelihood is a quadratic functional. We want to maximize this modified log likelihood $\tilde{L}_n$ (instead of L_n). We denote by $\mu(\theta_1,\theta_2)$ the lowest eigenvalue of the latter matrix. Under the condition $|\theta_1-\theta_2|\geq c_0>0$ the expectation of this matrix is positive definite, uniformly with respect to θ_1 and θ_2.

Using Lemma 1 and Lemma 2 we get that for all positive ε there exist $K>0$, $\varepsilon'>0$ and n_0 such that

$$\Pr\left[\sup_{(\theta_1,\theta_2)\in\mathcal{E}} n^{-1/2}\left\|\sum_{i=1}^{n} Y_i(\theta_1,\theta_2)\right\| < K, \text{ and for all } (\theta_1,\theta_2)\in\mathcal{E}\ \mu(\theta_1,\theta_2)>\varepsilon'\right]$$
$$>1-\varepsilon.$$

Hence there exists an $M>0$ such that for $n>n_0$, and with probability larger than $1-\varepsilon$ the supremum of $\tilde{L}_n$ is attained for $\left\|\left(p_1\theta_1 n^{\frac{1}{2}}, p_1\theta_1 n^{\frac{1}{2}}\right)\right\|<M$. Denoting by $\widehat{D}$ the quantity

$$\widehat{D} = \left[\frac{1}{n}\sum_{i=1}^{n} y_{i1}^2(\theta_1)\right]\left[\frac{1}{n}\sum_{i=1}^{n} y_{i2}^2(\theta_2)\right] - \left[\frac{1}{n}\sum_{i=1}^{n} y_{i1}(\theta_1)y_{i2}(\theta_2)\right]^2$$

this supremum is given by :

$$L_n(x_1,...,x_n,0) + \sup_{(\theta_1,\theta_2)\in\mathcal{E}} \frac{1}{2} D^{-1} \left\{ \frac{1}{n}\sum_{i=1}^{n} y_{i2}^2(\theta_2) \left[\frac{1}{\sqrt{n}}\sum_{i=1}^{n} y_{i1}(\theta_1)\right]^2 \right.$$

$$+\frac{1}{n}\sum_{i=1}^{n} y_{i1}^2(\theta_1) \left[\frac{1}{\sqrt{n}}\sum_{i=1}^{n} y_{i2}(\theta_2)\right]^2 -2\left(\frac{1}{n}\sum_{i=1}^{n} y_{i1}(\theta_1)y_{i2}(\theta_2)\right)\left[\frac{1}{\sqrt{n}}\sum_{i=1}^{n} y_{i1}(\theta_1)\right]$$

$$\left.\times\left[\frac{1}{\sqrt{n}}\sum_{i=1}^{n} y_{i2}(\theta_2)\right]\right\} 1\!\!1_{\left\{\sum_{i=1}^{n}\theta_1 y_{i1}(\theta_1)\geq 0;\sum_{i=1}^{n}\theta_2 y_{i2}(\theta_2)\geq 0\right\}}$$

$$= L_n(x_1,...,x_n,0)$$
$$+\sup_{(\theta_1,\theta_2)\in\mathcal{E}} \frac{1}{2}\left[T_n^2(\theta_1,\theta_2)\right] 1\!\!1_{\left\{\sum_{i=1}^{n}\left(e^{(\theta_1 X_i-\frac{\theta_1^2}{2})}-1\right)\geq 0;\sum_{i=1}^{n}\left(e^{(\theta_2 X_i-\frac{\theta_2^2}{2})}-1\right)\geq 0\right\}} + o_p(1)$$

where, using Lemma 2, $o_p(1)$ is a quantity which tends to 0 as n tends to infinity uniformly with respect to $\theta_1,\theta_2\in\mathcal{E}_1$. The log likelihood function can be written for every $(\theta_1,\theta_2)\neq(0,0)$ as

$$L_n(x_1,...,x_n,p_1,p_2,\theta_1,\theta_2) = L_n(x_1,...,x_n,0) + (p_1\theta_1,p_2\theta_2)\begin{pmatrix}\sum_{i=1}^{n}\dfrac{e^{x_i\theta_1-\frac{\theta_1^2}{2}}-1}{\theta_1}\\ \sum_{i=1}^{n}\dfrac{e^{x_i\theta_2-\frac{\theta_2^2}{2}}-1}{\theta_2}\end{pmatrix}$$

$$-\frac{1}{2}(p_1\theta_1,p_2\theta_2)\left(\begin{matrix}\sum_{i=1}^{n} y_{i1}^2(\theta_1)+\sum_{i=1}^{n}\left[g_{11}(x_i,\alpha p_1,\alpha p_2,\theta_1,\theta_2)-y_{i1}^2(\theta_1)\right]\\ \sum_{i=1}^{n} y_{i1}(\theta_1)y_{i2}(\theta_2)+\sum_{i=1}^{n}\left[g_{12}(x_i,\alpha p_1,\alpha p_2,\theta_1,\theta_2)-y_{i1}(\theta_1)y_{i2}(\theta_2)\right]\end{matrix}\right.$$

$$\left.\begin{matrix}\sum_{i=1}^{n} y_{i1}(\theta_1)y_{i2}(\theta_2)+\sum_{i=1}^{n}\left[g_{12}(x_i,\alpha p_1,\alpha p_2,\theta_1,\theta_2)-y_{i1}(\theta_1)y_{i2}(\theta_2)\right]\\ \sum_{i=1}^{n} y_{i2}^2(\theta_2)+\sum_{i=1}^{n}\left[g_{22}(x_i,\alpha p_1,\alpha p_2,\theta_1,\theta_2)-y_{i2}^2(\theta_2)\right]\end{matrix}\right)(p_1\theta_1,p_2\theta_2)^T.$$

If $\mathcal{D} = \left\{(p_1, p_2, \theta_1, \theta_2) \in [0,1]^2 \times \mathcal{E} \mid f(x, p_1, p_2, \theta_1, \theta_2) = f_0\right\}$ and $W(\mathcal{D})$ is an arbitrary but fixed neighbourhood of $\mathcal{D}$, Redner's result (1981, p.226; Theorem 3) implies that

$$\sup_{(p_1,p_2,\theta_1,\theta_2)\in[0,1]^2\times\mathcal{E}} L_n - \sup_{(p_1,p_2,\theta_1,\theta_2)\in W(\mathcal{D})} L_n = o_p(1).$$

An arbitrary neighbourhood of $\mathcal{D}$ can be included in a set $A_2(\delta) \cup A_3(\delta) \cup A_4(\delta)$ defined above. Uniformly with respect to $(p_1, p_2, \theta_1, \theta_2)$ on these sets, for $1 \leq k, l \leq 2$ the averages

$$\frac{1}{n}\sum_{i=1}^{n}[g_{kl}(X_i, \alpha p_1, \alpha p_2, \theta_1, \theta_2) - Y_{ik}(\theta_k)Y_{il}(\theta_l)]$$

converge in probability to zero as n tends to infinity and δ tends to zero. Then by Lemma 1, Lemma 2 and Lemma 3, for a sufficiently small δ, for all $\varepsilon > 0$, there exist n_0, $\varepsilon' > 0$ and $K > 0$ such that

$$\Pr\left(\begin{array}{c}\sup_{\theta_1,\theta_2\in\mathcal{E}_1} \frac{1}{\sqrt{n}}\left\|\sum_{i=1}^{n} Y_i(\theta_1,\theta_2)\right\| < K \\ \text{and for all } (\theta_1,\theta_2)\in\mathcal{E} \ : \ \mu(\theta_1,\theta_2) > \varepsilon'\end{array}\right) > 1-\varepsilon,$$

uniformly with respect to $(p_1, p_2, \theta_1, \theta_2)$ on $A_2(\delta) \cup A_3(\delta) \cup A_4(\delta)$. Thus with probability greater than $1 - \varepsilon$ there exists an $M > 0$ such that for $n > n_0$

$$\sup_{(p_1,p_2,\theta_1,\theta_2)\in A_2(\delta)\cup A_3(\delta)\cup A_4(\delta)} L_n(x_1, ..., x_n, p_1, p_2, \theta_1, \theta_2)$$

is attained on $\|(p_1\theta_1, p_2\theta_2)\sqrt{n}\| < M$. On this latter set, we get

$$\left|L_n - \tilde{L}_n\right| = \tfrac{1}{2}(\sqrt{n}p_1\theta_1, \sqrt{n}p_2\theta_2)$$

$$\begin{pmatrix} \frac{1}{n}\sum_{i=1}^{n}[Y_{i1}^2(\theta_1) - g_{11}] & \frac{1}{n}\sum_{i=1}^{n}[Y_{i1}(\theta_1)Y_{i2}(\theta_2) - g_{12}] \\ \frac{1}{n}\sum_{i=1}^{n}[Y_{i1}(\theta_1)Y_{i2}(\theta_2) - g_{12}] & \frac{1}{n}\sum_{i=1}^{n}[Y_{i2}^2(\theta_2) - g_{22}] \end{pmatrix} \begin{pmatrix} \sqrt{n}p_1\theta_1 \\ \sqrt{n}p_2\theta_2 \end{pmatrix}$$

$$\leq \frac{1}{2}M^2\left[\frac{1}{n}\sum_{i=1}^{n}\left|Y_{i1}^2(\theta_1) - g_{11}\right| + 2\frac{1}{n}\sum_{i=1}^{n}|Y_{i1}(\theta_1)Y_{i2}(\theta_2) - g_{12}| + \frac{1}{n}\sum_{i=1}^{n}\left|Y_{i2}^2(\theta_2) - g_{22}\right|\right].$$

Hence for $(p_1, p_2, \theta_1, \theta_2) \in [0,1]^2 \times \mathcal{E}$

$$\sup_{(p_1,p_2,\theta_1,\theta_2)_{\|(p_1\theta_1,p_2\theta_2)\|<\frac{M}{\sqrt{n}}}} \left|L_n - \tilde{L}_n\right| \leq \frac{1}{2}M^2$$

$$\left[\sup_{(p_1,p_2,\theta_1,\theta_2)_{\|(p_1\theta_1,p_2\theta_2)\|<\frac{M}{\sqrt{n}}}} \frac{1}{n}\sum_{i=1}^{n}\left|Y_{i1}^2(\theta_1) - g_{11}\right| + \right.$$

$$\sup_{(p_1,p_2,\theta_1,\theta_2)_{\|(p_1\theta_1,p_2\theta_2)\|<\frac{M}{\sqrt{n}}}} \frac{1}{n}\sum_{i=1}^{n}\left|Y_{i1}(\theta_1)Y_{i2}(\theta_2) - g_{12}\right| +$$

$$\left.\sup_{(p_1,p_2,\theta_1,\theta_2)_{\|(p_1\theta_1,p_2\theta_2)\|<\frac{M}{\sqrt{n}}}} \frac{1}{n}\sum_{i=1}^{n}\left|Y_{i2}^2(\theta_2) - g_{22}\right|\right].$$

Now, using Lemma 3 we get:

$$\sup\left\{\left|L_n - \tilde{L}_n\right| \mid (p_1, p_2, \theta_1, \theta_2) \text{ with } \|(p_1\theta_1, p_2\theta_2)\| < \frac{M}{\sqrt{n}}\right\} = o_p(1).$$

which leads to the result of Theorem 1.

Acknowledgments. The authors thank the editors and the referees for many helpful comments and suggestions.

References

BICKEL, P. & CHERNOFF, H. (1993). Asymptotic distribution of the likelihood ratio statistic in a prototypical non regular problem. *Statistics and Probability.* J.K. Ghosh et al. (eds). Wiley Eastern Limited, New-York, 83-96.

BILLINGSLEY, P. (1968). Convergence of probability measures. Wiley, New York.

BOCK, H. H., (1996). Probability models and hypotheses testing in partitioning cluster analysis. In *Clustering and classification*, P. Arabie, L. J. Hubert & G. De Soete (Eds.), World Scientific, New Jersey, 377-453.

DAVIES, R.B. (1977). Hypothesis testing when a nuisance parameter is present only under the alternative. *Biometrika* 64, 247-254.

GAREL, B (2001). Likelihood ratio test for univariate Gaussian mixture. *Journal of Statist. Planning and Inference* 96, 325-350.

GHOSH, J.K. & SEN, P.K. (1985). On the asymptotic performance of the log likelihood ratio statistic for the mixture model and related results, *Proc. Berkeley Conf. in honor of Jerzy Neyman and Jack Kiefer* (Vol II), L.M. Le Cam and R.A. Olshen (eds), Monterey, Wadsworth, 789-806.

GOUSSANOU, Y.L.F. (2001). Asymptotics of the generalized likelihood ratio test for Gaussian mixtures. *Unpublished Ph. D. Thesis*, University Toulouse 3.

HARTIGAN, J.A. (1985). A failure of likelihood asymptotics for normal mixtures. *Proc. Berkeley Conf. in honor of Jerzy Neyman and Jack Kiefer* (Vol.II), L.M. Le Cam and R.A. Olshen (eds), Monterey, Wadsworth, 807-810.

LINDSAY, B.G., (1995). Mixture models: theory, geometry and applications. *NSF -CBMS Regional Conference Series in Probability and Statistics*,USA, vol. 5.

McLACHLAN, & G.J., PEEL, D., (2000). Finite mixture models, *Wiley, New-York.*

REDNER, R.A. (1981). Note on the consistency of the maximum likelihood estimate for non identifiable distributions. *Ann. Statist.* 9, 225-228.

Obtaining Partitions of a Set of Hard or Fuzzy Partitions

Allan D. Gordon[1] and Maurizio Vichi[2]

[1] University of St Andrews, North Haugh, St Andrews, KY 16 9 SS, Scotland
[2] 'La Sapienza' University of Rome, Italy

Abstract. The paper presents methodology for obtaining a (secondary) partition of a set of (hard or fuzzy) primary partitions, specifying a consensus fuzzy partition for each of the classes in the secondary partition.

1 Introduction

The aim of classification studies is to summarize the relationships within a set of 'objects', often by obtaining a partition of them into disjoint classes of similar objects. In recent years, increased attention has been paid to the investigation of more general types of 'object'. In this paper, each 'object' is a partition of the same set of items. An example of such data is provided by the classic kinship terms data of Rosenberg and Kim (1975), obtained when each of a set of individuals was instructed to sort a set of fifteen kinship terms (grandfather, grandmother, grandson, granddaughter, father, mother, son, daughter, brother, sister, uncle, aunt, nephew, niece, cousin) into categories 'on the basis of some aspect of meaning'. Thus, each 'object' is a so-called *primary* partition of the set of kinship terms. Gordon and Vichi (1998) describe methodology for obtaining a so-called *secondary* partition of the set of primary partitions, in order to assess differences between individuals' views about the relationships between the kinship terms; associated with each of the classes of the secondary partition is a consensus partition that summarizes the set of primary partitions belonging to that class.

In the kinship terms example, each of the primary partitions is a hard partition, *i.e.* each item is assigned to a single class in the partition. In a *fuzzy* partition, each item has a set of membership functions (summing to 1) indicating the extent to which it belongs to each of the classes. In addition to a partition being provided directly, as was the case with the kinship terms data, it could have been obtained as the result of applying a clustering algorithm to a set of items. Such items could have been observed on several different occasions or under different experimental conditions, each data set having been classified separately.

This paper presents methodology for obtaining a (hard) secondary partition of a set of (hard or fuzzy) primary partitions, together with a consensus fuzzy partition for each of the classes in the secondary partition.

2 Methodology

The following terminology is used:

n — number of items;

r — number of primary partitions of the n items;

P_h — h-th primary partition of the set of n items $(h = 1, \dots, r)$;

g_h — number of classes in P_h $(h = 1, \dots, r)$;

w_h — weight associated with P_h $(h = 1, \dots, r)$;

$\mathbf{U}_h \equiv [u_{iph}]$ — $n \times g_h$ membership function matrix of P_h, where u_{iph} denote membership function of the i-th item for the p-th class of P_h $(i = 1, \dots, n; p = 1, \dots, g_h; h = 1, \dots, r)$;

s — number of classes in the secondary partition;

C_j — consensus partition of the j-th class of the secondary partitic $(j = 1, \dots, s)$;

c_j — number of classes in C_j $(j = 1, \dots, s)$;

$\mathbf{M}_j \equiv [m_{ilj}]$ — $n \times c_j$ membership function matrix of C_j, where m_{ilj} denote membership function of the i-th item for the l-th class of C_j $(i = 1, \dots, n; l = 1, \dots, c_j; j = 1, \dots, s)$;

$\mathbf{Z}_{hj} \equiv [z_{hpjl}]$ — $g_h \times c_j$ matrix, where z_{hpjl} =1 (resp., 0) if the p-th class of P contained (resp., is not contained) in the class which is matcl with the l-th class of C_j $(p = 1, \dots, g_h; h = 1, \dots, r; l = 1, ..$ $j = 1, \dots, s)$;

$\mathbf{B} \equiv [b_{hj}]$ — $r \times s$ matrix, where $b_{hj} = 1$ (resp., 0) if P_h is assigned (resp., is not assigned) to the j-th class of the secondary part $(h = 1, \dots, r; j = 1, \dots, s)$.

The membership functions are required to satisfy the following conditions:

$$u_{iph} \geq 0 \quad (i = 1, \dots, n; p = 1, \dots, g_h; h = 1, \dots, r), \tag{1}$$

$$\sum_{p=1}^{g_h} u_{iph} = 1 \quad (i = 1, \dots, n; h = 1, \dots, r), \tag{2}$$

$$m_{ilj} \geq 0 \quad (i = 1, \dots, n; l = 1, \dots, c_j; j = 1, \dots, s), \tag{3}$$

$$\sum_{l=1}^{c_j} m_{ilj} = 1 \quad (i = 1, \dots, n; j = 1, \dots, s). \tag{4}$$

If condition (1) is replaced by

$$u_{iph} \in \{0, 1\} \quad (i = 1, \dots, n; p = 1, \dots, g_h; h = 1, \dots, r), \tag{5}$$

each membership function matrix $\mathbf{U}_h$ specifies a hard partition.

In order to obtain a secondary partition, it is necessary to establish the relationship between each class of a primary partition and each class of each consensus partition. Gordon and Vichi (2001) describe three ways of defining a consensus partition of a given set of fuzzy primary partitions; the following methodology is based on their second definition. Under the assumption that $g_h \geq c_j$, the matrix $\mathbf{Z}_{hj}$ specifies how the classes of P_h are to be (possibly combined and) matched with the classes of C_j. The aim is to identify $\mathbf{Z} \equiv (\mathbf{Z}_{hj})$, $\mathbf{M} \equiv (\mathbf{M}_j)$ and $\mathbf{B} \equiv (b_{hj})$ which minimize the least-squares objective function

$$F(\mathbf{Z},\mathbf{M},\mathbf{B}) \equiv \sum_{h=1}^{r} w_h \sum_{j=1}^{s} b_{hj} \sum_{l=1}^{c_j} \sum_{i=1}^{n} (\sum_{p=1}^{g_h} u_{iph} z_{hpjl} - m_{ilj})^2, \tag{6}$$

subject to conditions (3), (4) and

$$z_{hpjl} \in \{0,1\} \quad (p = 1, ..., g_h; h = 1, \ldots, r; l = 1, \ldots, c_j; j = 1, \ldots, s) \tag{7}$$

$$\sum_{p=1}^{g_h} z_{hpjl} \geq 1 \quad (h = 1, \ldots, r; l = 1, \ldots, c_j; j = 1, \ldots, s) \tag{8}$$

$$\sum_{l=1}^{c_j} z_{hpjl} = 1 \quad (p = 1, \ldots, g_h; h = 1, \ldots, r; j = 1, \ldots, s). \tag{9}$$

A modification to cope with the situation in which $g_h < c_j$ is described by Gordon and Vichi (2001).

For given values of s and $\{c_j(j = 1, ..., s)\}$, a minimum of $F(\mathbf{Z},\mathbf{M},\mathbf{B})$ can be sought using the following iterative algorithm.

1. An initial secondary partition into s classes is obtained of the r primary partitions.
2. The consensus fuzzy partition C_j into c_j classes is found for the j-th class of the secondary partition ($j = 1, \ldots, s$), using the methodology described in Table 3 of Gordon and Vichi (2001).
3. Each of the r primary partitions is assigned to a class of the secondary partition to whose consensus fuzzy partition it is closest.
4. Steps 2 and 3 are iterated until there is no change in the secondary partition.

To increase confidence that the process has not become trapped in an inferior local optimum solution, it is advisable to repeat the iterations from several different initial secondary partitions.

At present, there does not seem to be an objective manner of specifying the number of classes in the secondary partition, s, and the number of classes in each of the consensus partitions, $\{c_j(j = 1, ..., s)\}$. One approach is to investigate a range of different values, and to assess the results by their interpretability.

3 Example

The methodology was applied to a subset of Rosenberg and Kim's (1975) kinship terms data comprising partitions provided by 85 female undergraduates at Rutgers University; the data are given in Rosenberg (1982, Table 7.1) with the 'nephew' and 'niece' columns interchanged, and are also available via the Internet at http://www-solar.mcs.st-and.ac.uk/~allan/classif.html. Solutions were obtained for a range of different values of s and $\{c_j(j = 1, ..., s)\}$. Table 1 shows the consensus fuzzy partitions obtained for the instance $s = 2$, $c_1 = 3 = c_2$; these results were obtained from a range of different random initial secondary partitions, and also from an initial partition obtained from a classification of the primary partitions.

Table 1. Fuzzy consensus partitions of Rosenberg and Kim's (1975) kinship terms data when the data have been partitioned into two classes, the fuzzy consensus partition of each of which comprises three classes. The results in the body of the Table are $100(m_{ilj})$, where m_{ilj} denotes the membership function of the i-th kinship term for the l-th class of the consensus partition associated with the j-th class of the secondary partition.

	j = 1			j = 2		
Kinship term	l = 1	l = 2	l = 3	l = 1	l = 2	l = 3
Grandfather	100	0	0	100	0	0
Grandmother	100	0	0	4	82	14
Grandson	89	7	4	91	9	0
Granddaughter	89	7	4	0	100	0
Father	1	92	7	100	0	0
Mother	1	92	7	4	91	5
Son	1	99	0	91	9	0
Daughter	1	99	0	0	100	0
Brother	0	99	1	100	0	0
Sister	0	99	1	23	73	4
Uncle	0	0	100	80	0	20
Aunt	0	0	100	3	73	24
Nephew	0	5	95	80	0	20
Niece	0	4	96	0	81	19
Cousin	0	7	93	10	0	90

The first consensus fuzzy partition is close to a hard partition into three classes which can be labelled 'grands' (grandfather, grandmother, grandson, granddaughter), 'nuclear family' (father, mother, son, daughter, brother, sister), and 'collateral relatives' (uncle, aunt, nephew, niece, cousin). The second partition is close to a hard partition into three classes specified by gender: male terms, female terms, and a singleton class containing the gender-ambiguous term 'cousin'. These results are consistent with previous analyses of these data reported by Carroll and Pruzansky (1980), Carroll and Arabie (1983) and Gordon and Vichi (1998).

References

CARROLL, J. D. and ARABIE, P. (1983): INDCLUS: An Individual Differences Generalization of the ADCLUS Model and the MAPCLUS Algorithm. *Psychometrika, 48*, 157–169.

CARROLL, J. D. and PRUZANSKY, S. (1980): *Discrete and Hybrid Scaling Models.* In: E. D. Lantermann and H. Feger (Eds.): *Similarity and Choice* Huber, Bern, 108–139.

GORDON, A. D. and VICHI, M. (1998): Partitions of Partitions. *Journal of Classification, 15*, 265–285.

GORDON, A. D. and VICHI, M. (2001): Fuzzy Partition Models for Fitting a Set of Partitions. *Psychometrika, 66*, 229–247.

ROSENBERG, S. (1982): *The Method of Sorting in Multivariate Research with Applications Selected from Cognitive Psychology and Person Perception.* In: N. Hirschberg and L. G. Humphreys (Eds.): *Multivariate Applications in the Social Sciences* Erlbaum, Hillsdale, NJ, 117–142.

ROSENBERG, S. and KIM, M. P. (1975): The Method of Sorting as a Data-gathering Procedure in Multivariate Research. *Multivariate Behavioral Research, 10*, 489–502.

Clustering for Prototype Selection using Singular Value Decomposition

A.K.V.Sai Jayram[1] and M.Narasimha Murty[2]

[1] Department of E.C.E, Indian Institute of Science, Bangalore 560012, India
[2] Department of C.S.A, Indian Institute of Science, Bangalore 560012, India

Abstract. Data clustering is an important technique for exploratory data analysis. The speed, reliability and consistency with which a clustering algorithm can organize large amounts of data constitute reasons to use it in applications like data mining, document retrieval, signal compression, coding and pattern classification. In this paper, we use clustering for efficient large-scale pattern classification; more specifically, we achieve it by selecting appropriate prototypes and features using Singular Value Decomposition (SVD). It is found that the SVD based clustering not only selects better prototypes, but also reduces the memory and computational requirements by 98% over the conventional Nearest Neighbour Classifier (NNC) (T.M.Cover and P.E.Hart (1967)), on OCR data.

1 Introduction

Clustering techniques can be broadly classified into two categories: partitional and hierarchical (A.K.Jain et al. (1999)). Hierarchical techniques organize data in a nested sequence of groups which can be displayed in the form of a dendogram or a tree.
A scheme for clustering documents based on Singular Value Decomposition (SVD) was proposed in (P.Drincas et al. (1999)). SVD can be used for both prototype selection and feature extraction. In this paper, we consider the use of SVD based clustering scheme (SVDBCS) for selecting prototypes and features for large scale handwritten pattern recognition. We study, experimentally, the performance of the proposed algorithm on a large handwritten digit data set with/without bootstrapping (Yoshihiko Hamamoto et al. (1997)). We show that the results obtained using the SVD based algorithm are superior to results reported in the literature (M.Prakash and M.Narasimha Murty (1997)) on the handwritten OCR data. We also compare the performance of the algorithm on VOWEL dataset.

We use the well-known bootstrapping technique (Yoshihiko Hamamoto et al.(1997)) for resampling to circumvent the curse of dimensionality (Richard O. Duda et al. (2nd ed. 2000)). K-Means Algorithm (KMA) (Shori Z. Selim and M.A.Ismail (1984)) and SVDBCS are then applied on bootstrapped data for prototype selection. The prototypes thus obtained are seen to be effective for classification.

We have observed that SVDBCS gives better prototypes. The technique (SVD) is also effective in reducing the dimension of the data, thus reducing both memory and computation requirements.

2 Prototype selection

2.1 Clustering by KMA

The most commonly used partitional clustering strategy is based on sum of squared error criterion. The objective of this scheme is to obtain a partition which, for a fixed number of clusters, minimizes the sum of squared errors. KMA is computationally efficient and gives good results if the clusters are compact, hyperspherical in shape and well-separated in feature space.

Case 1: Prototypes are selected by keeping the dimensions fixed and varying the number of clusters.

2.2 Clustering by SVDBCS

We consider the problem of dividing a set of m points in Euclidean n dimensional space into k clusters (m,n are variable while k is fixed), so as to minimize the sum of square of the distances from each point to its "cluster center". The above problem can be solved by Singular Value Decomposition. SVD is used to approximate a given matrix A, having rank M with a matrix D, having rank K, less than M. SVD gives the best reduced rank matrix D^* that is an approximation to A.

Algorithm for SVDBCS

As Singular value decomposition is used for both prototype selection and feature extraction, clustering is performed in three different ways.

Case 2: Having the dimension fixed and varying the number of clusters.

a. Obtain the best K rank approximation D^*, of the pattern matrix A.

b. Use D^* for clustering (P.Drincas et al. (1999)).

Case 3: Having the number of clusters fixed and changing the number of dimensions.

a. For each class 'i', find the mean vectorB_i.

b. Perform SVD on mean vector matrix B.

c. Find the first K vectors of the SVD of B.

d. Take the space V_{svd} spanned by first K singular values in the row space of B.

e. Project the patterns to V_{svd}, which is a K dimensional subspace.

Case 4: Varying both the number of clusters and number of dimensions.

a. In the subspace V_{svd} obtained by case 3, do clustering (P.Drincas et al. (1999)).

In each of the above four cases, each cluster centroid forms a prototype.

2.3 Bootstrapping

The performance of a classifier depends on the interrelationship between feature dimension, number of patterns, and classifier complexity. The number of training patterns is to be a function of the feature dimension for optimal training. This phenomenon is termed as "curse of dimensionality", which leads to the "peaking phenomenon" in classifier design. It has been observed in practice that the added features may degrade the performance of a classifier if the number of training samples that are used to design a classifier is small relative to the number of features. We use the algorithm reported and successfully used by Hamamoto et al. to circumvent the curse of dimensionality. This technique generates bootstrap samples by locally combining original training samples.

The algorithm for bootstrapping is as follows

a. Select one sample x_{k_0} from a set of N samples, X_N.

b. Find the 'r' nearest neighbor samples x_{k_1}, x_{k_2},.....x_{k_r} of x_{k_0}, based on Euclidean distance.

c. Compute the bootstrap sample $\mathrm{x}_{k_b} = \frac{1}{r+1}\sum_{j=0}^{r} x_{k_j}$

d. Repeat steps a, b, and c until each of the N samples is selected once.

In step a, the samples are chosen so that no sample is selected more than once. The bootstrap samples are added to the training set and clustering is performed on the resulting set.

Case 5:

Clustering using KMA on bootstrap + training samples. (refer case 1).

Case 6:

Clustering by SVDBCS on bootstrap + training samples. (refer case 2).

3 Experiments and Results

3.1 Data set used

OCR: The dataset consists of handwritten digits, and was used earlier (M.Prakash and M.Narasimha Murty (1997)). Each digit is a binary image of size 16 x 12 pixels. There are 6670 training (667 training samples per class) and 3333 labeled test samples. The value of every feature varies from 0 to 4, and there are a total of 192 features.

VOWEL: The dataset consists of eleven steady state vowels of British English using a specified set of LPC derived log area ratios. There are 10 such features associated with each vowel. Fifteen speakers have spoken each of the vowel six times, resulting in a total of 990 samples. These samples are split into 528 and 462 test and training samples respectively. For each class, the number of samples in the training and the test sets are 48 and 42 respectively. This data set is available in public domain as part of neural network bench mark data collection at Carnegie-Mellon University, USA.

3.2 Experiments

Experiments are carried out by

a. Having the dimension of the feature vector fixed (at 192 for OCR data and 10 for VOWEL data) and varying the number of clusters. (from 4 to 160 for OCR data and 1 to 48 for VOWEL data) (Case 1,Case 2, Case 5 and Case 6). Results obtained for Case 1 and Case 2 are shown below.

Table 1. Results on OCR data

Clusters/class	*KMA*	*SVDBCS*
4	87	87.16
8	88.25	88.66
12	90	90.37
16	90.82	91
20	91.06	91.39
40	92.02	92
80	92.59	92.71
120	92.35	92.74
140	92.83	92.92
160	**92.86**	**93.16**

Table 2. Results on VOWEL data

Clusters/class	*KMA*	*SVDBCS*
1	48.26	43.23
2	56.49	53.34
3	**58.87**	**54.59**
5	56.49	52.59
10	54.54	51.23
20	52.16	49.79
30	50.86	50.54
40	52.59	51.29
48	51.29	51.29

Bootstrapping is performed on training data to generate bootstrap samples. The generated bootstrapped samples are combined with training samples and are used to generate the cluster representatives. Results obtained for Case 5 and Case 6 on OCR data, are shown in Table 3.

Table 3. Results on OCR data

Clusters/class	*KMA*	*SVDBCS*
4	86.41	86.35
12	90.28	90.00
20	91.06	91.12
40	92.06	92.47
80	92.92	93.43
110	93.52	**94.00**
120	93.28	93.46
160	93.76	93.79

b. Having the number of prototypes fixed (at 667 for OCR data and 10 for VOWEL data) and varying the dimensionality (from 10 to 190 for OCR data and 1 to 10 for VOWEL data) (Case 3).
Results obtained for Case 3 are shown in Table 4 and Table 5.

Table 4. Results on OCR data

Dimension	*%Accuracy*
10	85.47
30	89.32
60	*91*
90	91.5
120	92
150	92.14
160	**92.3**
190	92.1

c. Varying the number of clusters from 10 to 150 and dimensions from 10 to 160 (Case 4).
Experimental results for Case 4 are plotted in Fig.1 and Fig.2.

3.3 Results and discussion

SVDBCS is applied on OCR and VOWEL data and the classification accuracy obtained by selecting K prototypes per class is compared to that obtained by same number of prototypes (K) using KMA.

SVDBCS, when used for prototype selection, performs better compared to KMA on OCR data, and is comparable to KMA on VOWEL data. (Case 1,2,5,6)

Table 5. Results on VOWEL data

Dimension	*%Accuracy*
1	18
2	41.77
4	46.10
5	55.68
6	53.64
7	**55.84**
8	53.03
10	53.03

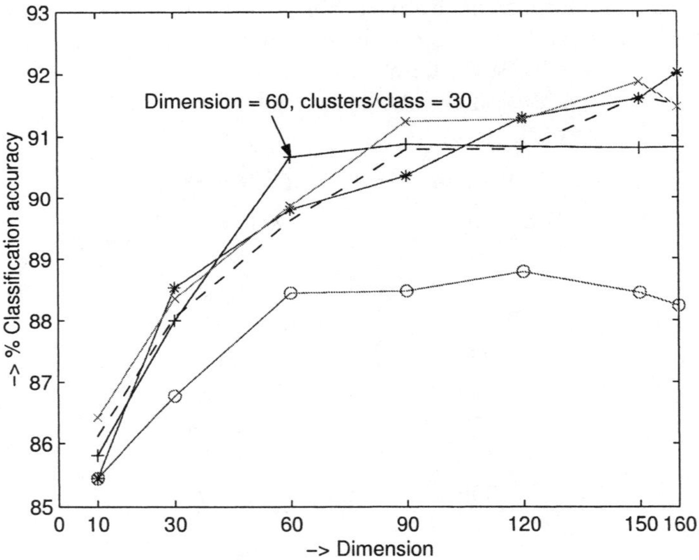

Legend : Clusters/class o : 10, + : 30, $*$: 70, $\star$: 110, $-$: 150

Fig. 1.

When used for feature extraction, SVDBCS reduces the dimensionality and increases the classification accuracy, compared to NNC (Case 3). For OCR data, the classification accuracy for a dimensionality of 60 and 30 prototypes per class is marginally less (by 1.3%) to that obtained by NNC (92%) with a dimensionality of 192 and 667 prototypes per class and hence is comparable. For VOWEL data, the classification accuracy for a dimensionality of 7 and 40 prototypes per class is equal to that obtained by NNC (56%) with a dimensionality of 10 and 48 prototypes per class.

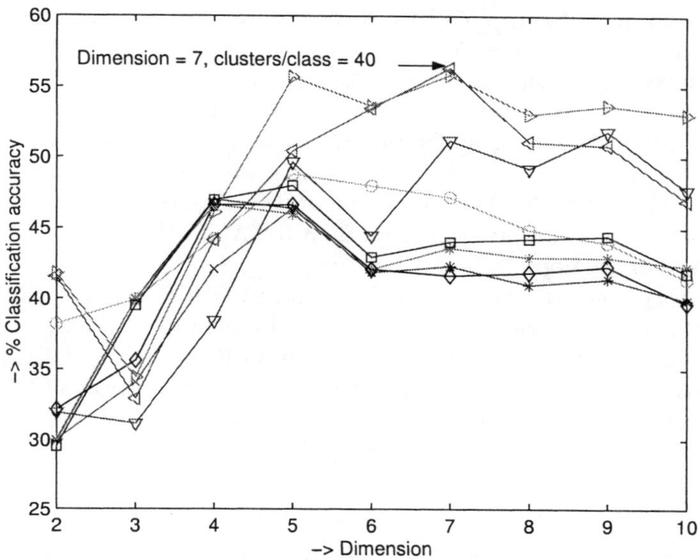

Legend : Clusters/class ∘ : 1, ⋆ : 2, + : 3, ∗ : 5, ⊓ : 10, ⋄ : 20, ∇ : 30, ◁ : 40, ▷ : 48

Fig. 2.

The classification accuracy obtained by SVDBCS on OCR data is found to be superior (94%) to that reported in the literature so far based on other algorithms (M. Prakash and M. Narasimha Murty (1997)), while it is comparable with other algorithms (M. Prakash and M. Narasimha Murty (1997)), on VOWEL data.

For OCR data, the classification accuracy for a dimensionality of 60 and 30 prototypes per class is comparable to that obtained by having a feature dimension of 192 and 667 prototypes per class, using NNC. So, we achieve a reduction in computation time and memory requirements by 98%. For VOWEL data, the classification accuracy for a dimensionality of 7 and 40 prototypes per class is equal to that obtained by having a feature dimension of 10 and 48 prototypes per class, using NNC. So, we achieve a reduction in computation time and memory requirements by 41%.

4 Summary and Conclusions

The SVD of a matrix is a powerful technique because it provides a best approximation to rank deficient matrices, and exposes the geometric properties of the matrix. SVD hence can be used as a generic techniqe for many applications like noisy signal filtering, time series analysis, etc.
In this paper we used SVD for both prototype selection and feature extraction on OCR and VOWEL data.

The proposed scheme, SVDBCS has the following advantages.

a. obtains better prototypes

b. reduces the dimensionality effectively, and

c. reduces the computation time and memory requirements.

SVDBCS performs better than KMA in terms of providing a good set of prototypes. The prototypes can be used along with NNC to realize increased classification accuracy and reduced computational resources. However, SVDBCS requires more computation time than KMA, during prototype selection and feature extraction.

Since prototype selection is an offline process, SVDBCS can be used for large-scale pattern classification. It is also observed that bootstrapping of the data is helpful in increasing the performance of both KMA and SVDBCS.

Acknowledgements: I would like to thank B. R. Suresh and Anand B. Jyoti for reviewing and helping me in improving the clarity of the paper.

References

ANIL K. JAIN, ROBERT P.W. DULIN and JIANCHANG MAO. (2000): Statistical Pattern Recognition: A Review. *IEEE Trans. Pattern Analysis and Machine Intelligence.* Vol.22, No.1, pp.4-37.

COVER, T.M. and HART, P.E. (1967): Nearest Neighbour Pattern Classification. *IEEE Trans. Information Theory.* vol.13, no. 1, pp.21-27.

DEWILDE, P. and DEPRETTERE, ED.F. (1988): Singular Value Decomposition: An introduction. In: *Ed. F. Deprettere, editor, SVD and Signal Processing: Algorithms, Applications, and Architectures.* Elsevier Science Publishers, North Holland, pp.3-41.

DRINCAS, P., ALAN FRIEZE, RAVI KANNAN, SANTOSH VEMPALA, VINAY, V. (1999): Clustering in large graphs and matrices. *Proc. of the symposium on Discrete Algorithms, SIAM*

JAIN, A.K., MURTY, M.N. and FLYNN, P.J. (1999): Data clustering: a review. *ACM computing surveys.* Vol 31, Issue 3, Nov pp-264-323.

JAIN, A.K. and CHANDRASEKARAN, R. (1982): Dimensionality and sample size considerations in pattern recognition practice, in: *Handbook of Dimensionality.* P.R.Krishnaiah and L.N.Kanal, Eds. New York: North-Holland

PRAKASH, M. and NARASIMHA MURTY, M. (1997): Growing subspace pattern recognition methods and their neural-network models. *IEEE Trans. Neural Networks.* Vol.8, No.1, pp.161-168.

RICHARD O. DUDA, PETER E. HART and DAVID G. STORK. (2000): *Pattern Classification (2nd ed.)*

YOSHIHIKO HAMAMOTO, SHUNJI UCHIMURA and SHINGO TOMITA. (1997): A Bootstrap Technique for Nearest neighbour Classifier. *IEEE Trans.Pattern Analysis and Machine Intelligence.* Vol 19, no 1, Jan pp.73-79.

Clustering in High-dimensional Data Spaces

Fionn Murtagh

School of Computer Science, Queen's University Belfast, Belfast BT7 1NN, Northern Ireland, f.murtagh@qub.ac.uk

Abstract. By high-dimensional we mean dimensionality of the same order as the number of objects or observations to cluster, and the latter in the range of thousands upwards. Bellman's "curse of dimensionality" applies to many widely-used data analysis methods in high-dimensional spaces. One way to address this problem is by array permuting methods, involving row/column reordering. Such methods are closely related to dimensionality reduction methods such as principal components analysis. An imposed order on an array is beneficial not only for visualization but also for use of a vast range of image processing methods. For example, clustering becomes in this context image feature detection.

1 Introduction

Bellman's (1961) "curse of dimensionality" refers to the exponential growth of hypervolume as a function of dimensionality. Many problems become tougher as the dimensionality increases. Nowhere is this more evident than in problems related to search and clustering.

In Murtagh and Starck (1998) (see also Starck, Murtagh, and Bijaoui (1998)), a constant computational time or $O(1)$ approach to cluster analysis was described. The computational complexity was, as is usual, defined in terms of the number of observations. This work related to problem spaces of dimensionality 2, with generalization possible to 3-dimensional spaces Chereul, Crézé, and Bienaymé (1997). Byers and Raftery (1996) proposed another very competitive approach.

It may be helpful to distinguish this work from clustering understood as mixture distribution modeling. A characterization follows which will describe the broad picture. Banfield and Raftery (1993) discuss algorithms for optimal cluster modeling and fitting. On the other hand, the work on clustering of Murtagh and Starck (1998), and Byers and Raftery (1996), is based on noise modeling. Mixture modeling and cluster modeling are essentially signal modeling. Given that observed data can be considered as a mixture of signal and of noise, one can approach data analysis from either of two perspectives: accurately model the signal, as in mixture modeling, with perhaps noise components included in the mixture; or accurately model the noise.

The latter lends itself well to the problem representation to be described in this article. In general, it lends itself well to situations when we consider data arrays as images. We will next look at when and how we can do this.

2 Matrix Sequencing

We take our input object-attribute data, e.g. document-term or hyperlink array, as a 2-dimensional image. In general, an array is a mapping from the Cartesian product of observation set, I, and attribute set, J, onto the reals, $f : I \times J \longrightarrow \mathbb{R}$, while an image (single frame) is generally defined for discrete spatial intervals X and Y, $f : X \times Y \longrightarrow \mathbb{R}$. A table or array differs from a 2-dimensional image, however, in one major respect. There is an order relation defined on the row- and column-dimensions in the case of the image. To achieve invariance we must induce an analogous ordering relation on the observation and variable dimensions of our data table.

A natural way to do this is to seek to optimize contiguous placement of large (or nonzero) data table entries. Note that array row and column permutation to achieve such an optimal or suboptimal result leaves intact each value x_{ij}. We simply have row and column, i and j, in different locations at output compared to input. Methods for achieving such block clustering of data arrays include combinatorial optimization (McCormick, Schweitzer, and White (1972), Lenstra (1974), Doyle (1988)) and iterative methods (Deutsch and Martin (1971), Streng (1991)). In an information retrieval context, a simulated annealing approach was also used in Packer(1989). Further references and discussion of these methods can be found in Murtagh (1985), March (1983), Arabie et al. (1988). Treating the results of such methods as an image for visualization purposes is a very common practice (e.g. Gale, Halperin and Costanzo (1984)).

We now describe briefly two algorithms which work well in practice.

Moments Method: Deutsch and Martin (1971) Given a matrix, $a(i, j)$, for $i = 1, 2, \ldots, n$, and $j = 1, 2, \ldots, m$. Define row moments as $m(i) = (\sum_j ja(i, j)) / (\sum_j a(i, j))$. Permute rows in order of nondecreasing row moments. Define column moments analogously. Permute columns in order of nondecreasing column moments. Reiterate until convergence.

This algorithm results (usually) in large matrix entries being repositioned close to the diagonal. An optimal result cannot be guaranteed.

Bond Energy Algorithm: McCormick, Schweitzer, and White (1972) Permute matrix rows and columns such that a criterion, $\text{BEA} = \sum_{i,j} a(i, j)(a(i - 1, j) + a(i + 1, j) + a(i, j - 1) + a(i, j + 1))$ is maximized.

An algorithm to implement the BEA is as follows: Position a row arbitrarily. Place the next row such that the contribution to the BEA criterion is maximized. Place the row following that such that the new contribution to the BEA is maximized. Continue until all rows have been positioned. Then do analogously for columns. No further convergence is required in this case. This algorithm is a particular use of the traveling salesperson problem, TSP, which is widely used in scheduling. In view of the arbitrary initial choice of row or column, and more particularly in view of the greedy algorithm solution, this is a suboptimal algorithm.

Matrix reordering rests on (i) permuting the rows and columns of an incidence array to some standard form, and then data analysis for us in this context involves (ii) treating the permuted array as an image, analyzed subsequently by some appropriate analysis method.

2.1 Matrix Permutation and Singular Value Decomposition

Dimensionality reduction methods, including principal components analysis (suitable for quantitative data), correspondence analysis (suitable for qualitative data), classical multidimensional scaling, and others, is based on singular value decomposition. It holds:

$$AU = \Lambda U$$

where we have the following. A is derived from the given data – in the case of principal components analysis, this is a correlation matrix, or a variance/covariance matrix, or a sums of squares and cross products matrix.

Zha et al. (2001) formalize the reordering problem as the constructing of a sparse rectangular matrix

$$W = \begin{pmatrix} W_{11} & W_{12} \\ W_{21} & W_{22} \end{pmatrix}$$

so that W_{11} and W_{22} are relatively denser than W_{12} and W_{21}. Permuting rows and columns according to projections onto principal axes achieves this pattern for W. Proceeding recursively (subject to a feasibility cut-off), we can further increase near-diagonal density at the expense of off-diagonal sparseness.

2.2 Lerman's Theorem for Ultrametric Matrices

As is well-known, a geometric space has an induced metric. The Euclidean metric is widely used. The Euclidean metric ($L = 2$) is just one of the Minkowski metrics, with others including the Hamming or city-block metric ($L = 1$), and the Chebyshev metric ($L = \infty$):

$$d_p(x, y) = \sqrt[p]{\sum_j \mid x_j - y_j \mid^p} \qquad p \geq 1.$$

A metric satisfies the property of triangular inequality $d(x, y) \leq d(y, z) + d(z, y)$. The ultrametric inequality is a more restrictive condition: $d(x, y) \leq \max\ (d(y, z), d(z, y))$. Consider a classification hierarchy defined on the object set. We can represent the tree with the objects at the base, and the embedded clusters extending upwards. If one defines the distance between objects as the lowest level in the tree in which the two objects first find themselves associated with the same cluster, then the resulting distance is an ultrametric one. Inducing a tree on an object-set is the transforming of a metric space into an ultrametric one.

Ultrametric distance matrices can be represented, subject to an appropriate ordering of objects, with quite particular relations between values as we move away from the diagonal. Lerman (1981) discusses ultrametric spaces in detail. Lerman's Theorem 2 (Lerman (1981), p. 45) describes properties of ultrametric distance matrices. The result we are most interested in is in regard to matrix reordering: an order can be found such that array elements are necessarily non-increasing as we move away from the diagonal, and row and column array elements have a number of such inequality properties.

Lerman's Theorem for the Form of Ultrametric Matrices: An $n \times n$ matrix of positive reals, symmetric with respect to the diagonal, is a matrix of distances associated with an ultrametric on the object-set iff a permutation can be applied to the matrix such that the matrix has the following form:

1. Beyond the diagonal term equaling 0, values in the same row are non-decreasing.
2. For each index k, if (condition b1) $d(k,k+1) = d(k,k+2) = \cdots = d(k,k+l+1)$ then (implication b2) $d(k+1,j) \leq d(k,j)$ for $k+1 < j \leq k+l+1$ and (implication b3) $d(k+1,j) = d(k,j)$ for $j > k+l+1$.

Therefore $l \geq 0$ is the length of the section starting, beyond the principal diagonal, the interval of columns containing equal values in row k.

We will exemplify Lerman's theorem using the Fisher iris data. The iris data of Anderson used by Fisher (1936) is a very widely-used benchmark dataset. The data consists of 3 varieties of iris flower, each providing 50 samples. There are measurements on 4 variables, petal and sepal length and breadth. The data matrix is therefore one of dimensions 150×4.

To derive ultrametric distances, we took the Fisher iris data, in its original 150×4 form. We constructed a complete link hierarchical clustering, using the Euclidean distance between the observation vectors. We read off the 150×150 ultrametric distances (ranks were used, rather than agglomeration criterion values) from this dendrogram. Fig. 1 (left) shows this ultrametric matrix. (The greyscale values have been histogram-equalized for better contrast.) When we reorder the rows and columns (the matrix is symmetric of course) in accordance with the ordering of singletons used by the dendrogram representation we get the visualization shown in Fig. 1 (right). Again contrast stretching through histogram-equalization was used. Note that the origin is in the lower left, i.e., following the image convention rather than the matrix convention.

2.3 Permuting Large Sparse Arrays

A few comments on the computational aspects of array permuting methods when the array is very large and very sparse follow Berry, Hendrickson, and

Fig. 1. Left: ultrametric matrix of 150 observations, in given order – clusters 1, 2 and 3 correspond to sequence numbers 1–50, 51–100, 101–150. Right: ultrametric matrix of these same observations, with the rows and columns permuted in accordance with a non-crossover representation of the associated dendrogram.

Raghavan (1996). Gathering larger (or nonzero) array elements to the diagonal can be viewed in terms of minimizing the envelope of nonzero values relative to the diagonal. This can be formulated and solved in purely symbolic terms by reordering vertices in a suitable graph representation of the matrix. A widely-used method for symmetric sparse matrices is the Reverse Cuthill-McKee (RCM) method.

The complexity of the RCM method for ordering rows or columns is proportional to the product of the maximum degree of any vertex in the graph represented by the array and the total number of edges (nonzeroes in the matrix). For hypertext matrices with small maximum degree, the method would be extremely fast. The strength of the method is its low time complexity but it does suffer from certain drawbacks. The heuristic for finding the starting vertex is influenced by the initial numbering of vertices and so the quality of the reordering can vary slightly for the same problem for different initial numberings. Next, the overall method does not accommodate dense rows (e.g., a common link used in every document), and if a row has a significantly large number of nonzeroes it might be best to process it separately; i.e., extract the dense rows, reorder the remaining matrix and augment it by the dense rows (or common links) numbered last.

One alternative approach is based on linear algebra, making use of the extremely sparse incidence data which one is usually dealing with. The execution time required by RCM may well require at least two orders of magnitude (i.e., 100 times) less execution time compared to such methods. However such methods, including for example sparse array implementations of correspondence analysis, appear to be more competitive with respect to bandwidth (and envelope) reduction at the increased computational cost.

Elapsed CPU times for a range of arrays are given in Berry, Hendrickson, and Raghavan (1996), and as an indication show performances between 0.025 to 3.18 seconds for permuting a 4000×400 array.

3 Incidence Data and Image Models

Consider co-occurrence data, or document-term dependence data. Contiguity of links, or of data values in general, is important if we take the 2-way data array as a 2-dimensional image. It is precisely this issue which distinguishes a data array from an image: in the latter data type, the rows and columns are permutation invariant.

We can define permutation invariance by some appropriate means. We can use the output of some matrix permuting method, such as the bond energy algorithm McCormick, Schweitzer, and White (1972) or a permuting method related to singular value decomposition Berry, Hendrickson, and Raghavan (1996).

The non-uniqueness of such solutions is not unduly important in this article and will not be discussed in detail. However we must justify our approach since it does rely on an array permutation method selected by the user. The resulting non-unique solution is acceptable because our ultimate goals are related to data visualization and exploratory data analysis. Our problem-solving approach is *unsupervised* rather than *supervised*, to use terms which are central in pattern recognition. We seek *an* interpretation of our data, rather than *the* unique interpretation. Of course, the unsupervised data analysis may well precede or be otherwise very closely coupled to supervised analysis (discriminant analysis, statistical estimation, exact database match, etc.) in practice.

4 Conclusion

A range of examples and case studies will be used in the presentation to exemplify the methodology described in this article.

The methodology developed here is fast and effective. It is based on the convergence of a number of technologies: (i) data visualization techniques; (ii) data matrix permuting techniques; and (iii) appropriate image analysis methods, if feasible of linear computational cost. We have discussed its use for large incidence arrays. We introduced noise modeling of such data, and showed how noise filtering can be used to provide as output a set of significant clusters in the data. Such clusters may be overlapping. Further development of this work would be to investigate hierarchical clusters, possibly overlapping, derived from the multiple scales.

We have also discussed this innovative methodology using a number of different datasets. It is clearly related to other well-established data analysis methods, such as seriation (one-dimensional ordering of observations), and nonparametric density estimation (the wavelet transform can be viewed as performing such density estimation).

We can note also the potential use of our new methodology for use in graphical user interfaces. The Kohonen self-organizing feature map, by now quite widely used for support of clickable user interfaces (Poinçot, Lesteven and Murtagh (1998), Poinçot, Lesteven, and Murtagh (2000)), presents a map of the documents (say), but not as explicitly of the associated index terms. Our maps cater equally for both documents and index terms. Furthermore, the way is open to the exploration of what can be offered by recent developments in client-server based image storage and delivery (see some discussion in Chapter 7 of Starck, Murtagh, and Bijaoui (1998) e.g. progressive transmission and foveation (i.e. progressive transmission in a local region) strategies. This perspective opens up onto a line of inquiry which could be characterized as *multiple resolution information storage, access and retrieval.* This is particularly relevant in the context of current international initiatives on the computational and data Grid infrastructure of the future.

References

BELLMAN, R. (1961) *Adaptive Control Processes: A Guided Tour.* Princeton University Press, Princeton.

MURTAGH, F. and STARCK, J.L. (1998) "Pattern clustering based on noise modeling in wavelet space", *Pattern Recognition*, **31**, 847–855.

STARCK, J.L., MURTAGH, F. and BIJAOUI, A. (1998) *Image and Data Analysis: The Multiscale Approach.* Cambridge University Press, Cambridge.

CHEREUL, E., CRÉZÉ, M. and BIENAYMÉ, O. (1997) "3D wavelet transform analysis of Hipparcos data", in Maccarone, M.C., Murtagh, F., Kurtz, M. and Bijaoui, A. (eds.). *Advanced Techniques and Methods for Astronomical Information Handling*, Observatoire de la Côte d'Azur, Nice, France, 41–48.

BYERS, S. and RAFTERY, A.E. (1996) "Nearest neighbor clutter removal for estimating features in spatial point processes", Technical Report 305, Department of Statistics, University of Washington.

BANFIELD, J.D. and RAFTERY, A.E. (1993) "Model-based Gaussian and non-Gaussian clustering", *Biometrics*, **49**, 803–821.

MCCORMICK, W.T., SCHWEITZER, P.J. and WHITE, T.J. (1972) Problem decomposition and data reorganization by a clustering technique, *Operations Research*, **20**, 993–1009.

LENSTRA, J.K. (1974) "Clustering a data array and the traveling-salesman problem", *Operations Research*, **22**, 413–414.

DOYLE, J. (1988) "Classification by ordering a (sparse) matrix: a simulated annealing approach", *Applied Mathematical Modelling*, **12**, 86–94.

DEUTSCH, S.B. and MARTIN, J.J. (1971) "An ordering algorithm for analysis of data arrays", *Operations Research*, **19**, 1350–1362.

STRENG, R. (1991) "Classification and seriation by iterative reordering of a data matrix", in Bock, H.-H. and Ihm, P. (eds.). *Classification, Data Analysis and Knowledge Organization Models and Methods with Applications*, Springer-Verlag, Berlin, pp. 121–130.

PACKER, C.V. (1989) "Applying row-column permutation to matrix representations of large citation networks", *Information Processing and Management*, **25**, 307–314.

MURTAGH, F. (1985) *Multidimensional Clustering Algorithms*. Physica-Verlag, Würzburg.

MARCH, S.T. (1983) "Techniques for structuring database records", *Computing Surveys*, **15**, 45–79.

ARABIE, P., SCHLEUTERMANN, S., DAWES, J. and HUBERT, L. (1988) "Marketing applications of sequencing and partitioning of nonsymmetric and/or two-mode matrices", in Gaul, W. and Schader, M. (eds.), *Data, Expert Knowledge and Decisions*. Springer-Verlag, Berlin, pp. 215–224.

GALE, N., W.C. HALPERIN and COSTANZO, C.M. (1984) "Unclassed matrix shading and optimal ordering in hierarchical cluster analysis", *Journal of Classification*, **1**, 75–92.

HONGYUAN ZHA, DING, C., MING GU, XIAOFENG HE and SIMON, H. (2001), "Bipartite graph partitioning and data clustering", preprint.

LERMAN, I.C. (1981) Classification et Analyse Ordinale des Données. Dunod, Paris.

FISHER, R.A. (1936) The use of multiple measurements in taxonomic problems, *Annals of Eugenics*, **7**, 179–188.

BERRY, M.W., HENDRICKSON, B. and RAGHAVAN, P. (1996) Sparse matrix reordering schemes for browsing hypertext, in *Lectures in Applied Mathematics (LAM) Vol. 32: The Mathematics of Numerical Analysis*, Renegar, J., Shub, M. and Smale, S. (eds.). American Mathematical Society, pp. 99–123.

POINÇOT, P., LESTEVEN, S. and MURTAGH, F. (1998), "A spatial user interface to the astronomical literature", *Astronomy and Astrophysics Supplement Series*, **130**, 183–191.

POINÇOT, P., LESTEVEN, S. and MURTAGH, F. (2000), "Maps of information spaces: assessments from astronomy", *Journal of the American Society for Information Science*. **51**, 1081–1089.

MURTAGH, F., STARCK, J.L. and BERRY, M. (2000), "Overcoming the curse of dimensionality in clustering by means of the wavelet transform", *The Computer Journal*, **43**, 107-120.

Quantization of Models: Local Approach and Asymptotically Optimal Partitions

Klaus Pötzelberger

Department of Statistics, Vienna University of Economics and Business Administration
Augasse 2-6, A-1090 Vienna, Austria

Abstract. In this paper we review algorithmic aspects related to maximum-support-plane partitions. These partitions have been defined in Bock (1992) and analyzed in Pötzelberger and Strasser (2001). The local approach to inference leads to a certain subclass of partitions. They are obtained from quantizing the distribution of the score function. We propose a numerical method to compute these partitions $\mathcal{B}$ approximately, in the sense that they are asymptotically optimal for increasing sizes $|\mathcal{B}|$. These findings are based on recent results on the asymptotic distribution of sets of prototypes.

1 Introduction

Let $E = (\Omega, \mathcal{F}, (P_\theta)_{\theta\in\Theta})$ denote a statistical model. A possibility to reduce the complexity of the model is to replace the σ-field $\mathcal{F}$ by the σ-field generated by a finite measurable partition $\mathcal{B} = (B_1, \dots, B_m)$. We call the model $(\Omega, \sigma(\mathcal{B}), (P_\theta)_{\theta\in\Theta})$ a quantization of the model E.

The quantization of a single probability distribution P is familiar in s-tatistics. It includes topics such as cluster analysis, principal points or unsupervised learning. The χ^2-test of homogeneity is based on a quantization of a model. Here typically the laws of metrically scaled random variables are replaced by multinomial distributions. The power of the test depends on the chosen partition of the sample space.

Let us briefly recapitulate the concept of principal points and introduce some notation. Suppose P is a distribution on $\mathbb{R}^d$ with finite second moment. For a partition $\mathcal{B} = (B_1, \dots, B_m)$ define the conditional means

$$p_i = \int_{B_i} x dP/P(B_i). \tag{1}$$

A partition $\mathcal{B}$ is optimal if it minimizes

$$\Delta_m(||,||^2, \mathcal{B}, P) = \sum_{i=1}^{m} \int_{B_i} ||x - p_i||^2 dx \tag{2}$$

among partitions of size at most m. The conditional means of an optimal partition are called prototypes, centroids or principal points (Flury (1990)).

Equation (1) defines the prototypes if the partition is given. On the other hand, if a set of prototypes $\{p_1, \ldots, p_m\}$ is given, then $\mathcal{B}$ is the Voronoi-partition, defined by the property that if $x \in B_i$, then

$$\|x - p_i\| = \min\{\|x - p_j\| \mid 1 \leq j \leq m\}.$$

Note that (2) is minimized by a partition $\mathcal{B}$, if and only if it maximizes the information

$$I^f(\mathcal{B}) = \sum_{i=1}^{m} f(p_i)P(B_i) \tag{3}$$

with $f(x) = \|x\|^2$. Bock (1992) and Pötzelberger and Strasser (2001) analyze quantizations based on general convex functions f. A partition is f-optimal if it maximizes the f-information (3), with prototypes defined by (1). These partitions are called MSP-partitions (Maximum-Support-Plane partitions) and are the corresponding generalizations of Voronoi-partitions.

The results on the quantization of a single probability distribution on $\mathbb{R}^d$ may be directly applied to models E, when Θ is finite. We have to identify the model with its standard measure and consider quantizations of the standard measure. For details see Strasser (2000). Within the framework of decision theory Pötzelberger (2002) showed that all admissible quantizations correspond to limits of MSP-partitions.

Admissibility of a partition is, in a certain sense, a minimal claim. It does not necessarily help to choose a specific convex function f and the corresponding optimal partition. Loosely speaking, there are too many admissible partitions. This is the case if Θ is infinite. Here, a characterization of admissible partitions involves a suitable definition of a MSP-partitions. Again, these partitions maximize certain functionals defined by convex functions. Admissible partitions are limits of MSP-partitions. However, if the model is large enough, all partitions are admissible.

The first aim of this paper is to give an example of, how a specific approach to inference defines the convex function f. We show that testing a one-point hypothesis against local alternatives leads to the quadratic function f and a quantization problem for the distribution of the score function.

The second theme is a method to compute partitions which are asymptotically optimal for $m \to \infty$, so-called regular partitions. These concepts are based on an analysis of the asymptotic law of the uniform distribution on the set of prototypes and of

$$\Delta_m(f, P) = \min\{\Delta_m(f, \mathcal{B}, P) \mid |\mathcal{B}| = m\},$$

the quantization error of the distribution P.

2 Local optimality

An example that leads to quantization problems with quadratic functions $f(x)$ is a local asymptotic approach to testing problems, when the procedures are based on partitions. Let a family of absolutely continuous probability distributions $(P_\theta)_{\theta\in\Theta}$ on $\mathbb{R}$ be given. Let p_θ denote the density of P_θ with respect to Lebesgue measure.

Procedures based on partitions $\mathcal{B} = (B_1, \dots, B_m)$ of $\mathbb{R}$ into m measurable sets lead to models $\mathcal{E}(\mathcal{B}) = (P_\theta(B_1), \dots, P_\theta(B_m))_{\theta\in\Theta}$ and thus to the analysis of multinomial distributions indexed by θ. We assume that Θ is an open subset of the d-dimensional Euclidian space and that for all measurable sets B the function $\theta \mapsto P_\theta(B)$ is twice differentiable. Models $\mathcal{E}(\mathcal{B})$ are locally asymptotic normal: Let $\theta_0 \in \Theta$, $t \in \mathbb{R}^d$ and $P_{n,t}$ the law of $X_1, \dots, X_n$ $\sim^{iid} P_{\theta_0 + tn^{-1/2}}$. Then

$$\log\left(\frac{dP_{n,t}}{dP_{n,0}}\right) = t^\top Z_n - \frac{1}{2} t^\top I_{\theta_0}(\mathcal{B}) t + R_n,$$

where under $P_{n,t}$, $R_n \to 0$ in probability and Z_n converges in distribution to a d-dimensional normal variable with covariance identity. Under $P_{n,0}$ its asymptotic mean is 0. Under local alternatives $P_{n,t}$, its asymptotic mean is $I_{\theta_0}(\mathcal{B})t$. $I_{\theta_0}(\mathcal{B})$ is the Fisher-information,

$$I_{\theta_0}(\mathcal{B}) = \sum_{i=1}^{m} P_{\theta_0}(B_i) \frac{\partial \ell(B_i, \theta_0)}{\partial \theta} \frac{\partial \ell(B_i, \theta_0)}{\partial \theta}^\top, \tag{4}$$

where $\ell(B_i, \theta) = \log P_\theta(B_i)$. We assume that $I_{\theta_0}(\mathcal{B})$ is invertible.

θ_0 represents the null-hypothesis. Local alternatives are parametrized by t, the deviation from the null-hypothesis. Since the model is locally asymptotic normal, procedures corresponding to the normal limit are asymptotically optimal. Asymptotically, the power of a test based on the partition $\mathcal{B}$ under local alternatives in the direction t is an increasing function of the norm of

$$I_{\theta_0}(\mathcal{B})^{-1/2} I_{\theta_0}(\mathcal{B}) t$$

and therefore of

$$t^\top I_{\theta_0}(\mathcal{B}) t. \tag{5}$$

A partition is therefore asymptotically optimal if it maximizes (5).

Let us assume that $\theta \mapsto p_\theta(x)$ is differentiable and that for all measurable sets B,

$$\frac{\partial P_\theta(B)}{\partial \theta} = \int_B \frac{\partial p_\theta}{\partial \theta}(x) dx.$$

We define the score function s_θ as

$$s_\theta(x) = \frac{\partial}{\partial \theta} \log p_\theta(x), \tag{6}$$

which is defined with P_θ-probability 1. Then

$$\frac{\partial \ell(B,\theta_0)}{\partial \theta} = \mathbb{E}_{\theta_0}[s_{\theta_0}(X) \mid B]. \tag{7}$$

Let us denote by P_θ^s the law of $s_\theta(X)$ for $X \sim P_\theta$ and let $f(x) = (t^\top x)^2$. Thus we have shown that

$$t^\top I_{\theta_0}(\mathcal{B})t = I^f(\mathcal{B}, P_{\theta_0}^s). \tag{8}$$

Furthermore, the task of choosing a partition that maximizes (5) is equivalent to choosing an f-optimal partition for the distribution of the score. For fixed t it may be viewed as a quantization problem for the quadratic function and the distribution of $t^\top s_{\theta_0}(X)$. Cases, when not only one single direction t, but a multidimensional set of directions has to be considered, lead to local Bayesian or to local minimax partitions.

3 Asymptotically optimal partitions

Except from rare cases, f-optimal partitions have to be computed numerically. Results on the asymptotic behaviour of the quantization error, when m, the size of the partition, goes to infinity, are valuable tools to assess the quality of the derived solutions. Furthermore, they allow to identify methods to compute partitions, which are, at least for large m, suitable choices. We call a sequence of partitions $\mathcal{B}^m$, with $|\mathcal{B}^m| = m$, asymptotically optimal, if

$$\lim_{n\to\infty} \frac{\Delta_m(f, \mathcal{B}^m, P)}{\Delta_m(f, P)} = 1. \tag{9}$$

The central result on the asymptotics of the quantization error is Zador's Theorem (1964). Let us denote by $\mathcal{P}_d$ the set of distributions P, which are absolutely continuous with respect to d-dimensional Lebesgue measure and which satisfy

$$\int ||x||^{2+\epsilon} dP(x) < \infty$$

for a $\epsilon > 0$.

Theorem 1. *There are constants $C(d)$ such that for all distributions $P \in \mathcal{P}_d$ with density p,*

$$\lim_{m\to\infty} m^{2/d} \Delta_m(||.||^2, P) = C(d) \left(\int p(x)^{d/(2+d)} dx \right)^{1+2/d}. \tag{10}$$

One approach to identify asymptotically optimal partitions $(\mathcal{B}^m)$ is the following. Let $\{p_1, \dots, p_m\}$ denote a set of prototypes and let $\hat{G}_m$ denote the uniform distribution on $\{p_1, \dots, p_m\}$. Suppose $\hat{G}_m$ converges weakly to a distribution G. Then the uniform distribution on the set of centroids of $(\mathcal{B}^m)$ should do likewise. Therefore, in a first step, identify the asymptotic distribution G of the prototypes. Then, in a second step, choose a set of approximate prototypes $\{\tilde{p}_1, \dots, \tilde{p}_m\}$, in the sense that its law (the uniform distribution on $\{\tilde{p}_1, \dots, \tilde{p}_m\}$) is approximately G.

Let $B(x,r)$ denote the open ball of radius r centered at x. Let G be a probability distribution and $c > 1$. We define

$$\eta_{G,c}(x) = \inf \left\{ \frac{G(B(x,r))}{(2r)^d} \mid 0 < r < 2c\|x\| \right\}. \tag{11}$$

Note that $G(\{x \mid \eta_{G,c}(x) = 0\}) = 0$.

Theorem 2. *Let $P \in \mathcal{P}_d$ with density p, let G be absolutely continuous with density g proportional to $p^{d/(2+d)}$. If there exists a $c > 1$ such that*

$$\int \eta_{G,c}(x)^{-2/d} dP(x) < \infty, \tag{12}$$

then, for any set of prototypes, $\hat{G}_m \to G$ weakly.

Theorem 2 holds in more general cases. For a proof see Pötzelberger (2000).

The computationally most simple approach is to take a sample of independent G-distributed random variables as set of prototypes. In this case the quantization error is random. Let $\Delta_m^G(f, P)$ denote its expectation. It can be shown that given regularity conditions, it converges at the rate $m^{-2/d}$. However, since G is not a one-point distribution,

$$\liminf_{m \to \infty} \frac{\Delta_m^G(f,P)}{\Delta_m(f,P)} > 1.$$

It is advisable to generate several samples and to take the sample with the smallest quantization error, i.e. the sample with the largest f-information. Random partitions are often used as starting values for numerical algorithms. However, this application is questionable. The f-information of the local optimum, to which certain procedures converge, seems to be independent of the f-information of the starting value, see Steiner (1999).

In the one-dimensional case, quantiles of G are asymptotically optimal. We call a Voronoi-partition $\mathcal{B} = (B_1, \dots, B_m)$ of $\mathbb{R}$ a regular partition generated by G, if the centroids $\{p_1, \dots, p_m\}$ satisfy

$$G((-\infty, p_i]) = \frac{i}{m+1}.$$

For the proof of the following theorem regularity conditions for the density p have to hold. A sufficient condition is that p is Hölder continuous.

Theorem 3. *Let $f = \|.\|^2$, $d = 1$ and let $P \in \mathcal{P}_d$ with density p. Assume that Q is absolutely continuous with density q. If there exists a $c > 1$ such that* (12) *holds for $\eta_{Q,c}$, then for the regular partitions $\mathcal{B}^m$ generated by Q,*

$$\lim_{m\to\infty} m^2 \Delta_m(\|.\|^2, \mathcal{B}^m, P) = C(1) \int q(x)^{-2} dP(x). \tag{13}$$

Let the density of G be proportional to $p^{1/3}$. If $\eta_{G,c}$ satisfies (12) *for a $c > 1$, then the sequence of regular partitions generated by G is asymptotically optimal.*

Example 1. Let $X \sim N(\theta, \sigma^2)$ with σ^2 fixed. Then the asymptotically optimal regular partition is generated by a normal distribution with mean θ and variance $3\sigma^2$.

Theorem 3 makes it possible to analyze the asymptotic efficiency of regular partitions. For a distribution Q with density q we call

$$\frac{(\int p(x)^{1/3} dx)^3}{\int q(x)^{-2} dP(x)}$$

the relative asymptotic efficiency. Let us briefly comment on the relative efficiency of two popular methods of quantization. The first is the use of equiprobable intervals. This partition is regular, generated by P. Unless $\int p(x)^{-1} dx < \infty$, the relative efficiency is 0. A second method for distributions with bounded support is the use of intervals of equal length. Let for instance $P([0,1]) = 1$. Partitions into intervals of equal length correspond to the uniform distribution. Its relative efficiency is $(\int p(x)^{1/3} dx)^3$. Figure 1 depicts the relative efficiency of the uniform distribution, when P is beta-distributed, i.e. $p(x) = B(\alpha, \beta) x^{\alpha-1}(1-x)^{\beta-1}$ for $0 < x < 1$.

Note that there are only two facts that lead to a characterization of asymptotically optimal partitions: The first is the knowledge of the asymptotic distribution of the prototypes. The second is the fact that the partition consists of convex sets.

A straightforward generalization of the concept of a regular partition to higher dimensions is not possible. The sole knowledge of the asymptotic distribution of the prototypes does not help to identify geometrical features of the partition into convex sets. For instance, in dimension 2, the asymptotically optimal regular partition is obtained by a tesselation into regular hexagons. However, partitions into rectangles may have the same asymptotic distribution of the centroids.

Intervals are obtained from the unit interval by scaling and translation. Scaling is essential if the distribution P is not uniform. For dimensions more that 2 even in the case of a uniform distribution the set, which is space-filling by translation and leads to asymptotically optimal partitions, is not known. For a discussion of regular partitions and of lattice partitions, see Graf and Luschgy (2000).

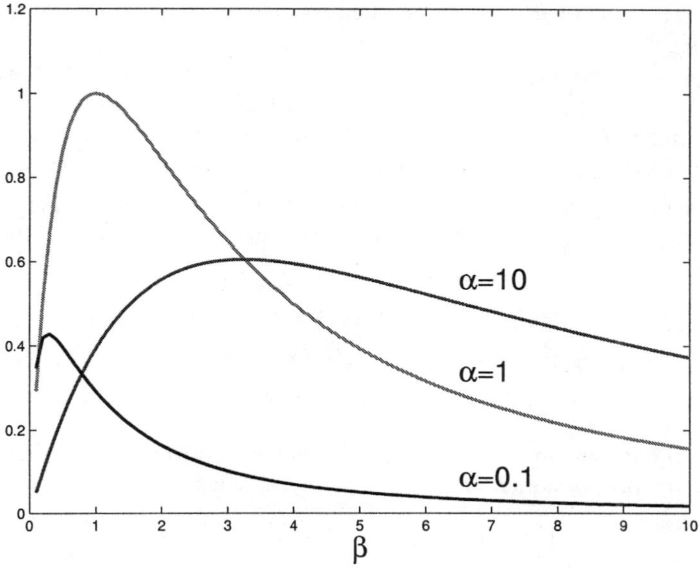

Fig. 1. Relative efficiency of uniform quantization

4 Asymptotically optimal Bayesian partitions

In section 2 we have defined the appropriate quantization problem for the local inference in the case of a fixed alternative parametrized by t. Without loss of generality let $\|t\| = 1$. In the case a partition has to be chosen simultaneously for a set of alternatives T, a Bayesian approach would specify a prior distribution π on T and choose the partition $\mathcal{B}$ that maximizes

$$\int t^\top I_{\theta_0}(\mathcal{B}) t d\pi(t).$$

Let

$$M = \int tt^\top d\pi(t) \tag{14}$$

and

$$f(x) = x^\top M x. \tag{15}$$

Then the Bayesian solution to the quantization problem is a partition that maximizes $I^f(\mathcal{B}, P^s_{\theta_0})$.

Assume that $x \mapsto s_{\theta_0}(x)$ is continuously differentiable. Since the data is one-dimensional and the parameter d-dimensional, the support of $P^s_{\theta_0}$ is a one-dimensional subset of $I\!R^d$. We show how the results on asymptotically optimal partitions of $\mathbb{R}$ may be applied to distributions on one-dimensional manifolds.

To simplify the exposition, we assume that $d = 2$ and that $s_{\theta_0}(x) = (s_{\theta_0,1}1(x), s_{\theta_0,2}(x))^\top$ with $s_{\theta_0,1}$ invertible. Let P^1 denote the distribution of $s_{\theta_0,1}(X)$ under $X \sim P_{\theta_0}$. $p^1(x)$ denotes its density. We abbreviate $s_{\theta_0,2} \circ s_{\theta_0,1}^{-1}$ by h. $P_{\theta_0}^s$ is then the distribution of $(X, h(X))^\top$ with $X \sim P^1$. Analogously to section 3, a distribution Q generates prototypes $(p_i, h(p_i))^\top$, where $Q((-\infty, p_i]) = i/(m+1)$. Assume that h is continuously differentiable. Since for $x_1 \to p_i$,

$$\begin{aligned} & m_{11}(x_1 - p_i)^2 + 2m_{12}(x_1 - p_i)(h(x_1) + h(p_i)) + m_{22}(h(x_1) + h(p_i))^2 \\ \approx \quad & m_{11}(x_1 - p_i)^2 + 2m_{12}(x_1 - p_i)^2 h'(x_1) + m_{22}(x_1 - p_i)^2 h'(x_1)^2 \\ = \quad & (x_1 - p_i)^2 \left(m_{11} + 2m_{12}h'(x_1) + m_{22}h'(x_1)^2\right), \end{aligned}$$

we have to choose the regular partition, which is optimal for a distribution with density proportional to $\tilde{p}(x) = \tau(x)p^1(x)$, where

$$\tau(x) = m_{11} + 2m_{12}h'(x) + m_{22}h'(x)^2.$$

Since M is positive-semidefinite, we have $\tau(x) \geq 0$.

Theorem 4. *Let M and f be given by* (14) *and* 15)*. Let p^1 denote the density of $s_{\theta_0,1}$ and let $h = s_{\theta_0,2} \circ s_{\theta_0,1}^{-1}$ be continuously differentiable. Assume that Q is absolutely continuous with density q. If there exists a $c > 1$ such that* (12) *holds for $\eta_{Q,c}$, then for the regular partitions $\mathcal{B}^m$ generated by Q,*

$$\lim_{m\to\infty} m^2 \Delta_m(f, \mathcal{B}^m, P_{\theta_0}^s) = C(1) \int q(x)^{-2}\tau(x)p^1(x)dx. \tag{16}$$

Let the density of G be proportional to $(\tau p^1)^{1/3}$. If $\eta_{G,c}$ satisfies (12) *for a $c > 1$, then the sequence of regular partitions generated by G isasymptotically optimal for the Bayesian quantization problem.*

Example 2. Let $X \sim N(\mu, \sigma^2), \theta = (\mu, \sigma^2)^\top$ and $\theta_0 = (0, 1)^\top$. Then $s_{\theta_0,1}(x) = -x$, $s_{\theta_0,2}(x) = (x^2 - 1)/2$, $h'(x) = x$ and $\tau(x) = m_{11} + 2m_{12}x + m_{22}x^2$. The asymptotically optimal regular partition is generated by a distribution with density

$$g(x) \propto (m_{1,1} + 2m_{1,2}x + m_{2,2}x^2)^{1/3} \exp\{-x^2/6\}.$$

References

BOCK, H.H. (1992): A clustering technique for maximizing ϕ-divergence, noncentrality and discriminating power. In M. Schader (ed.): *Analyzing and Modeling Data and Knowledge*, Springer, Heidelberg, 19–36.

FLURY, B.A. (1990): Principal points. *Biometrika, 77,* 33–41.

GRAF, S. and LUSCHGY, H. (2000):*Foundations of Quantization for Probability Distributions.* Lecture Notes in Mathematics 1730, Springer, Berlin Heidelberg.

PÖTZELBERGER, K. (2000): The general quantization problem for distributions with regular support. *Math. Methods Statist., 2,* 176–198.

PÖTZELBERGER, K. (2002): Admissible unbiased quantizations: Distributions without linear components. To appear in: *Math. Methods Statist.*

PÖTZELBERGER K. and STRASSER, H. (2001): Clustering and quantization by MSP-partitions. *Statistics and Decisions, 19,* 331–371.

STEINER, G. (1999): *Quantization and clustering with maximal information: Algorithms and numerical experiments*, Ph.D. Thesis, Vienna University of Economics and Business Administration.

STRASSER, H. (2000): Towards a statistical theory of optimal quantization. In W. Gaul, O. Opitz, M. Schader (eds.): *Data Analysis: Scientific Modeling and Practical Application*, Springer, Berlin Heidelberg, 369–383.

ZADOR, P.L. (1964): *Development and evaluation of procedures for quantizing Multivariate distributions*, Ph.D. Thesis, Stanford University.

The Performance of an Autonomous Clustering Technique

Yoshiharu Sato

Division of Systems and Information, Hokkaido University,
Kita-13, Nishi-8, Kita-ku, Sapporo, 060-8628 Japan

Abstract. Recently, a multi-agents system has been discussed in the field of adaptive control. An autonomous clustering is considered to be a multiple agents system which constructs clusters by moving each pair of objects closer or farther according to their relative similarity to all of the objects.

In this system, the objects correspond to autonomous agents, and the similarity relation is regarded as the environment. Defining a suitable action rule for the multi-agents, the clusters are constructed automatically.

In this paper, we discuss the ability of the detection of clusters by the autonomous clustering technique through the concrete examples.

1 Introduction

An autonomous agents system is a self-governing system capable of perceiving and acting in environments, which are complex or dynamic. In this system, the autonomous agents attain a given goal by the interactive actions, which are given by the information concerning its environment. This process is called action-perception cycle, where actions allow the agent to perceive information concerning its environment, which may lead to changes in the agent's internal state, which may in turn affect further actions.

For example, in the robot arms control, the agents are the actuators of the joints and sensor, ccd camera. By the interactive action of these agents, flexible behavior can be achieved. The other example is a berthing control. Namely, a ship comes alongside the pier. In this case, the agents are tugboats and main propeller. It is shown that this multi-agents system produces robustness and flexibility of the system. (Itoh, 1998)

In our clustering problem, the agents are regarded as the objects, and the environment is the observed similarity relation. The action is moving closer or further according to the relative position of the agents.

A standpoint of the autonomous clustering seems to be the same with 'self-organizing maps' (Kohonen, 1995) or 'generalized Kohonen maps' (Bock, 1998). The objective is to construct the clusters automatically, in some sense. But it is difficult to get clusters, in a self-organizing way, without any prior knowledge. In the autonomous clustering, the clusters can be constructed under the condition that the control parameter α is given.

2 Autonomous Clustering

We suppose that the observed similarity between n objects is given by

$$S = (s_{ij}), \quad 0 \leq s_{ij} \leq 1, \quad s_{ij} = s_{ji}, \quad s_{ii} = 1.$$

We assume that the points in a configuration space generated by the similarity give the initial state of the agents. The dimension of the configuration space will be less than n. We also assume that the agents can move to any directions. But without any restriction of the behavior of the agents, it is impossible to construct the expected clusters. So we introduce an action rule for the agents. The actions of the agents are determined as follows. Looking over the configuration from each agent, when two agents have the similar relative positions, they move closer. Otherwise they go away from each other. The relative position of the two objects,o_i and o_j, in the configuration space corresponds to the two column vectors, $\boldsymbol{s}_i$ and $\boldsymbol{s}_j$, of the similarity matrix S,

$$\boldsymbol{s}_i : (s_{1i}, s_{2i}, \cdots, 1, \cdots, s_{ji}, \cdots, s_{ni}), \quad \boldsymbol{s}_j : (s_{1j}, s_{2j}, \cdots, s_{ij}, \cdots, 1, \cdots, s_{nj}).$$

If these two vectors are similar, then two agents, o_i and o_j, get nearer to each other. This moving is implemented by increasing the similarity between o_i and o_j. Repeating this action, we can get the clusters.

Formally, this action is denoted as follows: When we denote the observed similarity and the similarity at t-step as $S^0 = (s_{ij})$ and $S^{(t)} = (s_{ij}^{(t)})$, respectively, the action from t-step to $(t+1)$-step is defined by

$$s_{ij}^{(t+1)} = \frac{\sum_{k=1}^{n} (s_{ik}^{(t)})^\alpha (s_{jk}^{(t)})^\alpha}{\sqrt{\sum_{\ell=1}^{n} (s_{i\ell}^{(t)})^{2\alpha}} \sqrt{\sum_{m=1}^{n} (s_{jm}^{(t)})^{2\alpha}}}, \quad (t = 0, 1, 2, \cdots, T), \tag{1}$$

where the parameter α is assumed to be greater than 1, and it play the important role to get non-trivial clusters. Using matrix notations, (1) is expressed as follows: Putting

$$S^{(t)} = [\boldsymbol{s}_1^{(t)}, \boldsymbol{s}_2^{(t)}, \cdots, \boldsymbol{s}_n^{(t)}], \quad S^{(t)\alpha} = [\boldsymbol{s}_1^{(t)\alpha}, \boldsymbol{s}_2^{(t)\alpha}, \cdots, \boldsymbol{s}_n^{(t)\alpha}]$$

and

$$D^{(t)\alpha} = \mathrm{diag}[\boldsymbol{s}_1^{(t)\alpha\prime} \boldsymbol{s}_1^{(t)\alpha}, \boldsymbol{s}_2^{(t)\alpha\prime} \boldsymbol{s}_2^{(t)\alpha}, \cdots, \boldsymbol{s}_n^{(t)\alpha\prime} \boldsymbol{s}_n^{(t)\alpha}].$$

Then (1) is denoted as

$$\begin{aligned} S^{(1)} &= \{D^{(0)\alpha}\}^{-\frac{1}{2}} S^{(0)\alpha\prime} S^{(0)\alpha} \{D^{(0)\alpha}\}^{-\frac{1}{2}} \\ S^{(2)} &= \{D^{(1)\alpha}\}^{-\frac{1}{2}} S^{(1)\alpha\prime} S^{(1)\alpha} \{D^{(1)\alpha}\}^{-\frac{1}{2}} \\ &\vdots \\ S^{(t+1)} &= \{D^{(t)\alpha}\}^{-\frac{1}{2}} S^{(t)\alpha\prime} S^{(t)\alpha} \{D^{(t)\alpha}\}^{-\frac{1}{2}} \\ &\vdots \end{aligned} \tag{2}$$

The convergence of the sequence $\{S^{(1)}, S^{(2)}, \cdots, S^{(t)}, \cdots\}$ is shown in Sato(2000).

3 The Features and Performance of Clustering

In order to make the properties of the autonomous clustering clear, we shall consider the feature of the clustering under the change of the parameter α by using the finite grid data. Since it is known that the initial configuration affects the clustering result, we consider the uniform structure for the initial configuration. The initial data is given in figure 3.1, where there are $12 \times 12 = 144$ grid points. For each calue of α, the number of clusters which are constructed by the autonomous technique are shown in table 3.1.

	parameter α				
	2	3	4	5	6
No. of clusters	1	4	64	100	144

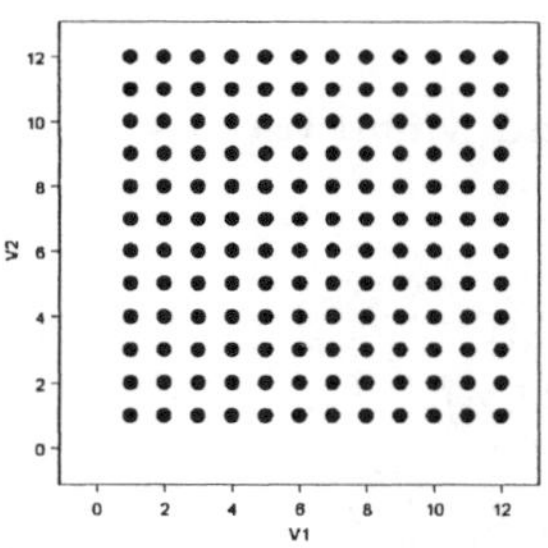

Figure 3.1 Original Configuration

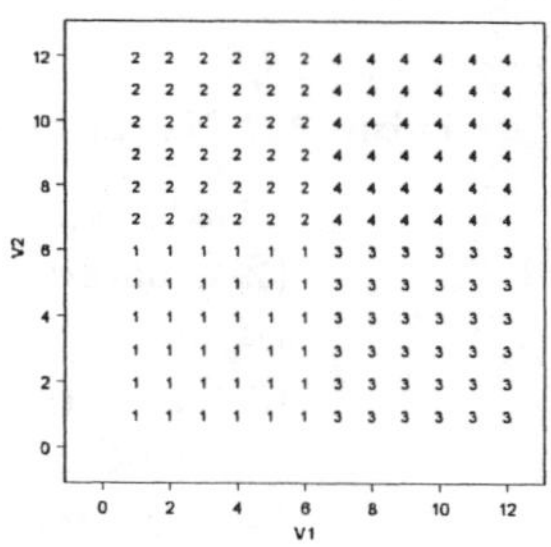

Figure 3.2 Clustering for $\alpha = 3$

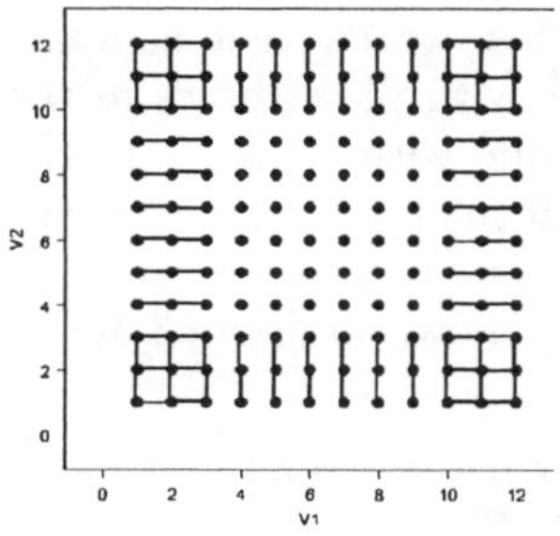

Figure 3.3 Clustering for $\alpha = 4$
Clusters are shown by the linkage with line

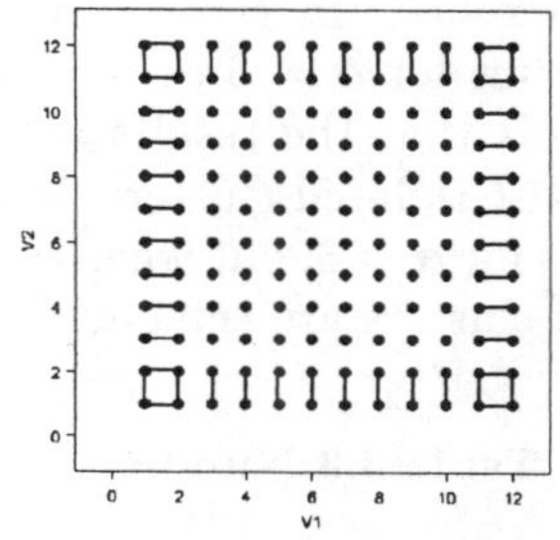

Figure 3.4 Clustering for $\alpha = 5$
Clusters are shown by the linkage with line

When $\alpha = 2$, all the objects merge to one cluster. In the case of $\alpha = 6$, each object makes one cluster. On the other hand, when $\alpha = 3$, the region of the data is divided to 4 equal subregions. (figure 3.2) For $\alpha = 4, 5$, we get the interesting results which show one of the feature of this method. In the clustering process of this method, the merging starts from the boundary and progress to the inner points. But for the inner points, the merging force acts from every directions. (in this case, four directions) Eventually, these inner points remain as the isolated points.(figure 3.3, 3.4) The interpretation of these results will depend on the situations. In some case, this grid data should be one cluster. But in other situation, the result for $\alpha = 3$ or $\alpha = 6$ may be accepted. This results suggest that the range of α may be sufficient from 2 to 6.

Next, we shall consider the case where there exist clusters clearly. In this case, we discuss the feature of clustering comparing with k-means method. For any case, the result of k-means method may not be the best, but it is well-known and it gives one of a guideline. The data are generated by the use of two-dimensional normal random numbers, where covariance matrix is an unit matrix. In this simulation, we consider the following three cases which shown in table 3.2.

Table 3.2 Number of objects and Mean vectors

Case A :	60	60		
(2-Clusters)	(0,0)	(5,0)		
Case B :	40	40	40	
(3-Clusters)	(0,0)	(5,0)	(2.5,4)	
Case C :	30	30	30	30
(4-Clusters)	(0,0)	(5,0)	(5,5)	(0,5)

In each case, we repeated 50 times. The initial mean vectors for k-means method are given as the original mean vector in table 3.2.

Table 3.3 shows the number of results which are the same with the k-means method. In these results, we know that we can get comparatively good result for $\alpha = 3$ or $\alpha = 4$. The number with parentheses in the table 3.3 means that the results are almost the same with k-means method but there exist isolated clusters. In the Case A, all the results are the same with k-means for $\alpha = 3$, but when $\alpha = 4$, the results are almost the same but some isolated clusters are generated. An example is shown in figure 3.5(a),(b).

Table 3.3 Number of same results with k-means method

Case \ α	2	3	4	5	6
A	0	50	(43)	0	0
B	0	38+(12)	(46)	0	0
C	0	46+(4)	(49)	0	0

In the Case B, when $\alpha = 3$, 38 results are same with k-means. 12 results are almost same but some isolated clusters are constructed. This example is shown in figure 3.6(a),(b).

In the Case C, the results are almost the same with k-means for $\alpha = 3$ and $\alpha = 4$. But some results are generated isolated clusters, for example, figure 3.7(a),(b) shows when $\alpha = 4$.

In k-means method, the object which has the same distance from several mean vectors is usually assigned to one of the clusters. The autonomous clustering is inclined to remain such a object as an isolated cluster. It seems that such a isolated cluster sometimes gives a useful information in the data analysis.

In any cases, when the data has coherent clusters, the parameter α will be given as 3. However, if we set $\alpha = 4$, the result seems to be rather safe, because number of generated clusters is larger than the case of $\alpha = 3$.

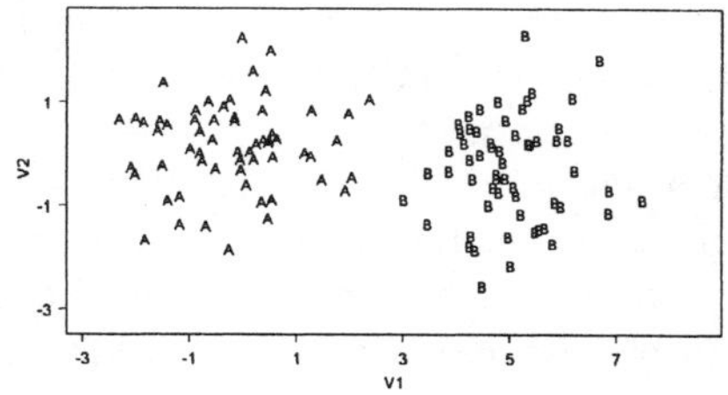

Figure 3.5(a) The result of k-means

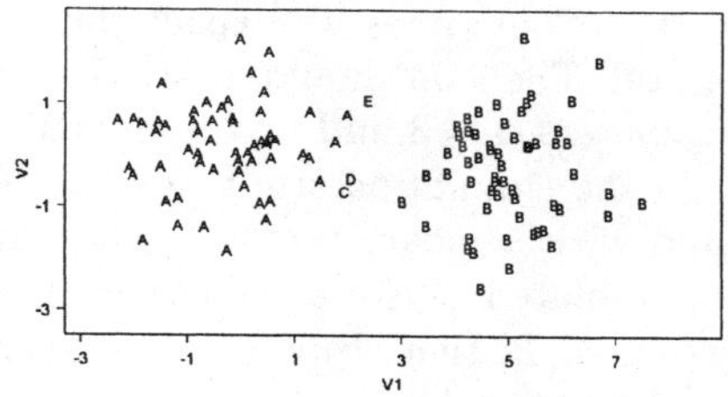

Figure 3.5(b) The result for $\alpha = 4$

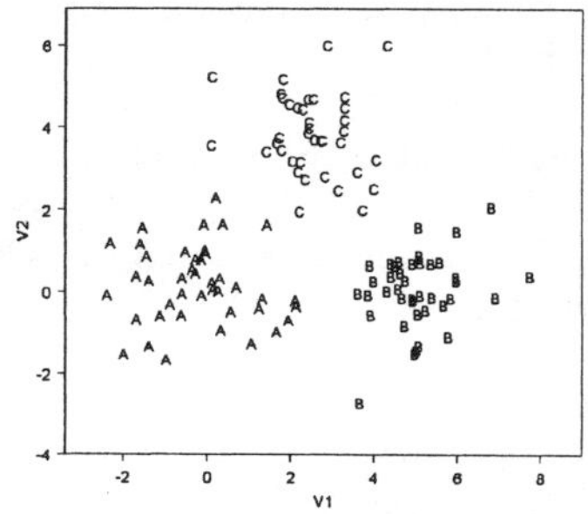

Figure 3.6(a) The result of k-means

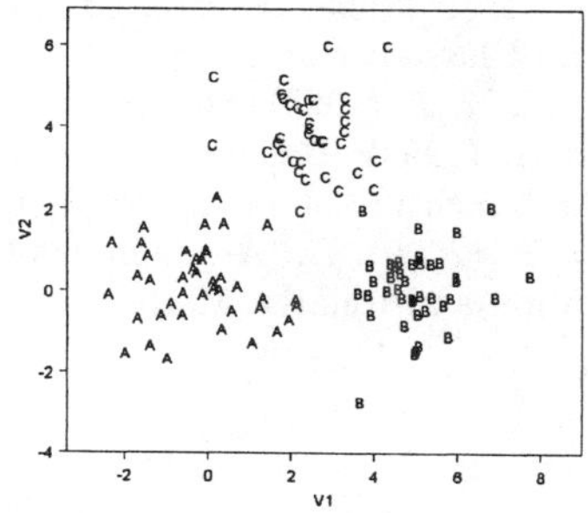

Figure 3.6(b) The result for $\alpha = 3$

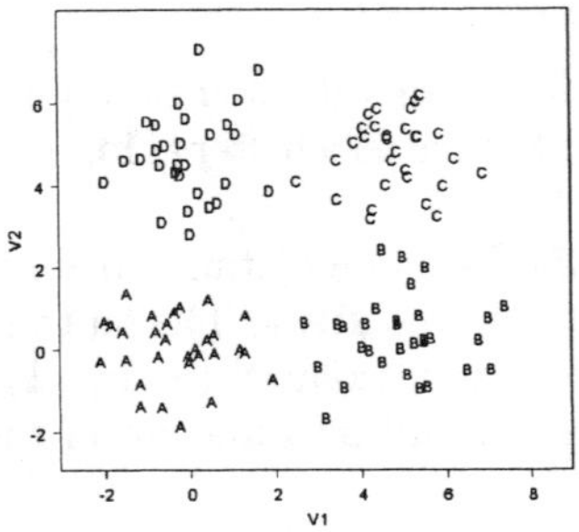

Figure 3.7(a) The result of k-means

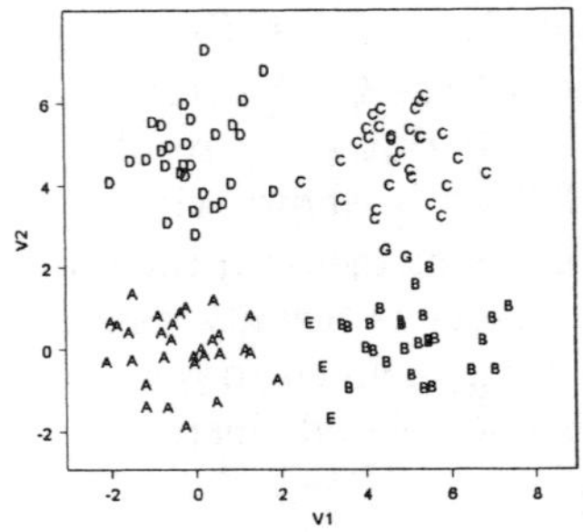

Figure 3.7(b) The result for $\alpha = 4$

4 Concluding remarks

When we apply the clustering method, the data does not always exist clumply clusters. Sometimes we expect the division of the given region like k-means method. Then we should apply the autonomous clustering method using the parameters $\alpha = 3$ and $\alpha = 4$ for both types of data.

In the simulation study, it is not clear yet which is better result as compared with k-means method. The autonomous clustering is inclined to generate isolated clusters. Whether this point works well or not must be investigated further. But it seems to be useful to detect the point which has intermediate state for belongingness.

References

ARBIB, M.A. (ed.)(1995). The Handbook of Brain Theory and Neural Networks. The MIT Press.

BOCK, H. H. (1998). Clustering and Neural Networks. Advances in Data Science and Classification (A. Rizzi et al. ed.), 265–277.

KOHONEN, T. (1995). Self-organizing maps. Springer, New York.

ITOH, H. (1998). Berthing Control with Multi-Agent System. Jour. Soc. Naval Architechture of Japan, Vol.184, 639–647

SATO, Y. (2000). An Autonomous Clustering Technique. Data Analysis, Classification and Related Methods (H.A.L.Kiers et al. ed.), 23–28.

Cluster Analysis by Restricted Random Walks

Joachim Schöll[1] and Elisabeth Paschinger[2]

[1] Institut für Statistik und Wahrscheinlichkeitstheorie
TU Wien, Wiedner Hauptstr. 8-10, A-1040 Wien
e-mail: j.schoell@tuwien.ac.at
[2] Institut für Theoretische Physik
TU Wien, Wiedner Hauptstr. 8-10, A-1040 Wien

Abstract. A new graph theoretical method is introduced to analyze a data set of objects described by a dissimilarity matrix d_{ij}. This method is based on the generation of a series of random walks in the data set. We define a random walk in a data set by moving in each time step from one object to another one at random. In order that the random walk depends on the pattern of the data, a restriction depending on the previous moves is imposed during its generation, so that the random walk is attracted by clusters formed of similar objects. We define a hierarchical set of graphs consisting of all connections of a series of random walks at a certain time step to detect the structure of the data and to derive an similarity measure between the objects. In an example an application of the method is shown.

1 Introduction

Cluster analysis is designed to classify a set of objects with known characteristics into subsets, such that in some sense members of the same subset resemble each other more than members of different subsets. One attempt to clustering is to establish linking models between groups based on graph theoretical concepts. For survey and review articles see, e.g., Ling (1973), Zahn (1971), Godehardt (1990) and Bock (1996).

Alpert and Kahng (1993) and Hagen and Kahng (1992) developed the concept of random walks (RWs) for circuit graph clustering. They realize RWs on a graph and detect clusters as cycles of distinct nodes which are more likely to be chosen between highly linked nodes of a graph.

We have developed a similar concept which makes random walks suitable for clustering a set of data described by a dissimilarity matrix. A random walk is chosen with restriction, i.e., in each step the RW is restricted to move only to more similar objects than in the former step. This defines a so called *restricted random walk* (RRW) which has a finite length which is, on the average, of order $\log(n)$ for a data set of size n. Such an RRW corresponds to a directed path that is linking objects with increasing similarity and is trapped by a group of similar objects. Since a single RRW corresponds to a similarity chain that investigates only a local subset of the data several RRWs have to be started. Using the union of all connections chosen at a certain time step in a series of RRWs we generate a hierarchical set of graphs whose components can be used to define an similarity measure between the objects.

Finally, in an example, we test our method and compare the resulting partition with those obtained by other methods. Furthermore, we show that RRWs can also be used for validating given clusters of a data set according of the connectivity of the thresholdgraph.

2 Restricted random walk in a data set

We investigate a data set consisting of n objects $X = \{1, \ldots, n\}$. The relationship between the objects i and j is assumed to be described by a dissimilarity measure d_{ij}. A random walk in the data set is a discrete time process: it starts at some object i_0 and at each time step an object is chosen at random. All objects are equally likely to be chosen. Such a random walk will not provide any information about the structure of the data set since it does not depend on the dissimilarity between the objects. In this work we modify this concept and introduce a new idea which we call *restricted random walk* (RRW). Such a walk is constructed as follows.

Assume that, for some $m \geq 0$, we have already chosen the first $m + 1$ objects $(i_0, \ldots, i_m)$ of the RRW. The following object i_{m+1} is not chosen at random from the entire data set $\{1, \ldots, n\}$, but rather from the subset

$$T_{m+1} = \{j \in X | d_{ji_m} < s_m\} \backslash \{i_m\}, \quad m = 1, 2, \ldots, \tag{1}$$

where $T_1 = X \backslash \{i_0\}$ and $s_m = d_{i_{m-1}, i_m}$ is called the step length. The object i_m of the previous step is excluded since immediately repeated consideration of this object only yields trivial information.

The number of steps of an RRW is called the *life-time* which is on average $O(\log(n))$. The deriviation is given in Schöll and Paschinger (2001). We found in examples that $\log_2(n)$ can be used as a rule of thumb to estimate the expected life-time.

The most important feature of an RRW in a set of grouped data is the following: At the beginning objects are chosen at random, while during the subsequent steps the set T_m restricts the RRW to objects which are more similar than the preceding ones since from the definition (1) its follows that s_m is a monotonically decreasing function. If the RRW has visited a cluster and the step length is already less than the dissimilarity to objects of other clusters, the RRW will end inside this cluster. The last connection chosen in an RRW is an element of the minimal spanning tree (MST).

An RRW corresponds to a *similarity chain* (SC). The definition of a SC in a data set X is as follows: a SC is an ordered subset of objects $\{i_0, \ldots, i_f\}$ fulfilling the condition $d_{i_{k-1}i_k} > d_{i_k i_{k+1}}$ for all $k = 1, \ldots, f-1$ and the length of a SC is the number of objects, i.e. f in this case. A SC of length f can be continued by choosing an object of the set $T_{f+1} = \{j \in X | d_{ji_f} < d_{i_{f-1}i_f}\}$ and is terminated if T_{f+1} is the empty set.

The ordering sheme of objects by RRWs, which is used in the following section to define a cluster method, is based on two properties:

1. The realization of an RRW or the corresponding SC can be regarded as an ordered *local* information retrieval process about the structure of the data, since the probability that a set of objects continuing a SC is part of a cluster is increasing with every further step.
2. A combination of a set of different SCs generated by a series of RRWs describes the *global* structure of the data set X.

3 Hierarchy

RRWs can be used for learning an ultrasimilarity matrix between objects of a data set. One single RRW determines a local similarity measure depending on the time scale: two successively chosen objects have to be more similar than two consecutively chosen objects before. In order to be able to guarantee that each object is involved at least once in the learning process, a series of RRWs have to be started. The most obvious way to create a series fulfilling this condition is to start an RRW at each object. Now we can define a set of graphs $G_k = (X, E_k)$ with vertices X and edges E_k based on a series in the following way: an edge between two objects is an element of E_k, if both were chosen in the kth step of any of the RRWs. Correctly speaking the graphs G_k are 'multigraphs' which we denoted in brief as 'graphs'. Since the first connections are sampled with equal probability the graph G_1 is with high probability a connected one while the graphs G_k with increasing index k get more and more dependent on the dissimilarity structure of the data, become unconnected and their components form clusters.

Unfortunately, this divisive construction of clusters will be overlapping, therefore we union the graphs G_k to a monotone decreasing sequence of graphs H_k as

$$H_k = \bigcup_{i=k}^{\infty} G_i, \quad k = 1, 2, \dots, \tag{2}$$

where we use the common notations of graph theory (e.g. see Behzad and Chartrand, 1971). This series of graphs fulfills $H_1 \supseteq H_2 \supseteq \cdots$.

The set of clusters V of our method is defined as the set of vertices of the components[1] of all graphs H_k, $k = 1, 2, \dots$ The clusters contained in V satisfy the conditions (i)-(iv) of an *n-tree* (Bobisud and Bobisud, 1972):

(*i*) $X \in V$,

(*ii*) $\emptyset \notin V$,

(*iii*) $\{i\} \in V$ for all $i \in X$, and

(*iv*) if $A, B \in V$, then $A \cap B \in \{\emptyset, A, B\}$.

[1] A component is a connected subgraph.

The condition (i) is not in general, but with high probability fullfilled since a number of edges greater than the threshold value $(n/2)\log n$ guarantees the connectedness for almost every random graph of size n and H_1 contains on average $n\log n$ edges (see e.g. Bollobás, 1985).

To obtain a similarity measure between the objects of X we associate to each cluster a 'height', as the maximal index of the graph where all objects of the cluster are still contained in one component. For each pair of objects i and j, h_{ij} is defined to be the height of the smallest cluster containing both the ith and the jth object. The symmetric matrix h_{ij} satisfies the ultrametric inequality $h_{ij} \geq \min_k(h_{ik}, h_{kj})$ for all objects and implies a self-similarity $h_{ii} = \infty$, $i = 1, \ldots, n$ which can be redefined to any proper value.

The set of clusters V with the associated heights determines an indexed hierarchy on the set of objects X and defines an unique graph theoretical cluster method called the *RRW method*. The hierarchy between the objects of X can be drawn in a dendrogram and should be interpreted as dendrograms of other hierarchical methods. In the following the capabilites of the method are demonstrated in a challenging application: we investigate a data set consisting of an elongated cluster that surrounds a compact cluster.

4 Application

We investigate a data set consisting of two clusters as shown in Fig. 1 where singeltons marked with label 's' belong to the compact group in the middle. A compact spherical cluster is surrounded by a ring-shaped cluster. Points of the inner cluster that are far away from the cluster center act as bridges between the two clusters. We sampled the objects from the random variables

$$\begin{pmatrix} X_i \\ Y_i \end{pmatrix} = R_i \begin{pmatrix} \cos\phi \\ \sin\phi \end{pmatrix}, \quad i = 1, 2 \tag{3}$$

with $R_1 \sim N(4, 0.5)$, $R_2 \sim N(0, 0.9)$ and $\phi \sim U(0, 2\pi)$. 150 objects sampled according to the first distribution form the a priori group 1 and 50 objects sampled according to the second distribution form the group 2, respectively.

The dendrogram in Fig. 2 is the result of a realization of fifty series of RRWs. We have chosen the number of series in the following way: if one constructs the dendrogram from an increasing number of series its structure will change for lower numbers but then tends to remain nearly constant (apart from being moved as a whole to higher levels). At this stage the method was stopped.

We interpret the hierarchy from the root at the top to the leaves at the bottom. The dendrogram in the upper part shows two main splits and additionally the separation of some outliers. The first interesting split takes place at level 9. The ring-shaped group matches completely with the bigger cluster A, while the compact group in the center is classified correctly apart from two singletons (cluster B). It becomes visible in the dendrogram, that objects in

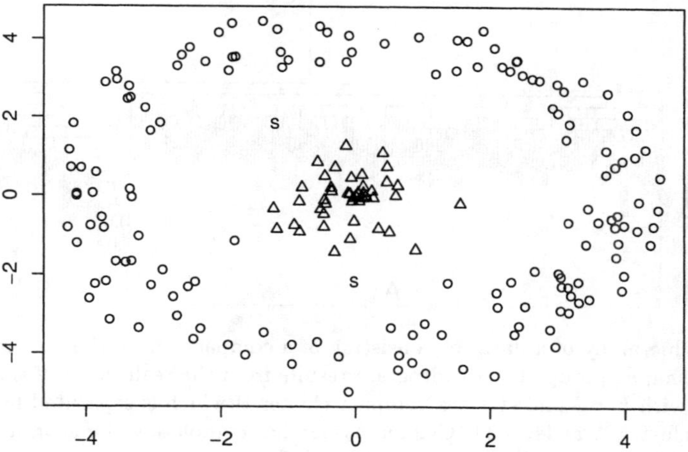

Fig. 1. The classified data set of the example: cluster A is marked with circles and cluster B with triangles and the two singeltons are marked with label 's'.

the center of the compact cluster B are very similar to one another, so that the substructure becomes obvious at much higher levels while cluster A is split into several compact clusters already in level 10.

For comparison, we have investigated the data set with the single-linkage method which is a hierarchical method that is well suited for the detection of elongated clusters. However, it was not able to detect the structure of the ring-shaped group completely: the substructure of the ring-shaped group was recognized, but, instead of merging these subgroups together, so that the group 1 becomes visible, some of them are merged with the compact group in the center via the bridges we have mentioned before.

It is well known that clustering methods may produce an inappropriate classification. So there is one outstanding question: how can we be sure that the split of the data set into the clusters A and B is the best one? In our simple two-dimensional example we are able to look at the plot and use our perception to check the resulting clustering. But a more general and more profound check is of course desirable. In the following we propose one that is based on RRWs.

Bailey and Dubes (1982) defined validity profiles to assess clusters depending on the threshold-graph (see also Gordon, 1999, Cha. 7). We use its methodology of raw profiles. Let $G = (X, E)$ be a graph and A a set of objects proposed as one cluster and B a set of objects proposed as another one.

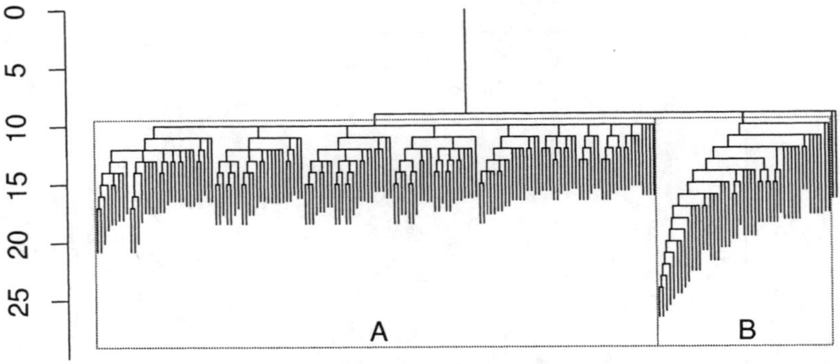

Fig. 2. A hierarchy of a data set consisting of a compact group that is surrounded by a ring-shaped group. The dendrogram results from the realization of three series of RRWs with $k = 3$ and shows a compact cluster B which is separated from a less compact cluster A at level 11. Cluster A matches completely the a priori group 1 and cluster B contains all except two objects of group 2.

Depending on the cluster A the set of edges E can be partitioned into

$$\begin{aligned} D_{in}(A) &= \{ \text{ edges within } A \ \} \\ D_{bet}(A) &= \{ \text{ edges linking } A \text{ with other clusters } \} \\ D_{out}(A) &= \{ \text{ external edges } \}. \end{aligned}$$

We define the *linking probability* between the two clusters A and B depending on the graph G as

$$p_G(A, B) = \frac{|D_{bet}(A) \cap D_{bet(B)}|}{|D_{in}(A) + D_{in}(B) + D_{bet}(A) \cap D_{bet(B)}|}. \tag{4}$$

A plot of these values for the hierarchy of graphs H_k as a function of the hierarchy level can be regarded as an estimate of the connectivity between the two clusters A and B and will be called a *linking probability profile* in the following.

We show that the validity of cluster A and B obtained with the RRW method in our example is superior to alternative classifications proposed, e.g., by the single linkage method, that first links the compact subgroups of the ring-shaped group with the compact group in the middle. For this reason we have plotted 7 linking probability profiles in Fig. 3 for the classifications obtained by 7 different methods: one obtained for the RRW classification, four for hierarchical cluster methods that use the single, complete, group average and Ward's link criterion, one for k-means, and the last one obtained by grouping the data into objects with x-coordinate less than zero and the remaining objects which is called splitting strategy in the following. The

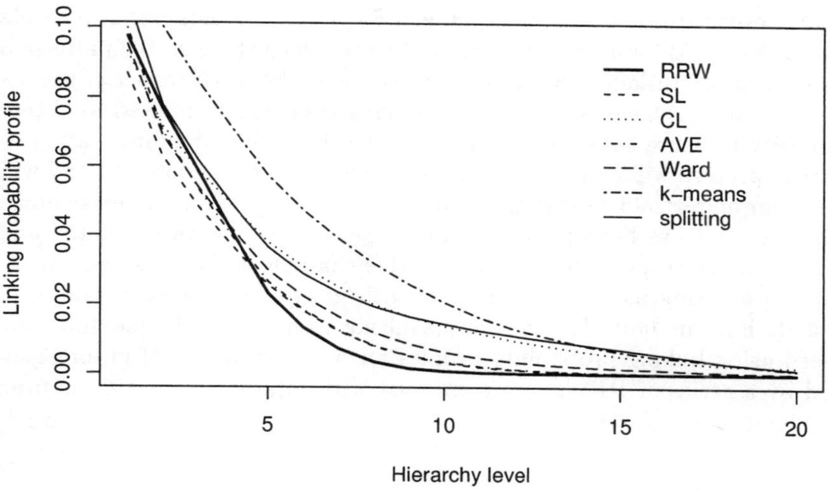

Fig. 3. The linking probability profile for 7 different partitions, that where obtained by the methods as indicated in the text. From the 5th level the RRW clusters are linked with lower probability than the clusters of other methods.

results were obtained by generating a hierarchical set of graphs $H_k, k = 1, 2, \ldots$ using a total number of 60 series of RRWs.

A validity of a classification is the better the lower the linking probability at a certain level is. In the first four levels, the profile shows that the single linkage method yields the best results, i.e., the SL cluster solution has the lowest connectivity at the beginning of RRWs. However, this result is not significant, since with the RRW method a two-cluster solution in a lower than the 5th level can only be achieved with a smaller number of series compared to the one we have used. From the 5th level upwards the classification into a ring-shaped and a compact cluster is superior to the other classifications. An astonishingly bad result is obtained by the complete link method, while the performance of the k-means algorithm is nearly the same as that of the splitting strategy. Two nearly indistinguishable profiles, which are in fact the closest ones to the RRW profile, are those of the single and the group average link method.

5 Summary and outlook

We have presented a new clustering method based on the realization of random walks in a data set. Each random walk is chosen under restrictions that depend on the dissimilarity between objects of the previous moves. This sampling process generates a similarity chain of objects and terminates if a move can only return to the penultimate object. Typically, the number of possible moves is $O(\log(n))$ for a data set with n objects. Using the directed paths of a series of restricted random walks we have defined a hierarchical set of graphs which components have been identified as clusters and can be used to obtain a similarity measure between two objects. We think that the formulation as a Markov process will gain insight into the properties of the method and will confirm our empirical investigations. Work in this direction is currently done.

The method has been tested on a data set containing an elongated and a close compact group. The challenge of this example is due to some objects which are acting as bridges between both groups and cause a failure of the single link method. The clusters resulting from the RRW method were assessed using linking probability profiles for a hierarchic set of graphs generated by a series of RRWs and compared with alternative partitions from other clustering methods.

References

ALPERT, C.J. and KAHNG, A.B. (1993): Geometric embedding for faster (and better) multi-way netlist partitioning. In: *Proceedings of the ACM/IEEE Design Automation Conference*, 743–748.

BAILEY, T.A. and DUBES, R. (1982): Cluster validity profiles. *Pattern Recognition*, 15(2), 61–83.

BEHZAD, M. and CHARTRAND, G. (1971): *Introduction to the theory of graphs.* Allyn and Bacon Inc., Boston.

BOBISUD, H.M. and BOBISUD, L.E. (1972): A metric for classifications. *Taxon*, 21, 607–613.

BOCK, H.-H. (1996): Probalistic models in cluster analysis. *Computational Statistics & Data Analysis*, 23, 5–28.

BOLLOBÁS, B. (1985): *Random Graphs.* Academic Press, London.

GODEHARDT, E. (1990): *Graphs as structural models: the application of graphs and multigraphs in cluster anaysis.* Advances in system analysis 4. Vieweg, Braunschweig, 2nd edition.

GORDON, A.D. (1999): *Classification.* Chapman & Hall, 2nd edition.

HAGEN, L. and KAHNG, A.B. (1992): A New Approach to Effective Circuit Clustering. In: *Proceedings of the IEEE Intl. Conf. on Computer-Aided Design*, 422–427.

LING, R.F. (1973): A probability theory of cluster analysis. *J. Am. Stat. Assoc.*, 68, 159–164.

SCHÖLL, J. and PASCHINGER, E. (2001): Classification by restricted random walks. *Pattern Recognition*, accepted.

ZAHN, C.T. (1971): Graph-theoretical methods for detecting and describing gestalt clusters. *IEEE Trans.Comp*, C-20(1), 68–86.

Missing Data in Hierarchical Classification of Variables - a Simulation Study *

Ana Lorga da Silva[1], Helena Bacelar-Nicolau[2], and Gilbert Saporta[3]

[1] Department of Mathematics, ISEG, Tecnic University, Lisbon, Portugal
[2] FPCE LEAD - Lisbon University, Lisbon, Portugal
[3] Statistics Department, CNAM, Paris, FRANCE

Abstract. Here we develop from a first work the effect of missing data in hierarchical classification of variables according to the following factors: amount of missing data, imputation techniques, similarity coefficient, and aggregation criterion. We have used two methods of imputation, a regression method using an OLS method and an EM algorithm. For the similarity matrices we have used the basic affinity coefficient and the Pearson's correlation coefficient. As aggregation criteria we apply average linkage, single linkage and complete linkage methods. To compare the structure of the hierarchical classifications the Spearman's coefficient between the associated ultrametrics has been used. We present here simulation experiments in five multivariate normal cases.

1 Introduction

The missing data problem has been dealt in a large number of papers and books where several methods to minimise missing data effect have been developed (Rubin(1974), Little and Rubin(1987), Dempster, Laird and Rubin(1977), Orchard and Woodbury(1972), Beale and Little(1975) among others).

When one wants to classify variables, for instance in marketing analysis and social sciences, one frequently finds missing data. We are interested in analysing the effect of missing data in some particular (originally complete) hierarchical classification structures of variables, as well the results of imputation methods in those cases. In the present work we consider hierarchical clustering models based on two similarity coefficients - basic affinity, (Bacelar-Nicolau(1981, 1988, 2000), Matusita(1955), Nicolau(1998) among others) and Pearson's correlation - and three classical aggregation criteria. We use two types of imputation methods in simulation studies with different percentage

* This work has been partially supported by the Franco-Portuguese Scientific Programme MSPLDM-542-B2 (Embassy of France and Portuguese Ministry of Science and Technology – ICCTI) and the Multivariate Data Analysis research team of CEAUL/FCUL.

of missing data at random. The data are issued from multinormal populations (Saporta(1990)).

2 Hierarchical cluster analysis

In this work we are interested in the classification of variables. We use the following hierarchical aggregation criteria:
Average linkage (AL): $C(A,B) = \frac{1}{(\#A)\times(\#B)} \sum c(X_j, X_{j'}), X_j \in A, X_{j'} \in B$
Single linkage (SL): $C(A,B) = max\{c(X_j, X_{j'}), X_j \in A, X_{j'} \in B\}$ and
Complete linkage (CL): $C(A,B) = min\{c(X_j, X_{j'}), X_j \in A, X_{j'} \in B\}$
where A and B represent two clusters and c is a similarity coefficient between two variables ($X_j, X_{j'}$ are $(n \times 1)$ variables) which can be one of the two following:

The (unweighted) basic affinity coefficient $\sum_{i=1}^{n} \sqrt{\frac{x_{ij}}{x_{.j}} \frac{x_{ij'}}{x_{.j'}}}$, where $x_{.j} = \sum_{i=1}^{n} x_{ij}$ and $x_{.j'} = \sum_{i=1}^{n} x_{ij'}$, as defined in Bacelar-Nicolau(2000).
The Bravais-Pearson correlation coefficient $c_p = \frac{\sum_{i=1}^{n}(x_{ij}-\overline{x}^j)(x_{ij'}-\overline{x}^{j'})}{s_{x^j} s_{x^{j'}}}$
and C is the respective extension to the clusters. In order to compare hierarchical classification models, we will use the Spearman's coefficient-c_s-between the ultrametric matrices, based on pairs of observations with the usual correction for ties.

3 The missing data - MAR

The data are said that missing at random if its missingness does not depend of the values assumed on the variables having missing values, but depends on the values observed in other completely observed variables. The expression of the general notion of MAR can be then written as: $Prob(R|X_{obs}, X_{miss}) = Prob(R|X_{obs})$, were X_{obs} represents the observed values of $X_{n \times p}, X_{miss}$ the missing values of$X_{n \times p}$ and $R = [R_{ij}]$ is a missing data indicator,

$$R_{ij} = \begin{cases} 1, \text{ if } x_{ij} \text{ observed} \\ 0, \text{ if } x_{ij} \text{ missing} \end{cases}$$

4 The imputation methods

An ordinary least square regression method (OLS) is used: is defined as usually, β_1, β_2 are estimated over the observed values of the dependent variable,$X_{obs} = \beta_1 + \beta_2 X'_{obs}$($X'_{obs}$ is a "sample"of X corresponding to the observed values of X - X_{obs}) and then the missing values of X (X_{miss}) are imputed by the regression on X'_{miss}(those are observed values coresponding to the missing dependent values of under the estimated model $X_{miss} = \beta_1 + \beta_2 X'_{miss}$.
An EM algorithm has been used as follows:
At the E step of the algorithm (at the tth iteration),

$$x_{ij}^t = \begin{cases} x_{ij}\ , & \text{if} \quad x_{ij} \quad \text{is} \quad \text{observed} \\ \hat{x_{ij}}, & \text{if} \quad x_{ij} \quad \text{is} \quad \text{missing} \end{cases}$$

"The E step imputes the best linear predictors of the missing values, using current estimates of the parameters available so that a suitable choice can be made. It also calculates the adjustments c_{jk} to the estimated covariance matrix needed to allow for imputation of missing values" Little and Rubin(1987) At the M step
$\mu^{(t+1)} = \left[n^{-1} \sum_{i=1}^{n} x_{ij}\right]$,
$\sigma_{jk}^{(t+1)} = (n-1)^{-1} E\left(\sum_{i=1}^{n} x_{ij} x_{ik} | X_{obs}\right) - \mu_j^{t+1} \mu_k^{t+1}, j, k = 1, ...p$. We consider missing values over a dependent variable.

5 Numerical experiments

In order to study the performance of the affinity and the Pearson's correlation coefficients as measures of similarity between variables, in hierarchical classification and in presence of missing data, we use here the three classical hierarchical clustering methods AL, SL and CL: in the cases of complete data; in MAR case - 10%, 15% and 20% (over the total of the data - each 1000×5 matrices); and when the missing data are filled-in using the two imputation methods as mentioned in **4.**.

One hundred samples have been generated of each type of simulated data set, from five normal multivariate populations: Cases A, B, C, D and E, such as: $X_i\tilde{\ }N(\mu_i, \Sigma_i), i = 1, ..., 5$,and are 1000×5 matrices($\mu_1 = \mu_2 = \mu_3 = \mu_4 = \mu_5$).
The values of the variance-covariance matrices have been chosen with the aim of obtaining specific hierarchical structures:

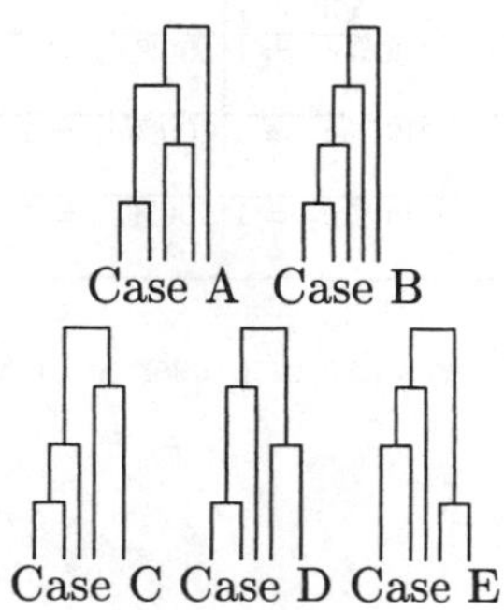

(order of variables:X_1, X_2, X_3, X_4, X_5)

Note that cases C, D and E, have the same topology but they have different agreggation levels. In order to have missing data at random MAR we have deleted (10%, 15% and 20%) values at random from variables X_1 and X_2.

In the following we present the results of the simulations, respectively in all the cases, by increasing order of the percentages of missing data (MD), according to the similarity coefficients, the agglomerative methods and the imputation methods. In each case we compare the ultrametrics associated to the originally complete data with the ultrametric matrices associated to the incomplete and the reconstructed data respectively(imputed data - ID). The comparison between ultrametrics is obtained using a 5% Spearman's bilateral test (the critical value is $c_s' = 0,684$, see for instance Saporta (1990)). In presence of MD, the classification is obtained by a listwise method i.e. we have only considered for the analysis the complete rows (by eliminating the rows with MD). In analysis of cases A,B,C,D and E,$c_s = 1$,$|c_s| > c_s'$,$|c_s| < c_s'$ mean that:
$c_s = 1$ the general "structure" of the two hierarchical classifications being compared is the same, that is the two associated ultrametrics are "ordinal equivalent" (each pair of ranked trees give the same "ordinal" structure).
$|c_s| > c_s'$the two hierarchical classification structures are not the same, but the two ultrametrics are "significantly correlated" (at 5%).
$|c_s| < c_s'$the two hierarchical classification structures are "significantly different"
The percentages of cases and $c_s = 1$,$|c_s| > c_s'$ and $|c_s| < c_s'$ are also indicated in each cell of the tables.

All the simulated complete data reproduced the same general hierarchical structure using both coefficients - Affinity and Pearson's correlation - and the three hierarchical methods, AL, SL and CL.

	Affinity coeficient			Pearson's coeficient		
MD	AL	SL	CL	AL	SL	CL
10%	100%$c_s = 1$	100%$c_s = 1$	100%$c_s = 1$	100%$c_s = 1$	89%$c_s = 1$ 11%$\|c_s\| > c_s'$	100%$c_s = 1$
15%	100%$c_s = 1$	100%$c_s = 1$	100%$c_s = 1$	100%$c_s = 1$	50%$c_s = 1$ 50%$\|c_s\| > c_s'$	100%$c_s = 1$
20%	100%$c_s = 1$	100%$c_s = 1$	100%$c_s = 1$	100%$c_s = 1$	69%$c_s = 1$ 31%$\|c_s\| > c_s'$	100%$c_s = 1$

Table 1. The results in presence of MD - Case A

6 Conclusions

In all the studied cases the affinity coefficient performs better than Pearson's correlation coefficient in presence of data missing at random.

	Affinity coeficient			Pearson's coeficient		
MD	AL	SL	CL	AL	SL	CL
10%	$100\% c_s = 1$	$69\% c_s = 1$ $31\% \lvert c_s \rvert > c'_s$	$100\% c_s = 1$	$99\% c_s = 1$ $1\% \lvert c_s \rvert > c'_s$	$92\% c_s = 1$ $8\% \lvert c_s \rvert > c'_s$	$99\% c_s = 1$ $1\% \lvert c_s \rvert > c'_s$
15%	$100\% c_s = 1$	$17\% c_s = 1$ $83\% \lvert c_s \rvert > c'_s$	$100\% c_s = 1$	$99\% c_s = 1$ $1\% \lvert c_s \rvert > c'_s$	$41\% c_s = 1$ $39\% \lvert c_s \rvert > c'_s$	$99\% c_s = 1$ $1\% \lvert c_s \rvert > c'_s$
20%	$79\% c_s = 1$ $21\% \lvert c_s \rvert > c'_s$	$3\% c_s = 1$ $97\% \lvert c_s \rvert > c'_s$	$97\% c_s = 1$ $17\% \lvert c_s \rvert > c'_s$	$12\% c_s = 1$ $88\% \lvert c_s \rvert > c'_s$	$3\% c_s = 1$ $97\% \lvert c_s \rvert > c'_s$	$12\% c_s = 1$ $88\% \lvert c_s \rvert > c'_s$

Table 2. results after using OLS method - Case A

	Affinity coeficient			Pearson's coeficient		
MD	AL	SL	CL	AL	SL	CL
10%	$98\% c_s = 1$ $2\% \lvert c_s \rvert > c'_s$	$81\% c_s = 1$ $19\% \lvert c_s \rvert > c'_s$	$98\% c_s = 1$ $2\% \lvert c_s \rvert > c'_s$	$93\% c_s = 1$ $7\% \lvert c_s \rvert > c'_s$	$70\% c_s = 1$ $30\% \lvert c_s \rvert > c'_s$	$93\% c_s = 1$ $7\% \lvert c_s \rvert > c'_s$
15%	$58\% c_s = 1$ $42\% \lvert c_s \rvert > c'_s$	$14\% c_s = 1$ $86\% \lvert c_s \rvert > c'_s$	$95\% c_s = 1$ $5\% \lvert c_s \rvert > c'_s$	$91\% c_s = 1$ $9\% \lvert c_s \rvert > c'_s$	$49\% c_s = 1$ $51\% \lvert c_s \rvert > c'_s$	$94\% c_s = 1$ $6\% \lvert c_s \rvert > c'_s$
20%	$75\% c_s = 1$ $25\% \lvert c_s \rvert > c'_s$	$8\% c_s = 1$ $92\% \lvert c_s \rvert > c'_s$	$80\% c_s = 1$ $20\% \lvert c_s \rvert > c'_s$	$31\% c_s = 1$ $69\% \lvert c_s \rvert > c'_s$	$1\% c_s = 1$ $99\% \lvert c_s \rvert > c'_s$	$31\% c_s = 1$ $69\% \lvert c_s \rvert > c'_s$

Table 3. Results after using EM method - Case A

	Affinity coeficient			Pearson's coeficient		
MD	AL	SL	CL	AL	SL	CL
10%	$100\% c_s = 1$	$100\% c_s = 1$	$100\% c_s = 1$	$100\% c_s = 1$	$100\% c_s = 1$	$100\% c_s = 1$
15%	$100\% c_s = 1$	$100\% c_s = 1$	$100\% c_s = 1$	$98\% c_s = 1$ $2\% \lvert c_s \rvert > c'_s$	$100\% c_s = 1$	$100\% c_s = 1$
20%	$100\% c_s = 1$	$99\% c_s = 1$ $1\% \lvert c_s \rvert > c'_s$	$100\% c_s = 1$	$94\% c_s = 1$ $6\% \lvert c_s \rvert > c'_s$	$97\% c_s = 1$ $3\% \lvert c_s \rvert > c'_s$	$81\% c_s = 1$ $18\% \lvert c_s \rvert > c'_s$ $1\% \lvert c_s \rvert < c'_s$

Table 4. Results in presence of MD - Case B

	Affinity coeficient			Pearson's coeficient		
MD	AL	SL	CL	AL	SL	CL
10%	$99\% c_s = 1$ $1\% \lvert c_s \rvert > c'_s$	$99\% c_s = 1$ $1\% \lvert c_s \rvert > c'_s$	$99\% c_s = 1$ $1\% \lvert c_s \rvert > c'_s$	$99\% c_s = 1$ $1\% \lvert c_s \rvert > c'_s$	$99\% c_s = 1$ $1\% \lvert c_s \rvert > c'_s$	$99\% c_s = 1$ $1\% \lvert c_s \rvert > c'_s$
15%	$100\% c_s = 1$	$100\% c_s = 1$	$100\% c_s = 1$	$100\% c_s = 1$	$100\% c_s = 1$	$100\% c_s = 1$
20%	$100\% c_s = 1$	$100\% c_s = 1$	$100\% c_s = 1$	$100\% c_s = 1$	$100\% c_s = 1$	$100\% c_s = 1$

Table 5. Results after using OLS method -Case B

	Affinity coeficient			Pearson's coeficient		
MD	AL	SL	CL	AL	SL	CL
10%	$100\% c_s = 1$	$100\% c_s = 1$	$100\% c_s = 1$	$100\% c_s = 1$	$100\% c_s = 1$	$100\% c_s = 1$
15%	$100\% c_s = 1$	$100\% c_s = 1$	$100\% c_s = 1$	$100\% c_s = 1$	$100\% c_s = 1$	$100\% c_s = 1$
20%	$100\% c_s = 1$	$100\% c_s = 1$	$100\% c_s = 1$	$100\% c_s = 1$	$100\% c_s = 1$	$100\% c_s = 1$

Table 6. Results after using EM method - Case B

Better results are obtained in presence of MD, than after the application of both imputation methods.

	Affinity coeficient			Pearson's coeficient		
MD	AL	SL	CL	AL	SL	CL
10%	$100\% c_s = 1$	$99\% c_s = 1$ $1\% \lvert c_s \rvert > c'_s$	$100\% c_s = 1$	$5\% c_s = 1$ $95\% \lvert c_s \rvert < c'_s$	$49\% c_s = 1$ $51\% \lvert c_s \rvert < c'_s$	$100\% \lvert c_s \rvert < c'_s$
15%	$100\% c_s = 1$	$96\% c_s = 1$ $4\% \lvert c_s \rvert > c'_s$	$99\% c_s = 1$ $1\% \lvert c_s \rvert > c'_s$	$5\% c_s = 1$ $95\% \lvert c_s \rvert < c'_s$	$41\% c_s = 1$ $59\% \lvert c_s \rvert < c'_s$	$100\% c_s < 1$
20%	$100\% c_s = 1$	$87\% c_s = 1$ $13\% \lvert c_s \rvert > c'_s$	$100\% c_s = 1$	$3\% c_s = 1$ $97\% \lvert c_s \rvert < c'_s$	$21\% c_s = 1$ $69\% \lvert c_s \rvert < c'_s$	$100\% \lvert c_s \rvert < c'_s$

Table 7. Results in presence of MD - Case C

	Affinity coeficient			Pearson's coeficient		
MD	AL	SL	CL	AL	SL	CL
10%	$100\% c_s = 1$	$100\% c_s = 1$	$100\% c_s = 1$	$98\% c_s = 1$ $2\% \lvert c_s \rvert < c'_s$	$100\% c_s = 1$	$98\% c_s = 1$ $2\% \lvert c_s \rvert < c'_s$
15%	$100\% c_s = 1$	$100\% c_s = 1$	$100\% c_s = 1$	$94\% c_s = 1$ $5\% \lvert c_s \rvert > c'_s$ $1\% \lvert c_s \rvert < c'_s$	$96\% c_s = 1$ $4\% \lvert c_s \rvert > c'_s$	$95\% c_s = 1$ $4\% \lvert c_s \rvert > c'_s$ $1\% \lvert c_s \rvert < c'_s$
20%	$94\% c_s = 1$ $2\% \lvert c_s \rvert > c'_s$ $4\% \lvert c_s \rvert < c'_s$	$94\% c_s = 1$ $2\% \lvert c_s \rvert > c'_s$ $4\% \lvert c_s \rvert < c'_s$	$94\% c_s = 1$ $2\% \lvert c_s \rvert > c'_s$ $4\% \lvert c_s \rvert < c'_s$	$21\% c_s = 1$ $58\% \lvert c_s \rvert > c'_s$ $21\% \lvert c_s \rvert < c'_s$	$21\% c_s = 1$ $58\% \lvert c_s \rvert > c'_s$ $21\% \lvert c_s \rvert < c'_s$	$21\% c_s = 1$ $58\% \lvert c_s \rvert > c'_s$ $21\% \lvert c_s \rvert < c'_s$

Table 8. Results after using both imputation methods - Case C

	Affinity coeficient			Pearson's coeficient		
MD	AL	SL	CL	AL	SL	CL
10%	$99\% c_s = 1$ $1\% \lvert c_s \rvert > c'_s$	$100\% c_s = 1$	$100\% c_s = 1$	$97\% c_s = 1$ $3\% \lvert c_s \rvert > c'_s$	$96\% c_s = 1$ $4\% \lvert c_s \rvert > c'_s$	$99\% c_s = 1$ $1\% \lvert c_s \rvert > c'_s$
15%	$98\% c_s = 1$ $1\% \lvert c_s \rvert > c'_s$ $1\% \lvert c_s \rvert < c'_s$	$99\% c_s = 1$ $1\% \lvert c_s \rvert > c'_s$	$99\% c_s = 1$ $1\% \lvert c_s \rvert > c'_s$	$95\% c_s = 1$ $5\% \lvert c_s \rvert < c'_s$	$90\% c_s = 1$ $10\% \lvert c_s \rvert > c'_s$	$95\% c_s = 1$ $5\% \lvert c_s \rvert > c'_s$
20%	$99\% c_s = 1$ $1\% \lvert c_s \rvert < c'_s$	$100\% c_s = 1$	$100\% c_s = 1$	$92\% c_s = 1$ $8\% \lvert c_s \rvert < c'_s$	$85\% c_s = 1$ $15\% \lvert c_s \rvert > c'_s$	$93\% c_s = 1$ $7\% \lvert c_s \rvert > c'_s$

Table 9. Results in presence of MD - Case D

	Affinity coeficient			Pearson's coeficient		
MD	AL	SL	CL	AL	SL	CL
10%	$100\% \lvert c_s \rvert > c'_s$	$99\% c_s = 1$ $1\% \lvert c_s \rvert > c'_s$	$96\% c_s = 1$ $4\% \lvert c_s \rvert > c'_s$	$59\% c_s = 1$ $40\% \lvert c_s \rvert > c'_s$ $1\% \lvert c_s \rvert < c'_s$	$47\% c_s = 1$ $63\% \lvert c_s \rvert > c'_s$	$60\% c_s = 1$ $40\% \lvert c_s \rvert > c'_s$
15%	$9\% c_s = 1$ $90\% \lvert c_s \rvert > c'_s$ $1\% \lvert c_s \rvert < c'_s$	$6\% c_s = 1$ $94\% \lvert c_s \rvert > c'_s$	$18\% c_s = 1$ $82\% \lvert c_s \rvert > c'_s$	$70\% c_s = 1$ $29\% \lvert c_s \rvert > c'_s$ $1\% \lvert c_s \rvert < c'_s$	$52\% c_s = 1$ $47\% \lvert c_s \rvert > c'_s$	$76\% c_s = 1$ $23\% \lvert c_s \rvert > c'_s$
20%	$99\% \lvert c_s \rvert > c'_s$ $1\% \lvert c_s \rvert < c'_s$	$100\% \lvert c_s \rvert > c'_s$	$100\% \lvert c_s \rvert > c'_s$	$31\% c_s = 1$ $68\% \lvert c_s \rvert > c'_s$ $1\% \lvert c_s \rvert < c'_s$	$18\% c_s = 1$ $82\% \lvert c_s \rvert > c'_s$	$68\% c_s = 1$ $32\% \lvert c_s \rvert > c'_s$

Table 10. Results after using OLS method - Case D

When using the imputation methods in case C both imputation methods gave the same results, and also that the affinity coefficient performs better than the Pearson's coefficient. In cases A, D and E the results are different when using the two imputation methods, sometimes the least squares method

	Affinity coeficient			Pearson's coeficient		
MD	AL	SL	CL	AL	SL	CL
10%	$98\% c_s = 1$ $2\% \lvert c_s \rvert > c'_s$	$83\% c_s = 1$ $12\% \lvert c_s \rvert > c'_s$ $1\% \lvert c_s \rvert < c'_s$	$92\% c_s = 1$ $7\% \lvert c_s \rvert > c'_s$ $1\% \lvert c_s \rvert < c'_s$	$93\% c_s = 1$ $7\% \lvert c_s \rvert > c'_s$	$99\% c_s = 1$ $1\% \lvert c_s \rvert > c'_s$	$93\% c_s = 1$ $7\% \lvert c_s \rvert > c'_s$
15%	$58\% c_s = 1$ $42\% \lvert c_s \rvert > c'_s$	$21\% c_s = 1$ $79\% \lvert c_s \rvert > c'_s$	$87\% c_s = 1$ $1\% \lvert c_s \rvert > c'_s$ $12\% \lvert c_s \rvert < c'_s$	$91\% c_s = 1$ $9\% \lvert c_s \rvert > c'_s$	$80\% c_s = 1$ $19\% \lvert c_s \rvert > c'_s$ $1\% \lvert c_s \rvert < c'_s$	$93\% c_s = 1$ $6\% \lvert c_s \rvert > c'_s$ $1\% \lvert c_s \rvert < c'_s$
20%	$75\% c_s = 1$ $25\% \lvert c_s \rvert > c'_s$	$1\% c_s = 1$ $99\% \lvert c_s \rvert > c'_s$	$97\% \lvert c_s \rvert > c'_s$ $3\% \lvert c_s \rvert < c'_s$	$31\% c_s = 1$ $69\% \lvert c_s \rvert > c'_s$	$19\% c_s = 1$ $81\% \lvert c_s \rvert > c'_s$	$98\% \lvert c_s \rvert > c'_s$ $2\% \lvert c_s \rvert < c'_s$

Table 11. Results after using EM method - Case D

	Affinity coeficient			Pearson's coeficient		
MD	AL	SL	CL	AL	SL	CL
10%	$100\% c_s = 1$	$100\% c_s = 1$	$100\% c_s = 1$	$100\% c_s = 1$	$100\% c_s = 1$	$100\% c_s = 1$
15%	$100\% c_s = 1$	$100\% c_s = 1$	$100\% c_s = 1$	$100\% c_s = 1$	$97\% c_s = 1$ $3\% \lvert c_s \rvert < c'_s$	$98\% c_s = 1$ $2\% \lvert c_s \rvert < c'_s$
20%	$100\% c_s = 1$	$98\% c_s = 1$ $1\% \lvert c_s \rvert > c'_s$ $1\% \lvert c_s \rvert < c'_s$	$99\% c_s = 1$ $1\% \lvert c_s \rvert > c'_s$	$100\% c_s = 1$	$87\% c_s = 1$ $13\% \lvert c_s \rvert < c'_s$	$92\% c_s = 1$ $8\% \lvert c_s \rvert < c'_s$

Table 12. Results in presence of MD - Case E

	Affinity coeficient			Pearson's coeficient		
MD	AL	SL	CL	AL	SL	CL
10%	$100\% c_s = 1$	$100\% c_s = 1$	$100\% c_s = 1$	$100\% c_s = 1$	$100\% c_s = 1$	$100\% c_s = 1$
15%	$100\% c_s = 1$	$100\% c_s = 1$	$100\% c_s = 1$	$100\% c_s = 1$	$100\% c_s = 1$	$100\% c_s = 1$
20%	$99\% c_s = 1$ $1\% \lvert c_s \rvert > c'_s$	$99\% c_s = 1$ $1\% \lvert c_s \rvert > c'_s$	$99\% c_s = 1$ $1\% \lvert c_s \rvert > c'_s$	$73\% c_s = 1$ $25\% \lvert c_s \rvert > c'_s$ $2\% \lvert c_s \rvert < c'_s$	$73\% c_s = 1$ $25\% \lvert c_s \rvert > c'_s$ $2\% \lvert c_s \rvert < c'_s$	$73\% c_s = 1$ $25\% \lvert c_s \rvert > c'_s$ $2\% \lvert c_s \rvert < c'_s$

Table 13. Results after using OLS method - Case E

	Affinity coeficient			Pearson's coeficient		
MD	AL	SL	CL	AL	SL	CL
10%	$100\% c_s = 1$	$100\% c_s = 1$	$100\% c_s = 1$	$100\% c_s = 1$	$100\% c_s = 1$	$100\% c_s = 1$
15%	$100\% c_s = 1$	$100\% c_s = 1$	$100\% c_s = 1$	$100\% c_s = 1$	$100\% c_s = 1$	$100\% c_s = 1$
20%	$99\% c_s = 1$ $1\% \lvert c_s \rvert > c'_s$	$99\% c_s = 1$ $1\% \lvert c_s \rvert > c'_s$	$99\% c_s = 1$ $1\% \lvert c_s \rvert > c'_s$	$78\% c_s = 1$ $22\% \lvert c_s \rvert > c'_s$	$76\% c_s = 1$ $24\% \lvert c_s \rvert > c'_s$	$76\% c_s = 1$ $24\% \lvert c_s \rvert > c'_s$

Table 14. Results after using EM method - Case E

performs better, others, is with the EM algorithm that we obtain better performance. We have obtained similar results in a study with Personality development data, in presence of seven variables(Silva et al.(2001))

The following developments of this work are related to other similarity coefficients and hierarchical structures, namely concerning a probabilistic classification approach,the use of the probabilistic weighted coefficient (without ignore objects or impute the missing data) and different types of missing data and imputation methods.

References

BACELAR-NICOLAU, H.(1981): Contributions to the Study of Comparison Coefficients in Cluster Analysis, Univ. Lisbon.

BACELAR-NICOLAU, H.(1988), Two probabilistic models for classification of variables in frequency tables, Classif. and Relat. Meth. of Data Analysis, H. .H. Bock (ed.), North Holland, pp. 181-186.

BACELAR-NICOLAU(2000) The Affinity Coefficient in Analysis of Symbolic Data Exploratory Methods for Extracting Statistical Information from Complex Data. H.H. Bock and E.Diday (Eds.), Springer,160-165.

BEALE, E. M. L. and LITTLE, R. J. A.(1975) Missing values in multivariate data analysis. J. R. *Statist. Soc.* B, **37**, 129-145.

DEMPSTER, A. P., LAIRD, N. M. and RUBIN, D. B.(1977) Maximum likelihood estimation from incomplete data via the EM algorithm (with discussion). *J. R. Statist. Soc.* B, **39**, 1-38.

LITTLE, R. J. A. and RUBIN, D. B.(1987) Statistical Analysis With Missing Data, John Wiley & Sons, New York.

MATUSITA,K.(1955) Decision rules, based on distance for problems of fit, two samples and estimation. Ann. Math. Stat., vol 26, n4,631-640.

NICOLAU F.C., BACELAR-NICOLAU, H. (1998), Some Trends in the Classification of variables, Data Science, Classification, and Related Methods, C. Hayashi, N. Ohsumi, K. Yajima, Y. Tanaka, H. H. Bock, Y. Baba (eds.), Springer, pp. 89-98

ORCHARD, T. and WOODBURY, M. A.(1972) A missing information principle: theory and applications. Proceedings 6th Berkley Symposium on Mathematical Statistics and Probability, **1**, 697-715.

RUBIN, D. B.(1974) Characterising the estimation of parameters in the estimation of parameters in incomplete-data problems. *JASA*, **69**,467-474.

SAPORTA, G.(1990) Probabilités, Analyse des Données et Statistique, Editions Technip, Paris.

SILVA,A.L, BACELAR-NICOLAU, SAPORTA, G. and GEADA, M.(2001) Missing Data in Hierarchical Classification – a study with Personality development data, - 32nd European Mathematical Psycology /EMPG 2001, 109-110.

Cluster Validation

Representation and Evaluation of Partitions

Alain Guénoche[1] and Henri Garreta[2]

[1] Institut de Mathématiques de Luminy,
[2] Laboratoire d'Informatique Fondamentale de Marseille,
163, avenue de Luminy, 13288 Marseille Cedex 9, France.

Abstract. Many methods lead to build a partition of a finite set, given a metric. In this text we propose some criteria to evaluate the quality of each class and other parameters to compare several partitions on the same data set. Then, we indicate how to represent graphically a partition as a tree of boxes, each one containing information about the quality of a class.

1 Introduction

Let X be a finite set of N elements and D a distance array containing the proximity values between any two elements of X. Let us recall that a p-partition P on X is a set of p separate classes, such that:

$$\forall i, j \quad X_i \cap X_j = \emptyset \quad \text{and} \quad \bigcup_{i=1}^{p} X_i = X$$

Many methods permit to build classes and/or partitions on X. The most classical ones optimize some criterion over the set of partitions with a given number of classes (Hansen and Jaumard (1997)). Criteria commonly used are the split of classes, the diameter of the partition, the sum of (squared) distances between elements in different classes or a function of inertia, computed as the sum of squared distances to a center (real or virtual) (Guénoche (2002)). Unfortunately, most of these criteria lead to NP-hard optimization problems (Brucker (1978)) and sub-optimal solutions are calculated. A projection into an euclidian space, using a multidimensional scaling or a factorial method is also commonly realized; then, a center method (k-means) is applied to the set of the projected points. More rarely, a sequential clustering method is applied (Hansen et al. (1995)); it consists in building a class as homogeneous as possible, then to iterate the procedure on X minus this class, until there is no class sufficiently homogeneous. In that case, we do not get a partition on X, but only on the clustered elements. Other very efficient techniques (Self Organisation Map (Kohonen (1982)), or Adaptative Quality-based Clustering (Thijs et al. (2001)) are often used, particularly for clustering large sets of proteins from their expression levels.

Whatever is the method, we are confronted with two problems, (i) the representation of the partitions, and (ii) the measurement of the quality of both classes and partition. They are often represented by lists of elements or by hypergraphs, that is a very poor and difficult way to understand partitions. This is a reason why many users prefer to apply a hierarchical clustering method and to have a look to the classes in the tree, because there are several very good softwares to display them.

In paragraph 1, we introduce several criteria to evaluate the quality of a class. In paragraph 2, we indicate other criteria for comparing partitions; they are not linked to those that are commonly used to build partitions (split, diameter or inertia) and they are also independent of the cardinality of the classes. In paragraph 3, we explain the representation of a partition by a "tree of boxes" , each box being characterized by the quality values of a class. An application, `QualiPart`, will be briefly presented.

2 Evaluating classes

The set X endowed with a distance D is a metric discrete space. In the Combinatorial Data Analysis approach (Arabie and Hubert (1996)) it is considered as a complete graph G, having X as its set of vertices and the edges being valued by D. We denote by ***threshold graph at level s*** the graph $G_{\leq s}$ with X as vertices and as edges all pairs (x, y) such that $D(x, y) \leq s$. Let Δ be the diameter of G that is the largest distance value.

For each class $X_k \subseteq X$ with $n_k = |X_k|$ elements, one can evaluate its homogeneity by calculating the following parameters:

- The *connectivity threshold* s_k. Each element x_i has a nearest neighbor at distance d_i. The threshold s_k of X_k is the largest value of d_i for x_i belonging to X_k.

$$s_k = \underset{x_i \in X_k}{Max} \ \underset{x_j \in X_k}{Min} \ D(x_i, x_j)$$

  It is also the length of the longest edge of a minimal spanning tree of the distance reduced to this class. If, in G, all the edges longer than or equal to s_k are removed, X_k is no more connected and this class is not consistent with the threshold.
- The *rate of links*. They correspond to pairs in X_k, having a length less than or equal to the connectivity threshold. The smallest value is $2/n_k$ (in the case where there is just a tree) and the largest one is 1 (when there is no value greater than s_k). The less it is, the weaker is the class, as a result of a chaining effect.
- The *diameter* δ_k. We indicate the ratio δ_k/Δ that must be as small as possible. All the links of X_k are less than or equal to this diameter and, at this threshold, X_k is a clique of $G_{\leq \delta_k}$. We shall remark that if the distance reduced to X_k fulfills the ultrametric condition, the connectivity threshold is equal to the diameter.

- The *size of the largest clique* of $G_{\leq s_k}$ included in X_k. The larger this cardinality, the stronger is the class. Unfortunately, the MaxClique-problem is NP-hard. Consequently, we just evaluate an interval. The lower bound is provided building a greedy clique: we start from a vertex having maximum degree and we add, as long as it exists, a vertex with maximum degree adjacent to all the previous elements in this clique. The upper bound q is the greatest number of elements of X_k having, in $G_{\leq s_k}$, a degree greater than or equal to $q-1$; it means that they have $q-1$ links not greater than s_k which might realize a clique (which is not tested).
- The *rate of well designed triples*. We only consider triples made of two elements x and y belonging to X_k and one element z out of it. Such a triple is well designed iff

  $$D(x,y) \leq Min\{D(x,z), D(y,z)\}$$

  This condition is not satisfied when an element out of the class is closer to an element in the class than to another element in the class.

In our approach, we consider a class to be homogeneous if it has a diameter δ_k clearly less than Δ and not too large compared to s_k; the ratio of links not greater than s_k would be high, close to 50%; the maximal clique would contain from third to half the number of elements; the rate of well designed triples will be greater than 50%. If the class concatenates elements just closed to a single one, that is when there is a *chaining effect*, the diameter is high, the ratio of links weak and there are many cliques with very few elements.

3 Comparing partitions

To compare the quality of several partitions build from the same data set, we must define criteria not identical to those that can be eventually optimized when constructing one of these partitions. Moreover, these quality criteria must be independent of the number of elements in the classes. They must neither favor balanced partitions i.e., having classes with approximatively the same number of elements, nor those that present many singletons. We have selected four parameters that fulfill these conditions. They are based on the principle that a partition P is good if large distance values are between-class links and if small values are within-class links.

Whatever the number of classes, a partition P on X induces a bipartition of the pairs of X elements: on the one hand the between-class pairs correspond the external edges, and on the other hand the within-class pairs correspond to internal edges. Let $(L_e|L_i)$ be this bipartition, and N_e and N_i be the number of elements of each part.

$$N_e = \tfrac{1}{2}\sum_{k=1}^{p} |X_k|\,(N - |X_k|) \qquad N_i = \tfrac{1}{2}\sum_{k=1}^{p} |X_k|\,(|X_k| - 1).$$

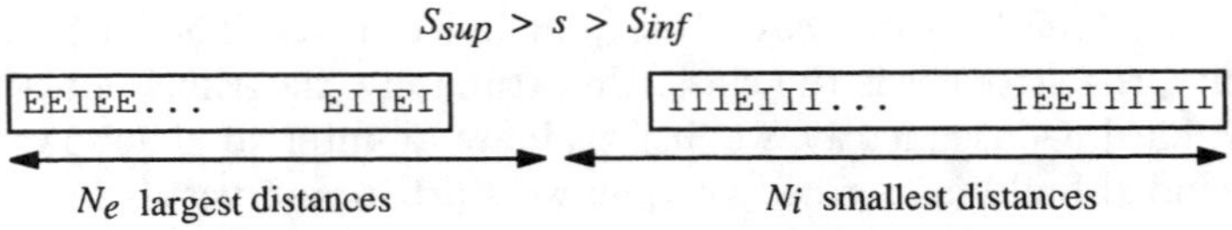

Fig. 1. Two bipartitions of the distance values ranked in decreasing order: external links (E) and internal links (I).

Evidently, we get $N_e + N_i = \frac{N(N-1)}{2}$, this quantity being denoted N_2 in the following.

These numbers induce another bipartition of the N_2 pairs in X (see Fig. 1). After ranking distance values in decreasing order, on the left side we have the N_e largest values and on the right side the N_i smallest values. This bipartition is denoted $(G_d|S_d)$. A perfect partition on X, according to the previous principle yields two identical bipartitions on pairs. Our two first parameters measure the similarity of these two bipartitions.

3.1 The rate of agreements on pairs

This is the percentage of pairs belonging to L_e and G_d or to L_i and S_d. They designate the largest distances used as external links and the smallest distances used as internal links, the difference between large (G_d) and small (S_d) being made at the threshold of the N_e-th distance value. As there can be some ties, we must define precisely what are the thresholds. First, we rank the distance values in decreasing order. Let s be the N_e-th largest distance value, and s' the next one ($s \geq s'$). If $s > s'$, we define $S_{sup} = s$ and $S_{inf} = s'$. Else, S_{sup} is the smallest value strictly greater than s and S_{inf} is the greatest value strictly lower than s. Between S_{sup} and S_{inf} either there is no further distance value, or there are values all equal to s. The *rate of agreements on pairs* is calculated by counting the pairs belonging to the intersection of the two bipartitions:

$Inter = 0$
For all pairs (x, y)
    If $Clas(x) \neq Clas(y)$ and $D(x, y) \geq S_{sup}$, $Inter = Inter + 1$
    If $Clas(x) = Clas(y)$ and $D(x, y) \leq S_{inf}$, $Inter = Inter + 1$
$\tau(a) = 1 - \frac{Inter}{N_2}$

3.2 The rate of weight

It is computed from the sums of distances on each of the four classes of pairs, respectively denoted by $\sum(L_e)$, $\sum(L_i)$, $\sum(G_d)$ and $\sum(S_d)$, as follows :

$$\tau(w) = \frac{\sum(L_e)}{\sum(G_d)} \times \frac{\sum(S_d)}{\sum(L_i)}.$$

These two ratios correspond to the weight of external links divided by the maximum that could be realized with N_e distance values and to the minimal weight of N_i edges divided by the weight of the internal links. Both are lower than or equal to 1 and so is the *rate of weight.* A rate close to 1 means that the between-class links belong to the largest edges and the intra-class links have been selected from the smallest ones. Consequently, this partition is one of the best possible ones.

Partitions maximizing the split have large values for this criterion, since they aim to put as external edges as many large distance values as possible. But it is not so for the rate of agreements on pairs, because large distance may link elements belonging to the same class.

3.3 The ratio of average lengths

Without depending on the cardinality of the classes, there is a very simple criterion to evaluate the quality of a partition; it is the ratio of the average lengths of the external edges and the average lengths of the internal edges:

$$\theta(l) = \frac{\frac{\sum(L_e)}{N_e}}{\frac{\sum(L_i)}{N_i}}$$

For a good partition, we can expect that it is greater than 1. The greatest feasible value is obtained replacing L_e and L_i by G_d and S_d in this formula. When $\theta(l)$ is large, the two types of edges are different and the partition is consistent. We have observed that if we select, for random euclidian or boolean distances, N points without cluster structure, this ratio is rarely larger than 1.3. But if the points are selected in balls centered on an axis and passing through the origin or in orthogonal hyperplans, this ratio is generally around 2 (Guénoche (2002)).

3.4 The ratio of well designed triples

This corresponding criterion for classes can be easily extended to partitions. Let T be the set of triples having just two elements in the same class.

$$|T| = \frac{1}{2}\sum_{k=1}^{p} |X_k| \left(|X_k| - 1\right) \left(N - |X_k|\right)$$

When the split of a class is lower than its diameter the rate of triples is lower than 1. Let x and y be in the same class and z in another one. We evaluate

$$\tau(t) = \frac{|\{x,y,z\} \in T \text{ such that } D(x,y) \leq Min\,\{D(x,z), D(y,z)\}|}{|T|}$$

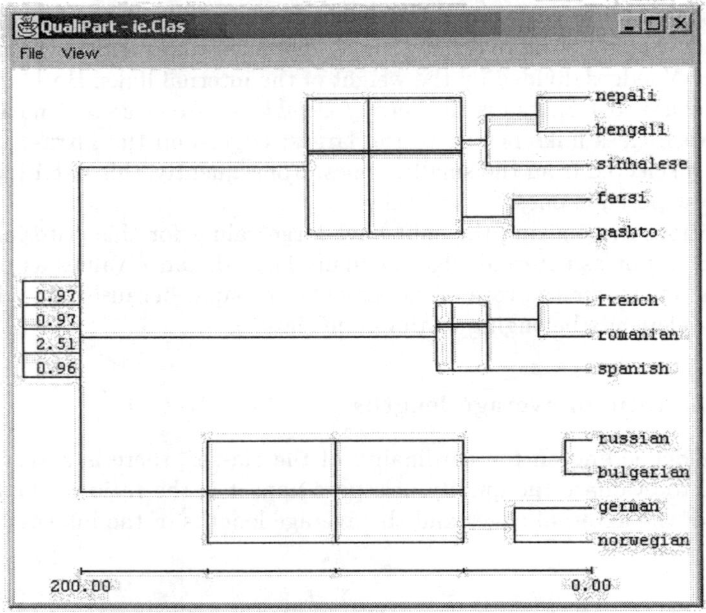

Fig. 2. Tree of boxes representing a partition in three classes of indo-european languages.

4 Representation of partitions

Classification trees or dendrograms are the most readable representation for hierarchies. They have a root, each leaf corresponding to an element has a label that can be written horizontally, if the tree expands vertically and if the horizontal axis is turned from large to small values (Cf. `NJPlot` [Perrière et al. 96]).

Partitions are also hierarchies with just one intermediate level between the elements and the whole set. So, in the tree style, but with much more information, we represent partitions as a tree of boxes. The root corresponds, as usual, to X to which as many rectangles (or boxes) as classes are linked. The horizontal axis is also scaled in the decreasing order. The root is placed at the position of Δ, the diameter of X, and all the leaves at position 0. Each class is included between two distance values, its threshold of connectivity s_k and its diameter δ_k. The larger the difference, the larger is the box and the class is less homogeneous. The height is just proportional to the number of elements.

We have seen that for a distance threshold lower than s_k, the class is disconnected. The subclasses can be represented by the single linkage hierarchy. It seems to be the best way to indicate their organization, and the level of the last union in this hierarchy is exactly s_k. So there remains, between levels

```
3
5 nepali bengali sinhalese farsi pashto
3 french romanian spanish
4 russian bulgarian german norwegian
```

Fig. 3. A partition of indo-european languages in three classes.

s_k and 0, the suitable place to draw it. The elements are ranked in an order compatible with this hierarchy.

Within the box, we draw two lines. Horizontally, it corresponds to the number of elements of the largest clique. When the lower and upper bounds are different, we take the average. A double line is placed near the top when it is high and near the bottom when it is small. Vertically, it corresponds to the rate of pairs having a distance not greater than s_k. This percentage varies along the width of the box. The line is drawn near the diameter when this rate is high and near the connectivity threshold in the opposite case.

Finally we display the four criteria values which qualify the partition in a box linked to the root.

5 Technical overview

An application, `QualiPart`, program has been written in Java. It uses two files ; the first one is for the classes and the second one for the distance array. The classes must be disjoint (it is tested) and their union form the complete set X. The first record in the file indicates the number of classes. Each class is described by its number of elements followed by the list of the labels of elements, separated by at least one space. The file given in Fig. 3 is the one represented by the tree of boxes in Fig. 2.

The distance file is in the PHILIP format. It must contain, in any order, all the labels referenced in the classes and labels must respect the same upper/lower case. In this file, there can be other elements absent in the classes. The useful distance array is extracted.

The program calculates first the criteria values for classes then those which evaluate the partition. All this values are shown in a text window, that can be saved as a text file. Simultaneously, it displays the tree of boxes. When clicking in a box, the corresponding class only remains on the screen and its criteria values are displayed in a message box. When clicking outside of the box, the full tree appears again. This drawings can be saved in PostScript format to be printed.

This application can be obtained from the authors, sending a e-mail to `garreta@luminy.univ-mrs.fr`.

Acknowledgements

This research was supported by the "Origines de l'homme, des langages et des langues" (OHLL) program. We also would like to thank an anonymous referee.

References

ARABIE, P. and HUBERT, L. (1996): An Overview of Combinatorial Data Analysis. In: P. Arabie, L. Hubert and G. de Soete (Eds.): *Clustering and Classification.* World Scientific Publ., River Edge, N.J., 5-63.

BRUCKER, P. (1978): On the complexity of clustering problems. In: M. Beckmann and H. Künzi (Eds.): *Optimization and Operations Research.* Lecture Notes in Economics and Mathematical Systems, 157, Springer-Verlag, Heidelberg, 45-54.

GUENOCHE, A. (2002): Partitions optimisées selon différents critères. *Mathématiques, Informatique et Sciences humaines*, to appear.

HANSEN, P. and JAUMARD, B. (1997): Cluster analysis and mathematical programming. *Mathematical Programming, 79*, 191-215.

HANSEN, P., JAUMARD, B., and MLADENOVIC, M. (1995): How to choose K entities among N. In: DIMAC Series in Discrete Mathematics and Theoretical Computer Science, 19, 105-115.

HUBERT, L.J. (1974): Spanning trees and aspects of clustering. *British J. of Math. Statist. Psychol.*, 27, 14-28.

KOHONEN, T. (1982): Self-organized formation of topologically correct feature maps. *Biol. Cybern.*, 43, 59-69.

PERRIERE, G. and GOUY, M. (1996): WWW-Query: An on-line retrieval system for biological sequence banks. *Biochimie*, 78, 364-369.

THIJS, G., MOREAU, Y., DE SMET, F., MATHYS, J., LESCOT, M., ROMBAUTS, S., ROUZE, P., DE MOOR, B., and MARCHAL, K. (2001): INCLUSive: INtegrated Clustering, Upstream sequence retrieval and motif Sampling. *Bioinformatics*, 2001, to appear.

Assessing the Number of Clusters of the Latent Class Model

François-Xavier Jollois[1], Mohamed Nadif[1] and Gérard Govaert[2]

[1] Laboratoire d'Informatique Théorique et Appliquée, Université de Metz, Ile du Saulcy, 57045 Metz Cedex, France
jollois@lita.univ-metz.fr - nadif@iut.univ-metz.fr

[2] Heudiasyc, UMR CNRS 6599, Université de Technologie de Compiègne BP 20529, 60205 Compiègne Cedex, France
govaert@utc.fr

Abstract. When partitioning the data is the main concern, it is implicitly assumed that each cluster can be approximately regarded as a sample from one component of a mixture model. Thus, the clustering problem can be viewed as an estimation problem of the parameters of the mixture. Setting this problem under the Maximum likelihood and Classification likelihood approaches, we first study the clustering of objects described by categorical attributes using the latent class model and we concentrate our attention on the problem of the number of components. To this end, we use three criteria derived within a Bayesian framework to tackle this problem. These criteria based on approximations of integrated likelihood and of integrated classification likelihood have been recently compared in Gaussian mixture. In this work, we propose to extend these comparisons to the latent class model.

1 Introduction

Most of the clustering methods used in practice are based on a distance or a dissimilarity measure and most of them available commercially are also of this type. However, basing cluster analysis on mixture models has become a classical and powerful approach (see for instance, McLachlan and Basford (1988), Banfield and Raftery (1993) and Celeux and Govaert (1992), (1995)). Thus, some available softwares, such as EMMIX (McLachlan and Peel (1998)) and MCLUST/EMCLUST (Fraley and Raftery (1999)), are based on this approach. Even if these methods are often focussed on continuous data, recently the clustering of categorical data has attracted much attention. We cite AUTOCLASS developed by Cheeseman and Stutz (1996), which became a popular tool in the data mining field where the clustering is an important process. Note that, to cluster categorical data, the authors used the latent class model for its simplicity and its performance.

One of major problems of interest in mixture models concerns the choice of the number of components. In recent years, a number of new criteria have been suggested, some of which, like the Cheeseman-Stutz (1996) criterion (CS) and the integrated classification likelihood criterion (ICL) proposed by Biernacki et al. (2000) have given encouraging results in Gaussian mixture context. In this paper, we discuss these criteria, along with the standard Bayesian information criterion (BIC) in the case of categorical data using the latent class model.

The paper is organized as follows. In Section 2, we present the mixture model and its relations with the cluster analysis. Section 3 is devoted to latent class model used for the categorical data. In Section 4, we briefly review some criteria derived within a Bayesian framework for assessing the number of clusters. In Section 5, we report some numerical experiments studying the behavior of these criteria.

2 Finite mixture model and clustering

In the mixture approach, it is assumed that the objects $\mathbf{x}_1, \ldots, \mathbf{x}_n$ to be clustered are from a mixture of a specified g of densities in some unknown proportions $\pi_1, \ldots, \pi_g$. That is, each object $\mathbf{x}_i$ is taken to be a realization of the mixture probability density function (p.d.f.),

$$f(\mathbf{x}_i; \theta) = \sum_{k=1}^{g} \pi_k \varphi_k(\mathbf{x}_i; a_k). \tag{1}$$

where $\varphi_k(\mathbf{x}_i; a_k)$ denotes the density of $\mathbf{x}_i$ from the kth component and a_k is the corresponding parameter. Here, the vector θ of unknown parameters consists of the mixing proportions $(\pi_1, \ldots, \pi_g) = \pi$ and the parameters of each component $(a_1, \ldots, a_g) = \mathbf{a}$. From the observed data $\mathbf{x}_1, \ldots, \mathbf{x}_n$, the log likelihood is given by

$$L(\mathbf{x}, \theta) = \log f(\mathbf{x}; \theta) = \sum_{i=1}^{n} \log \left(\sum_{k=1}^{g} \pi_k \varphi_k(\mathbf{x}_i; a_k) \right).$$

2.1 The complete data

In the clustering context, each $\mathbf{x}_i$ is conceptualized as having arisen from one of the components of the mixture model (1) being fitted, $\mathbf{z}_i$ is a g-dimensional vector with $\mathbf{z}_{ik} = 1$ or 0, according to whether $\mathbf{x}_i$ did or did not arise from the kth component. The complete data is therefore declared to be $(\mathbf{x}_1, \mathbf{z}_1), \ldots, (\mathbf{x}_n, \mathbf{z}_n)$. From this formulation the complete data log likelihood is given by

$$L(\mathbf{x}, \mathbf{z}; \theta) = \log f(\mathbf{x}, \mathbf{z}; \theta) = \sum_{i=1}^{n} \sum_{k=1}^{g} z_{ik} \log(\pi_k \varphi_k(\mathbf{x}_i; a_k)).$$

2.2 The maximum likelihood and classification approaches

In order to find an optimal partition $\hat{\mathbf{z}} = (\hat{\mathbf{z}}_1, \ldots, \hat{\mathbf{z}}_g)$, we used two approaches: the maximum likelihood (ML) approach and the classification maximum likelihood (CML) approach that we review briefly. The ML approach (Day (1969)) consists in estimating the parameters of the mixture and the partition is derived from these parameters using the maximum a posteriori principle (MAP). An iterative solution is provided by the expectation-maximization (EM) algorithm of Dempster et al. (1977). In the E-step, we compute the a posteriori probabilities $t_{ik} = \pi_k f(\mathbf{x}_i; \mathbf{a}_k)/\sum_{\ell=1}^{g} \pi_\ell f(\mathbf{x}_i; \mathbf{a}_\ell)$ and in the M-step, we compute $\theta^{(c+1)} = (\pi^{(c+1)}, \mathbf{a}^{(c+1)})$ that maximizes the conditional expectation $\mathbf{E}[L(\mathbf{x}, \mathbf{z}; \theta)|\mathbf{x}; \theta^{(c)}]$.

The second approach (Symons (1981)), sometimes called classification approach is based on the complete data. With this approach, θ and the unknown component-indicator vectors $\mathbf{z}_1, \ldots, \mathbf{z}_n$ of the observed data $\mathbf{x}_1, \ldots, \mathbf{x}_n$ are chosen to maximize the complete data log likelihood $L(\mathbf{x}, \mathbf{z}; \theta)$. This optimization can be done by the Classification EM (CEM) algorithm, a variant of EM, which converts the probabilities t_{ik}'s to a discrete classification in a Classification step before performing the Maximization step. Note that when the data are continuous, using the Gaussian mixture model, the standard k-means can be shown to be a simple version of the CEM algorithm. Simplicity, fast convergence and the possibility to process large data sets are the major advantages of the CEM algorithm. In our opinion, in several situations, it is interesting to start EM with the best partition obtained with CEM. This is a way to combine the advantages of both algorithms.

3 Categorical data and latent class model

The background of the latent class model, which was proposed by Lazarfeld and Henry (1968) is the existence of a latent categorical variable. In this model, the association between any pair of variables should disappear once the latent variable is held constant. This is the basic model of latent class analysis with its fundamental assumption of local independence. This hypothesis is commonly chosen when the data are categorical or binary (Celeux and Govaert (1992), Cheeseman and Stutz (1996)). That is, the density (frequency function) of an observation $\mathbf{x}_i$ can be expressed as

$$f(\mathbf{x}_i, \theta) = \sum_{k=1}^{g} \pi_k \prod_{j=1}^{m} \prod_{s=1}^{r_j} (a_{kjs})^{x_i^{js}} (1 - a_{kjs})^{1-x_i^{js}}, \text{ with } \sum_{s=1}^{r_j} a_{kjs} = 1$$

where every attribute $j = 1, \ldots, m$ is categorical, having a finite number of states r_j and $x_i^{js} = 1$ if the $j s$th state is observed and 0 otherwise. The assumption of local independence, sometimes called the naive Bayes, permit to estimate the parameters separately; this hypothesis greatly simplifies the computing, especially when the number of attributes is large. Although this assumption is clearly false in most real data, naive Bayes often performs clustering very well. This paradox is explained by Domingos and Pazzani (1997).

4 Assessing the number of components

Consider the problem of comparing a collection of models $M_1, ..., M_K$ where M_g is the latent class model with g components $(1 \leq g \leq K)$. Estimating the number of components is a difficult problem for which several approaches are in competition. Here, we consider this problem from a Bayesian perspective. Among the criteria proposed to this end, we will describe and use some criteria BIC, CS and ICL. The two first are based on an approximation of the integrated likelihood and the last is based on an approximation of the integrated classification likelihood. Next, we review these criteria that can be applied with the latent class model.

From the observed data, the integrated likelihood is given by

$$f(\mathbf{x}|M_g) = \int f(\mathbf{x}|M_g;\theta)p(\theta|M_g)d\theta \tag{2}$$

with $f(\mathbf{x}|M_g;\theta) = \prod_{i=1}^{n} f(\mathbf{x}_i|M_g;\theta)$ and $p(\theta|M_g)$ is the prior density for θ. A classical way to approximate (2) consists in using the Bayesian information criterion (BIC) (Schwarz (1978))

$$BIC(M_g) = \log f(\mathbf{x}|M_g;\hat{\theta}) - \frac{d}{2}\log n \approx \log f(\mathbf{x}|M_g)$$

where $\hat{\theta}$ is the maximum of likelihood estimation (MLE) of θ, d is the number of the parameters to be estimated, and here, it is equal to $k(\sum_{j=1}^{m} r_j - m + 1) - 1$. The BIC criterion is interesting for several reasons. It depends neither on the prior (see for instance Kass and Raftery (1995)) nor on the coordinate system of the parameters. This approximation is quite intuitive. The second term on the right hand-side penalizes the complexity of the model. Other approximations are used to choose the number of clusters, such as the CS criterion implemented in the Autoclass software (see also Chickering and Heckerman (1997)). This criterion is given by

$$CS(M_g) = L(\mathbf{x},\hat{\theta}) - n\sum_{k=1}^{g}\hat{p}_k \log \hat{p}_k - \frac{d_1}{2}\log n + K(n\hat{p}_1, \dots, n\hat{p}_g)$$

with

$$K(n\hat{p}_1, \dots, n\hat{p}_g) = \sum_{k=1}^{g} \log \Gamma(n\hat{p}_k + \frac{1}{2}) - \log(n + \frac{g}{2}) - g\log\Gamma(\frac{1}{2}) + \log\Gamma(\frac{g}{2}),$$

where $\Gamma(.)$ is the Gamma function. The parameter $d_1 = k(\sum_{j=1}^{m} r_j - m)$ denotes the number of unknown parameters in $\mathbf{a}$.

To take into account the ability of the mixture model to give evidence for a clustering structure of the data, Biernacki et al. (2000) considered the integrated classification likelihood

$$f(\mathbf{x}, \mathbf{z}|M_g) = \int f(\mathbf{x}, \mathbf{z}|M_g; \theta)p(\theta|M_g)d\theta$$

and to approximate this expression, they proposed the ICL criterion

$$ICL(M_g) = L(\mathbf{x}, \hat{\mathbf{z}}; \hat{\theta}) - \frac{d}{2}\log n$$

where $\hat{\mathbf{z}}$=MAP($\hat{\theta}$). As pointed by the authors, the ICL criterion has a close link with the BIC criterion. Indeed, it can be expressed as

$$ICL(M_g) = BIC(M_g) - E(\hat{t})$$

where $\hat{t} = (\hat{t}_{11}, \ldots, \hat{t}_{ng})$; $\hat{t}_{ik}$ is the estimation of a posteriori probability with $\hat{\theta}$ and $E(\hat{t}) = -\sum_{i=1}^{n}\sum_{k=1}^{g} \hat{t}_{ik} \log \hat{t}_{ik}$ is the entropy term which measures the overlap of the mixture components.

4.1 Monte Carlo experiments and Discussion

In all experiments, the clustering have been derived from the MLE of θ obtained with the EM algorithm which is initiated in the following way: first, the CEM algorithm is ran r times (here, we took r=20) from random centers. When it provided no empty cluster partition, the EM algorithm is initialized with the parameter values derived from this partition. On the other hand, when the CEM algorithm provided partitions with at least one empty cluster the EM algorithm is initiated r times with random centers. To illustrate the behavior of the criteria CS, ICL and BIC, we studied their performances on some simulations data sets consisted of 500×20, 1000×20 and 5000×20 values and generated according to the latent class models with $k = 3$, $r_j = 4$ for $j = 1, \ldots, m$ and in varying parameters according to the following scheme:

- the proportions are supposed equal ($\pi = (0.33, 0.33, 0.34)$) or different ($\pi = (0.10, 0.30, 0.60)$),
- the clusters are supposed well separated (+), moderately separated (++) or ill-separated (+++). These situations depend on the parameter $\mathbf{a}$ not reported here and which is chosen such as the proportions of misallocated objects by comparing the partition we obtained with the simulated one belong to $[6\%, 8\%], [11\%, 13\%]$ or $[16\%, 18\%]$.

For each Monte Carlo simulations, we generated 20 samples from each type simulated data and we compute the mean value of each criterion CS, BIC and ICL. In our experiments, we have seen that CS and BIC gave the same results. Then, we decide to report only the results on CS and ICL in Figures 1, 2 and 3. And, when we comment CS, we refer to the couple (CS, BIC).

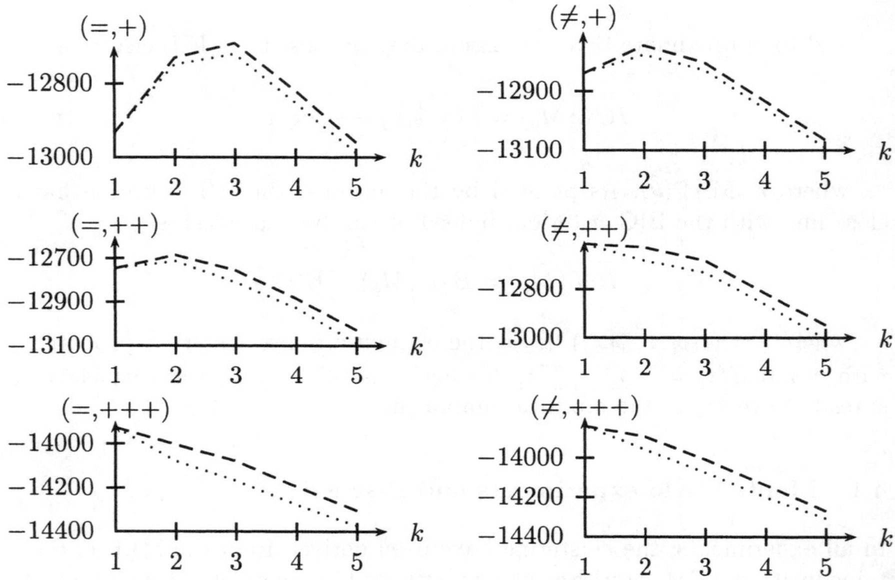

Fig. 1. $n = 500$, $d = 20$: behavior of CS (dashed line) and ICL (dotted line) for proportions equal (=) or different (≠) and clusters well separated (+), moderately separated (++) and ill-separated (+++).

The performances of CS and ICL to find the right number of components depend on the number of objects (n), on the proportions (π_k) and on the confusion degree of clusters. From the results, we deduce the following statements:

- CS and ICL underestimate the number of clusters when $n = 500$ but this deficiency disappears when n increases,
- ICL criterion works particularly better when the proportions are supposed equal,
- even if the proportions are dramatically different, CS criterion gives good results,
- the more the clusters are ill-separated, the more CS and ICL criteria encounter difficulties to find the right k.

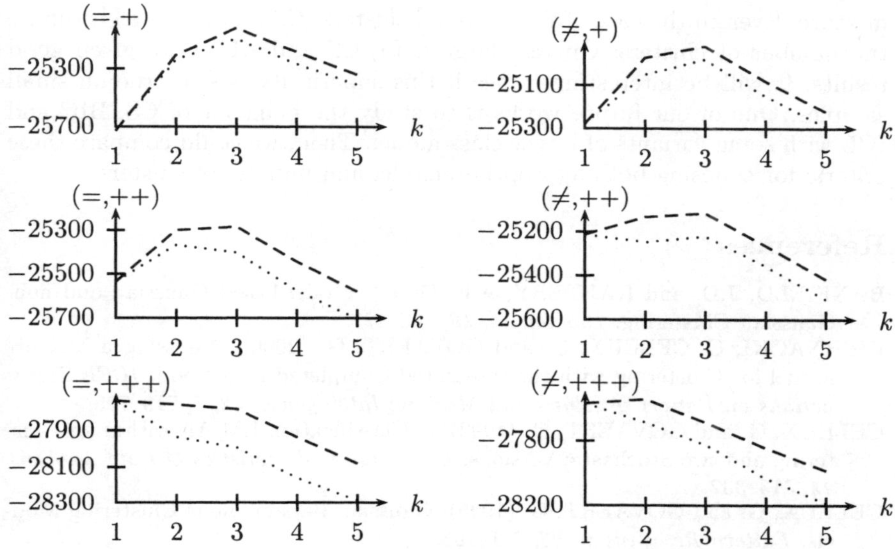

Fig. 2. $n = 1000$, $d = 20$: behavior of CS (dashed line) and ICL (dotted line).

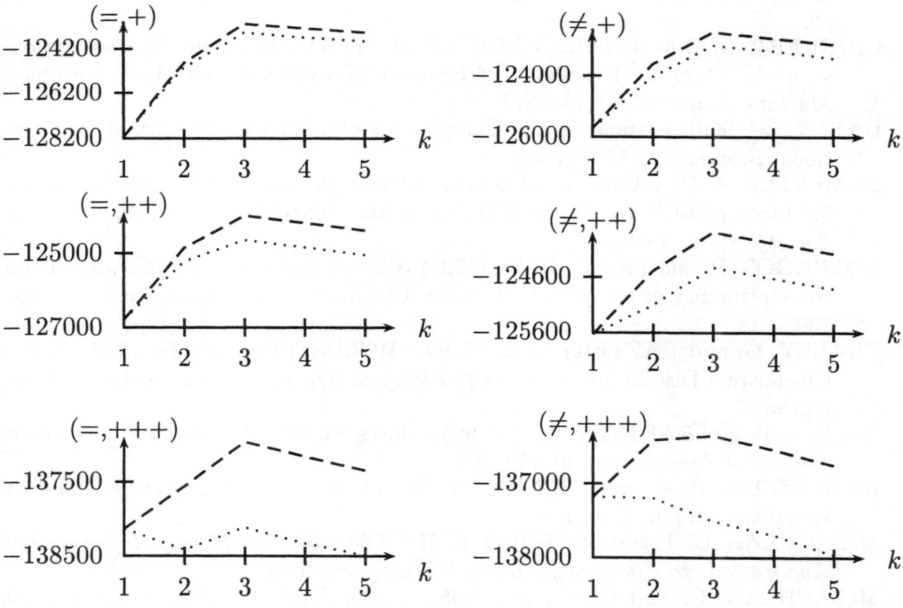

Fig. 3. $n = 5000$, $d = 20$: behavior of CS (dashed line) and ICL (dotted line).

Our experiments permit us to deduce that CS and BIC seem to outperform greatly ICL for the latent class model, contrary to the case of Gaussian mixture. Even in the case of ill-separated clusters, CS and BIC well estimate the number of clusters. On real large data, CS and BIC have given good results. It will be interesting to see if this superiority is still true on small samples. One of our future works is to study the behavior of CS, BIC and ICL with some variants of latent class model. Then, we could compare these criteria for choosing both appropriate model and number of clusters.

References

BANFIELD, J.D. and RAFTERY, A.E. (1993): Model-based Gaussian and non-Gaussian Clustering. *Biometrics, 49*, 803–821.

BIERNACKI, C. CELEUX, G. and GOVAERT, G. (2000): Assessing a Mixture Model for Clustering with the Integrated Completed Likelihood. *IEEE Transactions on Pattern Analysis and Machine Intelligence, 22*, 7, 719–725.

CELEUX, G. and GOVAERT, G. (1992): A Classification EM Algorithm for Clustering and two Stochastic Versions. *Computational Statistics & Data Analysis, 14*, 315–332.

CELEUX, G. and GOVAERT, G. (1995): Gaussian Parsimonious Clustering Models. *Pattern Recognition, 28*, 781–793.

CHEESEMAN, P. and STUTZ, J. (1996): Bayesian Classification (AutoClass): Theory and Results. In Fayyad, U. and Pitesky-Shapiro, G. and Uthurusamy, R. (Eds.): *Advances in Knowledge Discovery and Data Mining.* AAAI Press, 61–83.

CHICKERING, D.M and HECKERMAN, D. (1997): Maximum Approximations for the Marginal Likelihood of Bayesian Networks with Hidden Variables. *Machine Learning, 29*, 181–212.

DAY, N.E. (1969): Estimating the Components of a Mixture of Normal Distributions. *Biometrika, 56*, 464–474.

DEMPSTER, A. P., LAIRD, N. M. and RUBIN, D. B. (1997): Maximum Likelihood for Incomplete Data via the EM Algorithm. *Journal of the Royal Statistical Society, 39*, B, 1–38.

DOMINGOS, P. and PAZZANI, M. (1997): Beyond Independence: Conditions for the Optimality of the Simple Bayesian Classifier. *Machine Learning, 29*, 103–130.

FRALEY, C. and RAFTERY, A.E. (1999): MCLUST: Software for Model-Based Cluster and Discriminant Analysis. *Technical Report*, 342, University of Washington.

KASS, R. E. and RAFTERY, A. E. (1995): Bayes Factors. *Journal of the American Statistical Association, 90*, 773–795.

LAZARFELD, P. F. and HENRY, N. W. (1968): *Latent Structure Analysis.* Houghton Mifflin, Boston.

McLACHLAN, G. J. and BASFORD, K. E. (1988): *Mixture Models: Inference and Applications to Clustering.* Marcel Dekker, New York.

McLACHLAN, G. and PEEL, D. (1998): User's Guide to EMMIX-Version 1.0. *Technical Report*, University of Queensland.

SCHWARTZ, G. (1978): Estimating the Dimension of a Model. *Annals of Statistics, 6*, 461–464.

SYMONS, M. J. (1981): Clustering Criteria and Multivariate Normal Mixture. *Biometrics, 27*, 387–397.

Validation of Very Large Data Sets Clustering by Means of a Nonparametric Linear Criterion

Israel Lerman[1], Joaquim Pinto da Costa[2], and Helena Silva[3]

[1] IRISA, University of Rennes I,
France, e-mail: lerman@irisa.fr
[2] DMA/FCUP, Faculdade de Ciências, Universidade do Porto,
Portugal, e-mail: jpcosta@fc.up.pt
[3] ISEP, Universidade do Porto,
Portugal, e-mail: helenabras@clix.pt

Abstract. In this paper we present a linearization of the Lerman clustering index for determining the number of clusters in a data set. Our goal was to apply the linearized index to large data sets containing both numerical and categorical values. The initial index, which was based on the set of pairs of objects, had a complexity $O(n^2)$. In this work its complexity is reduced to $O(n)$, and so, we can apply it to large data sets frequently encountered in Data Mining applications. The clustering algorithm used is an extention of the k-means algorithm to domains with mixed numerical and categorical values (Huang (1998)). The quality of the index is empirically evaluated on some data sets, both artificial and real.

1 Introduction

Due to the size of the data sets often used in Data Mining, the time consuming to evaluate a clustering index is a crucial problem. Therefore, if n is the number of data points to be clustered, a complexity redution from $O(n^2)$ to $O(n)$ is a very important task. Our aim is to write the Lerman index in such a way that its complexity becomes $O(n)$. We will first begin by remembering the earlier Lerman index expression (Lerman 1973, 1981, 1983); after that we will present the general principle of the redution and then, we will supply the explicite adaptation in the case where the data points can have an euclidean representation. We will finish by extending the new index to the case of mixed categorical and numerical variables.

2 Classical form of the index

Let us consider the simplified classic formula of the Lerman index. Let E be a finite set of cardinal n and S a similarity measure defined on it. Let $F = P_2(E)$ represent the set of pairs or subset with two elements of E. The values of S can be defined by the table

$$\{S(p)|p \in F\} \tag{1}$$

of dimension $n(n-1)/2$.

Let $\pi(E)$ be a partition of E into k clusters

$$\pi(E) = \{E_1, E_2, \ldots, E_l, \ldots, E_k\}, \tag{2}$$

whose adequacy with the similarity measure S is to be evaluated.

Regarding the table (1), we can calculate its mean $\mu(S)$ and its variance $\sigma^2(S)$:

$$\mu(S) = \frac{2}{n(n-1)} \sum \{S(p) | p \in F\} \tag{3}$$

$$\begin{aligned} \sigma^2(S) &= \frac{2}{n(n-1)} \sum \left\{ [S(p) - \mu(S)]^2 \, | p \in F \right\} \\ &= \frac{2}{n(n-1)} \sum \left\{ [S(p)]^2 \, | p \in F \right\} - [\mu(S)]^2 \end{aligned} \tag{4}$$

On the other hand, the partition (2) can be represented as a subset of F:

$$R(\pi(E)) = \sum_{1 \leq l \leq k} P_2(E_l) \tag{5}$$

which defines the set of pairs put together by the partition $\pi(E)$. We can also designate by

$$S(\pi(E)) = \sum_{1 \leq l < l' \leq k} E_l * E_{l'} \tag{6}$$

the complementary set in F of $R(\pi)$. The expression $E_l * E_{l'}$ represents the set of non-ordered pairs $\{x, y\}$ where $x \in E_l$ and $y \in E_{l'}$, $1 \leq l < l' \leq k$.

Let us now designate by

$$r = card(R(\pi(E))) = \sum_{1 \leq l \leq k} \frac{n_l(n_l - 1)}{2} \tag{7}$$

and

$$s = card(S(\pi(E))) = \sum_{1 \leq l < l' \leq k} n_l \times n_{l'} \tag{8}$$

where $n_l = card(E_l), 1 \leq l \leq k$.

Before proceeding, the similarity S is normalized in the following way:

$$c(p) = \frac{S(p) - \mu(S)}{\sigma(S)} \tag{9}$$

In these conditions, the formula for the simplified form of the index is (see references above):

$$C(\pi, S) = \frac{1}{\sqrt{r \times s/f}} \sum \{c(p) | p \in R(\pi)\} \tag{10}$$

$$= \frac{1}{\sqrt{r \times s/f}} \sum \{\epsilon(p)c(p) | p \in F\} \tag{11}$$

where $f = r + s = card(F)$ and $\epsilon(p) = 1(p \in R(\pi))$.

3 Adaptation of the index for Euclidean data

Let us suppose that E can be represented by a set of points in an euclidean space. Let d be the metric distance, $I = \{1, 2, \ldots, i, \ldots, n\}$ the index set of E and I_l the subset of I containing the indexes of the cluster E_l, $1 \leq l \leq k$.

Instead of working with the table of similarity indices (1), we will consider from now on a table containing the distances between the elements of E:

$$\{d^2(i, i') | (i, i') \in I \times I\} \tag{12}$$

This table has dimension n^2. In these conditions, the expressions stated so far, which were based on unordered pairs, can be now expressed by using the ordered pairs.

$$\mu'(d^2) = \frac{1}{n^2} \sum \{d^2(i, i') | (i, i') \in I \times I\} \tag{13}$$

$$\sigma'(d^2) = \frac{1}{n^2} \sum \{\left[d^2(i, i') - \mu'(d^2)\right]^2 | (i, i') \in I \times I\} \tag{14}$$

The set of ordered pairs of objects that are clustered together is defined by:

$$R'(\pi(E)) = \sum_{1 \leq l \leq k} E_l \times E_l \tag{15}$$

and those pairs of objects that are in different clusters by:

$$S'(\pi(E)) = \sum_{1 \leq l \neq l' \leq k} E_l \times E_{l'} \tag{16}$$

In these conditions, we have:

$$r' = card(R'(\pi(E))) = \sum_{1 \leq l \leq k} n_l^2 \tag{17}$$

and

$$s' = card(S'(\pi(E))) = \sum_{1 \leq l \neq l' \leq k} n_l \times n_{l'} \tag{18}$$

For an ordered pair $q = (x, y)$, the normalized index related to $c(p)$ is:

$$c'(q) = \frac{d^2(x,y) - \mu'(d^2)}{\sigma'(d)} \tag{19}$$

and the index corresponding to (10) becomes:

$$C'(\pi, d^2) = \frac{1}{\sqrt{r' \times s'/n^2}} \sum \{c'(x,y) | (x,y) \in R'(\pi(E))\} \tag{20}$$

Let us remark that the index (20) can also be expressed by the formula:

$$C'(\pi, d^2) = \frac{1}{\sqrt{r' \times s'/n^2}} \sum_{1 \leq l \leq k} \sum \{c'(x,y) | (x,y) \in E_l \times E_l\} \tag{21}$$

4 Linear adaptation of the index for Euclidien data

We will start by expressing the basic formula that precisely allows the desired reduction in complexity. Let

$$\{M_i | i \in I\} \tag{22}$$

represent a cloud of n data points. We suppose, for simplification, that these points share the same weight; the generalization for different weights is immediate. Let G be the centroid of the cloud of points:

$$G = \frac{1}{n} \sum_{i \in I} M_i \tag{23}$$

It is known that

$$\frac{1}{n^2} \sum_{(i,i') \in I \times I} d^2(i, i') = \frac{2}{n} \sum_{i \in I} d^2(i, g) \tag{24}$$

where we replace M_i by i and G by g.

It is easy to discern that to calculate the left member takes $O(n^2)$ calculations of distances while the right member takes $O(n)$ calculations of distances.

The basic idea is then to replace the first two moments of the distribution

$$\{d^2(i, i') | (i, i') \in J \times J\} \tag{25}$$

by the first two moments of the distribution of

$$\{d^2(i, g_J) | i \in J\} \tag{26}$$

where g_J represents the centroid of the subset of data points indexed by J.

Now, in (21) let us consider the contribution of the cluster E_l for the value of the index. This is represented by:

$$\frac{1}{\sqrt{r's'/n^2}} \times \frac{1}{\sigma'(d^2)} \times \sum \{d^2(x, y) - \mu'(d^2) | (x, y) \in E_l \times E_l\} \tag{27}$$

But,

$$\sum \{d^2(x, y) - \mu'(d^2) | (x, y) \in E_l \times E_l\} = 2n_l \left\{ \sum_{x \in E_l} \left[d^2(x, g_l) - \frac{1}{n} \sum_{x \in E} d^2(x, g) \right] \right\} \tag{28}$$

Where g_l is the centroid of the cluster E_l. We remark that $d^2(x, g_l)$ is centered by the total moment of inertia. We can replace $\left[\sigma'(d^2)\right]^2$ by the variance of the distribution of $\{d^2(x,g)|x \in E\}$, which we designate by λ_g^2. The final expression is proportional to:

$$C_1(\pi, d^2) = \frac{1}{\sqrt{r's'}} \times \frac{1}{\lambda_g} \times \sum_{1 \leq l \leq k} n_l \sum \{d^2(x, g_l) - \mu_g | x \in E_l\} \tag{29}$$

where

$$\mu_g = \frac{1}{n} \sum_{x \in E} d^2(x,g)$$

and

$$\lambda_g^2 = \frac{1}{n} \sum_{x \in E} d^4(x,g) - \mu_g^2$$

5 The case of mixed numerical and categorical attributes

We are going to adapt our index for the metric case that is considered in (HUANG 1998) where the attributes may be numerical and/or categorical. Each one of the data objects can be characterized by $(v^1, \ldots, v^p, c^{p+1}, \ldots, c^m)$ where $v^1, v^2, \ldots, v^p$ represent the numerical attributes and $c^{p+1}, c^{p+2}, \ldots, c^m$ represent the categorical attributes $(m > p)$.

Here we consider $\{x_i | i \in I\}$ the data set $(I = \{1, 2, \ldots, n\})$. Let x_i^j be the value of the attribute j for the object x_i. This value is numerical if $1 \leq j \leq p$ and is categorical if $p+1 \leq j \leq m$. The distance between x_i and $x_{i'}$ can be evaluated using the following formula:

$$d^2(x_i, x_{i'}) = \sum_{1 \leq j \leq p} \left(x_i^j - x_{i'}^j\right)^2 + \sum_{p+1 \leq j \leq m} \delta\left(x_i^j, x_{i'}^j\right) \tag{30}$$

where $\delta(x_i^j, x_{i'}^j) = \begin{cases} 0 \; if \; x_i^j = x_{i'}^j \\ 1 \; if \; x_i^j \neq x_{i'}^j \end{cases}$ for $p+1 \leq j \leq m$

It is shown in (HUANG 1998) that the centroid of any set X is given by:

$$g_X = \left(g_X^1, \ldots, g_X^h, \ldots, g_X^p, f_X^{p+1}, \ldots, f_X^{p+j}, \ldots, f_X^m\right) \tag{31}$$

where g_X^h is the mean of the component h in X, $1 \leq h \leq p$, and where f_X^{p+j} represents the mode of the categorical variable $p+j$ (the most frequent category in X), $1 \leq j \leq m - p$.

Under these conditions, the adaptation of the index is immediate.

6 Application of the index

In this section the quality of the index is empirically evaluated on seven data sets. We tested on two artificial examples with two data sets each, and on three real data sets.

For each artificial example we have generated two data sets with 20,000 objects each. One of the data sets was defined on a bidimensional euclidean space, represented in Figures 1 and 5, and it contains five clusters. The second data set consisted on adding to the previous variables, four categorical attributes, having each four values. We determine in a random way the same distribution of these attributes on each cluster. But this common distribution of the four attributes is different from one cluster to another one. Thus, the probability distributions were the same for each categorical variable within a given cluster, but were different for different clusters. In Figure 2 and 7 it can be seen the five clusters identified by the k-prototype method (Huang 1998). The minimum value for our index is for $K = 5$ in both data sets of the first example , as can be seen in Figures 3 and 4. In Figures 8 and 9 it can be seen the variation of the index for the two data sets of the second example. For the mixed data set the number of clusters is correctly identified, whereas for the euclidean data set we got the minimum for $K = 3$ due to a weak efficience of the k-prototype identifying three and five clusters as it can be seen in Figure 6 and Figure 7.

In figures 10, 11 and 12 we can analyze the application of our index in the case of real data sets. These data sets belong to the Stalog database that are a subset of the datasets used in the European Statlog project. The data set , whose index values are represented in Figure 10 (Australian Credit Approval), concerns credit card applications. It has 690 objects with six numerical and eight categorical attributes and has 2 clusters. The data set, whose index values are represented in Figure 11, concerns image segmentation (Image Segmentation data). The instances were drawn randomly from a database of seven outdoor images, and the images were segmented to create a classification for every pixel. This data set has 2,310 objects with nineteen numerical attributes and has seven clusters. Finally in Figure 12 are the values of the index related to a data set (Shuttle Dataset) with 43, 500 objects with nine numerical attributes each and seven clusters.

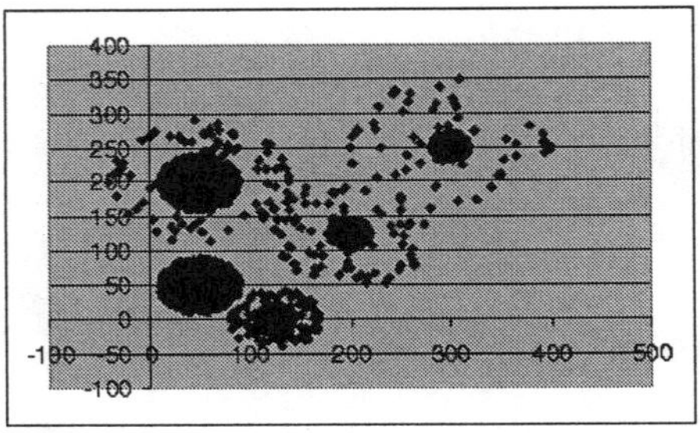

Fig. 1. Data set with 20,000 objects (two numerical attributes).

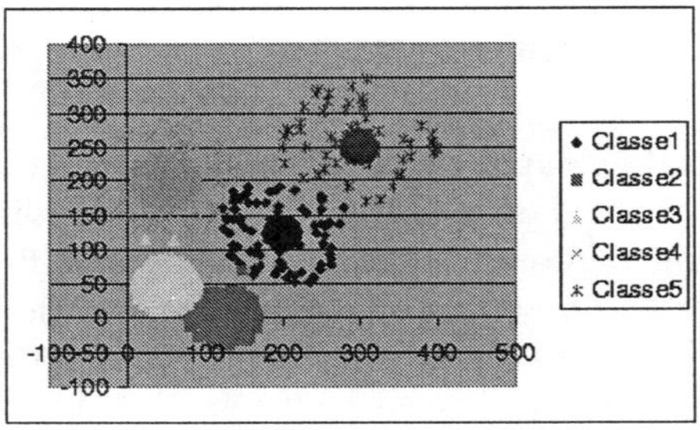

Fig. 2. Partition in five clusters identified by the k-prototype method, applied to the data set of Figure 1.

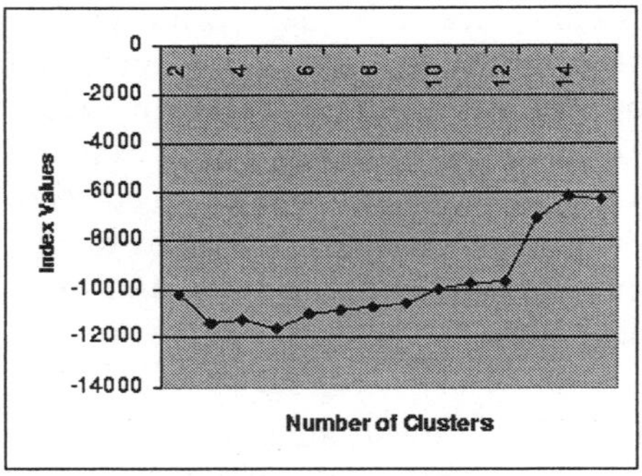

Fig. 3. Values of our index ($K = 2, \ldots, 15$) for the data set represented in Figure 1.

7 Conclusions and Future Work

We have developped a criterion for the identification of the number of clusters present in a data set; and it can be applied to both numerical and categorical variables. This criterion has a linear computational complexity, which makes it very interesting to use in large data sets, as is common in Data Mining. We have seen empirically the importance of the new criterion on seven data sets, four artificial and three real ones. The criterion has been applied in con-

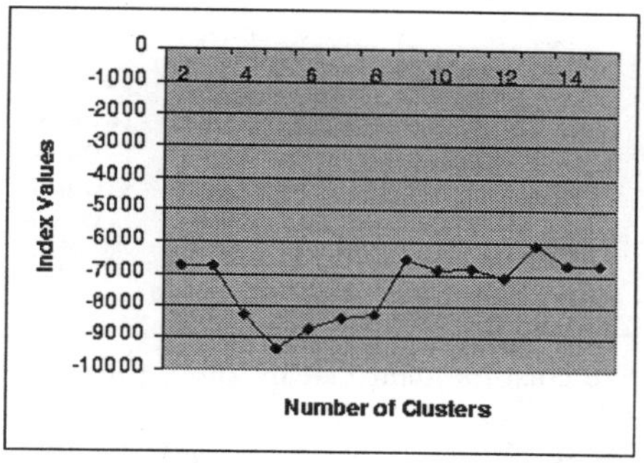

Fig. 4. Values of our index ($K = 2, \ldots, 15$) for the data set with two numerical and four categorical attributes and five clusters.

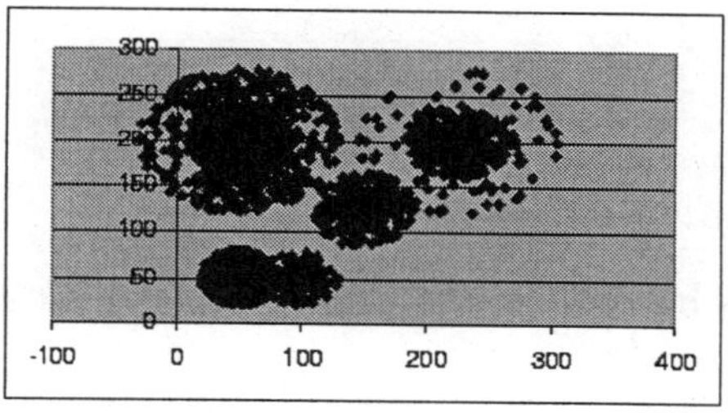

Fig. 5. Data set with 20,000 objects (two numerical attributes).

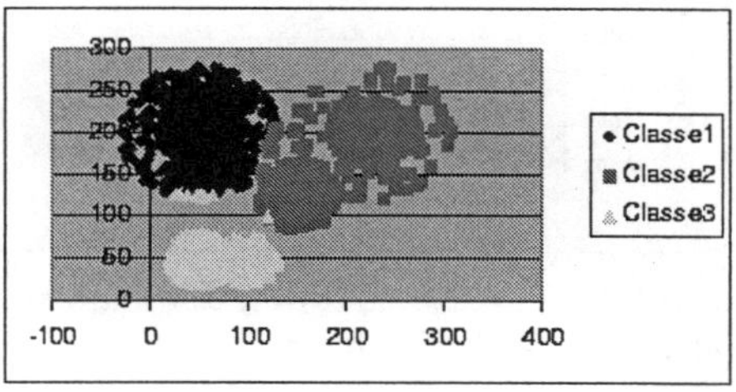

Fig. 6. Partition in three clusters identified by the k-prototype method, applied to the data set of Figure 5.

junction with the k-prototypes algorithm (Huang 1998), and because of that, some interesting partitions that could have been identified by our criterion were not, because k-prototypes didn't find them. We are developping a new clustering method that incorporates the new criterion into the construction of the partitions, and hope that in this way the above disadvantage will be solved. We are also planning to consider the adaption of the index for unstructured data, because as it is now it can not be applied to the case of just one cluster. Finally, in the case of numerical data, the comparison of the behaviour of our criterion with some classical ones (Milligan and Cooper 1985) will be studied. For one of the most important of these, the "Cubic

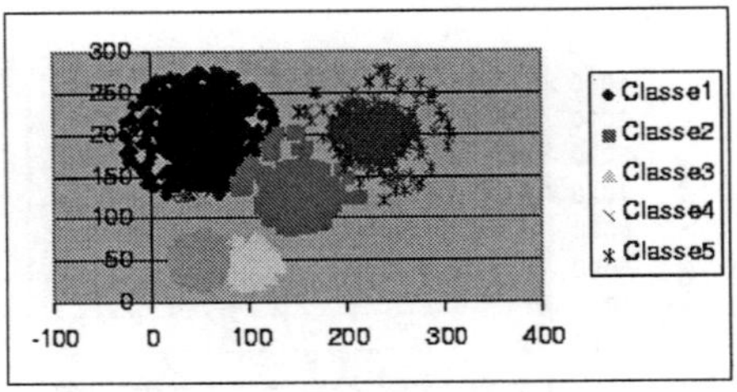

Fig. 7. Partition in five clusters identified by the k-prototype method, applied to the data set of Figure 5.

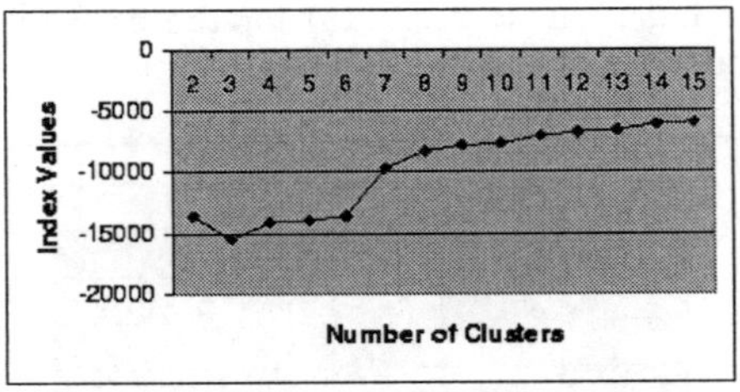

Fig. 8. Values of our index ($K = 2, \ldots, 15$) for the data set represented in Figure 5.

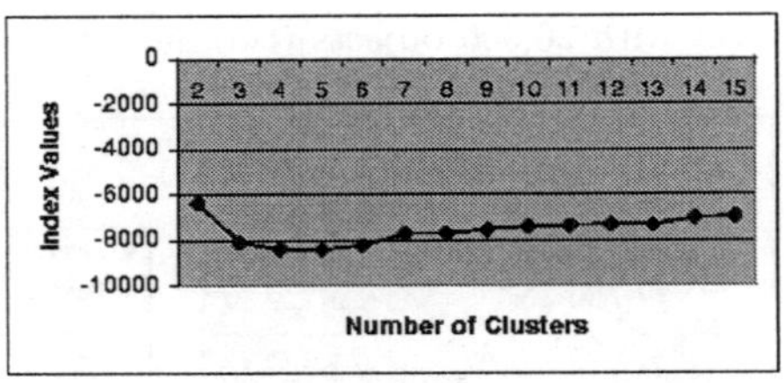

Fig. 9. Values of our index ($K = 2, \ldots, 15$) for the data set with two numerical and four categorical attributes and five clusters.

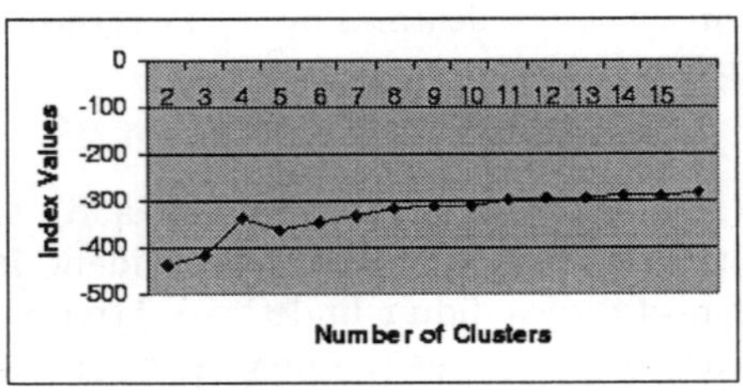

Fig. 10. Values of our index ($K = 2, \ldots, 15$) for the Australian Credit Approval data set.

Clustering Criterion (CCC)" such comparison has been performed relative to the initial quadratic form of our criterion (Mollière 1986).

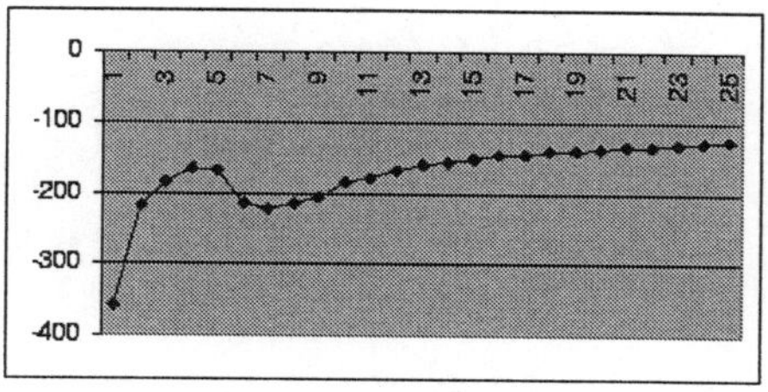

Fig. 11. Values of our index ($K = 2, \dots, 25$) for the Image Segmentation data set.

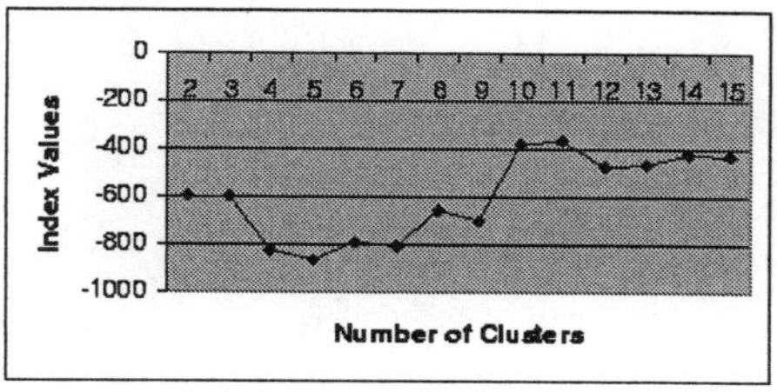

Fig. 12. Values of our index ($K = 2, \dots, 15$) for the Shuttle Dataset.

Acknowledgements

We thank Ross D. King for providing us with the real data sets.

References

HUANG, Z. (1998): Extensions to the K-Means Algorithm for Clustering Large Data Sets with Categorical Values. *Data Mining and Knowledge Discovery 2*, 283–304.

LERMAN, I.C.(1973): Étude distributionnelle de statisques de proximité entre structures finies de même type; application à la classification automatique. *Cahiers du Bureau Universitaire de Recherche Opérationnelle, 19*, Paris.

LERMAN, I.C.(1981): Classification et Analyse Ordinale des Données. Dunod, Paris.

LERMAN, I.C.(1983): Sur la signification des Classes Issues d'une Classiifcation Automatique de Données. In: J. Felsenstein (Eds.): *NATO ASI Series, Vol G1 Numerical Taxaromy.* Springer-Verlag.

MILLIGAN, G.W. and COOPER, M.C. (1985): An examination of procedures for determining the number of clusters in a data set. *Psychometrika*, 50, 159-179.

MOLLIÈRRE, J.L. (1986): What's the real number of clusters. In: W. Gaul and M. Schader (Eds): *Classification as a tool research.*. North Holland.

Discrimination

Effect of Feature Selection on Bagging Classifiers Based on Kernel Density Estimators

Edgar Acuña[1], Alex Rojas[2], and Frida Coaquira[3]

[1] Department of Mathematics, University of Puerto Rico at Mayaguez, Mayaguez, PR 00680, edgar@math.uprm.edu
[2] Department of Statistics, Carnegie Mellon University, Pittsburgh, PA 00680, arojas@stat.cmu.edu
[3] Department of Mathematics, University of Puerto Rico at Mayaguez, Mayaguez, PR 00680, frida_cn@math.uprm.edu

Abstract. A combination of classification rules (classifiers) is known as an *Ensemble*, and in general it is more accurate than the individual classifiers used to build it. One method to construct an *Ensemble* is *Bagging* introduced by Breiman, (1996). This method relies on resampling techniques to obtain different training sets for each of the classifiers. Previous work has shown that *Bagging* is very effective for unstable classifiers. In this paper we present some results in application of *Bagging* to classifiers where the class conditional density is estimated using kernel density estimators. The effect of feature selection in bagging is also considered.

1 Introduction

Many researches have investigated the technique of combining the predictions of multiple classifiers to produce a single classifier: Wolpert (1992), Breiman (1996, 1998), Quinlan (1996), Freund and Schapire (1996), Maclin and Optiz (1997), Bauer and Kohavi (1999) among others. The resulting classifier, also known as an *Ensemble*, is generally more accurate than the individual classifiers making up the *Ensemble*. Three popular methods for creating ensembles are: Bagging (Bootstrap aggregating) introduced by Breiman (1996), AdaBoosting by Freund and Schapire (1996) and Arcing (Adaptively resampling and combining) by Breiman (1998). These methods rely on resampling techniques to obtain different training sets for each of the classifiers. Stacking (Wolpert, 1992) is another method to combine classifiers but it does not use resampling, in this case the learning data set is divided in v parts similar to v-fold cross-validation and then a classifier is estimated in each of the part. The stacked classifier is a linear combination of the single classifiers.

Breiman (1996) heuristically defines a classifier as unstable if a small change in the learning data L can make large changes in the classification. Unstable classifiers have low bias but high variance, meanwhile the opposite occurs for stable classifiers. CART and neural networks are not stable classifiers, linear discriminant analysis and K-nearest neighbor classifiers are stable. It is expected a reduction of the bias and variance after the classifiers are combined.

Reference	Classifier	Relative Error Reduction (%)	# of datasets (w-l-t)
Breiman (1996)	CART	29.0	7 (7-0)
Freund & Schapire (1996)	C4.5	20.0	27(22-2-3)
Quinlan (1996)	C4.5	10.0	27(24-3)
Maclin & Opitz (1997)	C4.5	18.5	23(21-1-1)
Maclin & Opitz (1997)	Neural Nets	13.3	23(22-0-2)
Breiman (1998)	CART	36.0	11(11-0)
Bauer & Kohavi (1999)	MC4	14.5	14(14-0)
Dietterich (2000)	C4.5	16.5	33(30-0)

Table 1. Results of previous experiments in Bagging

Bagging and Adaboosting are very effective for unstable classifiers such as decision trees: CART, C4.5 and MC4 (see Breiman (B96, B98), Quinlan (Q96), Freund and Schapire (FS96), Bauer and Kohavi (BK99)) and neural networks (see Maclin and Optiz (MacO97)). A summary of previous results for Bagging is shown in Table 1. Adaboosting applied to decision trees and Naïve-Bayes performs generally better that Bagging, but not uniformly better, sometimes they degraded compared to the single classifier. The same conclusions were obtained for neural networks classifiers (MacO97). When tree classifiers are used, Bagging mainly reduces the variance, whereas boosting reduces, both the bias and the variance (B98 and BK99).

In the Table 2, we show the misclassification errors of applying Bagging to the ten datasets used in this paper (see also Table 3 in section 3). Note that in two of these datasets: *Vehicle* and *Iris* the best single classifier beats the combining techniques. In the *breastw* dataset the improvement is minimum and in the remaining seven, the combining techniques do a good job.

2 Classifiers based on kernel density estimators

From a Bayesian point of view, supervised classification is equivalent to compare estimates of the probabilities of belonging to each class with each other, assigning an object with measurement vector $\mathbf{x}$ to the class with the largest $\hat{f}(\mathrm{j}/\mathbf{x})$, j=1,2 .... .J. In order to obtain such estimates, one can estimate them indirectly via the class conditional density f($\mathbf{x}$/j) using the Bayes' theorem. Kernel density estimators can be used to carry out that task. For a given class j and a random sample $\mathbf{X}_1,\mathbf{X}_2,\ldots..\mathbf{X}_n$ of the p-dimensional random vector $\mathbf{X}$ with continuos components, the product kernel of the class conditional density at the point $\mathbf{x}$ is given by

$$\hat{f}(\mathbf{x}) = \frac{1}{nh_1h_2....h_p}\sum_{i=1}^{n}\prod_{v=1}^{p}K(\frac{\mathrm{x}_v-\mathbf{X}_{iv}}{h_v}) \tag{1}$$

where the kernel K will be usually a radially symmetric unimodal density function, for instance the multivariate normal density, and h_v represents the

Dataset	Q96 C4.5	FS96 C4.5	B96/98* CART	Mac097 C4.5	D00 C4.5	Best Single Classifier
Iris	5.13	5.0	NA	4.6	5.00	2.0 LDA
Sonar	23.80	24.3	NA	21.6	27.52	15.5 1-NN
Glass	27.01	25.7	24.05	28.4	27.23	23.8 1-NN
Heartc	21.52	20.9	NA	18.1	19.81	18.2 NN
Ionosphere	NA	6.2	8.25	6.0	NA	8.1 C4.5
Breastw	4.23	3.2	3.95	3.3	3.67	3.3 NN
Diabetes	23.63	24.4	21.35	21.9	NA	22.3 Log
Vehicle	25.54	26.1	NA	26.1	25.70	15.0 QDisc
German	25.81	24.8	NA	22.8	24.95	28.3 NN
Segmentation	2.74	2.7	NA	2.8	2.63	3.0 Ker

Table 2. Comparison of previous results of experiments on Bagging with the best single classifier for the data sets used in this paper.(*) Average of both studies, LDA: Linear discriminant analysis, NN: Neural Network,1-NN: 1-Nearest neighbour, Ker: Kernel classifier, Log: Logistic regression, Qdisc: Quadratic discrimination.

bandwidth of the v-th predictor. There are several approaches to select the optimal bandwidth.

In the Statlog Project (Michie, et. al. 1994), where 23 classifiers are compared in 22 datasets, classifiers based on kernel density estimators (ALLOC80) performed better than CART (a 13-8 victory) and tied with C4.5 (11-11). However, ALLOC80 appeared as the top 5 classifier for 11 datasets whereas C4.5 and CART appeared only for 5 and 3 datasets respectively. Classifiers based on kernel density estimation are unstable due to two reasons, first the presence of singularities in the log-likelihood function, and second the effect of outliers in the selection of the bandwidth.

In this paper we have used kernel density estimation with both fixed and adaptive bandwidth. The standard kernel density estimator uses a fixed bandwidth obtained by assuming that the class conditional density is a multivariate normal (see Silverman, p. 86). This is, $h_{opt} = (4/n(p+2))^{1/(p+4)}$, where p is the number of predictors and n is the number of instances. In the adaptive kernel, the bandwidth varies from one point to another, and its value will depend on the concentration level of data in the neighborhood of the point. The basic idea of this approach is to combine the standard kernel density estimation with the nearest neighbor approach (for more details see Silverman, p. 101).

We have also considered kernel density estimators for categorical predictors: binary, nominal and ordinal predictors. In the first case we have followed the Aitchison and Aitken's proposal (see Titterington, (1980)). That is, given n p-dimensional sample data points $\mathbf{x}_j$ the kernel density estimate of f at the point $\mathbf{x}$ is given by

$$\hat{f}(\mathbf{x}) = \frac{1}{n} \sum_{j=1}^{n} h^{p-d_j^2} (1-h)^{d_j^2} \tag{2}$$

where $\frac{1}{2} \leq h \leq 1$, and $d_j^2 = (\mathbf{x} - \mathbf{x}_j)'(\mathbf{x} - \mathbf{x}_j)$ is equivalent to the number of disagreements in corresponding components of $\mathbf{x}$ and $\mathbf{x}_j$. The smoothing parameter h has to be estimated, usually by cross validation. Titterington gave also a explicit formula for h and in addition he considered also kernel density estimates for nominal and ordinal variables. A kernel density estimator for a categorical variable X with k values and vector of sample proportions $\mathbf{r}$ can be written as $\hat{f}(x) = C'\mathbf{r}$, where C=I+(1-h)G is a square matrix of order k, and G is a matrix such that G_{ii}=-1, G$\mathbf{1}$'=$\mathbf{0}$ and $G_{ij} > 0$ for i$\neq$j. The smoothing parameter h is estimated by minimizing the mean squared error and, is given by

$$\hat{h} = 1 + \frac{tr(VG)}{tr(G'\mathbf{r}\mathbf{r}'G + VGG')} \tag{3}$$

where $V = n^{-1}(\Lambda - \mathbf{r}\mathbf{r}')$ and $\Lambda = diag(r_1, \ldots .r_k)$.

In the multivariate case if our vector of predictors $\mathbf{x}$ can be decomposed as $\mathbf{x}$=($\mathbf{x}^{(1)}$,$\mathbf{x}^{(2)}$), where $\mathbf{x}^{(1)}$ contains the p_1 categorical predictors and $\mathbf{x}^{(2)}$ includes the p_2 continuous predictors, then a mixed product kernel density estimator will be given by

$$\hat{f}(\mathbf{x}) = \frac{1}{nh_{21}, \ldots, h_{2p_2}} \sum K_{p_1}^{(cat)}(\mathbf{x}^{(1)}; \mathbf{x}_j^{(1)}, \mathbf{h}_1) K_{p_2}^{(cont)}\left(\frac{\mathbf{x}^{(2)} - \mathbf{x}_j^{(2)}}{h_2}\right) \tag{4}$$

where $K_{p_1}^{(cat)}$ is the kernel estimator for the vector $\mathbf{x}^{(1)}$ of categorical predictors and $K_{p_2}^{(cont)}$ is the kernel density estimator of the vector $\mathbf{x}^{(2)}$ of continuous predictors.

3 Experimental methodology

We chose 10 datasets coming from the Machine Learning Database at University of California Irvine (UCI) to evaluate the effect of combining KDE classifiers considering mixed type of predictors: continuous and categorical and fixed as well as adaptive bandwidth. In Table 3, we show the 10 datasets used in this paper including the number of instances, the number of classes and the number of each type of feature.

For each dataset we have performed the following procedure: The dataset is randomly divided in 10 parts, the first one is taken as the test sample and the remaining is considered as the learning sample. Next, 50 bootstrapped samples are taking from the learning sample and a KDE classifier is constructed with each of them. Finally, each instance of the test sample is assigned

Dataset	Instances	Classes	Features			
			continuous	binary	nominal	ordinal
Iris	150	3	4	-	-	-
Sonar	208	2	60	-	-	-
Glass	214	6	9	-	-	-
Heartc	303	2	13	-		
Ionosphere	351	2	33	-	-	-
Breastw	699	2	9	-	-	-
Diabetes	768	2	8	-	-	-
Vehicle	846	4	18	-	-	-
German	1000	2	20			
Segmentation	2310	2	19	-	-	-

Table 3. Datasets used in this paper

to a class by voting using the 50 classifiers previously constructed. The proportion of instances incorrectly assigned will be the bagged misclassification error. We repeat the steps considering now the second part as the test set and in this way we continue until the tenth part is considered as the test set. The procedure is repeated 10 times. The misclassification error of a single classifier is estimated by a 10-fold cross-validation and averaged over 50 runs. The bagged misclassification error is averaged on 100 repetitions. We also computed the ratio of the misclassification errors of the bagged classifier versus the single one. The misclassification errors and the ratios are shown in the Table 4.

The average of the error reduction for the 10 datasets after Bagging using the standard kernel was 2.24% whereas for the adaptive kernel was 4.39%. When C4.5 classifier was bagged (Quinlan (1996)) the average error reduction for the same datasets was 8.83%. Notice that we get improvement with the adaptive kernel for all datasets but *Glass*, although is very slight in some cases.

4 Bagging after feature selection

To deal with the curse of dimensionality problem we perform feature selection. A forward selection procedure, that belongs to the family of wrappers methods of feature selection (see Bauer and Kohavi (1998)), was used and repeated 10 times for datasets with less than 20 features and 20 times for datasets with more than 20 features. First we select the single feature that produces the highest classification rate estimated by 10 fold cross-validation using the classifier based on kernel density estimator. Once that this is done we search for the second feature that together with the first one yields the highest classification rate. The procedure continues until the classification

Dataset	Standard kernel			Adaptive kernel		
	Single	Bagged	Ratio	Single	Bagged	Ratio
Iris	3.53	3.60	1.0198	4.47	4.26	0.9530
Sonar	17.18	17.21	1.0017	16.37	15.70	0.9591
Glass	44.57	43.83	0.9834	35.46	35.51	1.0014
Heartc	22.86	21.78	0.9528	22.04	21.04	0.9546
Heartc*	22.26	21.75	0.9771	22.55	20.13	0.8927
Heartc**	22.30	20.80	0.9327	21.56	19.86	0.9212
Ionosphere	10.93	10.48	0.9588	10.33	10.08	0.9758
Breastw	3.64	3.77	1.0357	3.94	3.80	0.9645
Diabetes	26.38	26.21	0.9636	26.26	25.72	0.9794
Vehicle	35.15	34.61	0.9846	36.74	33.90	0.9227
German	34.61	33.81	0.9769	35.06	33.74	0.9624
German*	27.61	27.11	0.9819	27.34	25.12	0.9188
German**	28.71	27.95	0.9735	28.28	26.51	0.9374
Segmentation	15.86	15.32	0.9660	13.41	13.32	0.9933

Table 4. Comparison of misclassification error rates for single and bagged KDE classifiers.(*) Here the features are considered only as continuous and binary.(**) Here all the features are considered as continuous.

rate decreases. After that we compute the average number of selected features for each dataset, rounding it if it is necessary. Finally, for each dataset, we select the features appearing more frequently in the 10 replications. The Table 5 shows the features selected in each dataset for both the standard kernel and the adaptive kernel classifier.

Dataset	Standard kernel	Adaptive kernel
Iris	2,4,3	3,4
Sonar	10, 11, 12, 15, 16, 17 21, 22, 37, 46	11, 12, 16, 18, 20, 21 22, 36, 46, 49
Glass	2,4,7	4,5,7
Heartc	2,6,8,12,13	2,3,8,10,12,13
Ionosphere	1,2,3,4,12,13,14	1,3,4,5,6,30
Breastw	1,2,4,6	1,2,3,6,7
Diabetes	2,6,7,8	2,6,8
Vehicle	1,3,4,5,6,7,10,18	1,4,5,6,9,10,18
German	1,2,3,5,10	1,2,3,4,6
Segmentation	1,2,11,13,14,16,19	1,2,8,14,16,18,19

Table 5. Features selected for both types of kernel density estimators

Once that the predictors are selected we create two subsets of each of the datasets and then we perform bagging using the corresponding kernel classifier. In Table 6, we show the misclassification rates of the single and bagged kernel classifiers after feature selection

Dataset	Standard kernel			Adaptive kernel		
	Single	Bagged	Ratio	Single	Bagged	Ratio
Iris	3.58	3.60	1.0055	3.69	3.60	0.9756
Sonar	12.42	12.26	0.9871	15.64	14.66	0.9373
Glass	36.40	35.74	0.9818	32.05	32.33	1.0087
Heartc	21.16	20.63	0.9749	20.43	19.83	0.9706
Heartc*	20.20	19.15	0.9480	17.00	17.20	1.0117
Heartc**	19.04	18.85	0.9900	18.26	17.81	0.9753
Ionosphere	5.89	5.41	0.9185	10.01	9.40	0.9400
Breastw	3.45	3.40	0.9855	3.24	3.42	1.0555
Diabetes	24.75	22.81	0.9216	23.93	23.58	0.9853
Vehicle	29.53	29.20	0.9888	32.32	31.32	0.9690
German	27.72	27.83	1.0039	29.38	29.05	0.9887
German**	25.38	24.32	0.9582	25.17	23.74	0.9431
Segmentation	3.29	3.36	1.0212	4.42	4.43	1.0023

Table 6. Comparison of bagging and KDE classifiers after feature selection

5 Concluding remarks

Our experiments have lead us to the following conclusions:

1. Increasing the number of bootstrapped samples seems to improve the misclassification error for both types of kernels. We have tried experiments with samples of size 10 and 50 and on average we gained 2%. But the computing time increases more than three times.
2. Before feature selection *Bagging* the adaptive kernel performs better than the standard kernel, but it requires between 2 and 6 times more computing time.
3. After feature selection the performance of *Bagging* for both type of kernels is almost the same and there is very little improvement in most of the datasets. According to Table 6, the average of the error reduction for the 10 datasets after Bagging using the standard kernel was 2.84% whereas for the adaptive kernel was 2.13%.
4. Feature selection does a good job, because once that is done kernel density classifiers gives lower misclassification errors than decision trees classifiers. Only in three of the datasets: *Glass, Vehicle,* and *Segment* decision trees classifiers outperform kernel density classifiers.
5. Bagging classifiers get better results when it is applied to datasets where the single classifier performs poorly. In most of the datasets of this paper kernel density classifiers have performed quite well.

6. Titterington's proposal of kernel density estimators considering all type of variables does not help too much. In both datasets: *Heart* and *German*, that include all the types of features, sometimes the Titterington's proposal gives the greater misclassification error and besides that it takes longer to be computed in the single classifier as well in the bagged classifier.

We have written a S-Plus program (version 2000 release 3) to carry out all our tasks. The program was tested in a DELL workstation with a dual processor PENTIUM Xeon, and it is available from the authors upon request.

Acknowledgment

This work was supported by the Office of Naval Research grant N00014-00-1-0360 and by the PRECISE group of the Engineering School at the UPR-Mayaguez funded by NSF grant EIA 99-77071.

References

BAUER, E. and KOKAVI, R. (1999): An empirical comparison of voting classification algorithms: Bagging, Boosting and variants. *Machine Learning, 36*, 105–139.

BLAKE, C. and MERZ, C. (1998): UCI repository of machine learning databases. *Department of Computer Science and Information*, University of California, Irvine, USA.

BREIMAN, L. (1996): Bagging Predictors. *Machine Learning, 26*, 123–140.

BREIMAN, L. (1998): Arcing Classfiers. *Annals of Statistic, 26*, 801–849.

DIETTERICH, T.G (2000): An Experimental comparison of three methods for constructing Ensembles of decision trees: Bagging, Boosting, and randomization. *Machine Learning, 26*, 801–849.

FREUND, Y. and SCHAPIRE, R. (1996): Experiments with a new boosting algorithm. *In Machine Learning, Proceedings of the Thirteenth International Conference*, San Francisco, Morgan Kaufman, 148–156.

KOHAVI, R. and JOHN, G.H. (1997): Wrappers for feature subset selection. *Artificial Intelligence, 97*, 273–324.

MACLIN, R. and OPTIZ, D. (1997): An empirical evaluation of Bagging and Bosting. *Proceedings of the Fourteenth National Conference on Artificial Intelligence*, AAAI/MIT Press.

MICHIE, D., SPIGELHALTER, D.J. and TAYLOR, C.C. (1994): *Machine Learning, Neural and Statistical Classification.* London: Ellis Horwood.

QUINLAN, J.R. (1996): Bagging, Boosting and C4.5. *Proceedings of the Thirteenth National Conference on Artificial Intelligence*, AAAI/MIT Press, 725–730.

SILVERMAN, B.W. (1986): *Density Estimation for Statistics and Data Analysis.* Chapman and Hall. London.

TITTERINGTON, D.M. (1980): A comparative study of kernel–based density estimates for categorical data. *Technometrics, 22*, 259–268.

Biplot Methodology for Discriminant Analysis Based upon Robust Methods and Principal Curves

Sugnet Gardner and Niel le Roux

Department of Statistics, University of Stellenbosch, South Africa

Abstract. Biplots not only are useful graphical representations of multidimensional data,but formulating discriminant analysis in terms of biplot methodology can lead to several novel extensions. In this paper it is shown that incorporating both principal curves and robust canonical variate analysis algorithms in biplot methodology often leads to superior classification.

1 Introduction

The biplot was introduced by Gabriel (1971) as a graphical tool for simultaneously displaying the rows and columns of a matrix. In the traditional biplot any element of the matrix is represented by the inner product of the vectors corresponding to its row and column. The biplot can be related to principal component analysis and involves the singular value decomposition of a matrix. Although the traditional Gabriel biplot can be used - and is widely applied in practice - as a graphical aid in the important field of classification, the ideas of Gower (1995) have much to offer when applying biplot methodology in discriminant analysis and related classification procedures. These ideas culminated in the monograph by Gower and Hand (1996) introducing a new philosophy in biplot methodology by viewing biplots as the multivariate analogues of scatterplots. Calibrated biplot axes are used to relate the plotted points to the original variables, as is the case in ordinary scatterplots while the principal axes are used only as scaffolding. Gardner (2001) uses biplot methodoloby to propose several extensions of linear and non-linear classification procedures. In this paper extensions of biplot methodology by combining robust methods and principal curves into classification procedures are considered.

2 Classification based upon biplot methodology

The Gower and Hand philosophy allows well-known methods such as principal component analysis (PCA) and canonical variate analysis (CVA), as well as some newer and less well-known methods to be integrated into a unified presentation. The concept of inter-sample distance is central to all these methods of multidimensional scaling and forms the unifying concept. A PCA biplot is based on Pythagorean distance, also known as Euclidean distance, while a CVA biplot is based on Mahalanobis distance. Non-linear biplots are based on Principal Co-ordinate Analysis (PCO) using any Euclidean embeddable distance. An even further generalisation provides for categorical variables. All these biplots can be constructed with this unified method by simply using the appropriate distance measure. Gower and Hand (1996) use the terms interpolation and prediction for the relationships between the scaffolding and the original variables. Interpolation is defined as the process of finding the representation of a given (new) sample point in the biplot space in terms of the biplot scaffolding. On the other hand, prediction is inferring the values of the original variables for a point given in the biplot space in terms of the biplot scaffolding.

Since CVA and linear discriminant analysis (LDA) are equivalent, the CVA biplot is a straightforward application of a biplot in LDA. Flury (1997) describes the transformation to canonical variates as a non-singular transformation of the original variables $\underline{X}$ into a new p-variate vector $\underline{V} = B'\underline{X}$ where the eigen-vector matrix $B : p \times p$ contains the discriminant coefficients obtained by maximising the between class to with-in class variance ratio. This transformation has the following important consequence for the elements of the J canonical means: Although, in the original co-ordinate system, all elements of the p-dimensional class means may differ, the p-dimensional canonical means in the transformed co-ordinate system have the property that only their first $K = \min(p, J-1)$ elements can differ (*cf.* Flury, 1997). Furthermore, the variables $\underline{V}$ in the transformed canonical space are uncorrelated each having unit variance. It is therefore appropriate to use Pythagorean distance for classification in the transformed space. Applying the plug-in principal to observed data the matrix B is partitioned as $\underset{p\times p}{B} = \begin{bmatrix} \underset{p\times K}{B_{LDA}} & \underset{p\times(p-K)}{B_{(2)}} \end{bmatrix}$ and the canonical transformation is given by $U = XB_{LDA}$. Using the plug-in estimates of the parameters, classification is done using Pythagorean distance with classification regions defined as follows:

Definition 1. Classification region

Ignoring prior probabilities, a classification region for class j is defined as $C_j = \left\{ \underline{u} \in \mathcal{R}^K : \left\| \underline{u} - \overline{\underline{u}}_j \right\| < \left\| \underline{u} - \overline{\underline{u}}_h \right\| \text{for all } h \neq j \right\}, j = 1, 2, \ldots, J$ where $\mathcal{R}^K$ is the canonical space and $\overline{\underline{u}}_j$ the j-th canonical mean.

The CVA biplot is an r-dimensional representation of the class means after a transformation to the canonical space. The r-dimensional representation uses the first r columns of the matrix B, transforming the matrix of class means with the canonical discriminant function $\overline{Z}_{CVA} = \overline{X}B_{LDA,r}$. Since this transformation is a linear transformation the sample matrix X can be transformed into a matrix $Z_{CVA} = XB_{LDA,r}$ and the canonical means can also be obtained as the matrix of class means of the 'samples' given by the rows of Z_{CVA}. A more informative biplot can be obtained by plotting Z_{CVA} and $\overline{Z}_{CVA}$ simultaneously in the same plot. This will not only show the canonical means but also the spread of the sample points around their mean values. A tool is thus provided for visual appraisal of the degree of separation between classes.

Transforming the samples to a K-dimensional CVA biplot implies that: $r = K, Z_{CVA} = U = XB_{LDA}$and the biplot space $\mathcal{L}$ is the Pythagorean canonical space $\mathcal{R}^K$.

Therefore classification of an observation $\underline{x}^*$ is achieved as follows:

Interpolate $\underline{x}^* : p \times 1$, onto $\mathcal{L}$, giving $\underline{z}^* : r \times 1$;

Classify $\underline{z}^*$ to its nearest canonical mean using Pythagorean distance.

Often $r < K$ with the CVA biplot an approximation of the canonical space. In $\mathcal{R}^K$ the classification region for the j-th class will be the subspace that contains all points nearer to the j-th class mean than to any of the other class means. The space $\mathcal{R}^K$ will be divided into J disjoint classification regions, $C_1, C_2, \ldots, C_J$ with $C_h \cap C_j = \varnothing$ if $h \neq j$ and $\bigcup_{h=1}^{J} C_h = \mathcal{R}^K$. To classify a point $\underline{z}^*$ in the r-dimensional space $\mathcal{L}$ the representation of $\underline{z}^*$ in $\mathcal{R}^K$ given by $\underline{u}^* = \begin{bmatrix} \underline{z}^* : r \times 1 \\ \underline{0} : (K-r) \times 1 \end{bmatrix}$is needed. The distances between $\underline{u}^*$ and each of the class means are calculated and the predicted class membership corresponds to the nearest class mean. The intersections $C_j \cap \mathcal{L}$, $j = 1, 2, \ldots, J$ is a slice through the classification regions and form the representation of the classification regions in $\mathcal{L}$. The algorithms of Gower and Hand (1996) for obtaining prediction regions in a generalised biplot can be applied to construct these classification regions:

- Construct a two-dimensional $m \times m$ grid in $\mathcal{L}$, say E: $m^2 \times 2$;
- For each row of E, calculate which class mean is nearest;
- Label the point in $\mathcal{L}$ accordingly, *e.g.* by colouring the grid-points with different colours corresponding to the different category levels.

Apart from supplying a visual representation of the classification of observations, biplot methodology has the advantage of a reduction in dimension. Hastie, Tibshirani and Buja (1994) remark that the reduced space can be more stable and therefore the dimension reduction could yield better classification performance. An alternative to the above classification scheme is therefore to classify a 'new' observation to the nearest approximated canonical mean in the r-dimensional biplot space $\mathcal{L}$, where $r \leq K$.

3 Principal curves in discriminant analysis

Although biplot classification to the nearest class mean is very effective in LDA, a single (mean) value to represent a class of observations does not always provide an adequate summary of the behaviour of a class to separate it from other classes. Consider the simulated data set in Figure 1. Classification to the nearest class mean in cases like this is bound to fail. Indeed, the

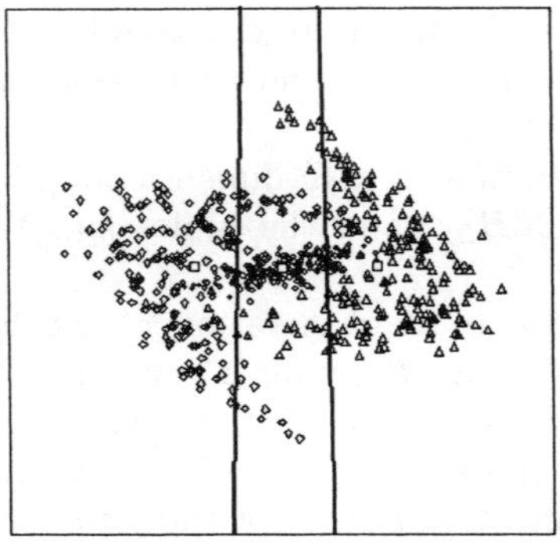

Fig. 1. CVA biplot of simulated data set with LDA classification regions

respective classification regions shown in this CVA biplot clearly bear witness to the failure of this method to provide an adequate classification into the three classes. What is needed, is extending the (single) summary measure to a measure that can adequately separate the classes from one another. Instead of using a straight line to summarise data a smooth curve that passes through the middle of the data in a smooth way, whether or not the middle of the data is a straight line could be used. Hastie and Stuetzle (1989) formally define principal curves as those smooth curves that are self-consistent for a distribution or data set. Self-consistency is based on conditional expectations viewed as random variables and is defined as follows:

Definition 2. Self-consistency (Flury, 1997; Tarpey and Flury, 1996)

Suppose $\underline{X}$ and $\underline{Y}$ are two jointly distributed random vectors, both of the same dimension, then $\underline{Y}$ is self-consistent for $\underline{X}$ if $E(\underline{X}|\underline{Y}) = \underline{Y}$. Since a principal curve is defined in terms of self-consistency which is based on a conditional expectation, Scott (1992) describes a principal curve as a local conditional mean. Replacing the point estimates (class means) with a principal curve for each class (conditional class mean) could lead to more flexibility in discrimination and classification. When using the class means for classification these mean values could be obtained in two ways: interpolation of the class means onto the biplot or calculating the class means of the interpolated biplot samples. These two possibilities of obtaining the class means are equivalent only in the case of linear transformations. Fitting a principal curve to a data set is a non-linear operation and therefore principal curves are not invariant with respect to the interpolation transformation. Interpolating the principal curve fitted to the observed data X will in general not correspond to the principal curve fitted to the interpolated sample points. Since the classification is done in the r-dimensional space $\mathcal{L}$ and the principal curves are used as conditional class means for these r-dimensional 'sample points', the principal curves are fitted to the interpolated sample points Z_{CVA}. A principal curve is fitted

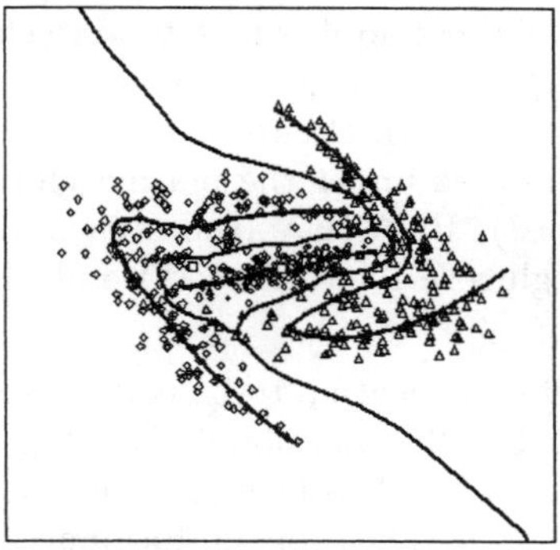

Fig. 2. CVA biplot of simulated data set with principal curve classification regions

for each class separately. A (new) sample point is interpolated onto the biplot and classified to the nearest principal curve. The biplot classification to the nearest principal curve as shown in Figure 2 has reduced the number of misclassified observations from 136 to 33 resulting in a substantial drop in error rate from 0.227 to 0.055. It is of interest to note that members of class 3 (the class in the centre of the display) are with the exception of a single observation, correctly classified with the principal curve procedure.

4 Robust biplot techniques

Hawkins and McLachlan (1997) remark that a single outlier can be enough to distort not only the class sample mean vector, but also the pooled sample covariance matrix. This distortion can lead to an arbitrarily poor classification rule, making a CVA biplot very sensitive to outliers. One possible solution is to exclude outlying observations. Two drawbacks are efficient identification of outliers and that an observation lying just outside the 'permissible cloud' is ignored while one just inside is admissible. Venables and Ripley (1999) remark that the sharp decision to keep or reject an observation is wasteful and that it is preferable to down weight dubious observations rather than rejecting them. The algorithm of Campbell (1982) which is based on robust M-estimation for CVA with a functional relationship formulation is used to obtain a robust CVA biplot similar to the ordinary CVA biplot with the matrix $B_{LDA,r}$ replaced by B_r^* obtained from Campbell's (1982) algorithm.

5 Example

The above biplot classification methodology is illustrated with the well-known diabetes data set from Reaven and Miller (1979) discussed *inter alia* in McLachlan (1992), Banfield and Raftery (1993) and Hawkins and McLachlan

(1997). A total of 145 non-obese adult subjects were classified, based on conventional clinical criteria, into three groups: patients suffering from chemical diabetes, overt diabetes and normal subjects. The study focussed on three variables: GL-AREA (the area under the plasma glucose curve for a 3-hour oral glucose tolerance test), IN-AREA (the area under the plasma insulin curve for a 3-hour oral glucose tolerance test) and SSPG (the steady-state plasma glucose response).

Assuming the clinical classification to be correct, an LDA analysis is performed on the data set. The 2-dimensional CVA biplot with corresponding classification regions is used for classification since the discrimination is performed on 3 classes. The classification performance is measured with the apparent error rate (APER) and the leave-one-out cross validation (CV) error rate. With ordinary LDA classification a total of 123 of the subjects were correctly classified, resulting in an 84.8% correct classification rate, with the corresponding CV rate of 82.8% with 120 subjects classified correctly. Inspection of this CVA biplot given in Figure 3(a) revealed that the observations that are incorrectly classified are far from their class mean and could therefore be viewed as outliers. Classification in a robust CVA biplot (*cf.* Figure 3(b)) which attempts to down weight the outlying values to minimise their effect on the estimation of the display space resulted in an improved CV classification performance of 89.7% with 130 of the 145 subjects now being correctly classified. The APER is exactly the same as the CV error rate. This is not surprising since the robust biplot should be less sensitive to the exclusion of a single sample point.

In both the CVA and robust CVA biplots the chemical and overt diabetes classes form elongated patterns instead of circular clusters around their class means. The result of this is that the sample points to the extremes of the elongated patterns are closer to the class mean of another class than to its own class mean. Replacing the class mean with a conditional mean (principal curve) which follows the observations for that particular class results in the biplot given in Figure 3(c). This improved the APER to 91.7% correct classifications (133 out of 145 subjects correctly classified). The algorithms for fitting the principal curves and the calculation of the distances from the sample point to the principal curves are computationally much more complex than the simple calculation of class means and Pythagorean distances, therefore, only the APER's are reported, which can be calculated from a single biplot compared to constructing n different biplots for measuring the CV error rate.

Although the principal curves follow the observations in the biplot representation this does not solve the inconsistency in the display space estimation as a result of outlier values. The improved classification obtained with the robust biplot confirmed that outliers did distort the accurate representation in the CVA biplot. Applying principal curve classification to the robust biplot should therefore eliminate the effect of the outliers and solve the classifica-

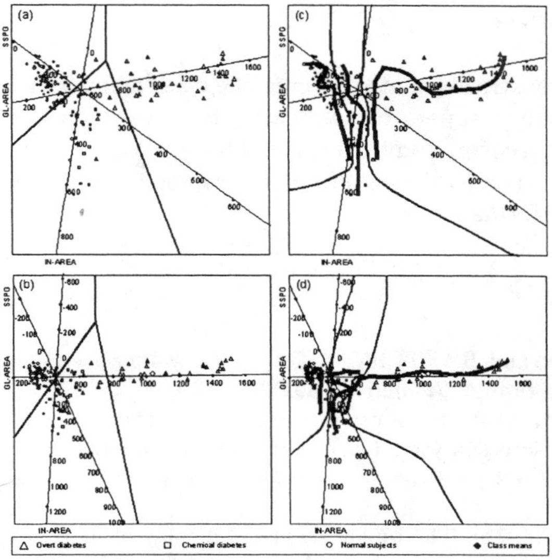

Fig. 3. CVA biplots of the diabetes data set

Table 1. LDA classification performance for the diabetes data set

Method	Correct resubstitution classification		CV classification	
CVA biplot	123	84.8%	120	82.8%
Robust CVA biplot	130	89.7%	130	89.7%
CVA biplot with principal curves	133	91.7%	—	—
Robust CVA biplot with principal curves	134	92.4%	—	—

tion shortcomings with classes forming long strips. Classification based upon principal curves in the robust CVA biplot shown in Figure 3(d), shows that this is indeed the case. Using both the robust and the principal curve extensions to the CVA biplot methodology resulted in a classification performance improvement from 123 to 134 correct classifications. This equals a 92.4% correct resubstitution classification rate. The classification performance is summarised in Table 1. Although the improvement from the ordinary CVA principal curve classification to the robust CVA principal curve classification is only marginal, with one more correct classification, a more stable display space is obtained which should lead to more accurate classifications of new sample points.

6 Conclusions

It has been demonstrated that replacing the class mean values with principal curves might result in superior classification in cases where linear discriminant analysis fails. Moreover, combining this biplot classification technique with robust methods leads to a more stable display space resulting in even better classification performance.

References

BANFIELD, J.D. and RAFTERY, A.E. (1993): Model-based Gaussian and non-Gaussian clustering. *Biometrics*, 49, 803 – 821.

CAMPBELL, N.A. (1982): Robust procedures in multivariate analysis. II: Robust canonical variate analysis. *Journal of Applied Statistics*, 31, 1 – 8.

FLURY, B. (1997): *A first course in multivariate statistics.* Springer-Verlag, New York.

GABRIEL, K.R. (1971): The biplot graphical display of matrices with application to principal component analysis. *Biometrika*, 58, 453 – 467.

GARDNER, S. (2001): *Extensions of biplot methodology to discriminant analysis with applications of non-parametric principal components.* Unpublished PhD thesis, University of Stellenbosch.

GOWER, J.C. (1995): A general theory of biplots. In: W.J. Krzanowski. (ed.): *Recent advances in descriptive multivariate analysis.* Clarendon Press, Oxford, 283 – 303.

GOWER, J.C. and HAND, D.J. (1996): *Biplots.* Chapman and Hall, London.

HASTIE, T. and STUETZLE, W. (1989): Principal curves. *Journal of the American Statistical Association*, 84, 502 – 516.

HASTIE, T., TIBSHIRANI, R., and BUJA, A. (1994): Flexible discriminant analysis by optimal scoring. *Journal of the American Statistical Association*, 89, 1255 – 1270.

HAWKINS, D.M. and McLACHLAN, G.J. (1997): High-breakdown linear discriminant analysis. *Journal of the American Statistical Association*, 92, 136 – 143.

McLACHLAN, G.J. (1992): *Discriminant analysis and statistical pattern recognition.* John Wiley, New York.

REAVEN, G.M. and MILLER, R.G. (1979): An attempt to define the nature of chemical diabetes using a multidimensional analysis. *Diabetologia*, 16, 17 – 24.

SCOTT, D.W. (1992): *Multivariate density estimation.* John Wiley, New York.

TARPEY, T. and FLURY, B. (1996): Self-consistency: A fundamental concept in statistics. *Statistical Science*, 11, 229 – 243.

VENABLES, W.N. and RIPLEY B.D. (1999): *Modern applied statistics with S-PLUS.* 3rd edition. Springer-Verlag, New York.

Bagging Combined Classifiers

Torsten Hothorn and Berthold Lausen

Department of Medical Informatics, Biometry and Epidemiology, University Erlangen-Nuremberg, Waldstraße 6, D-91054 Erlangen, Germany

Abstract. Aggregated classifiers have proven to be successful in reducing misclassification error in a wide range of classification problems. One of the most popular is bagging. But often simple procedures perform comparably in specific applications. For example, linear discriminant analysis (LDA) provides efficient classifiers if the underlying class structure is linear regarding the predictors.

We suggest bagging for a combination of tree classifiers and LDA. The out-of-bag sample is used as an independent learning sample for the computation of linear discriminant functions. The corresponding discriminant variables of the bootstrap sample are used as additional predictors for a classification tree. We illustrate the proposal by a glaucoma classification with laser scanning image data. Moreover, we analyse the properties with a simulation study and benchmark data sets. In summary, our proposal has misclassification error comparable to LDA when LDA performs best and comparable to bagged trees when bagged trees perform best.

1 Introduction

It is well known that classifiers like naive Bayes, linear discriminant analysis (LDA) or nearest neighbors perform comparably to more recent proposals in many applications, cf. Friedman(1997). Bootstrap aggregated classifiers Breiman(1996), Breiman(1998) reduce the misclassification error in many applications and artificial problems. The performance of aggregated rules depends on the performance of the unstable classifiers used. Standard classification trees (CTREE) are designed to identify partitions in the multivariate sample space. They perform poorly in problems where the classes are separated by linear combinations of the predictors. Consequently, Section 5.2.2 Breiman et al.(1984) suggested a modified tree classifier which allows for linear combinations of the original variables in a node. This requires the solution of a complex maximization problem in every node. Bagging can do better than single trees by approximating a hyperplane by multiple rectangular regions but can not be expected to do as good as LDA in this situation.

Various suggestions for combining different classification methods have been made in the past, usually by combining the different predictions or class probability estimates, for example see LeBlanc and Tibshirani(1996). Adapting proposals for regression trees cf. Ciampi(1991), Lausen et al.(1994), Lausen(1997), we propose linear discriminant variables as additional predictors in bagged classification trees.

2 Bagging a combination of LDA and classification trees

Let $\mathcal{L} = \{(y_i, \mathbf{x}_i), i = 1, \ldots, N\}$ denote a learning sample of N independent observations consisting of p-dimensional vectors of predictors $\mathbf{x}_i = (x_{i1}, \ldots, x_{ip}) \in \mathbb{R}^p$ and class labels $y_i \in \{1, \ldots, J\}$. The observations in the learning set are a random sample from some distribution function G

$$(y_1, \mathbf{x}_1), \ldots, (y_N, \mathbf{x}_N) \overset{\text{iid}}{\sim} G.$$

A classifier $C(\mathbf{x}, \mathcal{L})$ predicts future y-values for a vector of predictors $\mathbf{x}$ based on a learning sample $\mathcal{L}$. The aggregated classifier C_A is given by $C_A(\mathbf{x}) = \mathrm{E}_G C(\mathbf{x}, \mathcal{L})$, where the expectation is over learning samples $\mathcal{L}$ distributed according to G. Bagging estimates the aggregated rule $C_A(\mathbf{x})$ using the bootstrap by $\hat{C}_A(\mathbf{x}) = \mathrm{E}_{\hat{G}} C(\mathbf{x}, \mathcal{L}^*)$, where $\mathcal{L}^*$ is a random sample from the empirical distribution function $\hat{G}$

$$(y_1^*, \mathbf{x}_1^*), \ldots, (y_N^*, \mathbf{x}_N^*) \overset{\text{iid}}{\sim} \hat{G}.$$

Drawing a random sample of size N from the empirical distribution, a bootstrap sample of size N covers approximately 2/3 of the observations of the learning sample. The observations, which are not in the bootstrap sample, are called out-of-bag sample and may be used for estimating the misclassification error or for improved class probability estimates, see Breiman(1996), Breiman(1996).

In our framework, the out-of-bag sample is used to perform a LDA. The results are the coefficients of a $J - 1$ dimensional linear transformation of the original variables called linear discriminant or canonical variables for example Chapter 3.1. Ripley(1996). The discriminant variables are computed for the bootstrap sample and a new classifier is constructed using the original variables as well as the discriminant variables:

1) Draw B random samples $\mathcal{L}^{*^{(1)}}, \ldots, \mathcal{L}^{*^{(B)}}$ with replacement from $\mathcal{L}$ and let $X^{*^{(b)}}$ denote the matrix of predictors $\mathbf{x}_1^{*^{(b)}}, \ldots, \mathbf{x}_N^{*^{(b)}}$ from $\mathcal{L}^{*^{(b)}}$.
2) Compute a LDA using the out-of-bag sample $\mathcal{L} \setminus \mathcal{L}^{*^{(b)}}$, that gives a $p \times (J - 1)$ matrix $Z^{(b)}$, where the columns are the coefficients of the linear discriminant functions.
3) Construct the classifier C using the original variables as well as the discriminant variables of the bootstrap sample $(\mathcal{L}^{*^{(b)}}, X^{*^{(b)}} Z^{(b)})$.
4) Iterate steps 2) and 3) for all $b = 1, \ldots, B$ bootstrap samples.

A new observation $\mathbf{x}$ is classified by majority voting using the predictions of all classifiers

$$C\left((\mathbf{x}, \mathbf{x}Z^{(b)}), (\mathcal{L}^{*^{(b)}}, X^{*^{(b)}} Z^{(b)})\right) \text{ for } b = 1, \ldots, B.$$

Using the out-of-bag sample for the LDA, the coefficients of the discriminant function are estimated by an independent sample, thus avoiding overfitted discriminant variables in the tree growing process. Furthermore, it ensures that the training sample for the LDA is small and therefore the LDA becomes less stable.

In the following, we use classification trees as classifier C from above for bagging and will term the suggested procedure "LDA-CTREE-Bagging". Moreover, we compute bagged classification trees denoted "CTREE-Bagging" following Breiman(1996). The package `rpart` is used for the construction of classification trees with `R`, version 1.3.1. $B = 50$ bootstrap replications are used.

3 Classification of laser scanning image data

Glaucoma is an irreversible neuro-degenerative disease of the eye and requires methods for diagnosis before visual field defects are measurable. One method is based on a laser scanning image of the optic nerve head (ONH) by the Heidelberg Retina Tomograph HRT, Heidelberg Engineering (1997). This non-invasive confocal laser scanning system records a series of 32 images of the ONH. From the image series, a 2.5-dimensional topography image can be computed, with each pixel representing a depth value. The loss of retinal nerves is detected by various measurements of volumes and surfaces of the papilla derived from the topography images, see Section 4 for a detailed description.

The learning sample consists of 98 normal and 98 glaucomatous eyes and 62 variables derived from the HRT-measurements cf. Hothorn et al.(2002). Hothorn et al.(2002) demonstrate that bagging classification trees may improve glaucoma classification by HRT measurements. The .632+ bootstrap method by Efron and Tibshirani(1997) is used to estimate the misclassification error with 50 bootstrap replications. The estimated misclassification errors are: LDA 0.229, CTREE 0.191, CTREE-Bagging 0.158 and LDA-CTREE-Bagging 0.157. LDA-CTREE-Bagging seems to be at least comparable to CTREE-Bagging.

4 Simulation study

The performance of different classifiers can hardly be judged by using the learning sample only. We therefore compare the performance of the proposed procedure by a simulation of the predictors derived from laser scanning images of the ONH. The model introduced by Swindale et al.(2000) is used to describe the surface of the ONH:

$$\mu_\beta(u,v) = \\ -\left(\frac{w}{1+e^{(r-r_0)/s}} + a(u-u_0) + b(v-v_0) + c(u-u_0)^2 + d(v-v_0)^2 + w_0\right)$$

with $r = ||(u - u_0, v - v_0)||_2$, the usual 2-norm, and parameter vector

$$\beta = (w, w_0, r_0, s, u_0, v_0, a, b, c, d).$$

The arguments u and v vary between 0 and 3mm, which is the area scanned by the HRT. The parameter vector $\hat{\beta}_{norm}$ is estimated from the normal subjects in the study and $\hat{\beta}_{glau}$ is estimated from the glaucoma subjects. Progression in glaucoma is derived from the weighted average of the normal and glaucoma surfaces

$$\mu_\delta(u, v) = \delta\mu_{\hat{\beta}_{glau}}(u, v) + (1 - \delta)\mu_{\hat{\beta}_{norm}}(u, v) \text{ with } \delta \geq 0 \text{ and } (u, v) \in [0, 3]^2.$$

Figure 1 shows the surface of the ONH for a normal and a glaucomatous eye.

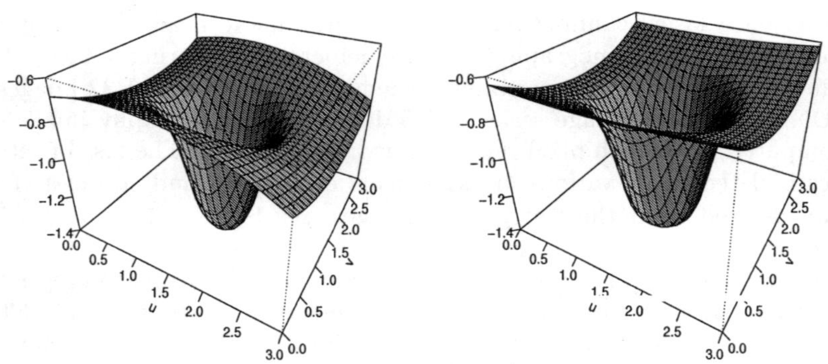

Fig. 1. Surface of the ONH model for normal (left, $\delta = 0$) and glaucomatous (right, $\delta = 1$) eyes with parameters estimated from the study population.

Given a surface with parameter vector β we simulate the predictors derived by the HRT as follows. The surface is computed on a grid $\mathbf{M}$ over $[0, 3]^2$:

$$\mathbf{M} = \left\{ i\frac{3}{k}, i = 0, \ldots, k \right\}^2, k = 40.$$

To model the measurement error by the HRT we add iid normal errors with variance σ^2 (white noise) to the surface at each point of the grid $\mathbf{M}$

$$x_\mu(u, v) \sim \mathcal{N}(\mu_\beta(u, v), \sigma^2), \quad (u, v) \in \mathbf{M},$$

where $\mathcal{N}$ denotes the normal distribution. The papilla is modeled as the set of all points on the grid $\mathbf{M}$ within a circle of radius ρ around the centre (u_0, v_0):

$$\mathbf{Pa} = \{(u,v) \in \mathbf{M} \big| \|(u-u_0, v-v_0)\|_2 \leq \rho\},$$

where ρ is estimated from the study population. The following subsets of the papilla are defined by four sectors of 90 degrees each (temporal, superior, inferior and nasal):

$$\begin{aligned}
\mathbf{Pa}_{temp} &= \{(u,v) \in \mathbf{Pa} \big| u > v \wedge (3-u) > v\} \\
\mathbf{Pa}_{nasal} &= \{(u,v) \in \mathbf{Pa} \big| u < v \wedge (3-u) < v\} \\
\mathbf{Pa}_{inf} &= \{(u,v) \in \mathbf{Pa} \big| u > v \wedge (3-u) < v\} \\
\mathbf{Pa}_{sup} &= \{(u,v) \in \mathbf{Pa} \big| u < v \wedge (3-u) > v\}.
\end{aligned}$$

The reference plane of the HRT x_{ref} is defined as the mean height on the contour line temporal minus 0.5mm:

$$x_{ref} = \left(|\mathbf{Pa}_c \cap \mathbf{Pa}_{temp}|^{-1} \sum_{(u,v)\in \mathbf{Pa}_c \cap \mathbf{Pa}_{temp}} x_\mu(u,v) \right) - 0.5.$$

$|\cdot|$ denotes the cardinality of a set and $\mathbf{Pa}_c$ is the contour line around the papilla including 2 cells of the grid

$$\mathbf{Pa}_c = \left\{ (u,v) \in \mathbf{Pa} \big| \rho - 2\frac{3}{k} \leq \|(u-u_0, v-v_0)\|_2 \leq \rho \right\}.$$

The following predictors are used by the HRT to describe the surface of the ONH: volume above reference x_{Var}, volume below reference x_{Vbr}, rim area x_{Ar}, area global x_A, mean height in contour x_{Mhc} and third moment x_{Tm}. Similar measures in the simulation model are defined as follows:

$$\begin{aligned}
x_A &= |\mathbf{Pa}| \,(3/k)^2 \\
x_{Ar}(\mathbf{Q}) &= \sum_{(u,v)\in\mathbf{Q}} \chi_{(x_{ref},\infty)}\left(x_\mu(u,v)\right)(3/k)^2 \\
x_{Var}(\mathbf{Q}) &= \sum_{(u,v)\in\mathbf{Q}} (x_\mu(u,v) - x_{ref})\chi_{(x_{ref},\infty)}\left(x_\mu(u,v)\right)(3/k)^2 \\
x_{Vbr}(\mathbf{Q}) &= \sum_{(u,v)\in\mathbf{Q}} (x_\mu(u,v) - \min_{(u,v)\in\mathbf{Pa}} (x_\mu(u,v))\chi_{(-\infty,x_{ref}]}\left(x_\mu(u,v)\right)(3/k)^2
\end{aligned}$$

$$x_{Mhc}(\mathbf{Q}) = |\mathbf{Q} \cap \mathbf{Pa}_c|^{-1} \sum_{(u,v)\in\mathbf{Q}\cap\mathbf{Pa}_c} (x_\mu(u,v) - x_{ref}) \text{ and}$$

$$x_{Tm}(\mathbf{Q}) = |\mathbf{Q}|^{-1} \sum_{(u,v)\in\mathbf{Q}} \left(x_\mu(u,v) - |\mathbf{Q}|^{-1} \sum_{(u,v)\in\mathbf{Q}} x_\mu(u,v) \right)^3,$$

where χ denotes the indicator function. The 26 predictors x_A, $x_{Ar}(\mathbf{Q})$, $x_{Var}(\mathbf{Q})$, $x_{Vbr}(\mathbf{Q})$, $x_{Mhc}(\mathbf{Q})$ and $x_{Tm}(\mathbf{Q})$ are computed from the noisy surface for either the papilla or one of the four sectors

$$\mathbf{Q} \in \{\mathbf{Pa}, \mathbf{Pa}_{temp}, \mathbf{Pa}_{nasal}, \mathbf{Pa}_{sup}, \mathbf{Pa}_{inf}\}.$$

δ	LDA	CTREE	CTREE-Bagging	LDA-CTREE-Bagging
			Setup 1	
0.1	0.376	0.470	0.447	0.428
0.5	0.027	0.223	0.168	0.029
0.9	0.000	0.110	0.080	0.000
			Setup 2	
0.1	0.443	0.459	0.436	0.435
0.5	0.246	0.216	0.168	0.152
0.9	0.147	0.137	0.093	0.067

Table 1. Model of glaucoma classification: Simulated error rates.

The first setup simulates the measurement error of the HRT itself for repeated examination of one normal and one glaucomatous eye. White noise with $\sigma = 0.2$ is added to the surface of a normal and a glaucomatous eye with different levels of progression ($\delta = 0.1, 0.5, 0.9$). This setup is not suitable for comparing the performance of classifiers in a human population with varying morphology of the ONH. Therefore the second setup simulates a clinical population with three subgroups with respect to the size of the area global (small, medium, large) incorporated. The classifiers are trained using a learning sample of size 200 (100 normal and 100 glaucoma surfaces) and tested using a test sample of the same size. The misclassification errors are averaged over 1000 learning and test samples.

Table 1 shows that LDA performs best in the first setup. CTREE-Bagging suffers a much higher error rate. It may be assumed that the two classes are separable by linear combinations of the predictors. LDA-CTREE-Bagging is comparable to the error rates of the LDA. Note that for progression $\delta = 0.9$ both LDA and LDA-CTREE-Bagging classify all observations correctly while CTREE-Bagging has misclassification error of 8%. In contrast, CTREE-Bagging outperforms LDA in the second setup. Classification trees are able to identify the subgroups with respect to the area global. LDA-CTREE-Bagging is at least comparable to CTREE-Bagging but has slightly lower estimates of misclassification error. It should be noted that both σ (standard deviation of the measurement error) and δ (level of progression) determine the distance between the two groups. Consequently, the results of Table 1 depend only on the ratio σ/δ.

Additionally, we compare the performance using three benchmark classification problems: Twonorm, Threenorm and Ringnorm for a description of the simulation setup see Breiman(1998). The results are listed in Table 2. In the Twonorm and Threenorm problem, LDA-CTREE-Bagging significantly improves CTREE-Bagging and is comparable to the best procedure (LDA). In the Ringnorm problem, LDA-CTREE-Bagging gives almost the same error rate as CTREE-Bagging does.

Dataset	Bayes-Error	LDA	CTREE	CTREE-Bagging	LDA-CTREE-Bagging
Twonorm	0.023	0.028	0.244	0.092	0.029
Threenorm	0.105	0.175	0.336	0.216	0.172
Ringnorm	0.013	0.382	0.229	0.118	0.111

Table 2. Benchmark problems: Simulated misclassification errors for three classification problems. 100 learning samples of size 300 are tested using a test sample of 18000 observations. The samples are created using the R-package `mlbench`.

5 Discussion

The method we suggest shows how linear classifiers can be incorporated in bagging. It is intuitively clear that bagged classification trees take advantage of the discriminant variables if there is linear structure in the data.

If it is not, for this proposal we pay the price of adding $J-1$ non-informative variables in the J classes case. But a combined method does not suffer model or method selection problems like over-optimism after implicit variable selection or method selection. The medical application and simulations discussed in Sections 3 and 4 indicate that LDA-CTREE-Bagging improves CTREE-Bagging significantly when linear hyperplanes are inherent in the data. It should be noted that our proposal is not restricted to the combination of CTREE and LDA. For example a possible application is bagging statistical models as building blocks of neural networks Ciampi and Lechevallier(2000).

Acknowledgments

Financial support from Deutsche Forschungsgemeinschaft, grant SFB 539-A4/C1, is gratefully acknowledged.

References

BREIMAN, L. (1996): Bagging predictors, *Machine Learning*, *Vol. 24*, 2, *123–140*.

BREIMAN, L. (1998): Arcing classifiers, *The Annals of Statistics*, *Vol. 26*, 3, *801–824*.

BREIMAN, L. (1996): Out-of-bag estimation, Tech. rep., Statistics Department, University of California Berkeley, Berkeley CA 94708.

CIAMPI, A. (1991): Generalized regression trees, *Computational Statistics and Data Analysis*, *Vol. 12*, *57–78*.

RIPLEY, B. D. (1996): Pattern Recognition and Neural Networks, Cambridge University Press, Cambridge, UK.

FRIEDMAN, J. H. (1997): On bias, variance, 0/1-loss, and the curse-of-dimensionality, *Data Mining and Knowledge Discovery*, *Vol. 1*, 1, *55–77*.

LEBLANC, M. and TIBSHIRANI, R. (1996): Combing estimates in regression and classification, *Journal of the American Statistical Association*, *Vol. 91*, 436, *1641–1650*.

EFRON, B. and TIBSHIRANI, R. (1997): Improvements on cross-validation: The .632+ bootstrap method, *Journal of the American Statistical Association*, *Vol. 92*, 438, *548–560*.

LAUSEN, B. (1997): Generalized regression trees applied to longitudinal nutritional survey data, in KLAR, R. and OPITZ, O. (Eds.): Classification and Knowledge Organization, Springer, Heidelberg, 467–474.

BREIMAN, L., FRIEDMAN, J. H., OLSHEN, R. A., and STONE, C. J. (1984): Classification and regression trees, Wadsworth, California.

LAUSEN, B., SAUERBREI, W., and SCHUMACHER, M. (1994): Classification and regression trees (CART) used for the exploration of prognostic factors measured on different scales, in DIRSCHEDL, P. and OSTERMANN, R. (Eds.): Computational Statistics, Physica-Verlag, Heidelberg, 483–496.

HEIDELBERG ENGINEERING (1997): Heidelberg Retina Tomograph: Bedienungsanleitung Software version 2.01., Heidelberg Engineering GmbH, Heidelberg.

HOTHORN, T., PAL, I., GEFELLER, O., LAUSEN, B., MICHELSON, G., and PAULUS, D. (2002): Automated classification of optic nerve head topography images for glaucoma screening, in Studies in Classification, Data Analysis, and Knowledge Organization (to appear), Springer, Heidelberg.

SWINDALE, N. V., STJEPANOVIC, G., CHIN, A., and MIKELBERG, F. S. (2000): Automated analysis of normal and glaucomatous optic nerve head topography images., *Investigative Ophthalmology and Visual Science*, *Vol. 41*, 7, *1730–42*.

CIAMPI, A. and LECHEVALLIER, Y. (2000): Constructing artificial neural networks for censored survival data from statistical models, in KIERS, H., RASSON, J.-P., GROENEN, P., and SCHADER, M. (Eds.): Data Analysis, Classification, and Related Methods, Springer, Heidelberg, 223–228.

Application of Bayesian Decision Theory to Constrained Classification Networks

Hans J. Vos

Department of Educational Measurement and Data Analysis, University of Twente, P.O. Box 217, 7500 AE Enschede, The Netherlands

Abstract. A classification network of classifying subjects as either suitable or unsuitable for a treatment followed by classifying accepted subjects as either a master or nonmaster will be formalized in the case of several relevant subpopulations. It will further be assumed that only a fixed number of subjects can be accepted for the treatment. The purpose of this paper is to optimize simultaneously this constrained classification network using Bayesian decision theory. In doing so, important distinctions will be made between weak and strong as well as monotone and nonmonotone decision rules. Also, a theorem will be given under what conditions optimal (weak) rules will be monotone.

1 Introduction

The application of Bayesian methods to statistical decision making (e.g., Lehmann, 1959) consists of two basic elements: A measurement model relating true scores to observed scores and a utility structure evaluating the total costs and benefits of all possible decision outcomes. The purpose of this paper is to optimize series of constrained classification decisions simultaneously within a Bayesian decision-theoretic framework by maximizing the overall expected utility (payoff) over all possible combinations of classification outcomes. To illustrate the simultaneous approach, a constrained classification network in industrial psychology, consisting of a quota-restricted selection decision for a training program followed by a mastery decision, is analyzed in the case of several relevant subgroups of employees. Subpopulations might be defined, for instance, by ethnicity or gender.

2 Formalizing the constrained classification network

A training program (i.e., the treatment) in an organization is often preceded by a pretest and followed by a posttest. The pretest (i.e., selection test) is administered to classify employees as either suitable or unsuitable for the program. Due to shortage of resources, however, only a fixed number of employees can usually be accepted for the program, that is, we are dealing with a quota-restricted selection decision. The posttest (i.e., mastery test) is used for classifying the accepted employees as either masters or nonmasters.

Typically, due to measurement and sampling error, the mastery test is an unreliable representation of the program objectives. The success criterion separating 'true masters' from 'true nonmasters' is supposed to be a threshold on the true score underlying the observed score on the mastery test.

For a randomly sampled employee, let the observed scores on the selection and mastery tests be continuous random variables X and Y, with realizations x and y, respectively. Let the true score for a randomly sampled employee be denoted by a continuous random variable T, with realization t. Furthermore, it will be assumed that the relation between X, Y, and T in subpopulation i $(i = 1, ..., g)$ can be represented by a joint density function $f_i(x, y, t)$. Finally, the standard denoting true mastery is a threshold value t_c on T, set in advance by the instructors of the program.

3 Different types of simultaneous decision rules

Let each of the possible actions be denoted by a_j $(j = 0,1,2)$, where $j = 0$, 1, and 2 stand for the actions to classify an employee as unsuitable, to classify an accepted employee as a nonmaster, and to classify an accepted employee as a master, respectively.

The decision rule for the mastery decision may or may not depend on the score of the selection test. Intuitively, one can imagine that a high performance on the selection test leads to a more lenient rule for the mastery decision because this prior information implies that a possible low score on the mastery test is more likely due to measurement error than to a true low performance. Simultaneous rules in which decisions are a function both of the current test score and previous test scores will be called *weak* rules in this paper. So, the compensatory character of weak rules can be interpreted as regression effects. If simultaneous rules are only a function of current test scores, the rules will be called *strong* rules. For our constrained classification network, a weak simultaneous rule δ can be defined as:

$$\begin{aligned}
\{(x,y) : \delta(x,y) = a_0\} &= A_i \times R \\
\{(x,y) : \delta(x,y) = a_1\} &= A_i^C \times B_i(x) \\
\{(x,y) : \delta(x,y) = a_2\} &= A_i^C \times B_i^C(x),
\end{aligned}$$

where A_i, A_i^C, $B_i(x)$, and $B_i^C(x)$ stand for, respectively, the sets of x and y values for which a random employee from subpopulation i is rejected or admitted for the program and accepted employees failed or passed the mastery test. R represents the set of real numbers.

Decision rules can take a monotone or nonmonotone form. A decision rule is *monotone* if cutting scores are used to partition the test scores into intervals for which different actions are taken. All other possible rules are *nonmonotone*. A weak monotone rule δ can now be defined as:

$$\delta(X,Y) = \begin{cases} a_0 \text{ for } X < x_{ci} \text{ and } Y \in R \\ a_1 \text{ for } X \geq x_{ci} \text{ and } Y < y_{ci}(x) \\ a_2 \text{ for } X \geq x_{ci} \text{ and } Y \geq y_{ci}(x), \end{cases}$$

with x_{ci} and $y_{ci}(x)$ being the cutting scores on X and Y for a random employee from subpopulation i, respectively.

4 Utility structure for the combined problem

Formally, a utility function $u_{ji}(t)$ describes the utility incurred when action a_j is taken for an employee from subpopulation i whose true score is t (see also, e.g., Vos, 2000). Mellenbergh and van der Linden (1981) proposed a linear utility function for determining optimal rules on the separate classifications. Here, their function is restated as a linear function in t for subpopulation i:

$$u_{ji}(t) = \begin{cases} b_{0i}(t_c - t) + d_{0i} \text{ for } j = 0 \\ b_{1i}(t - t_c) + d_{1i} \text{ for } j = 1 \\ b_{2i}(t - t_c) + d_{2i} \text{ for } j = 2. \end{cases}$$

5 Overall expected utility in a simultaneous approach

For the weak monotone rule δ and utility structure $u_{ji}(t)$, the expected utility for the unconstrained selection-mastery problem for a random employee from subpopulation i is equal to:

$$\begin{aligned} E[U(A_i^C, B_i^C(x))] \equiv & \int_{A_i} \int_R \int_R u_{0i}(t) f_i(x, y, t) dt dy dx \\ & + \int_{A_i^C} \int_{B_i(x)} \int_R u_{1i}(t) f_i(x, y, t) dt dy dx \\ & + \int_{A_i^C} \int_{B_i^C(x)} \int_R u_{2i}(t) f_i(x, y, t) dt dy dx. \end{aligned}$$

Taking expectations, completing integrals, and rearranging terms, this expression can be written as:

$$\begin{aligned} E[U(A_i^C, B_i^C(x))] = E[u_{0i}(T)] + \int_{A_i^C} \{ & E[u_{1i}(T) - u_{0i}(T) \mid x] \\ & + \int_{B_i^C(x)} E[u_{2i}(T) - u_{1i}(T) \mid x, y] k_i(y \mid x) dy \} q_i(x) dx, \end{aligned}$$

where $q_i(x)$ and $k_i(y \mid x)$ denote the p.d.f.'s of X and Y given $X = x$ in subpopulation i, respectively.

Then, when sampling randomly from the total population under consideration, the overall expected utility with g ($g \geq 2$) subpopulations becomes:

$$E[U(A_1^C, B_1^C(x), ..., A_g^C, B_g^C(x))] = \sum_{i=1}^{g} p_i E[U(A_i^C, B_i^C(x))],$$

where p_i, $\sum_{i=1}^{g} p_i = 1$, is the proportion of employees from subpopulation i in the total population of employees.

In the constrained selection-mastery problem, only a fixed number of employees can be accepted. The selection constraint can be expressed as:

$$p_0 = \sum_{i=1}^{g} p_i [P(x \in A_i^C] = \sum_{i=1}^{g} p_i \int_{A_i^C} q_i(x) dx,$$

where $0 < p_0 < 1$ represents the fixed proportion of all employees that can be accepted for the program.

6 Conditions sufficient for monotone rules

To find the conditions for monotonicity, first introduce the selection constraint into the function to be optimized through a Lagrange multiplier λ:

$$L(A_1^C, B_1^C(x), ..., A_g^C, B_g^C(x)) \equiv \sum_{i=1}^{g} p_i E[U(A_i^C, B_i^C(x)] + \lambda[-p_0 + \sum_{i=1}^{g} p_i \int_{A_i^C} q_i(x) dx].$$

Substitution of $E[U(A_i^C, B_i^C(x))]$, and rearranging terms, results in:

$$L(A_1^C, B_1^C(x), ..., A_g^C, B_g^C(x)) = \sum_{i=1}^{g} p_i \Big\{ E[u_{0i}(T)] + \int_{A_i^C} \{E[u_{1i}(T) - u_{0i}(T) \mid x] + \lambda + \int_{B_i^C(x)} E[u_{2i}(T) - u_{1i}(T) \mid x, y] k_i(y \mid x) dy\} q_i(x) dx \Big\} - \lambda p_0.$$

Using the well-known theorem that any integral is maximal for those values of the integration variable for which the integrand is nonnegative, an upper bound can be obtained to $L(A_1^C, B_1^C(x), ..., A_g^C, B_g^C(x))$. Applying this theorem to the inner integral of the above expression w.r.t. y, and using $q_i(x) \geq 0$, it follows that for all $B_i^C(x)$ and an arbitrary but fixed A_i^C:

$$L(A_1^C, B_1^C(x), ..., A_g^C, B_g^C(x)) \leq \sum_{i=1}^{g} p_i \Big\{ E[u_{0i}(T)] + \int_{A_i^C} \{E[u_{1i}(T) - u_{0i}(T) \mid x] + \lambda + \int_{B_i^{*C}(x)} E[u_{2i}(T) - u_{1i}(T) \mid x, y)] k_i(y \mid x) dy\} q_i(x) dx \Big\} - \lambda p_0,$$

with $B_i^{*C}(x) \equiv \{y : E[u_{2i}(T) - u_{1i}(T) \mid x, y] \geq 0\}$.

Again, applying this theorem to the outside integral in the right-hand side of the above inequality w.r.t. x, it follows that for all A_i^C:

$$L(A_1^C, B_1^{*C}(x), ..., A_g^C, B_g^{*C}(x)) \leq \\ \sum_{i=1}^{g} p_i \left\{ E[u_{0i}(T)] + \int_{A*_i^C} \{E[u_{1i}(T) - u_{0i}(T) \mid x] + \lambda + \\ \int_{B_i^{*C}(x)} E[u_{2i}(T) - u_{1i}(T) \mid x, y)]k_i(y \mid x)dy\} q_i(x)dx \right\} - \lambda p_0,$$

with

$$A_i^{*C} \equiv \{x : E[u_{1i}(T) - u_{0i}(T) \mid x] + \lambda + \\ \int_{B_i^{*C}(x)} E[u_{2i}(T) - u_{1i}(T) \mid x, y]k_i(y \mid x)dy \geq 0 \right\}.$$

Inserting the assumed utility function into $B_i^{*C}(x)$ and A_i^{*C}, results in:

$$\begin{aligned} B_i^{*C}(x) &= \{y : (b_{2i} - b_{1i})[E_i(T \mid x, y) - t_c] + d_{2i} - d_{1i} \geq 0\} \\ A_i^{*C} &= \{x : (b_{0i} + b_{1i})[E_i(T \mid x) - t_c] + d_{1i} - d_{0i} + \lambda + \\ &\int_{B_i^{*C(x)}} \{(b_{2i} - b_{1i})[E_i(T \mid x, y) - t_c] + d_{2i} - d_{1i}\} k_i(y \mid x)dy \geq 0 \right\}, \end{aligned}$$

where $E_i(T \mid x)$ and $E_i(T \mid x, y)$ denote the regression functions of T on x and T on x and y in subpopulation i, respectively.

For weak monotone rules, the sets A_i^{*C} and $B_i^{*C}(x)$ take the form $[x_{ci}, \infty)$ and $[y_{ci}(x), \infty)$, respectively. It follows that optimal rules take weak monotone forms if the left-hand sides of the inequalities for A_i^{*C} and $B_i^{*C}(x)$ are increasing functions in x and in y for all x, respectively.

7 Calculation of optimal cutting scores

Assuming the conditions for weak monotonicity are satisfied, optimal cutting scores can now be obtained for those values of x_{ci} and $y_{ci}(x)$ for which the inequalities for A_i^{*C} and $B_i^{*C}(x)$ turn into equalities. Since the sets A_i^{*C} and $B_i^{*C}(x)$ are defined for all x, and thus for x_{ci}, first the optimal (weak) cutting score on the selection test X in subpopulation i ($i = 1,..., g$) can be found by solving the $(2g + 1)$ equalities for x_{ci}, $y_{ci}(x_{ci})$, and λ subject to the selection constraint. Then, for each $x \geq x_{ci}$, optimal (weak) cutting scores on the mastery test Y in subpopulation i can be obtained by putting the left-hand side of the inequality for $B_i^{*C}(x)$ equal to zero and solving for $y_{ci}(x)$.

Since no analytical solutions for the system of equations could be found, Newton's iterative algorithm was used. Assuming a trivariate normal distribution for $f_i(x, y, t)$, from which $E_i(T \mid x, y)$ and $E_i(T \mid x)$ can be computed for known descriptive statistics of X and Y, a computer program called NEWTON was written to calculate optimal (weak) cutting scores.

8 An empirical example

Optimal rules were calculated for a constrained selection-mastery problem consisting of a sales training program, preceded and followed by respectively a selection and mastery test. Both tests consisted of 20 open-ended questions and had total scores ranging from 0-100. Data were available for a group of 76 employees. The threshold value t_c on T was fixed at 55. Due to having followed different training programs previously, the 76 employees could be divided into disadvantaged and advantaged subpopulations of 36 and 40 employees, denoted as respectively Subpopulation 1 and 2. Hence, p_1 and p_2 were equal to 36/76 and 40/76. The proportion p_0 that could be accepted for the training program was arbitrarily set equal to 0.75. Because Subpopulation 2 is considered more advantaged than Subpopulation 1, it was decided that $b_{01} > b_{02}$ and $b_{21} > b_{22}$. Furthermore, it is required that b_{0i}, $b_{2i} > 0$.

Using lottery methods, the utility parameters were empirically assessed as: $b_{01} = 3.5$, $b_{11} = b_{12} = 2$, $b_{21} = 3$, $b_{02} = 3$, $b_{22} = 2.5$, $d_{01} = d_{02} = -3$, $d_{11} = d_{12} = -7$, $d_{21} = d_{22} = -10$. For known descriptive statistics of X and Y, it then turned out that the conditions for weak monotonicity were satisfied.

The program NEWTON yielded the following results: $x_{c1} = 43.28$, $y_{c1}(x_{c1}) = 59.06$, $x_{c2} = 45.04$, and $y_{c2}(x_{c2}) = 64.49$. These results show that the optimal selection scores were lower for the disadvantaged than for the advantaged subpopulation. This makes sense since the disadvantaged employees should be accepted sooner than the advantaged employees.

As already explained, the weak rules introduce an element of compensation in the decision procedure: The employees can compensate low mastery scores by high scores on the selection test. For our data set, it appeared that the optimal mastery scores $y_{c1}(x)$ and $y_{c2}(x)$ had to be lowered by 0.541 and 0.906 for each score point above x_{c1} and x_{c2}, respectively.

9 Discussion

A final note is appropriate. Confining ourselves to monotone rules, computation of strong monotone rules with maximum overall expected utility (i.e., SMMEU cutting scores) might be useful if conditions for weak monotonicity do not hold. For the subclass of strong monotone rules, the sets A_i^C and $B_i^C(x) = B_i^C$ take the form $[x_{ci}, \infty)$ and $[y_{ci}, \infty)$, respectively. SMMEU cutting scores can then be obtained by inserting these intervals into $L(A_1^C, B_1^C(x), ..., A_g^C, B_g^C(x))$, differentiating w.r.t. x_{ci} and y_{ci}, setting the resulting expressions equal to zero, and solving the $(2g + 1)$ equalities for x_{ci}, y_{ci}, and λ subject to the selection constraint.

References

LEHMANN, E.L. (1959): *Testing Statistical Hypotheses.* Wiley, New York.

MELLENBERGH, G.J. and VAN DER LINDEN, W.J. (1981): The Linear Utility Model for Optimal Selection. *Psychometrika, 46,* 283–293.

VOS, H.J. (2000): A Minimax Solution for Sequential Classification Problems. In: H.A.L. Kiers, J.-P. Rasson, P.J.F. Groenen, and M. Schaders (Eds.): *Data Analysis, Classification, and Related Methods.* Springer, Heidelberg, 101–106.

Part II

Multivariate Data Analysis and Statistics

Multivariate Data Analysis

Quotient Dissimilarities, Euclidean Embeddability, and Huygens' Weak Principle

François Bavaud

Section d'Informatique et de Méthodes Mathématiques,
Faculté des Lettres, Université de Lausanne
CH-1015 Lausanne-Dorigny
e-mail: Francois.Bavaud@imm.unil.ch

Abstract. We introduce a broad class of categorical dissimilarities, the quotient dissimilarities, for which aggregation invariance is automatically satisfied. This class contains the chi-square, ratio, Kullback-Leibler and Hellinger dissimilarities, as well as presumably new "power" and "threshold" dissimilarity families. For a large sub-class of the latter, the product dissimilarities, we show that the Euclidean embeddability property on one hand and the weak Huygens' principle on the other hand are mutually exclusive, the only exception being provided by the chi-square dissimilarity D^χ. Various suggestions are presented, aimed at generalizing Factorial Correspondence Analysis beyond the chi-square metric, by non-linear distortion of departures from independence. In particular, the central inertia appearing in one formulation precisely amounts to the mutual information of Information Theory. [1]

1 Introduction and notations

Let n_{jk} be an $(|J| \times |K|)$ contingency table, with relative frequency $f_{jk} := n_{jk}/n$, row profiles $w_{jk} := n_{jk}/n_{j\bullet}$, column profiles $w^*_{kj} := n_{jk}/n_{\bullet k}$ and marginal (strictly positive) profiles $\rho^*_j := n_{j\bullet}/n = f_{j\bullet}$ and $\rho_k := n_{\bullet k}/n = f_{\bullet k}$, where $n_{j\bullet} := \sum_{k\in K} n_{jk}$ are the row marginals, $n_{\bullet k} := \sum_{j\in J} n_{jk}$ are the column marginals, and $n := n_{\bullet\bullet}$ is the grand total. By construction, $f_{jk} = \rho^*_j w_{jk} = \rho_k w^*_{kj}$; also, the row and column profiles transform as $w_{jk} = \rho_k w^*_{kj}/\rho^*_j$ and $w^*_{kj} = \rho^*_j w_{jk}/\rho_k$.

Independence quotients constitute a most convenient cell-by-cell representation of departures from independence. Surprisingly enough, they are not routinely used in standard Statistics, with the exception of quantitative Geography and Economics, where they are referred (when J or K enumerates regions) to as *location quotients*.

[1] Stimulating discussions with Martin Rajman and Jean-Cédric Chappelier are gratefully acknowledged. This work has benefited from a join grant from the University of Lausanne and the Swiss Federal Institute of Technology of Lausanne.

Definition 1 Independence quotients q_{jk} are the ratios of the observed counts to the expected counts under independence:

$$q_{jk} := \frac{n_{jk}\, n}{n_{j\bullet}\, n_{\bullet k}} = \frac{f_{jk}}{\rho_j^*\, \rho_k} = \frac{w_{jk}}{\rho_k} = \frac{w_{kj}^*}{\rho_j^*} \tag{1}$$

Property 1 *$q_{jk} = 1$ for all cells iff perfect independence holds. Also, in any case,*

$$\sum_{j \in J} \rho_j^*\, q_{jk} = 1 \quad \forall k \in K \qquad\qquad \sum_{k \in K} \rho_k\, q_{jk} = 1 \quad \forall j \in J \tag{2}$$

Conversely, independence quotients q_{jk} obeying (2) determine a contingency table unique up to a multiplicative constant by $n_{jk} := n\, \rho_j^*\, \rho_k\, q_{jk}$.

A dissimilarity D between J objects is a symmetric $(J \times J)$ non-negative matrix $D_{jj'}$ with a null diagonal. The most popular dissimilarity for contingency tables is the chi-square dissimilarity between rows j and j' expressing in terms of independence quotients as

$$D_{jj'}^{\chi} := \sum_{k \in K} \frac{n}{n_{\bullet k}} \left(\frac{n_{jk}}{n_{j\bullet}} - \frac{n_{j'k}}{n_{j'\bullet}}\right)^2 = \sum_{k \in K} \rho_k\, (q_{jk} - q_{j'k})^2 \tag{3}$$

D^{χ} is well known to be aggregation invariant (definition 3). Furthermore, the chi-square measure of row/column dependence obtains as half the weighted average of the dissimilarity between each pairs of rows, or *equivalently*, as the weighted average of the dissimilarity between each row and the average profile (remark 1):

$$\frac{\chi^2}{n} = \frac{1}{2} \sum_{jj'} \rho_j^*\, \rho_{j'}^*\, D_{jj'}^{\chi} = \sum_j \rho_j^*\, D_{j\rho}^{\chi} \qquad \text{where } D_{j\rho}^{\chi} := \sum_k \rho_k (q_{jk} - 1)^2$$

The equivalence emphasized above constitutes the *weak Huygens' principle* (definition 6). In the present case, the *strong Huygens' principle* $\sum_j \rho_j^* D_{ja}^{\chi} = \sum_j \rho_j^*\, D_{j\rho}^{\chi} + D_{a\rho}^{\chi}$ also holds (definition 6). Finally, the dissimilarities D^{χ} can be represented as Euclidean square distances in a space with at most $\min(|J|, |K|) - 1$ dimensions; this Euclidean embeddability property (definition 4) allows visualization in Factorial Correspondence Analysis.

What are the necessary implications, if any, between the above properties for a *general* dissimilarity D? In a previous publication (Bavaud 2000), we addressed in part the same question for *weights dissimilarities*, i.e. of the form $D_{jj'} = \sum_k G(\rho_k)\, F(w_{jk}, w_{j'k})$, for which necessary and sufficient conditions for aggregation invariance were established. In the present self-contained paper, we focus on *quotients dissimilarities* (definition 2), for which aggregation invariance is automatically satisfied (theorem 1)[2]. As far as dissimilarities

[2] to the best of our knowledge, this elementary but fundamental property has gone so far unnoticed.

are concerned, quotients dissimilarities appear as more natural objects than weights dissimilarities, and easier to manipulate as well.

Somewhat to our surprise, we realized that the Euclidean embeddability condition and the weak Huygens' principle seem to work in *competition* rather than in association. More specifically, we consider here a broad but strict subclass of quotients dissimilarities, namely the product dissimilarities (definition 7), covering all the (aggregation-invariant versions of) dissimilarities we remember having encountered in the literature. For this subclass, we prove (theorem 3, our main result) that Euclidean embeddability and the weak Huygens' principle *are mutually exclusive*, the only exception being provided by the chi-square dissimilarity D^{χ}.

2 Main results

Definition 2 Quotients dissimilarities $D_{jj'}$ between rows $j, j' \in J$ and quotients dissimilarities $D^*_{kk'}$ between columns $k, k' \in K$ are respectively defined as

$$D_{jj'} := \sum_k \rho_k \, F(q_{jk}, q_{j'k}) \qquad\qquad D^*_{kk'} := \sum_j \rho^*_j \, F(q_{jk}, q_{jk'}) \tag{4}$$

where $F(q, q') \geq 0$ is a non-negative function obeying $F(q, q') = F(q', q)$ and $F(q, q) = 0$, $\forall q, q' \geq 0$.

Remark 1 Definition (4) enables to compute dissimilarities such as D_{ja} where j is one of the original rows and a is a supplementary row, whose quotient profile $\{a_k\}_{k \in K}$ satisfies $a_k \geq 0$ and $\sum_k \rho_k \, a_k = 1$ in virtue of (2). Similarly, $D^*_{ka^*}$ is the dissimilarity between column k and supplementary column a^* of quotient profile $\{a^*_j\}_{j \in J}$ satisfying $a^*_j \geq 0$ and $\sum_j \rho^*_j \, a^*_j = 1$. As typical examples, consider the average row ρ whose associated quotient profile is $a_k = q_{\rho k} = \rho_k / \rho_k = 1$, as well as the average column ρ^* with associated quotient profile $a^*_j = q_{j\rho^*} = 1$.

Definition 3 A dissimilarity D is **aggregation invariant** if its values $D_{jj'}$ remain unchanged when two identical profiles $w^*_{k_1 j} = w^*_{k_2 j}$ (or equivalently $q_{jk_1} = q_{jk_2}$) associated to distinct columns k_1 and k_2 are further aggregated into a single column denoted $[k_1 \cup k_2]$ yielding the same profile $w^*_{[k_1 \cup k_2] j}$.

Theorem 1. *Quotients dissimilarities D and D^* are aggregation invariant.*

Proof: D of the form (4) is aggregation invariant iff

$$\rho_{k_1} \, F(q_{jk_1}, q_{j'k_1}) + \rho_{k_2} \, F(q_{jk_2}, q_{j'k_2}) = \rho_{[k_1 \cup k_2]} \, F(q_{j[k_1 \cup k_2]}, q_{j'[k_1 \cup k_2]})$$

for all $j \neq j'$, whenever $q_{jk_1} = q_{jk_2} =: b_j$ for all j. The above identity follows form $\rho_{[k_1 \cup k_2]} = \rho_{k_1} + \rho_{k_2}$ and

$$q_{j[k_1 \cup k_2]} = \frac{n_{j[k_1 \cup k_2]} \, n}{n_{j\bullet} \, n_{\bullet[k_1 \cup k_2]}} = \frac{(n_{jk_1} + n_{jk_2}) \, n}{n_{j\bullet} \, (n_{\bullet k_1} + n_{\bullet k_2})} = b_j = q_{jk_1} = q_{jk_2}$$

Aggregation invariance for D^* is proved similarly. □

Definition 4 A dissimilarity D is **Euclidean embeddable** if there exist coordinates x_{jl} for $j \in J$ and $l = 1, \ldots, r < \infty$ such that $D_{jj'} = \sum_{l=1}^{r}(x_{jl} - x_{j'l})^2$. [3]

Definition 5 Let $\{h_j\}_{j \in J}$ be a signed vector (not necessarily non-negative) normed to $\sum_{j \in J} h_j = 1$, and let a be a row of J or a supplementary row whose quotient profile obeys $\sum_k \rho_k \, a_k = 1$.

a) the **"type 1 products matrix"** B^h associated to dissimilarity D and center h is $B^h := -\frac{1}{2}(I - H)D(I - H')$, that is:

$$B^h_{jj'} := -\frac{1}{2}(D_{jj'} - \sum_{l \in J} h_l \, D_{lj'} - \sum_{l \in J} h_l \, D_{jl} + \sum_{l,l' \in J} h_l \, h_{l'} \, D_{ll'})$$

where $H := \mathbf{1}\, h'$, that is $(H)_{ll'} = h_{l'}$.

b) the **"type 2 products matrix"** ${}^a\!B$ associated to dissimilarity D and center a is ${}^a\!B := -\frac{1}{2}(D_{jj'} - D_{ja} - D_{aj'})$.

Note the type 2 products matrix ${}^{j_0}\!B$ associated to the center $j_0 \in J$ to be at same time a type 1 products matrix B^{h^0} with center $h^0_j = \delta_{jj_0}$. In general, the dissimilarities obtain from the "products matrices" as $D_{jj'} = B^h_{jj} + B^h_{j'j'} - 2B^h_{jj'}$, respectively $D_{jj'} = {}^aB_{jj} + {}^aB_{j'j'} - 2\, {}^a\!B_{jj'}$, and the well-known Schoenberg characterization holds:

Theorem 2. a) *D is Euclidean embeddable iff B^h is positive semi-definite (p.s.d.). In this case, $B^{\tilde{h}}$ is also p.s.d. for any signed $\tilde{h}$ obeying $\sum_j \tilde{h}_j = 1$.*

b) *D is Euclidean embeddable iff ${}^a\!B$ is p.s.d. In this case, ${}^{\tilde{a}}\!B$ is also p.s.d. for any supplementary row $\tilde{a}$.*

Proof: see e.g. Schoenberg (1935) or Gower (1982).

Definition 6 Huygens' principles relative to a quotient dissimilarity D are

$$\sum_j \rho^*_j \, D_{ja} = \sum_j \rho^*_j \, D_{j\rho} + D_{\rho a} \qquad \text{strong Huygens' principle} \qquad (5)$$

$$\sum_{jj'} \rho^*_j \, \rho^*_{j'} \, D_{jj'} = 2 \sum_j \rho^*_j \, D_{j\rho} \qquad \text{weak Huygens' principle} \qquad (6)$$

where $a_k \geq 0$ is any quotient profile obeying $\sum_k \rho_k \, a_k = 1$ and ρ is the average quotient profile, identically equal to 1 (remark 1).

Writing $a_k = q_{j'k}$ and applying the weighted average operator $\sum_{j'} \rho^*_{j'} \ldots$ shows the strong formulation (5) to entail the weak one (6).

---

[3] in presence of metric properties, $D_{jj'}$ thus behaves as the *square* of an Euclidean distance.

Definition 7 Product dissimilarities, f- and **g-dissimilarities** are quotient dissimilarities (definition 2) defined as

$$\begin{array}{ll} D_{jj'} = \sum_{k\in K} \rho_k \, (g(q_{jk}) - g(q_{j'k}))(h(q_{jk}) - h(q_{j'k})) & \text{(product dissimilarity)} \\ D_{jj'} = \sum_{k\in K} \rho_k \, (f(q_{jk}) - f(q_{j'k}))^2 & (f\text{-dissimilarity}) \\ D_{jj'} = \sum_{k\in K} \rho_k \, (g(q_{jk}) - g(q_{j'k}))(q_{jk} - q_{j'k}) & (g\text{-dissimilarity}) \end{array} \qquad (7)$$

where f, g and h are non-decreasing functions obeying $f(1) = g(1) = h(1) = 1$. When differentiable, we impose the normalization $f'(1) = g'(1) = h'(1) = 1$ to hold in addition.

Theorem 3. *The product dissimilarity (7)*

- *is Euclidean embeddable (definition 4) iff $g = h =: f$, that is iff the product dissimilarity is a f-dissimilarity.*
- *obeys Huygens' weak principle (6) iff $g(q) = q$ or $h(q) = q$, that is iff the product dissimilarity is a g-dissimilarity.*
- *is Euclidean embeddable and satisfies Huygens' weak principle iff $g(q) = h(q) = q$, that is iff the product dissimilarity is the chi-square dissimilarity D^χ. Also, D^χ is the only product dissimilarity obeying the strong Huygens' principle.*

Proof : the first part of the third point is a direct consequence of the first and second points. To prove the first point, observe (definitions 4, 7 and remark 1) that, if D is a product dissimilarity, then

$$^{\rho}\!B_{jj'} = \frac{1}{2} \sum_k \rho_k \left[(g(q_{jk}) - 1)\,(h(q_{j'k}) - 1) + (g(q_{j'k}) - 1)\,(h(q_{jk}) - 1) \right]$$

However, $^{\rho}\!B_{jj'}$ is p.s.d. for any quotient profile iff it is of the form $B_{jj';\rho} = \sum_k \rho_k \, G(q_{jk}) \, G(q_{j'k})$, that is iff $g = h =: f$.

To prove the second point, observe (definition 6) that, if D is a product dissimilarity, then

$$\sum_{jj'} \rho_j^* \, \rho_{j'}^* \, D_{jj'} - 2 \sum_j \rho_j^* \, D_{j\rho} = 2 \sum_k \rho_k \, (\hat{g}_k - 1)\,(\hat{h}_k - 1)$$

where $\hat{g}_k := \sum_j \rho_j^* \, g(q_{jk})$ and $\hat{h}_k := \sum_j \rho_j^* \, h(q_{jk})$. Then the weak Huygens principle holds iff $\hat{g}_k = 1$ or $\hat{h}_k = 1$ for any quotient profiles; but the only rule to be followed by all profiles is (2), which shows (with $h(1) = g(1) = 1$) that $g(q) = q$ or $h(q) = q$, or equivalently that D is a g-dissimilarity.

To prove the second part of the third point, suppose the product dissimilarity D to obey Huygens' strong principle, and therefore the weak principle as well. One finds Huygens' strong principle to hold iff $\sum_{jk} \rho_k \rho_j^* g(q_{jk})(g(a_k) - a_k) = 0$ for any quotient profiles q_{jk} and a_k, which implies $g(a_k) = a_k$, that is $D = D^\chi$. □

3 Examples and comments

a) Product dissimilarities do not exhaust the class of "interesting" quotient dissimilarities: as a counter-example, consider for instance the "symmetrized generalized power quotient dissimilarities"

$$D^{\lambda}_{jj'} := \frac{1}{2\lambda(\lambda+1)} \sum_k \rho_k \left(q_{jk}\,[(\frac{q_{jk}}{q_{j'k}})^{\lambda} - 1] + q_{j'k}\,[(\frac{q_{j'k}}{q_{jk}})^{\lambda} - 1]\right) \qquad \text{for } \lambda \text{ real}$$

whose name refers to similar functionals studied by Cressie and Read (1984) in the context of model selection.

b) For $s > 0$, g-dissimilarities with $g(q) := (q^s + s - 1)/s$ constitute the so-called "type s" dissimilarities, identified in Bavaud (2000) as an aggregation-invariant class of *weight* dissimilarities obeying Huygens's weak principle. Theorem 3 permits to show that this class is unique. The particular cases $g(q) = q$ (chi-square dissimilarity), $g(q) = \ln q + 1$ (logarithmic dissimilarity) and $g(q) = 2 - 1/q$ (ratio dissimilarity) obtain for $s = 1, 0, -1$ respectively.

c) Let us list some f-dissimilarities with promising properties: they are flexible, intuitively transparent regarding the non-linear distortion of departures of the independence, and thus potentially well-adapted and finely tunable for particular needs:

1) the *power dissimilarities* with $f(q) := (q^{\beta} + \beta - 1)/\beta$, restoring the chi-square dissimilarity for $\beta = 1$. The limit $\beta \to 0$ yields the logarithm $f(q) = \ln q + 1$. The case $\beta = 1/2$ gives the so-called Hellinger distance, whose aggregation invariance properties have been noticed and investigated by Escofier (1978). The limit of large positive β produces "caricature" dissimilarities in that object j tends to be characterized exclusively by its dominant feature k_0 (such that $q_{jk_0} \geq q_{jk}$). Inversely, low positive values of β overweight the effect of rare features.
2) the *presence-absence* dissimilarity with $f(q) := I(q > 0)$, where $I(A)$ is the indicator function for the event A. The resulting dissimilarity is aggregation invariant by construction (theorem 1), and so is the *weighted simple matching similarity*

$$S^{\text{simple, weighted}}_{jj'} := \sum_{k \in K} \rho_k \left[1 - (I(q_{jk} > 0) - I(q_{j'k} > 0))^2\right]$$

By contrast, the usual (unweighted or uniform) simple matching similarity

$$S^{\text{simple, unweighted}}_{jj'} := \sum_{k \in K} \left[1 - (I(q_{jk} > 0) - I(q_{j'k} > 0))^2\right]/|K|$$

(see e.g. Joly and Le Calvé (1994)) is not aggregation invariant. The same remarks apply to the unweighted and weighted dissimilarity of Jaccard defined as

$$S_{jj'}^{\text{Jaccard, unweighted}} = \frac{\text{number of features common to } j \text{ and } j'}{\text{number of features common to } j \text{ or } j'}$$

$$S_{jj'}^{\text{Jaccard, weighted}} := \frac{\sum_{k\in K} \rho_k \, I(q_{jk} > 0)\, I(q_{j'k} > 0)}{\sum_{k\in K} \rho_k \, [1 - I(q_{jk} = 0)\, I(q_{j'k} = 0)]}$$

3) the *major-minor* dissimilarity with $f(q) := I(q \geq 1)$. This dissimilarity only distinguishes whether quotients are above or below average (hence its name). Major-minor and presence-absence dissimilarities are particular cases of the *threshold* dissimilarities $f(q) := I(q \geq \theta) + I(\theta > 1)$ (obeying $f(1) = 1$ for any $\theta > 0$).
4) the *entropic* dissimilarity with $f(q) := 1 + \text{sgn}(q-1)\sqrt{2}\sqrt{q(\ln q - 1) + 1}$. Calculus shows $f(q)$ to be increasing with $f(1) = 0$ and $f'(1) = 1$. The resulting central (half-)inertia is

$$\frac{1}{2}\sum_j \rho_j^* \, D_{j\rho} = \frac{1}{2}\sum_{jk} \rho_j^* \, \rho_k \, (f(q_{jk}) - 1)^2 = \sum_{jk} \rho_j^* \, \rho_k \, (q_{jk}[\ln q_{jk} - 1] + 1) =$$
$$\sum_{jk} \rho_j^* \, \rho_k \, q_{jk} \ln q_{jk} = \sum_{jk} \rho_j^* \, w_{jk} \, \ln \frac{w_{jk}}{\rho_k} = \sum_{jk} \rho_j^* \, w_{jk} \, \ln w_{jk} -$$
$$\sum_k \rho_k \, \ln \rho_k = -H(K|J) + H(K) = I(J:K)$$

where $I(J:K) := H(J) + H(K) - H(J,K) \geq 0$ is the mutual information between rows $j \in J$ and columns $k \in K$, null iff J and K are independent, that is iff $q_{jk} \equiv 1$. The non-linear function $f(q)$ thus allows an *exact Euclidean representation for mutual information*, without having to expand the logarithm to the second order: mutual information $H(J) + H(K) - H(J,K)$ can be visualized as a particular instantiation of the central inertia, thus providing a direct link between Data Analysis and Information Theory.

d) The Euclidean embeddability property enjoyed by f-dissimilarities is obvious when considering the transformation $q_{jk} \to x_{jk} := \sqrt{\rho_k}\, f(q_{jk})$, transforming the *profile* $\{q_{jk}\}_{k\in K}$ of row j into *coordinates* $\{x_{jk}\}_{k\in K}$, since $D_{jj'} = \sum_{k\in K}(x_{jk} - x_{j'k})^2$ from definition 7. In this paper, we have defined the coordinates of the average profile by *first* averaging over the row profiles and *then* applying the above transformation. Proceeding the other way round, i.e. directly averaging the row coordinates, produces the same coordinates iff $f(q)$ is linear, which is the chi-square case. Distinguishing clearly between the above two procedures is therefore crucial: in terms of the average coordinates (rather than the average profiles) f-dissimilarities *do* indeed trivially satisfy the weak Huygens' principle. However, those average coordinates $\bar{x}_k$ are generally not invertible: the corresponding "quotient profile" $a_k := \rho_k^{-1/2} f^{-1}(\bar{x}_k)$ will not obey (2) in general.

4 Conclusion

The Euclidean embeddability condition is of course essential for validating visualization techniques in Data Analysis. On the other hand, the weak Huygens' principle justifies the definition of a local dissimilarity by splitting the double summation into a summation restricted on pairs satisfying a given relation (such as a contiguity relation in local variance formulations; see e.g. Lebart (1969)) and its complementary. Thus the theoretical results obtained in this paper should invite to reconsider a few practical aspects in Visualization, Classification and Factor Analysis, when dealing with generalized, non chi-square dissimilarities. In particular:

- usual (dis-)similarities indices should be modified into their aggregation-invariant versions (section 3.c.2).
- the distinction between representing clusters by averaging profiles *or* coordinates should be carefully addressed (section 3.d).
- to consider dissimilarities between binary profiles as particular cases of general categorical dissimilarities (sections 3.c.2 and 3.c.3) is more direct than the current practice, which operates the other way round by first dichotomizing categorical variables.

References

BAVAUD, F. (2000): On a class of Aggregation-invariant Dissimilarities obeying the weak Huygens' principle. In H.A.L. Kiers and al. (Eds.): *Data Analysis, Classification and Related Methods.* Springer, New York, 131-136.

CRESSIE, N. and READ, T.R.C. (1984): Multinomial goodness-of-fit tests. *J.R.Statist.Soc.B*, *46*, 440-464.

CRITCHLEY, F. and FICHET, B. (1994): The partial order by inclusion of the classes and dissimilarity on a finite set, and some of their basic properties. In B. Van Custem (Ed.): *Classification and Dissimilarity Analysis.* Lecture Notes in Statistics, Springer, New York, 5-65.

ESCOFIER, B. (1978): Analyse factorielle et distances répondant au principe d'équivalence distributionnelle. *Revue de Statistique Appliquée*, *26*, 29-37

GOWER, J.C. (1982): Euclidean distance geometry. *The Mathematical Scientist*, *7*, 1-14

JOLY, S. and LE CALVE, G. (1994): Similarity functions. In B. Van Custem (Ed.): *Classification and Dissimilarity Analysis.* Lecture Notes in Statistics, Springer, New York, 67-86.

LEBART, L. (1969): L'analyse statistique de la contiguïté. *Publications de l'ISUP*, *XVIII*, 81-112

SCHOENBERG, I.J. (1935): Remarks to Maurice Fréchet's article "Sur la définition axiomatique d'une classe d'espaces vectoriels distancés applicables vectoriellement sur l'espace de Hilbert". *Annals of Mathematics*, *36*, 724-732

Conjoint Analysis and Stimulus Presentation – a Comparison of Alternative Methods

Michael Brusch[1], Daniel Baier[1], and Antje Treppa[2]

[1] Institute of Business Administration and Economics
[2] Institute of Production Research
Brandenburg University of Technology Cottbus, D-03044 Cottbus, Germany

Abstract. The rapid development of the multimedia industry has led to improved possibilities to realistically present new product concepts to potential buyers even before prototypical realizations of the new products are available. Especially in conjoint studies - where product concepts are presented as stimuli with systematically varying features - the usage of pictures, sounds, animations, mock ups or even virtual reality should result in a reduction of respondent's uncertainty with respect to (w.r.t.) innovative features and (hopefully) to an improved validity of the collected preferential responses. This paper examines differences between three different stimulus presentation methods: verbal, multimedia, and real.

1 Introduction

Conjoint analysis is among the most frequently applied marketing research methods. Since its introduction to market segmentation and product design by Green and Rao (1971), a large number of problems in its application have become known, and it has continuously and methodically been improved resulting in a large number of specialized tools for data collection and analysis (for an overview see, e.g., Baier and Gaul (1999), Baier and Gaul (2000)).

In the data collection step of a conjoint study a sample of potential buyers (respondents) is asked to judge presented alternative product concepts (so-called stimuli, systematically varying, e.g., w.r.t. design attributes of research interest) as a whole. Then, in the data analysis step of the conjoint study, the contribution of each product concept attribute and its levels to the whole preference (so-called part worths) is analytically determined yielding a perfect database for designing products according to the wishes of the potential buyers and for the producer's commercial interests. So, it is not surprising that conjoint analysis has often been identified as an important method in the development process of new products (for further application fields see e.g., Baier (1999), Wittink et al. (1994), Wittink and Cattin (1989)).

2 Stimulus presentation methods and their discussion

2.1 Traditional stimulus presentation methods

Three different stimulus presentation methods are distinguished fundamentally: verbal, pictorial, and real presentation, elements of which have also already been used in combination (see e.g., Baier (1999), Wittink et al. (1994)).

Through progress in the field of information and communication technology, further developments in the combination of known elements (texts, pictures) as well as in the integration of further elements (sounds, tastes, smells etc.) can be expected. This progress is desired to obtain a more realistic presentation of product concepts. In this case, one however assumes that all qualities are not to be represented exclusively by new elements, but that still some of the qualities will be described verbally. [1]

The multimedia stimulus presentation, on account of its new nature and its increased importance, is discussed in the next section.

2.2 Multimedia stimulus presentation

Multimedia is a term of the 1990's that combines different media elements which address different senses, such as eyes and ears, and consequently affect individuals in many ways. The best aspect of multimedia comes from computer-assisted integration which allows the customers a time-independent information access. Another positive aspect of computer-assisted integration is that it allows interactions with the system. It is not surprising that multimedia has been used as a marketing tool in various fields (see e.g., Silberer (1995)).

In recent years also in conjoint analysis the usage of multimedia has increased in order to bypass the known disadvantages of verbal descriptions. So, e.g., stimulus presentations have utilized more pictures, video sequences as well as acoustic elements (voices, noises, sounds) instead of traditional written product concept attribute descriptions (see e.g., Baier (1999)).

The fundamental assumption of the model of cognition psychology implies that it is not the objectively observed, but rather the subjectively perceived cognitive reality, that is responsible for behavior of individuals (Zimbardo and Gerrig (1999)). From theoretical and empirical findings of imagery research (Ruge (1988), Kroeber-Riel (1993)), especially the dual coding approach (Paivio (1971), Paivio (1978)), it has been found that the perception of verbal-textual descriptions fundamentally differs from the perception of graphic and/or multimedia presentation forms. Humans process picture information with much smaller cognitive expenditure than with text information (Scharf et al. (1996)). Justified through the close combination of perception and judgement during information processing, this circumstance can have an influence on the validity of conjoint study results.

A summary of the advantages and disadvantages which are expected from the different stimulus presentation methods is given in Table 1.

[1] If the product concept of interest is, e.g., a radio, tone quality can be represented through acoustic elements, design through graphic stimuli and price through verbal stimuli. That means that an as realistic as possible representation can only be achieved by a combination of the different element alternatives (Stadie (1998)).

In order to analyze more precisely which effects these new types of stimulus presentations have on the validity of conjoint studies, in comparison with traditional stimulus presentation alternatives, the results of previous empirical studies are given below.

Table 1. Theoretical Advantages and Disadvantages of Stimulus Presentation Methods (Summarized aspects from different sources, e.g., Scharf et al. (1996), Weisenfeld (1989), Loosschilder et al. (1995), Ernst and Sattler (2000), Scharf et al. (1997))

Method	
Theoretical Advantages	*Theoretical Disadvantages*
Verbal	
• Easy preparation of the used stimuli • Simple to use	• Unwanted perception differences • Inadequately product concept attributes presentation
Pictorial	
• Facilitation of evaluation tasks	• Pictures' information cannot be controlled systematically • Evaluating pictures quality instead of the product concept
Multimedia	
• High participation motivation by a interesting and less tiring stimulus presentation method • Small cognitive load of the respondents through the easy and natural accessibility of presented information • Big reality content of purchase behavior of consumer • More realistic information	• Difficult ascertainment of preferences for single product concept attributes from the total judgement • Difficult decision-making for respondents because of the higher degree of complexity of product concepts • Evaluating individual relevant product concept attributes • Judgements about perception of the general product concept • Possibility of distraction of the actual judgement task • High costs
Real	
• Most realistic form of stimulus presentation	• Bad availability of production facilities, production materials and pre-production models in case of new products • Difficulties in presentation of service addones • Very high costs

3 Results of empirical comparisons

Numerous studies compare the different types of product concept attribute presentations with regard to reliability and validity (for an overview, see Ernst and Sattler (2000)).

Studies between verbal and pictorial as well as verbal and real presentations of stimuli have been carried out in the past. The results can be summarized as follows: stimulus presentation methods do not have an influence on the reliability of the results of a conjoint study. There may be effects on internal validity (comparing verbal stimulus presentations and real stimulus presentations for the benefit of the verbal presentation) and an influence on the direct results (e.g. part worth estimates or importances) or derived results (e.g. predicted purchases or market shares) of the different alternatives. As for the benefit of a specific presentation method (verbal, pictorial or real) no clear statement can be made (Ernst and Sattler (2000)).

In the study of Ernst and Sattler (2000) (the first comparison between verbal and multimedia stimulus presentation alternatives) the results with regard to reliability and validity did not clearly deviate. Finally, they indicate the need for further studies in order to confirm these results empirically.

A further investigation of this problem by a new empirical comparison is the purpose of this article. We compare three different presentation alternatives (verbal, multimedia, real).

4 A new empirical comparison

Door locking systems were chosen as product concepts for this study since they could be judged w.r.t. function as well as w.r.t. style. The results of a preliminary study determined the product concept attributes and levels which were used in the conjoint study (for further details of the door locking system see Treppa (2001)). The four attributes of the door locking system and their levels in the study are shown in Table 2.

Table 2. Door Locking Systems Attributes and Levels

<table>
<tr><td>Attributes</td><td colspan="3">Levels</td></tr>
<tr><td>Doorknob-Inside</td><td colspan="2">Permanent</td><td>Permanent and temporary</td></tr>
<tr><td>Doorknob-Outside</td><td>Emergency opening</td><td>Normal</td><td>Closing with key</td></tr>
<tr><td>Handle Form</td><td>Frankfurter model</td><td>Frankfurter U</td><td>Frankfurter arc</td></tr>
<tr><td>Price</td><td>Low</td><td>Middle</td><td>High</td></tr>
</table>

Three different presentation alternatives were used for this study: verbal presentation ("verbal conjoint"), multimedia presentation ("multimedia conjoint"), as well as real presentation ("real conjoint"). In the interview, the individual product concept attributes and levels first were explained to the respondents. Then, the respondents were asked in a conjoint task to sort nine door locking systems in order of their individual preference. In a last part of the interview, a so-called holdout task, nine additional door locking systems were used to collect purchase intentions (with scales ranging from "definitely would buy" to "definitely would not buy"). The two times nine stimulus cards were generated systematically using orthogonal plans w.r.t. the attributes and levels in Table 2. The stimulus cards of the conjoint and the holdout task had different product concept attribute level combinations.

Within the "verbal conjoint", the attributes and their levels were explained to the respondents only through written abstracts. On the stimulus card, the presentation of the attributes and their levels also occurred in writing. Only the cards shown in the holdout task contained illustrations of the locking systems as well as written descriptions in order to be able to simulate more realistically the purchase act. These holdout cards had the same design in all three types of interviews. "Multimedia conjoint" and "real conjoint" differ from "verbal conjoint" only w.r.t. the description of the attributes and the attribute levels and the stimulus cards used. The respondents of "multimedia conjoint" received the explanations of the attributes and their levels by the use of written as well as multimedia descriptions. The functions of the individual door locking systems were presented by means of 3-D animation, and the different handle forms were presented by means of pictures. The stimulus cards of the conjoint task had, in addition to written descriptions, pictorial illustrations of the door locking systems. A sample for the "multimedia conjoint" stimulus card used is shown in Fig. 1. At the same time, the illustration shows the appearance of the stimulus cards in the holdout task, which only differed by other attribute level combinations at the same time. The respondents of "real conjoint" received, in addition to the written abstracts, real preproduction models to practically examine the functions. The stimulus cards of the conjoint and the holdout task had the same design as in the case of "multimedia conjoint".

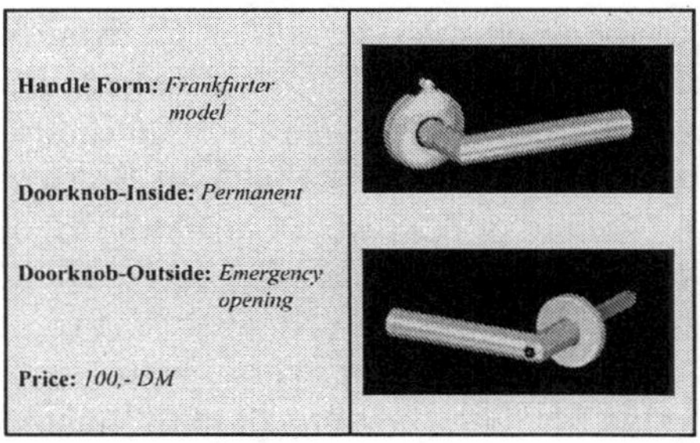

Fig. 1. Sample for a Multimedia Conjoint Stimulus Card

A between-subject design with three independent partial random samples with 35 ("verbal conjoint"), 35 ("multimedia conjoint"), and 38 ("real conjoint") respondents were chosen be able to test influence of the three examined stimulus presentations types on validity. In addition, the selection of independent partial random samples on the one hand guaranteed that the facility to supply information of the respondents is not overtaxed, and on the other hand that arrangement and/or learning effects, which influence answer behavior and therefore can lead to distortions in the results, can be avoided (Huber et al. (1993), Agarwal and Green (1991)).

In order to achieve comparability between the three partial random samples, the respondents were selected from a restricted homogenous population (students and employees of a German university). But this limitation brings the disadvantage that the results cannot be generalized easily (for a more comprehensive discussion compare, Sattler et al. (2001)). Furthermore, no market potential estimate is possible with the achieved results, but, this was not the subject of this comparison.

5 Comparison results and conclusion

The results of our comparison are summarized in Table 3 and Table 4. Table 3 includes the differences in part worth estimates depending on the type of product presentation. Table 4 shows the mean Spearman rank-order correlation and Kendall's tau as well as the first-choice-hit-rate for every type of stimulus presentation.

The part worths were estimated using ordinary least squares. The survey results are very similar for all three types of presentation. It is identifiably that the doorknob-inside and the price part worths have the same ordering within every method. Only "verbal conjoint" shows different orderings w.r.t. the doorknob-outside (closing with key) and the handle form (frankfurter u) to both other presentation methods.

The coefficients for the Spearman rank-order correlation and Kendall's tau were estimated between ranks from conjoint and holdout tasks. Both coefficients show the expected ordering results: "Real conjoint" has the highest coefficient values for Spearman as well as for Kendall, "verbal conjoint" has the lowest coefficient values and the coefficient values for "multimedia conjoint" are between both other presentation methods. The coefficients are significantly different with an F-value of 9.95 for Spearman with a probability of 0.0001 and an F-value of 8.93 with a probability of 0.0003 for Kendall. Only the first-choice-hit-rate (shows the hit rate, where the first selected stimulus in the holdout task is the one stimulus that represents the maximum of the sum of part worth estimates for every respondent) is surprising, which indicates the best value for "multimedia conjoint" instead for "real conjoint" as expected. Probably the reason is that the respondents of the "multimedia conjoint" have seen the multimedia attribute level presentation as type of product concept presentation, which were similar to the presentation of the stimulus cards in the holdout task and answered consequentially more consistent.

To conclude: In our new empirical comparison we could show that multimedia stimulus presentation can improve the validity of conjoint study results compared to verbal presentations. Of course, further comparisons are needed.

Table 3. Estimation Results: Mean Part Worth Estimates (Standard Deviation)

		"Verbal Conjoint"	"Multimedia Conjoint"	"Real Conjoint"
Doorknob-Inside	Permanent	0.127 (1.230)	0.204 (0.758)	0.027 (0.808)
	Permanent and temporary	-0.127 (1.230)	-0.204 (0.758)	-0.027 (0.808)
Doorknob-Outside	Emergency opening	-0.125 (1.639)	-0.532 (1.218)	-0.589 (1.839)
	Normal	0.005 (1.785)	0.767 (1.701)	0.903 (1.982)
	Closing with key	0.120 (1.665)	-0.235 (1.569)	-0.314 (1.860)
Handle Form	Frankfurter Model	0.121 (1.331)	1.000 (1.491)	0.259 (1.135)
	Frankfurter U	0.364 (1.144)	-0.480 (1.345)	-0.071 (0.987)
	Frankfurter arc	-0.486 (1.616)	-0.520 (1.761)	-0.188 (1.483)
Price	Low	0.432 (0.705)	0.452 (0.716)	0.549 (0.651)
	Middle	-0.105 (0.804)	0.174 (0.525)	0.200 (0.563)
	High	-0.327 (0.704)	-0.626 (0.689)	-0.750 (0.846)

Table 4. External Validity: Correlation Coefficients and First-choice-hit-rates

	"Verbal Conjoint"	"Multimedia Conjoint"	"Real Conjoint"
Spearman	0.349 (0.428)	0.601 (0.298)	0.675 (0.222)
Kendall	0.269 (0.358)	0.480 (0.277)	0.540 (0.205)
First-choice-hit-rate	31.43 %	48,57 %	39,47 %

References

AGARWAL, M.K., GREEN, P.E. (1991): Adaptive Conjoint Analysis versus Self Explicated Models: Some Empirical Results. *International Journal of Research in Marketing, 8*, 141–146.

BAIER, D. (1999): Methoden der Conjointanalyse in der Marktforschungs- und Marketingpraxis. In: Gaul, W., Schader, M. (Eds.): *Mathematische Methoden der Wirtschaftswissenschaften.* Physica, Heidelberg, 197-206.

BAIER, D., GAUL, W. (1999): Optimal Product Positioning Based on Paired Comparison Data. *Journal of Econometrics, 89*, Nos. 1-2, 365–392.

BAIER, D., GAUL, W. (2000): Market Simulation Using a Probabilistic Ideal Vector Model for Conjoint Data. In: Gustafsson, A., Herrmann, A., Huber, F. (Eds.): *Conjoint Measurement - Methods and Applications*. Springer, Berlin, 97–120.

ERNST, O., SATTLER, H. (2000): Multimediale versus traditionelle Conjoint-Analysen. Ein empirischer Vergleich alternativer Produktpräsentationsformen. *Marketing ZFP*, *2*, 161–172.

GREEN, P.E. and RAO, V.R. (1971): Conjoint Measurement for Quantifying Judgmental Data. *Journal of Marketing Research*, *8*, 355–363.

HUBER, J.C., WITTINK, D.R., FIEDLER, J.A., MILLER, R. (1993): The Effectiveness of Alternative Preference Elicitation Procedures in Predicting Choice. *Journal of Marketing Research*, *30*, 105–114.

KROEBER-RIEL, W. (1993): *Bildkommunikation. Imagerystrategien für die Werbung*. Vahlen, München.

LOOSCHILDER, G.H., ROSBERGEN, E., VRIENS, M., WITTINK, D.R. (1995): Pictorial Stimuli in Conjoint Analysis - to Support Product Styling Decisions. *Journal of the Market Research Society*, *37*, 17–34.

PAIVIO, A. (1971): *Imagery and Verbal Processes*. Holt, Rinehart and Winston, New York a.o.

PAIVIO, A. (1978): A Dual Coding Approach to Perception and Cognition. In: Pick, A., Saltzman, E. (Eds.): *Modes of Perceiving and Processing Information*. Lawrence Erlbaum Associates, Hillsdale, 39–51.

RUGE, H.D. (1988): *Die Messung bildhafter Konsumerlebnisse*. Physica, Heidelberg.

SATTLER, H., HENSEL-BÖRNER, S. and KRÜGER, B. (2001): Die Abhängigkeit der Validität von demographischen Probanden-Charakteristika: Neue empirische Befunde. *Zeitschrift für Betriebswirtschaft*,*7*, 771–787.

SCHARF, A., SCHUBERT, B., VOLKMER, H.P. (1996): Conjointanalyse und Multimedia. *Planung und Analyse*, 26–31.

SCHARF, A., SCHUBERT, B., VOLKMER, H.P. (1997): Konzepttests mittels bildgestützter Choice-Based Conjointanalyse. *Planung und Analyse*, *5*, 24–28.

SILBERER, G. (1995): Marketing mit Multimedia im Überblick. In: Silberer, G. (Eds.): *Marketing mit Multimedia. Grundlagen, Anwendungen und Management einer neuen Technologie im Marketing*. Schäffer-Poeschel, Stuttgart, 3–31.

STADIE, E. (1998): *Medial gestützte Limit Conjoint-Analyse als Innovationstest für technologische Basisinnovationen*. Springer, Münster.

TREPPA, A. (2001): *Konzeption eines integrativen Vorgehensmodells zur Unterstützung der Konstruktionsmethodik*, Dissertation, BTU Cottbus.

WEISENFELD, U. (1989): *Die Einflüsse von Verfahrensvariationen und der Art des Kaufentscheidungsprozesses auf die Reliabilität der Ergebnisse bei der Conjoint Analyse*. Duncker & Humblot, Berlin.

WITTINK, D.R., CATTIN, P. (1989): Commercial Use of Conjoint Analysis: An Update. *Journal of Marketing*,*53*, 91–96.

WITTINK, D.R., VRIENS, M., BURHENNE, W. (1994): Commercial Use of Conjoint Analysis in Europe: Results and Critical Reflections. *International Journal of Research in Marketing*, *11*, (1), 41–52.

ZIMBARDO, P.G., GERRIG, R.J. (1999): *Psychologie*. Springer, Berlin.

Grade Correspondence-cluster Analysis Applied to Separate Components of Reversely Regular Mixtures

Alicja Ciok[1,2]

[1] Institute of Computer Science PAS, ul. Ordona 21, 01-237 Warsaw, Poland
[2] Institute of Home Market and Consumption, ul. Al. Jerozolimskie 87, 02-001 Warsaw, Poland

Abstract. The paper presents how the method called grade correspondence-cluster analysis (GCCA) can extract data subtables, which are characterized by specifically regular, distinctly different data structures. A short review of basic ideas underlying GCCA is given in Sec. 1. A description of straight and reverse regularity of data tables transformed by the GCCA is given in Sec. 2. These concepts are illustrated on a real data example, which describes development factors and economic status of Polish small business service firms. In the next sections, this data table is analyzed and effects of the method are demonstrated.

1 Basic ideas of the grade correspondence-cluster analysis (GCCA)

The basic notions of the grade correspondence analysis and the clustering method, which is based on it, were presented in several papers, e.g. Ciok (1998), Ciok et al. (1995). Referring interested readers to them we recall now only a few ideas which are necessary for understanding this paper.

- The input data table, after appropriate normalization, must have the *form of a bivariate probability table.* Any two-dimensional table with non-negative values can be easily transformed into this form. The transformed table will be denoted by $P = (p_{i,j})$, $i = 1, ..., m$; $j = 1, ..., k$. Due to the universal form, data structures can be expressed in terms of stochastic dependence between marginal (row and column) variables, say X and Y, irrespective of the data table contents.
- Instead of categorical variables X and Y, the pair of *continuous* variables (X^*, Y^*) defined on $[0,1] \times [0,1]$ is considered. The density h of the new distribution is constant and equal to $h_{ij} = p_{ij}/(p_{i\bullet}p_{\bullet j})$ on any rectangle

$$\{(u,v) : S^X_{i-1} < u \leq S^X_i \text{ and } \mathrm{S}^{\mathrm{Y}}_{\mathrm{j-1}} < \mathrm{v} \leq \mathrm{S}^{\mathrm{Y}}_{\mathrm{j}}\} \tag{1}$$

where $S^X_i = \sum_{l=1}^{i} p_{l\bullet}$, $S^Y_j = \sum_{l=1}^{j} p_{\bullet l}$ and $p_{i\bullet} = \sum_{l=1}^{k} p_{il}$, $p_{\bullet j} = \sum_{l=1}^{m} p_{lj}$ for $i = 1, ..., m$; $j = 1, ..., k$. This density is called the randomized grade density of (X, Y). The distribution of X^* and Y^* is called the copula of (X, Y). Copulas are bivariate distributions on $[0,1] \times [0,1]$ with uniform marginals (literature on copulas is enormous - see eg. Nelsen (1991)). Evidently, X^* and Y^* are each uniform on $[0,1]$.

- Any change in permutations of rows and columns of the data table (categories of X and Y) affects values of S_i^X and S_j^Y and consequently changes the copula.
- The overrepresentation map serves as a very convenient tool for visualisation of data structures. Every cell in the data table is represented by the respective rectangle in $[0,1]\times[0,1]$ (cf. 1) and is marked by various shades of grey, which correspond to magnitudes of the randomized grade density. The value range of the grade density is divided into several intervals (five are used in the paper). Each colour represents a particular interval; the black corresponds to the highest values, the white - to the lowest. As grade density h measures deviation from *independence* of variables X and Y, dark colours indicate overrepresentation ($h > 1$), light colours show underrepresentation ($h < 1$). Widths of rows (columns) reflect respective marginal sums.
- The randomized grade correlation coefficient $\rho^*(X,Y) = cor(X^*, Y^*)$ measures dependence between variables X and Y; for discrete variables it is equal to Schriever's extension of Spearman's rho (cf. Schriever (1985)). Coefficient ρ^* may be expressed by various equivalent formulas. The one convenient in the correspondence and cluster analysis is the following:

$$\rho^*(X,Y) = 6\int_0^1 (u - C^*(Y:X)(u))du = 6\int_0^1 (u - C^*(X:Y)(u))du,$$

where $C^*(Y:X)(t) = 2\int_0^1 r^*(Y:X)(u)du$ is called the randomized grade correlation curve and $r^*(Y:X)(t) = E(Y^* \mid X^* = t)$ is the randomized grade regression function.
- The grade correspondence analysis (GCA) maximizes positive dependence between X and Y (measured by ρ^*) in the set of all permutations of rows and columns of the data table (categories of X and Y). Let note that due to the optimal permutations there is no need to assume any particular form of this dependency. The important property of the G-CA is that *similar rows as well as columns are always placed close to one another.* In this case, the values of the regression functions r^* can serve as a similarity measure, because both regressions: $r^*(Y:X)$ and $r^*(X:Y)$ are nondecreasing for the optimal permutations (Ciok et al. (1995)).
- The grade cluster analysis (GCCA) is based on optimal permutations provided by the GCA. Assuming that numbers of clusters are given, rows and/or columns of the data table (categories of X and/or Y) are optimally aggregated. The respective probabilities are the sums of component probabilities, and they form a new data table. In this case, optimal clustering means that $\rho^*(X,Y)$ is maximal in the set of these aggregations of

rows and/or columns, which are *adjacent in optimal permutations.* The rows and columns may be aggregated either separately (i.e. we maximize ρ^* for aggregated X and nonaggregated Y or for nonaggregated X and aggregated Y), or simultaneously. In this paper, only the first method is used. Details concerning the maximization procedure can be found in Ciok (1998).

2 Straight and reverse regularity of the GCA tables

Concepts of regularity and strength of bivariate positive dependence are not usually linked together. Data analysts pay attention to strength of dependence, which can be measured by the grade correlation ρ^*. As explained in Sec. 1, the GCA transforms any two-way probability table into a bivariate distribution on the unit square, which has the maximal value of ρ^*. This distribution has the possibly strongest positive dependence and non-decreasing regression functions. However, any set of two-way probability tables transformed by the GCA with a common value of ρ^* and the same pair of marginal distributions is still strongly differentiated according to what is being called *regularity* of positive dependence.

Orderings and classes of positive dependence are widely discussed in the statistical literature but not in regard to GCA tables (i.e. two-way probability tables transformed by the GCA), and the term "regularity" is rarely mentioned in this context. As a full review is not possible in this paper, only a short introduction of one aspect, which has important implications for the clustering based on the GCCA, will be discussed.

Well-known hierarchical classes with increasing regularity of positive dependence are called the total positivity classes (TP) of order 2,3,...etc. In particular, any discretized $m \times k$ binormal table is TP of order $min(m,k)-1$. Figure 1a presents an overrepresentation map of such a discretized binormal table for $\rho^* = 0.085$. Its regularity is expressed by the fact that overrepresentations (grade density values) form a saddle surface with a non-decreasing hummock from the upper left corner to the bottom right corner of the unit square. Overrepresentation tends to decrease when one proceeds from any point of the hummock towards sides of the square.

Let us look now on Figure 1b. It is an overrepresentation map of another GCA table, which has identical value of ρ^* and the same marginals as the table in Figure 1a. Intuitively speaking, the second table is also quite regular but looks very differently from the first. At the first sight, it is even difficult to be sure that it is a GCA table, since it seems that some other row permutation (in particular 2,1,4,3) might improve correlation ρ^*. This however is not true, because the permutation 1,2,3,4 was checked to be optimal for this table and any change can only reduce ρ^* (eg. ρ^* for row permutation 2,1,4,3 and column permutation 1,2,3,4,5 is about half of the maximal value of ρ^*).

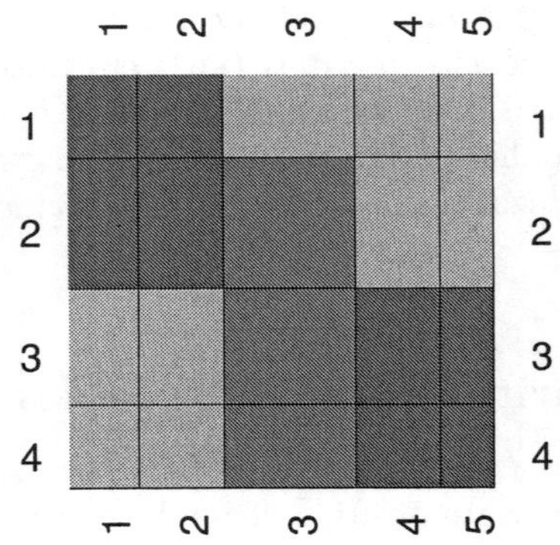

a)

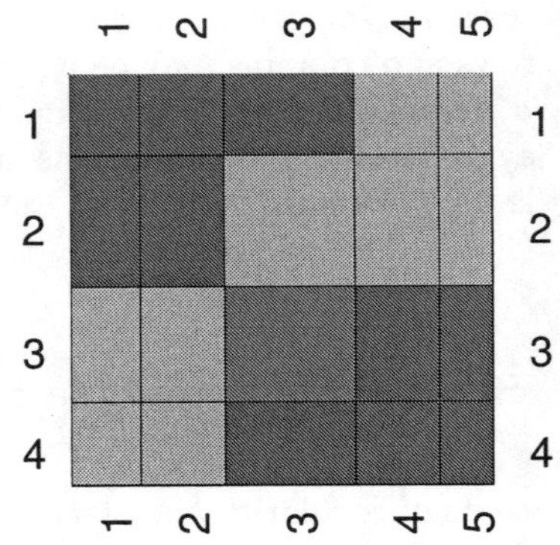

b)

Fig. 1. Overrepresentation maps for: a) a straightly regular structure, b) a reversely regular table (small business data after optimal GCCA aggregation of rows and columns).

A close examination of Figures 1b shows that Figure 1a is in a way *reverse* to Figure 1a. Instead of a hummock formed by darker rectangles in Figure 1a, there is a depression formed by lighter rectangles in Figure 1b; it goes from the bottom left to the upper right corner of the unit square; overrepresentation tend to increase from any point of the depression towards sides of the square. We propose to call structures of the type demonstrated in Figure 1a the GCA structures with high *straight regularity*, and structures of the type visible in Figure 1b - the GCA structures with high *reverse regularity.* The first type forms a unified "one-way" structure, which provides no doubt that the only way of potential aggregation is to join *adjacent* rows (and adjacent columns as well), and that there is just one latent variable according to which rows (columns) are ordered. For structures of the second type it is different: it is worth effort to separate in them two or more components which possess strong straight regularity, each representing another latent variable ordering rows (columns).

A GCA structure with high reverse regularity and very small maximal ρ^* was unexpectedly met in the research of survey data which concerned condition of small business service firms in Poland, described in Sec. 3. In fact, Figure 1b shows the overrepresentation maps for this data after GCA rearrangement and following aggregation according to the optimal clustering. Interpretation of the structures obtained in this research will be given in further sections.

This example illustrates the fact that GCA structures with high (straight or reverse) regularity should be carefully investigated even when the corresponding value of maximal ρ^* is very small. Attempts are made to evaluate statistical significance of regular structures using procedures described in Patefield (1981) or Ciok (2001) but this is beyond the frames of the paper.

Table 1. Analyzed questions from small business questionnaire

Number	Question
3	Has the territorial scope of activities declined?
5	Has the number of employees declined?
6	Is the activity unprofitable?
7	Has the level of profitability of services declined?
12	Is there a strong competition in the market?
13	What is the firm's position in the market?
How important are factors deciding the level of services?	
16.1	Personnel qualifications
16.2	Up-to-dateness of equipment
16.3	Raw materials and materials
16.4	Technologies for service provision
16.5	Location of the firm
How important are elements of the firm's environment?	
17.1	Bank services
17.2	Tax system
17.3	Possibilities of choice of suppliers
17.4	Possibilities of gaining employees
17.5	Governmental policy
17.6	Local government's policy
How important are elements affecting the firm's development?	
18.1	Ever increasing requirements of customers
18.2	Growing competition
18.3	Greater financial possibilities
18.4	Better conditions in terms of premises
18.5	Personnel qualifications development
21	Is the firm affiliated?

3 Survey of small business service firms

The main goal of the research which was carried on by dr L. Kuczewska from the Institute of Home Market and Consumption in 1999, was to detect factors which are conducive to development of the small business in Poland. The chosen sample included 89 Polish service firms of small and medium sizes and from four branches (1 - hairdressing and beauty saloons, 2 - travel offices, 3 - language courses, 4 - dentists). The used questionnaire included several questions; only 23 of them (presented in Table 1) were chosen here. The set of questions can be partitioned into two subsets. The first includes questions (3-13) which characterize an actual development status of the firm. High values mean that firm conditions are worsening; the exception is question 13, where the scale is reversed. The remaining questions (16.1-18.5) serve to assess various factors with respect to their significance to the firm activity. All factors are assessed on the same 5 point scale, where 5 corresponds to very important factors. The table of questionnaire answers was normalized before application of the grade procedures. In this case, the division by the

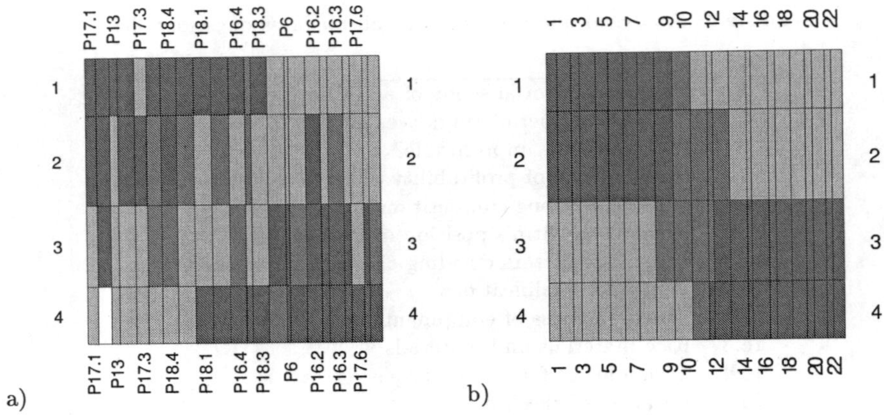

Fig. 2. Overrepresentation maps for: a) the small business data after GCA and row aggregation, b) discretized binormal probability table with the same value of ρ^* and the same marginals.

total sum of elements in the data table was used, what enables formally to treat the data as a probability table.

4 GCCA results for the whole sample

The GCA applied to the 89 × 23 table provided the following optimal permutation of columns (variables): 17.1 17.4 13 16.5 17.3 17.2 18.4 18.5 18.1 16.1 16.4 21 18.3 18.2 6 12 16.2 5 16.3 7 17.6 3 17.5, with the corresponding value of maximal ρ^* equal to 0.096. Figure 2a shows the overrepresentation map for the data table transformed by the GCA and aggregated according to the optimal clustering into 4 clusters of rows. Due to this aggregation the general structure of the data is better visible. This a GCA structure, where ρ^* cannot be improved by any permutation of columns and/or of aggregated rows. Moreover, it is evidently highly reversely regular. Its straightly regular, discretized binormal analogue (with the same value of ρ^* and the same marginals) is shown in Figure 2b.

It is easy to see that row clusters 1 and 3 form a natural pair in which characteristics of cluster 1 are exactly opposite to characteristics of cluster 3. In cluster 1 there are high overrepresentations for the first 13 questions in the optimal permutation (from 17.1 to 18.3). In cluster 3 high overrepresentations occur only in the case of remaining questions. Consequently in cluster 1 high values prevail in the first group of questions, in cluster 3 they prevail in the remaining questions. So the overrepresentation map for this 2 × 23 subtable is close to a straightly regular structure.

A similar situation occurs for the 2 × 23 subtable formed by clusters 2 and 4. Both clusters have also opposite characteristics, the only difference is in partition of the questions set, which determines the structure. In this case, the first question subset includes first 8 questions (from 17.1 to 18.5).

Using the probability language, one may say that *the analyzed data table (treated after normalization as a probability table) can be expressed as a mixture of two probability tables corresponding to straightly regular structures:*

$$P_{i,j} = a_1 P^1_{i,j} + a_2 P^2_{i,j},$$

where $P^k_{i,j} = 0$ if row i does not belong to the k-th structure, $k = 1, 2$.

Now the question arises: why a firm selects a particular structure. To answer this question, a new variable was generated, which characterizes the affiliation of a given firm to one of the four row clusters provided by the GCCA. This variable was found to be correlated with several variables from the same questionnaire which characterize the firms. The respective two-way contingency tables were calculated and the GCA was applied to them. The most interesting result was obtained for the variable which represented the branches. In the overrepresentation map for this table a very clear straightly regular structure is visible (not shown here for lack of place). This means that the GCA permutation of firms strongly correlates with firm branches. So the next hypothesis is that the detected pair of structures in the data is just an effect of mixing different branches (subsamples) in the chosen sample.

5 GCCA results for particular branches

The GCA procedure was applied to the data subtables restricted to particular branches. Table 2 shows the optimal permutations of variables for these subtables. As the particular branches must have specific requirements, it is not surprising that positions of particular variables in the optimal permutations differ significantly. The extremal positions are particularly important, because they practically define the detected structures. Particularly great differences are visible at one end of permutations. It is interesting that positions are changed not only by factors influencing firms development but also by variables which characterize their status. For example, the very important variable 13 (firm position on the market) may be meaningless in the detected structure when all answers are similar - this occurs in branch 4. So the conclusion is that important development factors as well as description of development status is specific for every branch. On the other hand, at the opposite end of permutations, three variables 17.5, 17.6 and 3 usually appear (invariantly from the branch). Consequently, this effect is also present in the whole data table. Moreover, the structures detected for particular branches appeared also reversely regular with characteristics similar to those described in the previous section. So the branch structures are also mixtures of structures. Why firms choose one or another structure is still open. The analyzed questionnaire did not provide the answer. Moreover, too small firm sample did not allow further partitions. Therefore further investigations are needed with a larger sample of firms and expanded questionnaire.

Table 2. Positions of analyzed variables in optimal permutations obtained for branch subtables.

Var. no	Whole data	Branches 1	2	3	4	Var. no	Whole data	Branches 1	2	3	4
17.1	1	19	2	3	19	18.3	13	12	12	12	15
17.4	2	10	21	1	1	18.2	14	14	14	16	4
13	3	2	3	4	13	6	15	17	7	7	12
16.5	4	5	13	2	3	12	16	18	19	6	18
17.3	5	1	1	17	5	16.2	17	16	8	21	6
17.2	6	20	6	14	2	5	18	13	18	18	21
18.4	7	6	5	9	8	16.3	19	8	16	20	11
18.5	8	7	4	8	7	7	20	15	22	15	20
18.1	9	9	11	10	14	17.6	21	23	23	22	16
16.1	10	11	10	11	9	3	22	21	17	19	22
16.4	11	4	15	5	10	17.5	23	22	20	23	23
21	12	3	9	13	17						

6 Final remarks

Another method of extracting subpopulations of objects, which is also based on the GCCA but does not take regularity into account is proposed in Szczesny (2001).

References

CIOK, A. (1998): Discretization as a tool in cluster analysis. In: A. Rizzi, M. Vichi, and H.-H. Bock (Eds.): *Advances in Data Science and Classification.* Springer, Heidelberg, 349–354.

CIOK, A., KOWALCZYK, T., PLESZCZYNSKA, E., and SZCZESNY, W.(1995): Algorithms of grade correspondence-cluster analysis. *The Collected Papers of Theoretical and Applied Computer Science*, *7*, no 1-4, 5–22.

CIOK, A. (2001): Significance testing in the grade correspondence analysis. In: M. A. Klopotek, M. Michalewicz, and S. T. Wierzchon (Eds.): *Intelligent Information Systems 2001*, Physica-Verlag, Heidelberg, 67–74.

PATEFIELD, W. M. (1981): Algorithm AS159. An efficient method of generating random RxC tables with given row and column totals. *Applied Statistics*, **30**, 91-97.

NELSEN, R.B. (1991): Copulas and association. In: G. Dall'Aglio, S. Kotz, G. Salinetti (Eds.): Advances in Probability Distributions with Given Marginals. Kluwer Academic Publishers, Dordrecht.

SCHRIEVER B. F. (1985): *Order dependence*, Ph.D. Dissertation, Free University of Amsterdam.

SZCZESNY, W. (2001): Grade correspondence analysis applied to contingency tables and questionnaire data. *Intelligent Data Analysis*, **5**, 1–35.

Obtaining Reducts with a Genetic Algorithm

José Luis Espinoza

School of Mathematics, Technological Institute of Costa Rica,
Cartago, Costa Rica.

Abstract. Given a data table associated with p qualitative attributes measured on n objects, grouped clasified in k classes or categories, a rough set of that table is a subset of the p attributes that produce the same classification of the individuals than all the attributes together. In this paper we present a genetic algorithm to calculate some reducts of such a data table. We propose a fitness function and an evolutive program, with the modifications to assure convergence. Finally, we present some numerical results.

1 Rough sets

The Rough Set theory (Pawlak (1982), Pawlak (1985)) explores the partitions associated with the subsets of attributes. When the data are classified in classes, we may measure the quality of the partition generated by a subset of attributes, in terms of its closeness to the original classification. Rough sets are considered nowadays as a tool for knowledge discovery in databases, pursuing cause-effect relationships in data (see Pawlak (1998)). So, the general purpose of rough sets is to find schemas of data with the minimum of variables of dependency. In this sense, this fits the main purpose of discriminant analysis (Trejos(1994)).

Let $\Omega = \{\omega_1, \omega_2, \ldots, \omega_n\}$ be a set of n objects and $X = \{x^1, x^2, \ldots, x^p\}$ the set of p qualitative attributes measured on Ω. Let Y_1, Y_2, ... ,Y_k the k categories or classes that represent the classification of the individuals in Ω.

Given a subset of attributes $P \subseteq X$, $P \neq \emptyset$, we define an equivalence relation $\wp$ in Ω, called *indiscernibility relation*, in the following way:

$$\forall \omega_i, \omega_h \in \Omega, \omega_i \wp \omega_h \iff x^j(\omega_i) = x^j(\omega_h) \ \forall x^j \in P.$$

where the symbols $x^j(\omega_i)$ and $x^j(\omega_h)$ denote, respectively, the values of attribute x^j on the individuals ω_i and ω_h. The indiscernibility relation produces a partition over Ω given by the equivalence classes $[\omega]_\wp$, with ω in Ω.

For any category Y_j, a *lower approximation* $L(Y_j)$ and an *upper approximation* $U(Y_j)$ are defined as:

$$L(Y_j) = \{\omega \in \Omega : [\omega]_\wp \subseteq Y_j\}$$

$$U(Y_j) = \{\omega \in \Omega : [\omega]_\wp \cap Y_j \neq \emptyset\}.$$

Two indexes of quality are defined: First, the *quality of representation to the classification*

$$W_Q = \frac{\sum_{i=1}^{k} \text{card}(L(Y_i))}{\sum_{i=1}^{k} \text{card}(U(Y_i))},$$

and the *exactitude of the approximation to the classification*

$$W_E = \frac{\sum_{i=1}^{k} \text{card}(L(Y_i))}{\text{card}(\Omega)}.$$

A *reduct* of the set of attributes is any subset P, $P \subseteq X$ such that P generates the same partition as X and none of the attributes in P is redundant, that is, we cannot eliminate an attribute x from P without changing the partition induced by the indiscernibility relation.

When we are interested in calculating reducts, we look for subsets of X with indexes W_Q and W_E as close as possible to 1, and at the same time with a minimum number of attributes.

For a table of n individuals and p attributes, the most elementary method to find reducts is the exhaustive one, that examinates all the subsets of X and then determines if that subset is a reduct or not. Although there are some implementations that avoid many unnecessary verifications, the amount of calculations that are required increases exponentially as n or p increase. In order to avoid the combinatorial problem, the search of reducts by means of iterative methods is very useful, such as genetic algorithms and others that (Goldberg (1989), Davis and Steenstrup (1987)), following the optimization of a criterion, allow us to obtain a global optimum.

2 A genetic algorithm

Typically, a genetic algorithm (GA) is an iterative procedure of genetic operations made over a population Q of individuals, in order to optimize some objective function over a feasible set. The individuals are genetic representations of the feasible solutions of a positive fitness function f to be maximized. There are two genetic operations wich are usually present in a genetic algorithm: crossover and mutation. Finally, when we are modeling a genetic algorithm, we give some parameters, such as the probability of mutation, p_m, the probability of crossing, p_c, and the size of the population. At each generation, we select an intermediate population following a mechanism of simulated roulette wheel in such a way that the individuals with the greatest values in the fitness function have a bigger probability to be selected than the others.

In order to establish conditions for the convergence of a genetic algorithm, Rudolph (1994) has proved that a canonical genetic algorithm does not necessarily converge to an optimum of the fitness function. But in the same paper, he proves that a modification of the algorithm assures the convergence. That modification consists of calculate the best individual ω_t^* at each generation t and does not modify it by mutation or crossover to keep it for the next generation. In the next iteration that best individual could be substituted by another with a better value in f and so on. The genetic algorithm that converges is as follows.

Modified GA

$t \longleftarrow 0$
Generate Q_0 (Initial population)
Evaluate f on Q_0
Calculate ω_0^* and keep it in a specific position of Q_0 (we may suppose the first position).
Repeat

1. With probability p_c, do crossover in couples from Q_t, except to the first one.
2. With probability p_m, do mutations in each individual from Q_t, except to the first one.
3. Evaluate f on Q_t.
4. Calculate ω_t^*.
5. Include ω_t^* as the first member of Q_{t+1}.
6. Select from Q_t a new population Q_{t+1}, maintaining the individual ω_t^* calculated in the previous step.

7 $t \longleftarrow t+1$
Until a criterium is valid.
End.

Genetic algorithms have also been used by Wróblewski (1998) in the obtaining of reducts. We made an independent implementation since 1997 (Espinoza (1998)) that was submitted in 1999 (Espinoza (1999)). The main differences between our work and Wróblewski's algorithm, are that he uses an hybrid algorithm that first calculates a reduct and uses a fitness function depending only of its number of attributes. On the other hand, our algorithm calculates until 25 reducts and we use the preceding modified GA, with the following choices.

Representation. A subset $P \subseteq X$ is represented as a binary vector $P = (e_1, e_2, \ldots, e_p)$, where $e_j \in \{0,1\}$ $j = 1, \ldots, p$ and $e_j = \begin{cases} 0 & \text{if } x^j \notin P \\ 1 & \text{if } x^j \in P \end{cases}$

Fitness function. Given $P \subseteq X$, let q be the number of zeros in P; that is q is the number of absent attributes. It is desirable to obtain sets of attributes with q as large as possible and quality indexes as large as possible too. For this, we propose the following fitness function

$$f(P) = \alpha + \frac{W_Q + W_E}{2}(1 + \alpha\frac{q}{p})$$

where α is a positive constant to assure the positiveness of f. The factor $(W_Q + W_E)/2$ is included to encourage an increasing on the average of the quality and exactitude of the representation. At the same time, the factor $(1 + \alpha\frac{q}{p})$ encourages a decreasing of the number of attributes in P.

We presented a first fitness function in Espinoza (1998) but it was not good enough. Later, after some changes, we obtained $f(P)$.

Selection and genetic operators In order to simulate the *selection* of individuals according to their fitness, the roulette algorithm was implemented, giving to each individual of the population a sector of the "roulette wheel" proportional to its relative value in f. The implementation of a *mutation* was very simple because of the binary representation of sets of attributes: given $P \subseteq X$, with probability p_m we substitute the value of a position randomly chosen. The value one is changed into zero and viceversa. The *crossover* of pairs of sets was implemented by choosing a random position and, with probability p_c, interchanging the blocks to form a new pair of sons.

3 Some numerical results

We programmed the genetic algorithm in C++, recording in a vector of 25 objects the best sets obtained along a running of the algorithm.

3.1 "Switching circuits"

Table 1 has been taken from Pawlak (1985) and corresponds to switching circuits, with five qualitative variables (a, b, c, d, and e) measured on 15 objects classified in two categories. The author presents a single reduct of this table: The set $\{a, b, d, e\}$.

We ran the GA with 2000 iterations, a population size of 100, and the probabilities $p_m = 0.005$ and $p_c = 0.5$. The value of α was 0.001 in all the cases. The best individual calculated in all the iterations was the only known reduct: the set of attributes $\{a, b, d, e\}$. The exactitude of representation calculated is 0.4 and the quality of approximation is 1.

3.2 "Felines"

A table called "felines" consists of 30 kinds of felines classified in four classes. The number of explanatory variables is fourteen. This table, recorded in the file *felines.txt* may be requested to the e-mail: jespinoza@itcr.ac.cr or downloadead from the web-site:

www.itcr.ac.cr/carreras/matematica/profesores/JoseL

	a	b	c	d	e	Class
1	0	0	0	0	0	0
2	0	0	0	0	1	0
3	0	0	1	1	1	0
4	0	1	1	0	1	0
5	1	1	1	1	1	0
6	1	0	0	0	1	0
7	0	0	0	1	1	1
8	0	1	0	0	0	1
9	0	1	1	0	0	1
10	1	1	0	0	0	1
11	1	1	0	1	0	1
12	1	1	1	1	0	1
13	1	1	1	0	0	1
14	1	0	0	1	1	1
15	1	0	0	1	0	1

Table 1. Switching circuits data table.

There are 131 known reducts of this table, and the GA has calculated the best 17 of them. In fact, the GA has obtained exactly the 17 reducts with three attributes (see Table 2). We ran the GA with 2000 generations, a population size of 500, $p_m = 0.05$, $p_c = 0.5$ and $\alpha = 0.001$. In all the reducts obtained, the exactitude of representation and the quality of approximation is 1 and the fitness function has obtained an optimal value of 1.0027.

We applied our method in some other data tables and the results were satisfactory. The program has been run 500 times with the "switching circuits" and "felines" tables and has found the best known value of the fitness function and some reducts in the 100 percent of the cases. It has been run 200 times with a table of zoological data, with 101 species of animals classified in seven classes, and 16 explanatory variables. The table *zoo.txt* may be unloaded or requested in the same way than the table *felines.txt*. In this experiment, in the 100 percent of the cases the algorithm has obtained the optimum and some reducts.
In all the operations we have used a value of $\alpha = 0.001$. At the same time which this level of the parameter α assures that f is positive, it gives the weight of the number of absent attributes in f.

4 Conclusions

We have presented a convergent genetic algorithm for the calculations of reducts from a table of data. The GA tends to obtain the reducts with the least number of variables. In the cases in which the algorithm has been run,

	a	b	c	d	e	f	g	h	i	j	k	l	m	n	Attributes
1	0	0	1	0	0	1	0	0	0	0	0	1	0	0	{c,f,l}
2	0	1	0	0	0	1	1	0	0	0	0	0	0	0	{b,f,g}
3	1	0	0	0	0	0	1	0	0	1	0	0	0	0	{a,g,j}
4	0	0	0	0	0	1	0	0	0	0	0	1	0	1	{f,l,n}
5	0	0	0	1	0	0	1	0	0	1	0	0	0	0	{d,g,j}
6	1	0	0	0	0	1	0	0	0	1	0	0	0	0	{a,f,j}
7	0	0	0	1	0	1	0	0	0	0	0	1	0	0	{d,f,l}
8	0	0	0	1	0	0	1	0	0	0	1	0	0	0	{d,g,k}
9	0	0	0	1	0	0	1	0	0	0	0	1	0	0	{d,g,l}
10	1	0	0	0	0	1	1	0	0	0	0	0	0	0	{a,f,g}
11	0	0	0	1	0	1	1	0	0	0	0	0	0	0	{d,f,g}
12	0	0	0	1	0	0	0	0	0	1	1	0	0	0	{d,j,k}
13	0	0	0	1	0	0	0	0	0	0	1	1	0	0	{d,k,l}
14	0	0	0	0	0	1	0	0	0	0	0	1	1	0	{f,l,m}
15	0	0	0	1	0	1	0	0	0	1	0	0	0	0	{d,f,j}
16	0	0	0	0	0	1	1	0	0	0	0	1	0	0	{f,g,l}
17	1	0	1	0	0	1	0	0	0	0	0	0	0	0	{a,c,f}

Table 2. Reducts of the feline data table with the genetic algorithm.

in all the cases some optimum reducts were obtained. So, the implemented genetic algorithm is efficient in solving the proposed problem.

References

DAVIS, L. and STEENSTRUP, M. (1987): Genetic algorithms and simulated annealing: an overview. In: L. Davis (Ed.): *Genetic Algorithms and Simulated Annealing.* Pitman, London.

ESPINOZA, J. L. (1998): Conjuntos Aproximados y Algoritmos Genéticos. In: W. Castillo, and J. Trejos (Eds.): *XI Simposio Internacional de Métodos Matemáticos Aplicados a las Ciencias.* Univ. of Costa Rica, Tech. Inst. of Costa Rica, 215-223.

ESPINOZA, J. L. (1999): *Obtención de Conjuntos Aproximados Mediante Algoritmos Genéticos.* Unpublished Magister Scientiae Dissertation, Tech. Inst. of Costa Rica.

GOLDBERG, D. E. (1989): *Genetic Algorithms in Search, Optimization and Machine Learning.* Addison-Wesley, Reading-Mass.

PAWLAK, Z. (1982): Rough sets. *International Journal of Computer and Information Sciences, 11(5),* 341–356.

PAWLAK, Z. (1985): *Rough Sets. Theoretical Aspects of Reasoning about Data.* Kluwer Publishing Co., Dordrecht.

PAWLAK, Z. (1998): Reasoning about Data - A Rough Set Perspective. In: L. Polkowski, and A. Skowron (Eds.): *Rough Sets and Current Trends in Computing.* Springer, Heidelberg, 25–34.

RUDOLPH, G. (1994): Convergence analysis of canonical genetic algorithms. *IEEE Transactions on Neural Networks*, Vol. 5, No. 1, January.

TREJOS, J. (1994): *Contribution à l'Acquisition Automatique de Connaissances à Partir de Données Qualitatives.* Thèse de Doctorat, Université Paul Sabatier, Toulouse.

WRÓBLEWSKI, J. (1998): Genetic Algorithms in Decomposition and Classification Problems. In: L. Polkowski, and A. Skowron (Eds.): *Rough Sets in Knowledge Discovery 2.* Springer, Heidelberg, 471-487.

A Projection Algorithm for Regression with Collinearity

Peter Filzmoser[1] and Christophe Croux[2]

[1] Department of Statistics and Probability Theory, Vienna University of Technology, Wiedner Hauptstr. 8-10, A-1040 Vienna, Austria
[2] Department of Applied Economics, K.U.Leuven Naamsestraat 69, B-3000 Leuven

Abstract. Principal component regression (PCR) is often used in regression with multicollinearity. Although this method avoids the problems which can arise in the least squares (LS) approach, it is not optimized with respect to the ability to predict the response variable. We propose a method which combines the two steps in the PCR procedure, namely finding the principal components (PCs) and regression of the response variable on the PCs. The resulting method aims at maximizing the coefficient of determination for a selected number of predictor variables, and therefore the number of predictor variables can be reduced compared to PCR. An important feature of the proposed method is that it can easily be robustified using robust measures of correlation.

1 Introduction

We consider the standard multiple linear regression model with intercept,

$$y_i = \beta_0 + x_{i1}\beta_1 + \ldots + x_{ip}\beta_p + \varepsilon_i \qquad i = 1, \ldots, n \tag{1}$$

where n is the sample size, $\boldsymbol{x}_i = (x_{i1}, \ldots, x_{ip})^\top$ are collected as rows in a matrix $\boldsymbol{X}$ containing the predictor variables, $\boldsymbol{y} = (y_1, \ldots, y_n)^\top$ is the response variable, $\boldsymbol{\beta} = (\beta_0, \beta_1, \ldots, \beta_p)^\top$ are the regression coefficients which are to be estimated, and $\boldsymbol{\varepsilon} = (\varepsilon_1, \ldots, \varepsilon_n)^\top$ is the error term.

Problems can occur when the predictor variables are highly correlated, this situation is called *multicollinearity.* The inverse of $\boldsymbol{X}^\top\boldsymbol{X}$ which is needed to compute the least squares (LS) estimator $\hat{\boldsymbol{\beta}}_{LS}$ of $\boldsymbol{\beta}$, becomes ill-conditioned and is numerically unstable. This matrix is also used for computing the standard errors of the LS regression coefficients and for the correlation matrix of the regression coefficients. In a near-singular case they can be inflated considerably and cause doubt on the interpretability of the regression coefficients.

A number of techniques have been proposed when collinearity exists among the predictors. One possibility is principal component regression (PCR) (see, e.g., Basilevsky, 1994) where principal components (PCs) obtained from the predictors $\boldsymbol{X}$ are used within the regression model. Most of the problems mentioned above are then being avoided. If all PCs are used in the regression model, the response variable will be predicted with the same precision as with the LS approach. However, one goal of PCR is to simplify the regression model by taking a reduced number of PCs in the prediction set. We could simply take those first $k < p$ PCs in the regression model having the largest variances (*sequential selection*) but often PCs with smaller variances are higher correlated with the response variable. Hence, it might be more advisable to set up the PCR model by a *stepwise selection* of PCs due to an appropriate measure of association with the response variable. In more detail, in the first step we could search for that PC having the largest *Squared Multiple Correlation* (SMC) with the response variable, in the second step search for an additional PC resulting in the largest SMC, and so on. In fact, due to the uncorrelatedness of the principal components, this comes down to selecting the k components having the largest squared (bivariate) correlations with the dependent variable.

PCR can deal with multicollinearity, but it is not a method which directly maximizes the correlation between the original predictors and the response variable. It was noted by Hadi and Ling (1998) that in some situations PCR can give quite low values for the SMC. PCR is a two-step procedure: in the first step one computes PCs which are linear combinations of the x-variables, and in the second step the response variable is regressed on the (selected) PCs in a linear regression model. For maximizing the relation to the response variable we could combine both steps in a single method. This method has to find $k < p$ predictor variables z_j ($j = 1, \ldots, k$) which are linear combinations of the x-variables and have high values of the SMC with the y-variable. Note that the linear combination of the predictor variables giving the theoretical maximal value of SMC with the dependent variable is determined by the coefficients of the LS-estimator. Of course, due to the multicollinearity problem mentioned before, we will not aim at a direct computation of this LS-estimator.

2 Algorithm

The idea behind the algorithm is to find k components $z_1, \ldots, z_k$ having the property that the Squared Multiple Correlation between y and the components is as high as possible, under the constraints that these components are mutually uncorrelated and have unit variance. Under these contraints, it is easy to check that

$$SMC = \mathrm{Corr}^2(y, z_1) + \mathrm{Corr}^2(y, z_2) + \ldots + \mathrm{Corr}^2(y, z_k).$$

We will try to optimize the above SMC in a sequential manner. First by selecting a z_1 having maximal squared correlation with the dependent variable, and then by sequentially finding the other components having maximal correlation with y while still verifying the imposed side restrictions. Below we propose an easy heuristical algorithm yielding a good approximation to the solution of the stated maximization problem under contraints.

For finding the first predictor variable z_1, we look for a vector $\boldsymbol{b}$ resulting in a high value of the function

$$\boldsymbol{b} \rightarrow |\text{Corr}(\boldsymbol{y}, \boldsymbol{X}\boldsymbol{b})|. \tag{2}$$

The correlation in (2) is the usual sample correlation coefficient between two column vectors. Since the value of the objective function in (2) is invariant with respect to scalar multiplication of $\boldsymbol{b}$, we add the constraint that $\text{Var}(\boldsymbol{X}\boldsymbol{b}) = 1$. The maximum of (2) would be obtained by choosing $\boldsymbol{b}$ as the LS estimator $\hat{\beta}_{LS}$ due to model (1). However, to avoid the multicollinearity problem, we are not looking at the global maximum of this function, but we restrict ourselves at evaluating (2) at the discrete set

$$B_{n,1} = \left\{ \frac{\boldsymbol{x}_i}{||\boldsymbol{x}_i||}; i = 1, \ldots, n \right\}.$$

(Similar as in the algorithm of Croux and Ruiz-Gazen (1996) for principal component analysis.) The scores of the first component are then simply given by the vector $\boldsymbol{z}_1 = \boldsymbol{X}\boldsymbol{b}_1$, where $\boldsymbol{b}_1$ is the value maximizing the function (2) over the set $B_{n,1}$. Afterwards, $\boldsymbol{b}_1$ is rescaled in order to verify the side restriction of having unit sample variance for the scores of the first component. The set $B_{n,1}$ is the collection of vectors pointing at the data, and can be thought of as a collection of potentially interesting directions. Note that finding $\boldsymbol{b}_1$ can be done without any numerical difficulty, and in $O(n^2)$ computation time. Even in the case of more variables than observations ($p > n$), which is of interest for example in spectroscopy, the variable $\boldsymbol{z}_1$ can easily be computed. For very large values of n, one could pass to a subset of $B_{n,1}$.

For finding the scores of the second component $\boldsymbol{z}_2$, we need to restrict to the space of all vectors having sample correlation zero with $\boldsymbol{z}_1$. Denote $\boldsymbol{X}_j$ the j-th column of the data matrix $\boldsymbol{X}$, containing the realizations of the variable x_j with $1 \leq j \leq p$. Herefore we regress all data vectors $\boldsymbol{y}, \boldsymbol{X}_1, \ldots, \boldsymbol{X}_p$ on the already obtained first component $\boldsymbol{z}_1$, just by means of a sequence of $p+1$ simple bivariate regressions. Since all these regressions are bivariate, they cannot imply any multicollinearity problems. We will continue then to work with the residual vectors obtained from these regressions, which we denote by $\boldsymbol{y}^1, \boldsymbol{X}_1^1, \ldots, \boldsymbol{X}_p^1$. Note that all these vectors are uncorrelated with $\boldsymbol{z}_1$. With $\boldsymbol{X}^1 = (\boldsymbol{X}_1^1, \ldots, \boldsymbol{X}_p^1)$, the second predictor variable is found by maximizing the function

$$\boldsymbol{b} \rightarrow |\text{Corr}(\boldsymbol{y}^1, \boldsymbol{X}^1\boldsymbol{b})|. \tag{3}$$

The maximum in (3) can in principle be achieved by taking the LS estimator for $\boldsymbol{b}$. Using the statistical properties of LS estimators, it can be seen that this would yield a SMC value between y and z_1, z_2 equal to the SMC between y

and the x-variables. But, again, since we are concerned with the collinearity problem, we will approximate the solution of (3) by searching only in the set $B_{n,2} = \left\{ \frac{\boldsymbol{x}_i^1}{\|\boldsymbol{x}_i^1\|}; i = 1, \ldots, n \right\}$ where $\boldsymbol{x}_i^1$ denotes the i-th row vector of $\boldsymbol{X}^1$. The vector $\boldsymbol{b}_2$ maximizing the function (3) over the set $B_{n,2}$ defines, after rescaling to get unit variance, the scores on the second component $\boldsymbol{z}_2 = \boldsymbol{X}^1 \boldsymbol{b}_2$. Since we passed to the space of residuals, the first two components will be uncorrelated, as required.

Note that if we would have worked with the theoretical maximum in (2) of the first step, then the objective function (3) would be equal to zero, since LS residuals are orthogonal to the explicative variables. In the latter case, there is no correlation left to explain after the first step. But since we are only approximating the solution, and not computing the LS-estimator directly, we will still have a non-degenerate solution to (3). A comparison of the numerical values of the maxima in (2) and (3) tells us how much explicative power is gained by adding z_2 to the model.

The other components $\boldsymbol{z}_3, \ldots, \boldsymbol{z}_k$ are now obtained in an analogous way as $\boldsymbol{z}_2$. Component $\boldsymbol{z}_l$ ($l = 3, \ldots, k$) is found by maximizing

$$\boldsymbol{b} \rightarrow |\mathrm{Corr}(\boldsymbol{y}^{l-1}, \boldsymbol{X}^{l-1}\boldsymbol{b})| \tag{4}$$

where $y^{l-1}, \boldsymbol{X}^{l-1} = (\boldsymbol{X}_1^{l-1}, \ldots, \boldsymbol{X}_p^{l-1})$ are obtained by regressing the previously obtained residual series $\boldsymbol{y}^{l-2}, \boldsymbol{X}_1^{l-2}, \ldots, \boldsymbol{X}_p^{l-2}$ on component $\boldsymbol{z}_{l-1}$. We approximate the solution of (4) by considering only the n candidate vectors of the set $B_{n,l} = \left\{ \frac{\boldsymbol{x}_i^{l-1}}{\|\boldsymbol{x}_i^{l-1}\|}; i = 1, \ldots, n \right\}$.

3 Example

We consider a data set from geochemistry which is available in form of a geochemical atlas (Reimann et al., 1998). An area of 188000 km^2 in the so-called Kola region at the boundary of Norway, Finland, and Russia was sampled. More than 50 chemical elements have been analyzed for all 606 samples. For some of the most interesting elements like gold (Au) it is not possible to obtain reliable estimations of the concentration because often the concentration is below the detection limit. It would thus be advantageous to estimate the contents of the "rare" elements by using the information of the other elements. Similar chemical structures of rocks and soil allow a dependency among the chemical elements which can sometimes be very strong. This leads to a regression problem with multicollinearity. Filzmoser (2001) used a robust PCR method to predict the contents of these "rare" elements. Here we apply our proposed algorithm to predict the concentration of cromium (Cr) using 54 other chemical elements as predictors. Cr belongs certainly not to the "rare" elements, but it is strongly related to many other elements, and hence it is suitable for testing our method and comparing it with conventional PCR. Figure 1 shows the comparison of (1) PCR with sequential selection

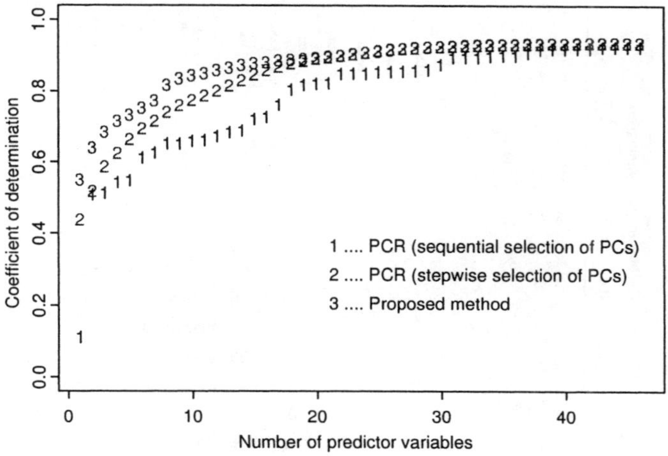

Fig. 1. Comparison of (1) PCR with sequential selection of PCs, (2) PCR with stepwise selection of PCs, (3) proposed method. The coefficient of determination is drawn against the number of predictor variables.

of PCs, (2) PCR with stepwise selection of PCs, and (3) the newly proposed method. The coefficient of determination is drawn against the number of predictor variables used in the linear model. Figure 1 shows that for each fixed number of predictor variables the coefficient of determination was highest for the new method. As already expected, the sequential selection of PCs according to their largest variances gives the lowest coefficient of determination. Our method would therefore allow a major reduction of the number of explanatory variables in the regression model.

4 Simulation study

We want to compare the proposed method with the classical multiple regression approach, and with PCR regression. Therefore, we generate a data set $\boldsymbol{X}_1$ with $n = 500$ samples in dimension $p_1 = 50$ from a specified $N(\mathbf{0}, \boldsymbol{\Sigma})$ distribution. For obtaining collinearity we generate $\boldsymbol{X}_2 = \boldsymbol{X}_1 + \boldsymbol{\Delta}$, and the columns of the noise matrix $\boldsymbol{\Delta}$ are independently distributed according to $N(0, 0.001)$. Both matrices $\boldsymbol{X}_1$ and $\boldsymbol{X}_2$ are combined in the matrix of independent variables $\boldsymbol{X} = (\boldsymbol{X}_1 | \boldsymbol{X}_2)$. Furthermore, we generate a dependent variable as $\boldsymbol{y} = \boldsymbol{X}\boldsymbol{a} + \boldsymbol{\delta}$. The first 25 elements of the vector $\boldsymbol{a}$ are generated from a uniform distribution in the interval $[-1, 1]$, and the remaining elements of $\boldsymbol{a}$ are 0. The variable $\boldsymbol{\delta}$ comes from the distribution $N(0, 0.8)$. So, $\boldsymbol{y}$ is a linear combination of the first 25 columns of $\boldsymbol{X}_1$ plus an error term.

In the simulation we consider PCR of $\boldsymbol{y}$ on $\boldsymbol{X}$ by sequentially selecting the PCs according to the magnitude of their eigenvalues and by stepwise

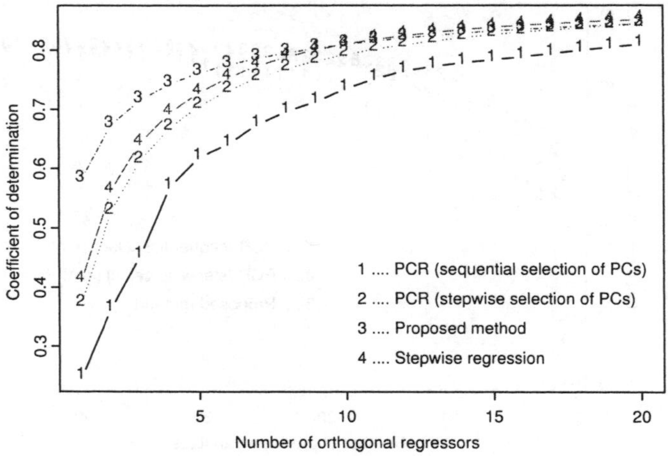

Fig. 2. Comparison of (1) PCR with sequential selection of PCs, (2) PCR with stepwise selection of PCs, (3) the proposed method, and (4) stepwise regression. The mean coefficient of determination is drawn against the number of predictor variables.

selecting the PCs according to the largest increase of the R^2 measure, our proposed regression method, and stepwise regression (forward selection of the predictor variables). We computed $m = 1000$ replications and a maximum number $k = 20$ of predictor variables.

We can summarize the resulting coefficients of determination by computing the average SMC over all m replications. Denote $R^2_{LS}(t, j)$ the resulting SMC coefficient of the t-th replication if j regressors are considered in the model. Then

$$R^2_{LS}(j) = \frac{1}{m} \sum_{t=1}^{m} R^2_{LS}(t, j), \tag{5}$$

for each number of regressors $j = 1, \ldots, k$. Figure 2 shows the mean coefficient of determination for each considered number j of regressors. We find that our method gives a higher mean coefficient of determination especially for a low number of predictor variables which is most desirable. PCR with sequential selection gives the worst results. For obtaining the same mean coefficient of determination, one would have to take considerably more predictor variables in the model than for our proposed method.

5 Discussion

The two-step procedure of PCR, namely computing the PCs of the predictor variables and performing regression of the response variable on the PCs can be reduced to a single step procedure. Like PCR, the proposed method is able to deal with the problem of multicollinearity, but the new predictor variables which are linear combinations of the original x-variables lead in general to a higher coefficient of determination compared to PCR. Since one usually tries to explain the main variability of the response variable by a possibly low number of predictor variables, the proposed method is preferable to PCR, and also to the stepwise regression technique as was shown by the simulation study.

Often it is important to find a simple interpretation of the regression model. Since PCR as well as our proposed method are searching for linear combinations of the x-variables, the resulting predictor variables will in general not be easy to interpret. Our example has shown that the interpretation of the predictor variables is not always necessary. However, if an interpretation is desired, one has to switch to other methods which can deal with collinear data, like ridge regression (Hoerl and Kennard, 1970). There are also interesting developments of methods in the chemometrics literature. Araújo et al. (2001) introduced a projection algorithm for sequential selection of x-variables in problems with collinearity and with very large numbers of x-variables.

An important advantage of our proposed method is that it can easily be robustified. It is well known that outliers in the y-variable and/or in the x-variables can have a severe influence on regression estimates, even for bivariate regressions. Hence, robust regression techniques like least median of squares (LMS) regression or least trimmed squares (LTS) regression (Rousseeuw, 1984) have been developed which can resist the effect of outliers. The classical correlations used in (2) and (3) can also be replaced by robust versions. Note that Croux and Dehon (2001) introduced robust measures of the multiple correlation.

References

ARAÚJO, M.C.U., SALDANHA, T.C.B., GALVÃO, R.K.H., YONEYAMA, T., and CHAME, H.C. (2001): The Successive Projections Algorithm for Variable Selection in Spectroscopic Multicomponent Analysis. *Chemometrics and Intelligent Laboratory Systems, 57*, 65–73.

BASILEVSKY, A. (1994): *Statistical Factor Analysis and Related Methods: Theory and Applications.* Wiley & Sons, New York.

CROUX, C. and DEHON, C. (2001): Estimators of the Multiple Correlation Coefficient: Local Robustness and Confidence Intervals, to appear in *Statistical Papers*, `http://www.econ.kuleuven.ac.be/christophe.croux`.

CROUX, C. and RUIZ-GAZEN, A. (1996): A Fast Algorithm for Robust Principal Components Based on Projection Pursuit. In: A. Prat (ed.): *Computational Statistics.* Physica-Verlag, Heidelberg, 211–216.

FILZMOSER, P. (2001): Robust Principal Compnent Regression. In: S. Aivazian, Yu. Kharin, and H. Rieder (Eds.): *Computer Data Analysis and Modeling.*

Robust and Computer Intensive Methods. Belarusian State University, Minsk, 132–137.

HADI, A.S. and LING, R.F. (1998): Some Cautionary Notes on the Use of Principal Components Regression. *The American Statistician, 1*, 15–19.

HOERL, A.E. and KENNARD, R.W. (1970): Ridge Regression: Biased Estimation for Nonorthogonal Problems. *Technometrics, 12*, 55–67.

REIMANN, C., ÄYRÄS, M., CHEKUSHIN, V., BOGATYREV, I., BOYD, R., DE CARITAT, P., DUTTER, R., FINNE, T.E., HALLERAKER, J.H., JÆGER, Ø., KASHULINA, G., LEHTO, O., NISKAVAARA, H., PAVLOV, V., RÄISÄNEN, M.L., STRAND, T., and VOLDEN, T. (1998): *Environmental Geochemical Atlas of the Central Barents Region.* NGU-GTK-CKE special publication. Geological Survey of Norway, Trondheim.

ROUSSEEUW, P.J. (1984): Least Median of Squares Regression. *Journal of the American Statistical Association, 79*, 871–880.

Confronting Data Analysis with Constructivist Philosophy

Christian Hennig

Seminar für Statistik,
ETH-Zentrum (LEO),
CH-8092 Zürich, Switzerland

Abstract. This paper develops some ideas from the confrontation of data analysis with constructivist philosophy. This epistemology considers reality only dependent of its observers. Objective reality can never be observed. Perceptions are not considered as representations of objective reality, but as a means of the self-organization of humans. In data analysis, this leads to thoughts about the impact of the gathering of data to the reality, the necessity of subjective decisions and their frank discussion, the nature of statistical predictions, and the role of probability models (frequentist and epistemic). An example from market segmentation is discussed.

1 Introduction

Some recent developments in epistemology, namely constructivist and post-modern theories, had a large impact in the social and educational sciences, but they are widely ignored up to now in the foundations and practice of many natural sciences including mathematics, statistics and data analysis. Data analysts are concerned with the generation of knowledge and with the question of how to learn about the reality from specific observations, which lies in the heart of constructivist epistemology. This is why I think it it fruitful to confront constructivist philosophy with data analysis, even though the rejection of the concept of "objective reality" by most of the constructivists seems off-putting to many researchers educated in the spirit of the natural sciences. This paper is meant as a short sketch of ideas to stimulate discussions. In Section 2 I give a brief introduction to constructivist philosophy. In Section 3, I argue that data based research should be considered as a conscious, subjective process which changes the observed reality rather than as a means to find out what the objective truth is. In Section 4, the role of probability models is discussed. This is followed by an illustrating example from market segmentation.

2 Constructivist philosophy

2.1 A short introduction

In the literature there are various interpretations of "constructivism". One may distinguish "radical" from "social" and "methodological" constructivism (introductions and anthologies are e.g. Berger and Luckmann (1966), Watzlawick (1984), Gergen and Davis (1985), von Glasersfeld (1995)). There are three principles common to constructivist approaches to epistemology:

There is no observation without observer. There is no means to go beyond a persons observations, except by observations of others, observations of observations, respectively. Every judgment about the validity of an observation (which can be a belief or an inference as well) depends upon the observer, and upon the person who judges. In this sense, no objectivity is possible.

Observations are constructed in social dependence. As human beings, we are bound by language and culture. We do not only learn how to *communicate* perceptions, we learn even to *perceive* by means of language and culture. We are bound by material constraints as well, but this can only be observed through language and culture. Individuals get to socially recognized and accepted perceptions by interaction between their actions and the actions of their social systems. This process, starting from the first perceptions in the earliest childhood to sophisticated scientific experiments, is called the "construction of observations." It can be analyzed on the personal and social level.

Perception is a means of self-organization, not of representation. Constructivist epistemology rejects the hypothesis that observations and perceptions should be analyzed as somewhat biased representations of objective reality, because it is not possible to assess the difference between reality and representation independently of observers. Instead, perceptions are thought as a means for an individual (or a social system) to organize itself to fit (more or less) successfully to the constraints of its environment, which are recognized to exist, but not to be accessible objectively. Note that there also cannot be objectivity in the attribution of "success" to a process of self-organization. Values like this must be culturally negotiated.

Because the construction of observations involves individual actions, we can ascribe (more or less) responsibility for it to the individual. In particular, a social system is constituted by the way in which its members communicate their observations and beliefs. That is, all members influence the constructions which are valid for the social system, and the other way round.

2.2 Consequences

How can an epistemology influence the practice of a science? I think that the main contribution of the constructivist philosophy can be a shift of the focus of interest from some problems (e.g. "What is objectively true?") to others:

1) How do data analysts construct their perception of reality by use of models, methods and communication?

2) What perception of reality gives rise to the models and methods? This is reversal to 1 and illustrates that the whole constructive process can be thought as circular.
3) In constructivism, alternative realities (personal as well as social) are possible. Given a construction of reality, how can alternatives be constructed, how and why is such a construction hindered, respectively?
4) What is the role of subjective decisions and how can the responsibility of the subjects be made visible?

According to its own standards, constructivism should not be seen as a "correct" or "wrong" philosophy. It is able to shift and broaden someone's view, but it has also been possible to develop many of the following ideas without an account to constructivism, as can be seen from some of the references given below.

3 Data based research as a constructive process

Data analysis deals with the transformation of observations: A phenomenon of interest (for which I use "nature" as a generic term) is transformed into data, usually numbers or categories. The data are analyzed by use of certain methods and models, which transform the raw data into statistical summaries and graphics. These results have to be transformed back to the nature by means of interpretation. An observer can distinguish the perceived nature from the gathered data and the chosen mathematical model as three different realities. The following discussion is to show that actions and decisions involved in the data analytic transformations affect all these realities, so that data analysis can be said to change observed realities.

3.1 Transforming nature into data

The transformation of nature into categories is fundamental for the development of language. Language is essential for human coordination and development, and at the same time it often tends to obscure the richness of nature.

The gathering of data can be thought as analogous. The definition of categories, quantities and measurements makes certain aspects of a problem important, while non-measured aspects tend to vanish from the consciousness (that is, from the observed reality) of the analysts.

As an example consider the comparison of the quality of schools. If such a comparison should result in a ranking, it has to be carried out on the base of a one-dimensional ordinal criterion and usually on the base of numerical data. For example, unified tests resulting in a number of points can be performed.

This may have a strong impact on the considered reality and its perception. If the content of such a test is known at least approximately, schools and teachers will try to train their students to optimize the test results, no matter if the tested items correspond to the needs and talents of their particular groups of students. Further, not every capacity is equally easy to measure. This results in a down-weighting of abilities which are more difficult to assess by tests or to the invention of more or less questionable measures for them.

3.2 Data analysis: subjective and responsible

From a positivist point of view, the aim of data analysis is to make reproducible statements about the objective reality. For this purpose, it is desirable to find a unique optimal method to prevent a subjective impact. Because non-standard treatments of data always are suspect to be subjective, it is preferred to use methods which optimize widely recognized criteria such as minimum variance. It lies in the nature of such criteria that they often are not very well adapted to the individual circumstances. The sensitivity of the variance criterion with respect to the outlier problem may serve as an illustration: It is recognized that robust methods can be used with advantage in most situations, but many data analysts think that robust statistics suffers from offering the user too many different estimation methods.

On the other hand, the constructivist approach stresses the active, responsible role of the data analyst. In agreement, Tukey (1997) justifies that different experts may draw different, equally reasonable conclusions from the same dataset, and Hampel (1998) argues that almost every data problem can give rise to a closely adapted, new and idiosyncratic treatment superior to the application of standard methods, when time and resources suffice.

It is easy to recognize the subjectivity, which is necessarily involved in choosing a model, a class of methods, an optimality criterion and tuning constants (if required). Positivists like to present statistics as a unified bundle of rules stating clearly the method to apply under given circumstances, despite the fact that their assumptions can never be verified. This tends to hide the responsibility of the analyst for her results.

The constructivist viewpoint does not mean that the choice of methods and models is arbitrary. Researchers are responsible for giving confidence in the import and benefit of a study to its addressees. It is crucial that the background and the arguments for subjective decisions are given as detailed as possible to gauge them. Receivers (constructivist or not) want to know what the exact reasons for choosing a method have been and how well the analysis has been connected to the subject matter. The confidence may be enlarged by the presentation of a variety of well explained solutions, e.g. defined by different tuning constants. Different solutions stimulate discussions and sharpen the perception.

Lack of time or computing power, concentration on other aspects of a study, routine use of familiar methods or even lack of statistical knowledge are legitimate, honest reasons for such a choice. However, they should be made transparent and may be criticized with every right by somebody who thinks to be able to do better. "Legitimate" does not mean "uniquely true". Therefore, it is useful to consider the result of a data analytic study as a responsible construction of a perception, as opposed to the revelation of a hidden truth.

3.3 Transforming results into nature: prediction

I focus on only a single aspect of interpretation. The results of a data analysis are often interpreted as if they enable somewhat uncertain predictions about the future behavior of a setup, the further development of a times series or the expected error rate of the classification of future observations, say.

My main point about such predictions is that they always need to assume that the future equals the past in terms of the underlying model (probabilistic or not). This means that every possible difference between future and past has to be judged as non-essential by the researcher (corresponding to the interpretation of the term "random" in Section 4.1). This may be reasonable in some controlled technical experiments, but is usually a very restricted view in every setup where human decisions are involved. Often, e.g. in stock markets, the prediction itself influences the future. In Germany, in the seventies the need of nuclear power was advertised by the use of over-pessimistic predictions for the electricity consumption, neglecting totally the possibility to influence the reasons for the consumption instead of simply providing more electricity. From a constructivist viewpoint, there is the danger that an uncritically adapted model for prediction may obscure the perception of possibilities to change the behavior reflected in the model. It is more constructive to use the outcome of the model as an illustrative scenario which we may want to prevent, or, in other cases, to reach.

4 Probability models and reality

Every formal method of data analysis can be considered as more or less explicitly model based. In this section I concentrate on probability models, because their relation to the real world has been discussed most intensively. Most of the remarks apply to other formal models for data as well.

4.1 The role of model assumptions

For model based methods in the frequentist sense it is usually assumed that the data is generated by some random mechanism which can adequately be described by some probability model. While most statisticians do not believe that such models are exactly true, the usual communication of statistics is based on the argument that there is a true model in nature generating the data and the assumed model should match this at least approximately.

Such communication sometimes leads to an ignorance of the essential non-randomness of the data. Consider, e.g., the use of the term "error" for deviations from the expectation. Such deviations are declared pathological in some psychological theories, often before taking their individual non-random reasons into account.

No probability model, how general it may be without getting completely useless, can ever be verified by observations. This holds even for approximate validity and for verification in a statistical sense (as opposed to logical):

- The usual logic of a statistical test implies that only a rejection of the null hypothesis is meaningful. The practice of goodness-of-fit tests to check model assumptions is reversal.
- Accepting a model conditional on the outcome of a goodness-of-fit test is a sure way to violate the model assumptions, because the chance for significance is, say, 5% under the model, but 0% for the resulting data. Graphical assessment of model assumptions shares, less formally, the same problems.
- Only simple error variance and dependency structures can be distinguished from each other by observations, while it is never possible to exclude less regular non-i.i.d. models.

Thus, true probability models should not be treated as existent in any scientifically observable reality. Instead, models can be interpreted as concepts of the human mind which help to structure perceptions.

A probability model may formalize a regular structure which a researcher perceives or presumes about the observed phenomena. To assume a model while not believing in its objective truth means to assess possible deviations from this structure as *non-essential* with respect to the research problem of interest. Thus, "random" in a frequentist sense could mean that an observer judges the sources of variation as non-essential.

This is a subjective decision which can be made transparent. Insofar, a model can be utilized to discuss different perceptions of a phenomena and to compare them with observations. But if the model assumptions are accepted without discussion, the sources of deviations vanish from the perception of the researchers, and this leads to a narrow view of reality.

Further, it makes sense to use models as "test beds" for methods (Davies and Kovac (2001), note also Davies (1995) for a concept of probability models avoiding reference to a non-observable objective reality). The true answer to an interesting real data analytic problem cannot be known (benchmark data are not "interesting" in this sense, because the truth about them must be assumed as known), and so it cannot be tested whether a data analytic method is able to find it. This means that formal models - not necessarily probabilistic - are useful to compare the quality of methods.

Model-based methods are often rejected for the reason that they should not be applied if the model assumptions are not fulfilled. This viewpoint rests on a misunderstanding about such assumptions, which are not meaningful about objective reality, but about the perception of researchers. The advantage of model based methods is that the circumstances at which they work (or not) are made at least partially transparent. This can also be achieved by proceeding the other way round: Finding a good model for a given method, as suggested by Tukey (1962).

4.2 Concepts of probability: subjectivism is not constructivist

It was stated in the previous section that probability models only reflect the perceptions and attitudes of the researchers. This could lead to the thought that probabilities should always be interpreted as epistemic, as done in the subjectivist Bayesian approach. But while an aleatory interpretation requires non-verifiable assumptions about the material reality, the epistemic interpretation requires non-verifiable assumptions about the states of mind of the individuals. For example, to observe behavior corresponding to epistemic probabilities, it is necessary to postulate a linear scale of utility valid for different interacting individuals. Further, it is excludes that individuals change their a priori opinions during experiments in reaction to events of any kind, unless they were modeled in advance.

In conclusion, epistemic probabilities as models for beliefs of individuals are subject to analogous objections as those raised in the previous section against the frequentist models, as long as they are meant to approximate objectively true states of mind. And they share the same advantages if they are meant to illustrate the ground on which researchers act.

The decision between aleatory and epistemic probabilities should be a decision between *interests* of the researchers, namely the interest in modeling a single reality shared by all involved individuals or the interest to model individually differing but internally consistent points of views.

If reality is accepted as non-objective and observer-dependent, the missing verifiability of the classical concepts does not make it necessary to reject them in favor of more elaborate concepts like imprecise probabilities (see, e.g., Walley (1991)). Addition of complexity does model more complex perceptions to the price that analysis gets more difficult (which is sometimes useful), but it cannot move the models nearer to objective reality.

5 An example

A classification-related example should serve to illustrate some of the discussed issues. Carroll and Chaturvedi (1998) apply their proposed method of k-midranges clustering to a segmentation of the mail order insurance market. "k-midranges clustering" means that a partition of cases is computed in order to minimize the maximum distance of a case to the nearest cluster center (these are shown in the paper to be "midranges" of the clusters) in the worst variable.

Data were collected from 600 potential purchasers. A conjoint analysis was used to derive importances (non-negative and summing up to 1 for every subject) for nine financial services attributes such as price of basic service, general advice/information and bill payment, i.e., there are 600 cases and 9 variables. The authors present a k-means solution with 5 clusters, chosen "on the grounds of highest Variance-Accounted-For and interpretability". Further they compute k-midrange clusterings for $k = 1, \ldots, 14$ and discuss the 3-cluster solution because of the steepest fall of the criterion function and the 13-cluster solution because of best interpretability. They observe that "the k-midranges procedure and the k-means procedure offer very different solutions in terms of segment membership". No decision about a unique best solution is offered.

The data could alternatively be analyzed by assuming that there is a real segmentation following a stochastic model, a mixture of Normal distributions with arbitrary covariance structure, say. The clusters could then be estimated by a maximum likelihood method. In fact, k-means is a likelihood maximizer for a Normal partition model with equal spherical covariance matrices and k-midranges leads to a (non-unique) maximizer of the likelihood for a partition model with uniform distributions on suitable defined hypercubes.

What is the real clustering, the correct model, the best method? From the constructivist viewpoint, the more interesting question is how to utilize the data to generate a convincing perception which supports the successful self-organization of the addressee, a company, say.

This aim does not require a unique partition of the data. One or few clusters, may be overlapping and from different partitions, can be used for product design or to tailor a marketing campaign to a specific segment. An interplay of a graphical analysis and knowledge of the essential properties of the methods (often derived from model assumptions) may serve to decide about the most convincing clusters for this purpose. It is helpful to transfer the formalized objective functions as directly as possible into perceptions of the subject matter: Should accessible market segments rather have low value ranges (k-midrange) or low variance (k-means)? Because the two sound similar, it is interesting if the difference in the solutions can be translated to a certain meaning for the market structure. Presumably, the company is interested in the main bulk of the purchasers of the segment, which may be better described by k-means, while the property of k-midranges to prevent the inclusion of single extreme purchasers may not be too helpful here. The stability of the solutions should be taken into account as well as aspects outside the data such as the profile of the company and the behavior of competitors. The researchers could further try to formalize a well accessible market segment directly in terms of a new objective function. A lot of non-standard subjective treatment seems to be reasonable here.

While I stress that no method can be judged as objectively optimal, it may be wondered if the subsequent earnings of the company would be a quality measure of a data based decision. But this measure (given that it could be accurately observed) is obviously affected by many other aspects, in particular by the ability of the company to address the selected target group. A successful campaign influences the attitudes of the customers, and it may generate the segment aimed for. Thus, under favorable circumstances even a poor data analysis may construct the reality proving its success.

6 Conclusion

I gave a short introduction to constructivist philosophy and derived some, may be provoking, ideas connected with data analysis.

Human beings are dependent of structuring their thoughts, and they can organize themselves often well by inventing models, generating data, and analyzing them. They should, however, not forget that this changes their thoughts and perceptions and the thoughts and perceptions of others. Lots of interesting and important processes between individuals vanish if we only concentrate on looking at the data. Data analysis might benefit from taking the construction processes into account more consciously.

References

BERGER, P. L. and LUCKMANN, T. (1966): *The Social Construction of Reality.* Anchor Books, New York.

CARROLL, J. D. and CHATURVEDI, A. (1998): K-Midranges Clustering. In: A. Rizzi, M. Vichi, and H.-H. Bock (Eds.): *Advances in Data Science and Classification.* Springer, Heidelberg, 3–14.

DAVIES, P. L. (1995): Data Features. *Statistica Neerlandica, 49*, 185–245.

DAVIES, P. L. and KOVAC, A. (2001): Local extremes, runs, strings and multiresolution. *Annals of Statistics, 29*, 1–47.

FEYERABEND, P. (1988): *Against Method* (revised version). Verso, London.

GERGEN, K. J. and DAVIS, K. E. (1985) (Eds.): *The Social Construction of the Person.* Springer, New York.

HAMPEL, F. (1998): Is statistics too difficult? *Canadian Journal of Statistics, 26*, 497–513.

TUKEY, J. W. (1962): The future of data analysis. *Annals of Mathematical Statistics, 33*, 1–67.

TUKEY, J. W. (1997): More honest foundations for data analysis. *Journal of Statistical Planning and Inference, 57*, 21–28.

VON GLASERSFELD, E. (1995): *Radical Constructivism: A Way of Knowing and Learning.* The Falmer Press, London.

WALLEY, P. (1991): *Statistical Reasoning with Imprecise Probabilities.* Chapman and Hall, London.

WATZLAWICK, P. (1984) (Ed.): *The Invented Reality.* Norton, New York.

Statistical Methods

Maximum Likelihood Clustering with Outliers

María Teresa Gallegos

Fakultät für Mathematik und Informatik, Universität Passau,
D-94030 Passau, Germany

Abstract. Suppose that we are given a list of n $\mathbb{R}^p$–valued observations and a natural number $r \leq n$. Further, assume that r of them arise from any one of g normally distributed populations, whereas the other $n-r$ observations are assumed to be contaminations. We develop estimators which simultaneously detect $n-r$ outliers and partition the remaining r observations in g clusters. We analyze under which conditions these estimators are maximum likelihood estimators. Finally, we propose algorithms that approximate these estimators.

1 Introduction

Maximum likelihood methods for the clustering problem under the assumption of normally distributed clusters are nowadays classic. Depending on how much information is (a priori) available on the parameters (mean vector and covariance matrix) of the various generating statistical laws, three main situations are to be discerned. Whereas the first one assumes that the clusters are normally distributed with *unknown* mean vectors and the same spherical covariance matrix of *unknown* size, the second situation (Section 2) is less restrictive, since it replaces the latter condition on the scatter matrices with the assumption of *unknown* but *equal* covariance matrices. Finally, the third situation (Section 3) drops any requirement of a prior information on the parameters, and handles the general case of *unknown, general* mean vectors and covariance matrices of the different clusters. The maximum likelihood estimators corresponding to the first and second situations are better known in the literature as the trace and determinant criterion (Friedman and Rubin (1967)), respectively. In both of them, the *pooled within–groups sum of squares and products matrix, SSP matrix* $\mathbf{W}$, a weighted sum over the sample variances of the various clusters, plays a central role. The trace and the determinant criteria postulate as estimators the partitions which minimize the trace and the determinant of $\mathbf{W}$, respectively. The maximum likelihood estimator for the general situation (the third one above) is the clustering $\mathcal{R} = \{R_1, \dots, R_g\}$ of the data set in g clusters which minimizes the objective function $\prod_{j=1}^{g} \det \mathbf{V}_{\mathcal{R}}(j)^{|R_j|}$ ($\mathbf{V}_{\mathcal{R}}(j)$ is the sample covariance of the jth cluster). We will refer to it as the *General Determinant Criterion.*

It is known that the methods mentioned above are not robust to *outliers.* Since almost all real data sets contain outliers, it is natural to assume that a proportion of the observations are contaminations or spurious observations. Therefore, attempts to robustify these methods are desirable.

With the aim of robustifying the trace criterion, Cuesta-Albertos, Gordaliza and Matrán (1997) introduced *impartial trimming*: given a trimming level $\alpha \in]0,1[$, find the subset of the data of size $\lfloor n(1-\alpha) \rfloor$ which is optimal for the corresponding trace criterion.

In this communication, we develop robust versions of the determinant criterion and the general determinant criterion. Both are natural extensions of Rousseeuw's (1984) minimum covariance determinant (MCD) estimator to cluster analysis. The MCD looks for a subset of *given* size of the data set whose covariance matrix has lowest determinant. The robust estimates of mean vector and covariance matrix are the sample mean and the sample covariance matrix, respectively, of this subset. Similarly, for a given natural number r our robust extensions of the determinant and the general determinant criterion, seek a subset R of the data set with r elements and a partition $\mathcal{R}$ of this subset in g clusters which optimize the corresponding determinant and general determinant criterion, respectively. We call these extensions the *Determinant Criterion with Outliers* and the *General Determinant Criterion with Outliers*, respectively.

These estimators are the object of study of this communication. Our first main result shows that, besides being meaningful descriptive tools on their own, both criteria are maximum likelihood estimators of well–defined statistical models. As a consequence of this fact both methods perform well whenever the data set is a realization of these underlying statistical laws. A remarkable fact of this result is its validity under very mild conditions on the statistical laws which generate the outliers.

Calculation of the determinant and general determinant criteria with outliers requires computation of a subset of size r out of the n observations and its subsequent partitioning in g clusters. Hence, as in the case of the determinant and general determinant criterion, their computation is infeasible for most real data sets and approximation algorithms are desirable. Our second main result deals with the design of algorithms which calculate good approximations for both determinant criteria with outliers. To our knowledge, the proposed algorithms are also new for the non robust versions of these criteria. We must however mention the local search due to Späth (1985), which also computes good approximations for the determinant criterion (without outliers). Our algorithms borrows some ideas of Rousseeuw and Van Driessen's (1999) algorithm for the MCD.

The interested reader finds proofs of the results presented here in Gallegos (2000 and 2001).

1.1 General notations

We denote the set of (strictly positive) natural numbers by $\mathbb{N} = \{0, 1, 2, \cdots\}$ ($\mathbb{N}_> = \{1, 2, \cdots\}$). Let $p \in \mathbb{N}_>$. We write $\mathbb{A} > 0$ in order to indicate that a matrix $\mathbb{A} \in \mathbb{R}^{p \times p}$ is positive definite. We denote by $\det \mathbb{A}$ its determinant. The symbol $\mathbf{I}_p$ stands for the identity matrix ($\in \mathbb{R}^{p \times p}$). Let $g \in \mathbb{N}_>$ and let

o_i, $i \in 1..g$, be points of a set E. We denote the tuple $(o_1, \ldots, o_g)$ by o_1^g. For a list or set R the notation $|R|$ stands for its cardinality.

Let R be a nonempty list of elements in $\mathbb{R}^p$. Let us denote by $\mathcal{C}_g(R)$ the set of all partitions (or configurations) of the list R in N clusters. We say that a cluster is *admissible* if it contains $p+1$ or more elements. Let $\mathcal{R} = \{R_1, \ldots, R_N\}$ be a partition in $\mathcal{C}_g(R)$. We call it *admissible* if all its clusters are admissible. We will often identify an element of $\mathcal{C}_g(R)$ with the integral interval $1..g$. Furthermore, for a nonempty cluster $R_j \in \mathcal{R}$, $j \in 1..g$, we denote by $\mathbf{m}_{\mathcal{R}}(j)$ and $\mathbf{V}_{\mathcal{R}}(j)$ the sample mean vector and the sample covariance matrix of the jth cluster, respectively. Finally, we denote by $\mathbf{W}_{\mathcal{R}}$ the SSP matrix of the partition $\mathcal{R}$, i.e.,

$$\mathbf{W}_{\mathcal{R}} = \sum_{j=1}^{g} \sum_{\mathbf{x} \in R_j} (\mathbf{x} - \mathbf{m}_{\mathcal{R}}(j))(\mathbf{x} - \mathbf{m}_{\mathcal{R}}(j))^t = \sum_{j=1}^{g} |R_j| \mathbf{V}_{\mathcal{R}}(j), \tag{1}$$

where both sums above range over all nonempty clusters of the partition $\mathcal{R}$.

The symbol $N_p(\boldsymbol{\mu}, \boldsymbol{\Sigma})$ denotes both the p–dimensional multinormal distribution with mean vector $\boldsymbol{\mu} \in \mathbb{R}^p$ and covariance matrix $\boldsymbol{\Sigma} \in \mathbb{R}^{p \times p}$ and its Lebesgue density. $N_p(\boldsymbol{\mu}, \boldsymbol{\Sigma})(\mathbf{x})$ denotes the value of this density at $\mathbf{x} \in \mathbb{R}^p$.

Let $n \in \mathbb{N}_>$, let $S = (\mathbf{x}_1, \mathbf{x}_2, \ldots, \mathbf{x}_n)$ be a sublist of $\mathbb{R}^p$ of n observations in general position, i.e., any $p+1$ elements in S are affinely independent. Let $r \in 1..n$, and let $g \in 1..r$ such that $r \geq gp + 1$. We will often identify the observation $\mathbf{x}_i$ in S with the corresponding index i ($\in 1..n$) and the list S with the integral interval $1..n$. Throughout this paper we refer to the just given elements as the input (S, n, r, g). If not otherwise specified, a (sub)list has as elements observations in $\mathbb{R}^p$ and a (sub)set has as reference set the integral interval $1..n$.

2 The determinant criterion with outliers

The classical determinant criterion of cluster analysis can be robustified, i.e., extended in order to deal with the presence of outliers in the the set of observations; cf. Gallegos (2000). As a byproduct, one obtains a clean estimation of the parameters (here, mean vectors and common covariance matrix) of the distinct underlying statistical laws. To be more precise, given the input (S, n, r, g), the goal is to find a *sublist* R of S of r elements, the list of *regular* observations, and a *partition* $\mathcal{R}$ of this list in g clusters such that $\mathbf{W}_{\mathcal{R}}$ minimizes the *objective function* $\det \mathbf{W}_{\mathcal{R}}$, i.e.,

$$\min \det \mathbf{W}_{\mathcal{R}} = \min_{R \in \binom{S}{r}} \ \min_{\mathcal{R} \in \mathcal{C}_g(R)} \det \mathbf{W}_{\mathcal{R}}; \tag{2}$$

here, the minimum on the left side ranges over all partitions in the set $\mathcal{R} \in \bigcup_{R \in \binom{S}{r}} \mathcal{C}_g(R)$. We call (2) the *Determinant Criterion with Outliers.*

In other words, first find the optimal partitions of all sublists R of S with r observations and, subsequently, choose the sublist with least value of the objective function of its optimal partition. Note that the condition $r \geq gp+1$ assures that at least one cluster has $p+1$ or more elements, a condition which in turn implies $\mathbf{W}_{\mathcal{R}} > 0$. For $g = 1$, the determinant criterion with outliers is Rousseeuw's MCD: a robust method which seeks for a subset of k elements of the n observations whose covariance matrix has lowest determinant. For $r = n$, the criterion (2) is the determinant criterion of cluster analysis.

It is well known that the determinant of the SSP matrix $\mathbf{W}_{\mathcal{R}}$ of the grouping $\mathcal{R}$ of a sublist of observations R in a given number of clusters (here g) is a measure of the scatter about the corresponding mean vectors. Large values indicate a high degree of scatter about the sample mean of at least one cluster, whereas low values represent concentration about all clusters. In this sense, those observations not belonging to the optimal set R may be considered as contaminations: their presence dramatically increases this measure of scatter. This observation makes also the Determinant Criterion with Outliers a plausible descriptive tool *per se*.

Recall that the determinant criterion (without outliers) turns out to be (also) a maximum likelihood estimator, if the various underlying statistical subpopulations are normally distributed with general mean vectors but the same covariance matrix. Therefore, it should not surprise us that this fact is also true for the determinant criterion with outliers, provided that the contaminations come from "suitable" statistical populations. For this sake, we first extend, adapt and unite both the usual statistical clustering setup, cf. Mardia et al. (1979), Section 13.2, and Mathar's (1981), Section 5.2, outlier model to the present situation. To be concrete, let us denote by

$$\Lambda := \{\ell : 1..n \to 0..g \mid |\ell^{-1}(0)| = n - r\}$$

the set of assignments of observations to clusters. The parameter set of our statistical experiment is

$$\Theta := \Lambda \times (\mathbb{R}^p)^g \times \{\boldsymbol{\Sigma} \in \mathbb{R}^{p \times p} | \boldsymbol{\Sigma} > 0\} \times (\mathbb{R}^p)^n, \tag{3}$$

where the first component of Θ stands for the unknown assignment of observations to clusters. The next two components stand for the unknown parameters of the g underlying Gaussian populations which generate the regular observations. And finally, the last component of Θ stands for the unknown parameters of the statistical laws generating the outliers. These are assumed to be Gaussian with unknown mean vector and the identity as covariance matrix. Let $X_1, X_2, \dots, X_n : (\Omega, P) \to \mathbb{R}^p$ be n independent random variables such that the p.d.f of X_i, $i \in 1..n$, under the parameter $\theta = (\ell, \boldsymbol{\mu}_1^g, \boldsymbol{\Sigma}, \boldsymbol{\nu}_1^n)$ is given by

$$f_{X_i,(\ell,\boldsymbol{\mu}_1^g,\boldsymbol{\Sigma},\boldsymbol{\nu}_1^n)}(\mathbf{x}) = \begin{cases} N_p(\boldsymbol{\mu}_{\ell(i)},\boldsymbol{\Sigma})(\mathbf{x}), & \ell(i) \in 1..g, \\ N_p(\boldsymbol{\nu}_i,\mathbf{I}_p)(\mathbf{x}), & \ell(i) = 0. \end{cases} \tag{4}$$

Clearly, these elements and assumptions define a statistical experiment. The list of observations S is a realization of these random variables. In the notations of the first paragraph of this section, we have $R_j = \ell^{-1}(j)$, $j \in 1..g$, and $S \setminus R = \ell^{-1}(0)$. Furthermore, we denote by $\mathcal{R}^\ell$ the partition of the list R induced by the assignment ℓ.

2.1 Theorem

The MLE of ℓ for the statistical model given above is the Determinant Criterion with Outliers. Moreover, if we denote its induced partition by $\mathcal{R}^\star$ then the MLE of $\boldsymbol{\mu}_1^g$ is $(\mathbf{m}_{\mathcal{R}^\star}(1),\ldots,\mathbf{m}_{\mathcal{R}^\star}(g))$ and the MLE of the common covariance matrix $\boldsymbol{\Sigma}$ is $\frac{1}{r}\mathbf{W}_{\mathcal{R}^\star}$.

Computation of the determinant criterion with outliers is infeasible for most real data sets and approximation algorithms are desirable. We now proposed such an algorithm. Starting from an initial configuration (in the set $\bigcup_{R\in\binom{S}{r}} \mathcal{C}_g(R)$) our algorithm consists of the successive iteration of the building block 2.2 below until convergence. The key idea of this step is to use the Mahalanobis distances from the observations to the different clusters (of a given grouping) in order to construct (if possible) a configuration with smaller SSP determinant than the current one. The point of convergence is our approximation to the determinant criterion of outliers.

2.2 The building block of our algorithm

// Input: The mean vectors $\mathbf{m}_\mathcal{R}$ and the SSP matrix $\mathbf{W}_\mathcal{R}$ of a configuration $\mathcal{R}$.
// Output: A configuration $\mathcal{R}_{\text{new}}$ such that $\det \mathbf{W}_{\mathcal{R}_{\text{new}}} \le \det \mathbf{W}_\mathcal{R}$.

(i) Compute the (squared) distances

$$d_\mathcal{R}(j,i)^2 := (\mathbf{x}_i - \mathbf{m}_\mathcal{R}(j))^t \mathbf{W}_\mathcal{R}^{-1}(\mathbf{x}_i - \mathbf{m}_\mathcal{R}(j)), \qquad j \in 1..g,\ i \in 1..n.$$

(ii) For each $i \in S$, determine $c_i := \operatorname{argmin}_{j\in 1..g} d_\mathcal{R}(j,i)^2$.
(iii) Determine the permutation $\pi : 1..n \to 1..n$ that satisfies

$$d_\mathcal{R}(c_{\pi(1)},\pi(1))^2 \le d_\mathcal{R}(c_{\pi(2)},\pi(2))^2 \le \ldots \le d_\mathcal{R}(c_{\pi(n)},\pi(n))^2.$$

(iv) Put $R_{\text{new}} = \{\pi(1),\pi(2),\ldots,\pi(r)\}$ and, for each $j \in 1..g$, put $R_{\text{new},j} = \{i \in 1..r | c_{\pi(i)} = j\}$. Finally, let $\mathcal{R}_{\text{new}} := \{R_{\text{new},1},\ldots,R_{\text{new},g}\}$.
(v) Return $\mathcal{R}_{\text{new}}$.

Two consecutive building blocks yielding partitions with the same value of the objective function mean convergence of the algorithm. The just proposed method is an extension of Rousseeuw and Van Driessen's algorithm for the MCD ($g = 1$). Furthermore, for this value of g the former reduces to the latter.

2.3 A simulation study

In order to evaluate the performance of the algorithm we simulate the following situation: we consider three normally distributed populations whose mean vectors are equidistant and aligned on the x–axis in $\mathbb{R}^p$, $p \in \{2, 4, 8\}$. Whereas the mean vector of the middle cluster is the origin, the norms of the mean vectors of the second and third clusters are both equal to $\chi^2_p(.999)$. The common covariance matrix $\boldsymbol{\Sigma}$ is a diagonal matrix whose entries are given by the first p entries of the list $(1.0, 9.0, 1.2, 1.4, 1.6, 1.8, 2.0, 2.2)$. We generate 2500 observations from the first cluster and 1500 and 1000 from the second and third cluster, respectively. The number of contaminations is 500. For each dimension $p \in 2..8$ and for the input parameters $n = 5500$, $g = 3$ and $r = 5000$ we want to reconstruct from scratch the simulated clustering. For this sake, we apply a multistart method and construct 3000 random initial configurations and iterate CC–steps until convergence is reached. Our approximation to the MCDC is that limit configuration which (among all the trials) has least value of the determinant of the corresponding SSP matrix. The rate of misclassified regular observations achieved by our approximation is .048 for the dimension 2 and .025 and .003 for the dimensions 4 and 8, respectively.

3 The general determinant criterion with outliers

This section deals with the general determinant criterion with outliers, a robust version of the general determinant criterion, cf. Gallegos (2001). Although the results achieved for this estimator are analogous to the ones of the determinant criterion with outliers, the technical tools used for their derivation are more complex than in the former case. (This is particularly true for the proposed algorithm, which computes good approximations for the general determinant criterion with outliers).

Given the input (S, n, r, g) $(r \geq g(p+1))$, each admissible partition $\mathcal{R}$ of a *sublist* R of S with r elements in g clusters which minimizes the *objective function* $\prod_{j=1}^{g} \det \mathbf{V}_{\mathcal{R}}(j)^{|R_j|}$ is a General Determinant Criterion with Outliers. Note that the additional condition $r \geq g(p+1)$ assures the existence of admissible partitions. (By restricting our choice to the set of admissible partitions we avoid degeneracy of scatter matrices.)

Plainly, the general determinant criterion with outliers is a useful descriptive statistic. Recall that the non robust version of the general determinant criterion ($r = n$) is a maximum likelihood estimator, whenever the g underlying statistical subpopulations are normally distributed with (unknown) general mean vectors and general covariances matrices. Taking into account this fact together with the conditions that assure optimality (in the maximum likelihood sense) of the determinant criterion with outliers, cf. Theorem 2.1, it is plausible that if we first redefine Λ as $\{\ell : 1..n \to 0..g \,|\, |\ell^{-1}(j)| \geq p+1, j \in 1..g, |\ell^{-1}(0)| = n - r\}$, second we replace the third component of the parameter set Θ, cf. (3), by $\{\boldsymbol{\Sigma} \in \mathbb{R}^{p\times p} | \boldsymbol{\Sigma} > 0\}^g$ and, finally we replace the various densities (4) by

$$f_{X_i,(\ell,\boldsymbol{\mu}_1^g,\boldsymbol{\Sigma}_1^g,\boldsymbol{\nu}_1^n)}(\mathbf{x}) = \begin{cases} N_p(\boldsymbol{\mu}_{\ell(i)}, \boldsymbol{\Sigma}_{\ell(i)})(\mathbf{x}), & \ell(i) \in 1..g, \\ N_p(\boldsymbol{\nu}_i, \mathbf{I}_p)(\mathbf{x}), & \ell(i) = 0, \end{cases} \tag{5}$$

then we obtain the following analogous to Theorem 2.1. (The notations given in the previous section concerning the relationship between an assignment ℓ and the induced admissible partition hold here, too).

3.1 Theorem

The MLE of ℓ for the statistical model given above is the General Determinant Criterion with Outliers. Moreover, if we denote it by $\mathcal{R}^*$ then the MLE of $\boldsymbol{\mu}_1^g$ and $\boldsymbol{\Sigma}_1^g$ is $(\mathbf{m}_{\mathcal{R}^*}(1), \ldots, \mathbf{m}_{\mathcal{R}^*}(g))$ and $(\mathbf{V}_{\mathcal{R}^*}(1), \ldots, \mathbf{V}_{\mathcal{R}^*}(g))$, respectively.

Note that if absence of outliers is assumed ($n - r = 0$) then the result just stated is well known. The situation is other, if we consider the proposed algorithm for the computation of approximations to the general determinant criterion with (or without) outliers. To our knowledge the method described below is new even in the pure clustering situation. The proposed algorithm works in the same way as the one given in Section 2, i.e., it iterates a basic step until convergence. The basic step has as input a given admissible partition (result of the random generation of an initial configuration or result of the previous iteration) and as output an admissible configuration with lower or equal value of the objective function. Now, computation of the general determinant criterion with outliers requires optimization of a more intricate objective function than the (simple) determinant criterion with outliers. This fact is reflected by a more complex building block than the one given in 2.2. To be more precise, the squared distance $d_{\mathcal{R}}(j,i)^2$ from the observation i to the jth cluster (of a given admissible configuration $\mathcal{R}$) is now given by

$$d_{\mathcal{R}}(j,i)^2 := (\mathbf{x}_i - \mathbf{m}_{\mathcal{R}}(j))^t \widetilde{\mathbf{V}_{\mathcal{R}}(j)}^{-1} (\mathbf{x}_i - \mathbf{m}_{\mathcal{R}}(j)), \tag{6}$$

$j \in 1..g$, $i \in 1..n$, where $\widetilde{\mathbf{V}_{\mathcal{R}}(j)} = \frac{p\mathbf{V}_{\mathcal{R}}(j)}{\det \mathbf{V}_{\mathcal{R}}(j)^{\frac{1}{p}}}$. Matrices of the type $\widetilde{\mathbf{V}_{\mathcal{R}}(j)}$ appear already in Bock (1996) when considering minimization of the objective function $\sum_{j=1}^{g} |R_j| \det \mathbf{V}_{\mathcal{R}}(j)^{\frac{1}{p}}$ in a pure clustering framework.

3.2 The building block of the algorithm

// Input: An admissible configuration $\mathcal{R}$, the mean vectors $\mathbf{m}_{\mathcal{R}}$ and the matrices $\widetilde{\mathbf{V}_{\mathcal{R}}}$.
// Output: A feasible configuration $\mathcal{R}_{\text{new}}$ such that its objective function
// is less than or equal to that of $\mathcal{R}$.

(i) Compute the matrix of squared distances $(d_{\mathcal{R}}(j,i)^2)_{\substack{j\in 1..g \\ i\in 1..n}}$, where $d_{\mathcal{R}}(j,i)^2$ is defined in (6).
(ii) For each $i \in 1..n$, determine $c_i := \operatorname{argmin}_{j\in 1..g} d_{\mathcal{R}}(j,i)$.
(iii) Determine the permutation $\pi : 1..n \to 1..n$ that satisfies

$$d_{\mathcal{R}}(c_{\pi(1)}, \pi(1))^2 \le d_{\mathcal{R}}(c_{\pi(2)}, \pi(2))^2 \le \ldots \le d_{\mathcal{R}}(c_{\pi(n)}, \pi(n))^2.$$

(iv) Put $R' = \{\pi(1), \pi(2), \ldots, \pi(r)\}$ and, for each $j \in 1..g$, put $R'_j = \{i \in 1..r | c_{\pi(i)} = j\}$. If all the R'_js are admissible then proceed with (vii).
(v) Put $L = \{\pi(r+1), \pi(r+2), \ldots, \pi(n)\}$;
Let $\mathcal{D}$ be the set of all clusters j such that $|R'_j| \le p$ (the *defective* clusters);
Let $l := \sum_{j\in\mathcal{D}}(p + 1 - |R'_j|)$;
Remove from the set $\bigcup_{j\in 1..N\setminus\mathcal{D}} R'_j$ the l observations $\pi(k)$ with largest distances to $c_{\pi(k)}$ without violating admissibility, add these to the set L, and remove these from the set R'.
(vi) while ($\mathcal{D}$ is nonempty) do
from the distances $d_{\mathcal{R}}(j,i)^2$, $j \in \mathcal{D}, i \in L$, find the smallest one, add the corresponding observation to both the corresponding defective cluster R'_j and the set R', and remove it from the set L;
If this cluster contains now $p + 1$ elements then remove it from $\mathcal{D}$.
od.
(vii) Let $\mathcal{R}' := \{R'_1, \ldots, R'_g\}$;
If $\mathcal{R}'$ satisfies

$$\prod_{j=1}^{g} \left(\frac{1}{|R'_j|} \sum_{\mathbf{x}\in R'_j} (\mathbf{x} - \mathbf{m}_{\mathcal{R}}(j))^t \widetilde{\mathbf{V}_{\mathcal{R}}(j)}^{-1} (\mathbf{x} - \mathbf{m}_{\mathcal{R}}(j)) \right)^{p|R'_j|} \le \tag{7}$$

$$\prod_{j=1}^{g} \left(\frac{1}{|R_j|} \sum_{\mathbf{x}\in R_j} (\mathbf{x} - \mathbf{m}_{\mathcal{R}}(j))^t \widetilde{\mathbf{V}_{\mathcal{R}}(j)}^{-1} (\mathbf{x} - \mathbf{m}_{\mathcal{R}}(j)) \right)^{p|R_j|}.$$

with "<" then $\mathcal{R}_{\text{new}} \longleftarrow \mathcal{R}'$;
else if $\mathcal{R}'$ satisfies (7) with "=" then $\mathcal{R}_{\text{new}} \longleftarrow$ the (admissible) configuration among $\mathcal{R}'$ and $\mathcal{R}$ with smaller value of the objective function; otherwise, $\mathcal{R}_{\text{new}} \longleftarrow \mathcal{R}$.

(viii) Return $\mathcal{R}_{\text{new}}$.

For $g = 1$ this step is Rousseeuw–Van Driessen's C–step.

References

CUESTA-ALBERTOS, J. A., GORDALIZA, A. and MATRÁN, C. (1997): Trimmed k–Means: An attempt to robustify quantizers, *The Annals of Statistics*, Vol. 25, No. 2, 553–576.

BOCK, H. H. (1996): Probabilistic models and statistical methods in partitional classifications problems. Tutorial session. Fifth Conference of International Federation of Classification Societies. April 2–3, 1996, Tokyo, Japan.

FRIEDMAN, H. P. and RUBIN, J. (1967): On Some Invariant Criteria for Grouping Data. *Journal of the American Statistical Association*, 62, 1159–1178.

GALLEGOS, M. T. (2000): A Robust Method for Clustering Analysis. Technical Report MIP–0013 October 2000. Fakultät für Mathematik und Informatik, Universität Passau.

GALLEGOS, M. T. (2001): Robust clustering under general normal assumptions. Technical Report MIP–0103 September 2001. Fakultät für Mathematik und Informatik, Universität Passau.

MARDIA, K. V., KENT, J. T. and BIBBY, J. M. (1979): Multivariate Analysis. Academic Press, London, New York, Toronto, Sydney, San Francisco.

MATHAR, R. (1981): Ausreier bei ein– und mehrdimensionalen Wahrscheinlichkeitsverteilungen. Dissertation, Mathematisch–Naturwissen schaftliche Fakultät der Rheinisch–Westfälichen Technischen Hochschule, Aachen.

ROUSSEEUW, P. J. (1984): Least Median of Squares Regression. *Journal of the American Statistical Association*, 79, 871–880.

ROUSSEEUW, P. J. and VAN DRIESSEN, K. (1999): A Fast algorithm for the Minimum Covariance Determinant Estimator, *Technometrics*, Vol. 41, 212–223.

SPÄTH, H. (1985): Cluster Dissection and Analysis. Theory, FORTRAN programs, examples. Ellis Horwood, Chichester.

An Improved Method for Estimating the Modes of the Probability Density Function and the Number of Classes for PDF-based Clustering

Michel Herbin[1] and Noel Bonnet[1,2]

[1] LERI, University of Reims,
IUT Leonard de Vinci,
Rue des Crayère BP 1035
51687 Reims cedex, France
[2] INSERM Unit 514 (UMRS, IFR 53),
45, rue Cognaq Jay
51092 Reims cedex, France

Abstract. This paper proposes an improvement of the estimation of the modes of the Probability Density Function (PDF) in clustering procedures. The k nearest neighbours are excluded when the PDF is estimated trying to avoid parasitic modes of the PDF estimation. The number of detected modes is analyzed using the bootstrap technique. The number of clusters is equal to the most frequent number of modes obtained when resampling the data (if the frequency is greater than 50%). The method to estimate the number of clusters is illustrated by an example in image processing.

1 Introduction

The nonparametric and unsupervised clustering procedures can be divided into two large classes: the hierarchical and the density based approaches. These two classes do not consider the model-based procedures using for instance Gaussian mixture models because of their parametric character. The hierarchical clustering approaches are outside the scope of this paper. The other clustering approaches use the notion of density which can be implicit or explicit in the procedures. Many density based clustering use prototypes (i.e. centers of clusters) and distances between the prototypes and the data. The classical k-means procedure is an example of these ones. Because of using distances to a center, these procedures assume that the clusters are convex. But real data do not always verify this assumption. In this paper, we only consider the density based clustering procedures that have neither explicit nor implicit assumptions on the shape of the clusters. Such procedures search the modes of the density (Herbin (1996), Cheng (1995), Touzani (1989)). We study these approaches assuming that each density mode is associated with a cluster.

Our goal is not to define the clusters but to determine the number of clusters detecting the modes of the density. We propose to estimate the Probability Density Function (PDF) and to search the modes. The Parzen-Rosenblatt (Parzen (1962)) method permits to estimate the PDF. We do not use the model approach of the PDF to avoid implicit assumptions on the shape of the clusters. This PDF estimation depends on a scale factor which is called smoothing parameter in the literature (Silverman (1986)). The higher the scale factor, the smoother the density estimation. Using a too large scale factor, the density can exhibit one mode, thus we detect only one cluster. Using a too small factor, the density can have one mode for each distinct data, thus we detect one cluster per distinct data (i.e. separate data). This scale approach permits us to estimate the number of clusters using the implicit convex shape of the clusters (Kothari (1999)). But this paper focuses on methods for estimating the number of clusters without this assumption on the shape of the clusters.

When resampling the data with the bootstrap technique, the number of modes of the estimated density can change. We select the scale factor values which minimize the variation of this number. If the variation is small enough, then the most frequent number of modes enables us to propose a number of clusters. In this paper, we propose to combine the multi-scale approach and the bootstrap technique to estimate the number of clusters.

Section 2 describes a method to estimate the Probability Density Function (PDF) and proposes an improvement of the classical Parzen-Rosenblatt approach. The goal is to avoid the parasitic modes induced by data which are alone outside clusters. Then we introduce the mode detection and the decrease of the number of modes when the scale factor increases. Section 3 proposes to study the variation of the number of modes when resampling the data. Then the numbers of clusters is proposed increasing the scale factor from a small value. Section 4 gives discussions and conclusions.

2 Estimation of the PDF and its modes

In this paper, we propose an explicit estimation of the PDF. Let X_1, X_2,... X_n be a set of n data samples in d-dimensional space. The classical Parzen-Rosenblatt PDF estimation is defined as:

$$\hat{f}_1(x) = \frac{1}{n} \sum_{i=1}^{n} \left(\frac{1}{h^d} K\left(\frac{x - X_i}{h} \right) \right) \tag{1}$$

where K is the kernel function and h is the smoothing parameter. In this paper we consider the classical Gaussian kernel.

Our goal is to estimate the modes (i.e. the local maxima) of the PDF to reveal the data clusters. But the Parzen-Rosenblatt method gives some parasitic modes which are induced by the sample noise. Thus this method

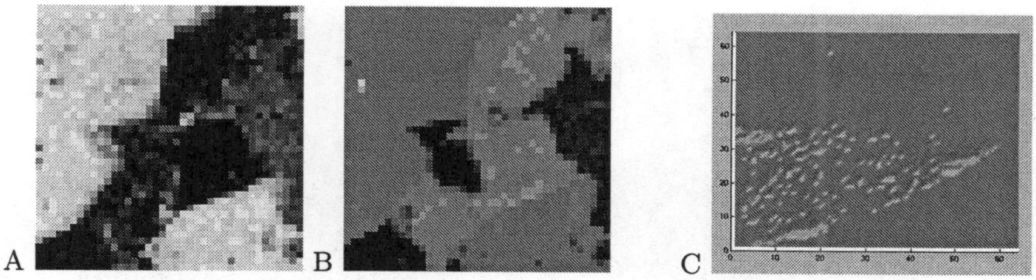

A, B : A bivariate images (39 × 40 pixels = 1560 samples).
C : 2D histogram (64 × 64 bins).

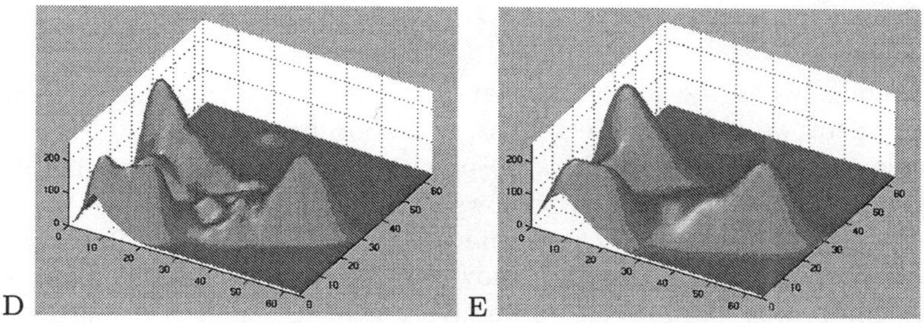

D, E : *PDF* estimated with Parzen method with $h = 2.5$ and $h = 3.5$.

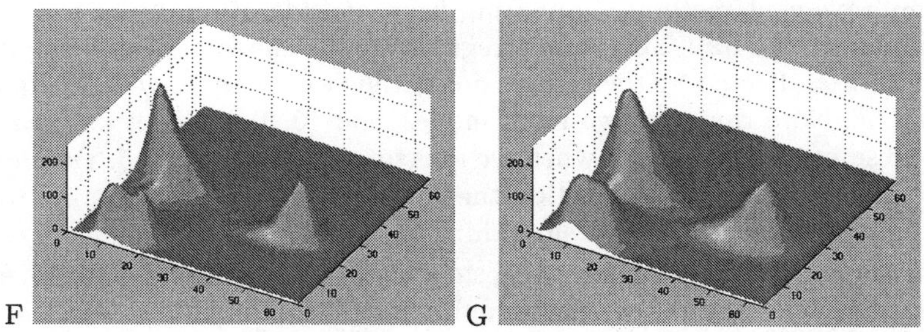

F, G : *PDF* estimated without the k nearest neigbours ($k = 32$).

Fig. 1. Estimation of the Probability Density Function.

leads to using large value of h to avoid the unwanted modes. To reduce sample noise, we propose to estimate the *PDF* at each point x without using the k nearest neighbour data. We define the *PDF* as:

$$\hat{f}_2(x) = \frac{1}{n-k} \sum_{j_x=k+1}^{n} \left(\frac{1}{h^d} K \left(\frac{x - X_{j_x}}{h} \right) \right) \tag{2}$$

where the data X_{j_x} with $j_x = 1, 2, ...n$ are ranked from the nearest from x to the farthest. In the example of this paper, the value k is chosen equal to $0.02 * n$ (2% of the samples).

We do not study the properties of this *PDF* estimation. We only give an example to display the gain of this method which avoids the increase of

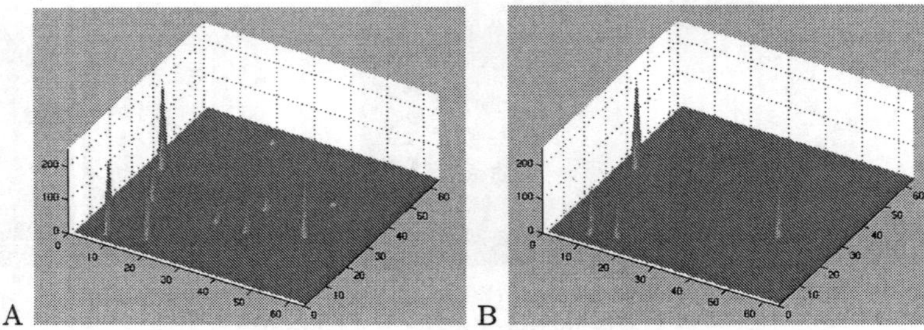

Fig. 2. Detection of PDF modes : 10 modes of Fig.1 D and 4 modes of Fig.1 F.

PDF modes when the smoothing parameter is small. Note that without this improvement we have to select a larger smoothing parameter. Fig.1 A and B display two images with 39×40 pixels. These images give 1560 data in a 2D space ($n = 1560$). Fig.1 C shows the 2D histogram with 64×64 bins. The Parzen-Rosenblatt PDF estimation is applied in this $[0, 64] \times [0, 64]$ discretized space. Fig.1 D and E show the PDF estimates using two values of the smoothing parameter, respectively $h = 2.5$ and $h = 3.5$. We can observe the smoothing effect of the parameter h. In Fig.1 F and G, we apply our improvement method eliminating from consideration 2% of the nearest neighbours ($k = 32$). The main irregularities disappear and the number of modes is not affected by the sample noise, thus we do not have to oversmooth the PDF. Note that the two main modes near $(0, 0)$ are kept using $h = 2.5$ and these modes disappear when we oversmooth with $h = 3.5$. Our method by eliminating from consideration the k nearest neighbours permits to keep the main modes without oversmoothing the PDF estimation.

In this paper, we consider the smoothing parameter as a scale factor. When the scale factor increases, small details disappear. The disappearance proceeds by stages according to the concepts of scale-space theory (Witkin (1983)).

The PDF modes are detected in the discretized space by comparing the density value of an element to the density values of its connected neighbours. In 1D space, each element has two connected neighbours. In 2D space, we use the 8-connect neighbourhood. In a space of dimension d, each element has $3^d - 1$ connected neighbours. The modes of the PDF estimation can reveal the clusters of data sample.

We consider the previous example of PDF estimation with a scale factor equal to 2.5 ($h = 2.5$). We detect the modes of the PDF. The PDF estimation with Parzen method (see Fig.1 D) has 10 modes displayed in Fig.2 A. The PDF estimation with our method (see Fig.1 F) has 4 modes displayed in Fig.2 B.

When increasing the scale factor, the number of modes decreases because of smoothing the PDF. The bold line in Fig.3 shows the decrease of the number of modes with the classical stages of the multi-scale approach. Intuitively,

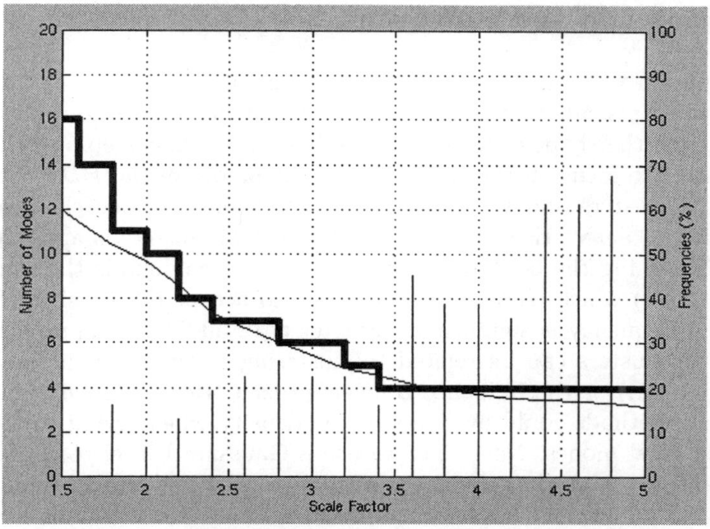

Bold line : number of modes of the initial data sample,
Thin line : mean number of modes using 30 bootstrap samples,
Stems : frequencies of the most frequent number between 0% and 100%.

Fig. 3. Number of modes as a function of the scale factor

the largest stage is selected to reveal the intrinsic number of clusters, but we propose another method to select the number of clusters.

3 Resampling with bootstrap and number of clusters

When resampling data with the bootstrap technique (Efron (1993)), the number of modes can vary. A bootstrap sample could give a number of PDF modes different from the one which is obtained with the initial sample. We propose to keep the most frequent number of modes when many bootstrap samples are used. In this work, we use 30 bootstrap samples. If the most frequent number of modes is greater than 50%, then this number is a candidate for the number of clusters. When the scale factor is too small, the most frequent number of modes is rejected (less than 50 %). Increasing the scale factor, the first accepted number of modes is proposed as the number of clusters. Note that the most frequent number of modes is equal to 1 when the scale factor is too large and 1 is always accepted. The mean number of modes is shown in Fig.3, the stems give the frequencies of the most frequent number. The first accepted number of modes is equal to four. Thus we propose four clusters in this 1560 data, this result is the same as in Herbin (1996).

4 Discussion and conclusion

We propose a new method to estimate the number of modes of the PDF. This method permits to estimate the number of clusters without implicit assumption on the shape of these clusters. This is the first step of a clustering procedure. We focus on the estimation of the modes of the PDF which can reveal the clusters and we propose an improvement of the classical PDF estimation to detect these modes. The number of clusters is approached by resampling using the bootstrap technique. This technique is the same as in Herbin (2001) which uses a chi-square test to accept the number of modes when the frequency is significantly greater than 50 %. Our approach of the number of clusters can be related to clustering methods as the mean shift (Cheng (1995)) or the influence zones approach (Herbin (1996)), but these clustering methods could oversmooth the density when trying to avoid undesirable local modes. Note that we use a Gaussian kernel to estimate the PDF, and we think that the shape of the kernel is not critical to determine the density modes. Our method is applied with success in simulation conditions and with real data. Our future work will study the detailed properties of this mode estimation approach.

References

CHENG, Y. (1995): Mean Shift, Mode Seeking, and Clustering. *IEEE Trans. on Pattern Annalysis and Machine Intelligence, 17*, 790–799.

EFRON, B., and TIBSHIRANI, R. (1993): *An introduction to the bootstrap*. Chapman and Hall, London.

HERBIN, M., BONNET, N., and VAUTROT, P. (1996): A Clustering method based on the estimation of the probability density function and on the skeleton by influence zones. *Pattern Recognition Letters, 17*, 1141–1150.

HERBIN, M., BONNET, N., and VAUTROT, P. (2001): Estimation of the number of clusters and influence zones. *Pattern Recognition Letters, 22*, 1557–1568.

KOTHARI, R., and PITTS, D. (1999): On finding the number of clusters. *Pattern Recognition Letters, 20*, 405–416.

PARZEN, E (1962): On estimation of a probability density function and mode. *Ann. Math. Statist., 33*, 1065–1076.

SILVERMAN, B.W. (1986): *Density Estimation for Statistics and Data Analaysis.* Chapman and Hall, London.

TOUZANI, A., and POSTAIRE, J.G. (1989): Clustering by mode boundary detection. *Pattern Recognition Letters, 9*, 1–12.

WITKIN, A.P. (1983): Scale space filtering: a new approach to multi-scale description. *Proc. Internat. Joint Conf. on Artificial Intelligence*, 1019–1022.

Maximization of Measure of Allowable Sample Sizes Region in Stratified Sampling

Marcin Skibicki

Department of Statistics, University of Economics in Katowice,
ul. Bogucicka 14, 40-226 Katowice, Poland

Abstract. This paper concerns the problem of optimal stratified sampling when the purpose is to simultaneously estimate the means of many variables. The goal is stratify the population and allocate sample in ways that minimize survey costs and ensure that all variances of the mean estimators are sufficiently small. The paper proposes a solution method for the case of imprecisely determined unit sampling costs. The method is based on maximization of the volume of a certain subset of the admissible sample sizes set.

1 Introduction

While conducting a sampling survey of a finite population in order to estimate mean values of many variables, we aim at minimizing of the sample size. This allows us to reduce survey costs. On the other hand, the sample size must ensure the demanded accuracy of estimation. In the case of sample selection according to the stratified sampling scheme, the problem lies in allocating the samples in the strata in such a way that the survey costs are minimal and the variances of the mean estimators of all variables do not exceed fixed values. This issue, formulated by Dalenius (1957), was considered by Greń (1963, 1966), Kokan and Khan (1967). The solution to the problem requires appropriate data on the variables in the population. The survey costs are usually defined as a linear function of the sample sizes from the strata with the stratum unit costs as coefficients. When the unit costs of surveying the elements in each stratum vary (the strata are heterogeneous with regard to the unit costs), there is a problem of adopting the proper coefficients of the cost function. Moreover, in practice we may reach a situation in which the unit costs of survey are calculated only with some accuracy and their exact values will only be noted during the survey.

In practice, the population is stratified in such a way that the variances of the variables in all the strata are generally as small as possible. This way of stratification does not always lead to an optimal solution to the described problem, although criteria which indirectly and properly decrease values of variances of the variables allow us to obtain solutions close to the optimal ones. The algorithm of population stratification and sample allocation in s-trata proposed below may be useful in the case of creating strata which are heterogeneous with regard to the unit costs and in the case of imprecisely determined unit costs. The generalization of the problem formulated by Skibicki and Wywiał (2001) and Wywiał (2000) will be presented.

2 Form of the problem

Let $\mathcal{U} = \{U_h,\ h = 1, \ldots, H\}$ be a partition of a given population of N units into the strata. The size of a stratum U_h is denoted by N_h. Let n_h denote the size of a simple sample s_h which is randomly selected without replacement from the h-th stratum. We observe m variables in the population. The statistic from the sample $s = \bigcup_{h=1}^{H} s_h$ has the form:

$$\overline{y}_{is} = \sum_{h=1}^{H} w_h \overline{y}_{is_h}.$$

This is the unbiased estimator of the population mean of an i-th variable, where

$$w_h = \frac{N_h}{N}, \quad \overline{y}_{is_h} = \frac{1}{n_h} \sum_{k \in s_h} y_{ik}, \qquad h = 1, \ldots, H.$$

The variance of this estimator can be written as follows:

$$D^2(\overline{y}_{is}) = \sum_{h=1}^{H} w_h^2 \frac{(N_h - n_h)}{N_h n_h} v_{ih}, \tag{1}$$

where

$$v_{ih} = \frac{1}{N_h - 1} \sum_{k \in U_h} (y_{ik} - \overline{y}_{ih})^2.$$

Suppose that the unit costs of surveying the elements in each stratum are the same. Let k_h be the unit cost in an h-th stratum and let us denote the fixed upper limits for the estimators variances by e_i for $i = 1, \ldots, m$. For a fixed stratification $\mathcal{U}$, the problem of minimization of the survey costs subject to a fixed precision of the estimation can be written:

$$\begin{cases} k(n_1, \ldots, n_H) = \sum_{h=1}^{H} k_h n_h = \min \\ D_i^2(\overline{y}_{is} \leqslant e_i, \qquad i = 1, \ldots, m \\ 2 \leqslant n_h \leqslant N_h, \qquad h = 1, \ldots, H. \end{cases} \tag{2}$$

Substituting $x_h = \frac{1}{n_h}$ for $h = 1, \ldots, H$, we obtain a problem corresponding to problem (2) of the form:

$$\begin{cases} f(x_1, \ldots, x_H) = \min \\ \sum_{h=1}^{H} a_{ih} x_h \leqslant b_i, \qquad i = 1, \ldots, m \\ \frac{1}{N_h} \leqslant x_h \leqslant \frac{1}{2}, \qquad h = 1, \ldots, H. \end{cases} \tag{3}$$

where

$$f(x_1, \ldots, x_H) = \sum_{h=1}^{H} \frac{k_h}{x_h} \tag{4}$$

and

$$a_{ih} = w_h^2 v_{ih}\,, \quad b_i = e_i + \sum_{h=1}^{H} \frac{w_h^2 v_{ih}}{N_h}\,, \qquad h = 1, \ldots, H,\ i = 1, \ldots, m\,.$$

The variances (1) are functions of the within-stratum variances of the variables v_{ih}. Therefore Skibicki and Wywiał (2001) formulated the problem of stratification of a finite population optimal in the sense of problem (3), which generally allows us to reduce the survey costs. However, the strata obtained in this way are heterogeneous with regard to the unit costs, which makes the cost function difficult to define. We may consider, for example, the expected survey costs, which boils down to accepting coefficients in function (4) of the form:

$$k_h = \frac{1}{N_h} \sum_{j \in U_h} c_j\,,$$

where c_j is the unit cost of surveying an i-th element of the population.

Let us consider the case of costs c_j that are not known precisely, but which are assumed to lie in the interval $[\alpha_{h,\mathcal{U}}, \beta_{h,\mathcal{U}}]$ if $j \in U_h$ ($k_h \in [\alpha_{h,\mathcal{U}}, \beta_{h,\mathcal{U}}]$), where $\alpha_{h,\mathcal{U}}, \beta_{h,\mathcal{U}}$ are fixed for $h = 1, \ldots, H$. The values $\alpha_{h,\mathcal{U}}$ and $\beta_{h,\mathcal{U}}$ are calculated as expected unit costs for a given stratification $\mathcal{U}$ and c_j respectively the most and the least optimistic one. Then, in the sense of problem (3) we obtain a set of functions of the form (4). Usually, there isn't any s-tratification optimal for all these cost functions, but it is possible to stratify the population so as to decrease the possible survey costs.

3 Method of solving the problem

Figure 1 represents a region of feasible solutions to problem (3) for $H = 2$, $m = 2$ and a stratification $\mathcal{U}$ (restrictions L_1, L_2) and $\mathcal{U}'$ (restrictions L_1', L_2'). Points a, a' are solutions to the problem for $\frac{k_1}{k_2} = 2.5$, and b, b' for $\frac{k_1}{k_2} = 0.4$. Stratification $\mathcal{U}'$ gives lower survey costs in comparison with $\mathcal{U}'$ for, in particular, $\frac{k_1}{k_2} \in (0.4, 2.5)$.

As point $(x_1, \ldots, x_H)$ which represents the solution to problem (3) lies on the boundary of the region fixed by the restrictions (see Kokan and Khan (1967)). Thus a proper increase of the measure of this region should lead to better solution to problem (3). Taking into consideration the fact that we search for a stratification which gives minimum survey costs only for $k_h \in [\alpha_{h,\mathcal{U}}, \beta_{h,\mathcal{U}}]$, we should limit oneself to some subregion which includes these solutions.

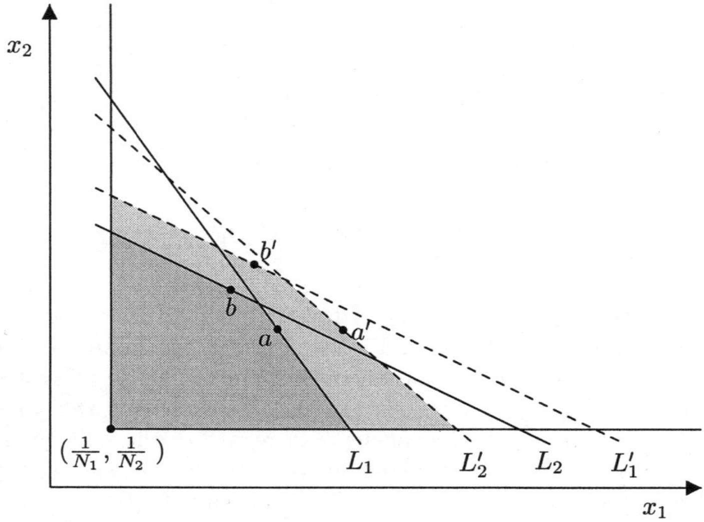

Fig. 1. Region of feasible solutions

Let, for a fixed stratification $\mathcal{U}$, $\mathrm{X_j} = (X_{1,j}, \ldots, X_{H,j})$ be the solution to problem (3) for $k_h = \alpha_{h,\mathcal{U}}$, $h \in \{1, \ldots, H\} \setminus \{j\}$ and $k_j = \beta_{j,\mathcal{U}}$. The set of points $(x_1, \ldots, x_H)$ fixed by the inequalities:

$$\sum_{h=1}^{H} a_{ih} x_h \leqslant b_i \,, \qquad i = 1, \ldots, m \,, \tag{5}$$

$$x_h \geqslant \frac{1}{N_h} \,, \qquad h = 1, \ldots, H \,, \tag{6}$$

$$x_h \leqslant X_{h,h} \,, \qquad h = 1, \ldots, H \,, \tag{7}$$

includes solutions to problem (3) for every $k_h \in [\alpha_{h,\mathcal{U}}, \beta_{h,\mathcal{U}}]$, $h = 1, \ldots, H$.

To prove this remark let $\mathrm{X}^* = (X_1^*, \ldots, X_H^*)$ and $\mathrm{X}^{**} = (X_1^{**}, \ldots, X_H^{**})$ be adequately solutions to problem (3) for some $k_h^*, k_h^{**} \in [\alpha_{h,\mathcal{U}}, \beta_{h,\mathcal{U}}]$, $h = 1, \ldots, H$. Since X^* and X^{**} are optimal for adequate unit costs we have

$$\sum_{h=1}^{H} \frac{k_h^*}{X_h^*} \leqslant \sum_{h=1}^{H} \frac{k_h^*}{x_h^{**}} \,.$$

$$\frac{k_j^{**}}{X_j^{**}} + \sum_{\substack{h=1 \\ h \neq j}}^{H} \frac{k_h^{**}}{X_h^{**}} \leqslant \frac{k_j^{**}}{X_j^*} + \sum_{\substack{h=1 \\ h \neq j}}^{H} \frac{k_h^{**}}{X_h^*} \,,$$

Hence, after adding the sides of the inequalities in appropriate way we have

$$\sum_{\substack{h=1\\h\neq j}}^{H}(k_h^* - k_h^{**})\left(\frac{1}{X_h^*} - \frac{1}{X_h^{**}}\right) \leqslant (k_h^{**} - k_j^*)\left(\frac{1}{X_j^*} - \frac{1}{X_j^{**}}\right).$$

Assuming that $k_h^* = k_h^{**}$ for $h \in \{1, \dots, H\} \setminus \{j\}$ and $k_j^* \leqslant k_j^{**}$ we obtain from the above inequality $X_j^* \leqslant X_j^{**}$. Hence, for any $j = 1, \dots, H$ in the case of change of the cost k_j adequately changes adequately X_j. Because X^*, X^{**} fulfill inequalities (5) and the form of function (4) implies that the values X_h^*, X_h^{**} should be generally as low as possible, we have

$$\sum_{\substack{h=1\\h\neq j}}^{H} a_{ih}X_h^{**} \leqslant \sum_{\substack{h=1\\h\neq j}}^{H} a_{ih}X_h^*.$$

It means that for the decreased value k_j we obtain in the solution the value X_j which is not higher and X_h which are no lower for $h \in \{1, \dots, H\} \setminus \{j\}$. Hence, if $k_h^* \leqslant k_h^{**}$ for $h \in \{1, \dots, H\} \setminus \{j\}$ and $k_j^* \geqslant k_j^{**}$, then $X_h^* \geqslant X_h^{**}$.

So the considered problem of the survey costs minimization is replaced by the problem of searching for a stratification $\mathcal{U}_0$ such that:

$$\mathrm{vol}(\mathcal{U}_0) = \max_{\mathcal{U}\in\mathbb{U}_{\mathbb{H}}} \mathrm{vol}(\mathcal{U}), \tag{8}$$

where $\mathbb{U}_{\mathbb{H}}$ is the set of all the population stratifications into H strata. Function $\mathrm{vol}(\mathcal{U})$ is the volume of the region fixed by inequalities (5, 6, 7) for the stratification $\mathcal{U}$. The problem of the maximization of $\mathrm{vol}(\mathcal{U})$ was formulated and solved by Wywiał(2000) under the assumption that $k_h = 1$ for all $h = 1, \dots, H$. Moreover, Wywiał mentioned that the criterion (8) is not necessary optimal in the sense of the stratification problem defined by the expression (3).

4 Algorithm of searching for optimum stratification

Because of a compound form of the objective function, the simplest method of searching for an optimum stratification seems to be an algorithm based on step by step modification of an initial stratification. The algorithm proposed below is similar to the k-means method.

Description: One iterative step of the algorithm relies on the selection of a following element in the population and checking if its shift to each of the other strata gives an increase of the objective function value. If true, the shift which gives the highest increase is implemented and this new stratification is adopted. If none of the shifts gives any increase, the stratification remains the same. All elements of the population are checked successively (after the last element the first one is checked, etc.) until the moment we do not observe any shifts which increase of the objective function value.

Methods of calculating the volume of an appropriate region are well known, but it is worth noticing that the simplex fixed by inequalities (5, 6, 7) is convex and also fixed by hiperplanes may be simply triangulated.

Example 3. We have the population of 7523 farms from one of the Polish administrative unit. The data used derives from the general agricultural census for 1996. Four variables have been selected to survey:
1. Arable land (in hectares),
2. Stock of cattle (in heads),
3. Stock of pigs (in heads),
4. Value of sale (in thousands of zlotys).

The population was divided into 2 strata on the basis of territorial division. Strata sizes, unit costs and mean values of the variables are presented in table 1.

Table 1. Stratification

strata	unit	unit	mean values of variables			
sizes	costs α_h	costs β_h	I	II	III	IV
3053	1.0	2.0	4.460	2.50	5.58	3.748
4470	1.5	3.0	4.321	2.13	4.90	2.842

Restrictions of the variances of the mean estimators amount to: $e_1 = 0.048$, $e_2 = 0.013$, $e_3 = 0.067$, $e_3 = 0.026$, and they were calculated as follows:

$$e_i = \left[\overline{y}_i \frac{\gamma_i}{100\%}\right]^2 .$$

The γ_i represents an assumed standard deviation of the estimator expressed as a percentage of the mean value. Here $\gamma_i = 5\%$ for $i = 1, \ldots, 4$.

Results obtained with the above mentioned algorithm are represented by the table 2. The columns contain adequately: vector of the strata sizes, vector of the unit costs α_h and β_h, vector of the sample sizes. The volume of the region considered here is: 1.2e-05.

Table 3 contains the matrices of the variances $[v_{ih}]$ for the initial and final stratification.

Table 2. Results

strata sizes	unit costs α_h	unit costs β_h	sample sizes
6942	1,30	2,61	358
581	1,20	2,41	189

Table 3. Variances of variables in strata

stratum	variances of variables			
no.	I	II	III	IV
1	7,650	6,78	87,72	50,389
2	8,350	4,06	200,91	63,019
1	6,068	3,54	17,57	6,111
2	15,920	11,37	1275,49	557,204

References

DALENIUS T. (1957), *Sampling in Sweden. Contribution to methods and theories of sample survey practice*, Almqvist & Wiksells, Stockholm.

GREŃ J. (1963), Sample allocation in multivariate stratified sampling (in Polish), *Przegląd Statystyczny*, vol. 10, pp. 291-302.

GREŃ J. (1966), On some usage of nonlinear programming in survey sampling (in Polish), *Przegląd Statystyczny*, vol. 13, pp. 203-217.

KOKAN A.R., KHAN S. (1967), Optimum allocation in multivariate surveys: an analytical solution, *Journal of the Royal Statistical Society*, vol. B 29, pp. 115-125.

SKIBICKI M., WYWIAŁ J. (2001), Examples of using clustering methods to optimization of stratified sampling (in Polish), *Wiadomości Statystyczne*, no. 8, pp. 4-11.

WYWIAŁ J. (2000), On optimal stratification in case of means vector estimation (in Polish), *Prace Naukowe Akademii Ekonomicznej we Wrocławiu*, no. 857, pp. 217-223.

On Estimation of Population Averages on the Basis of Cluster Sample

Janusz Wywiał

Department of Statistics, University of Economics in Katowice,
ul. Bogucicka 14, 40-226 Katowice, Poland

Abstract. Clustering methods are applied to improving precision of estimation of means in a fixed and finite population. The particular case of estimation of population means on the basis of sample drawn from a population partitioned into strata by means of clustering algorithms is considered by Huisman (2000) and Wywiał (1995, 1998, 2001). Here an estimator from cluster sample of a vector of population averages is considered. The accuracy of the estimator increases when the degree of the within-cluster spread increases. The precision of the vector estimator is compared with precision of simple sample means. The problem is finding such partition of a population into mutually disjoint clusters that the vector estimator from the cluster sample is more precise estimator of the population averages than the vector of simple sample means. Some clustering algorithm is proposed.

1 The basic properties of the vector of cluster means

A fixed population of size N is denoted by Ω. It is convenient to treat the population as a subset of the natural numbers: $\Omega = \{1, 2, \ldots, N\}$. Let us assume that the population Ω is divided into G mutually disjoint clusters Ω_p ($p = 1, \ldots, G$) that $\bigcup_{p=1}^{G} \Omega_p = \Omega$. If each cluster is of the same size denoted by M, the population Ω has size $N = GM$. A k-th ($k = 1, \ldots, N$) outcome of a variable $y = [y_1 \ldots y_m]$ is denoted by $y_k = [y_{k1} \ldots y_{km}]$. The sum of observations of an i-th variable in a p-th cluster is as follows: $z_{pi} = \sum_{k \in \Omega_p} y_{ki}$. The mean value of an i-th variable in the p-th cluster is: $\overline{y}_{ip} = \frac{1}{M} z_{pi}$. The mean value of an i-th variable over all clusters is: $\overline{z}_i = \frac{1}{G} \sum_{p=1}^{G} z_{pi}$. The population mean of an i-th variable takes the following form: $\overline{y}_i = \frac{1}{N} \sum_{p=1}^{G} z_{pi}$. The variance-covariance matrix is denoted by $\mathbf{C} = [c(y_i, y_j)]$ where $c(y_i, y_j) = \frac{1}{N-1} \sum_{p=1}^{G} \sum_{k \in \Omega_p} (y_{ki} - \overline{y}_i)(y_{kj} - \overline{y}_j)$. The variance-covariance matrix of cluster sums is denoted by $\mathbf{C}(z) = [c(z_i, z_j)]$ where $c(z_i, z_j) = \frac{1}{G-1} \sum_{p=1}^{G} (z_{pi} - \overline{z}_i)(z_{pj} - \overline{z}_j)$.

Let S be the cluster sample of size g. The random sample S is drawn without replacement according to the following design: $P(S) = \binom{G}{g}^{-1}$ for all possible S.

The estimator of the vector $\overline{\mathbf{y}} = [\overline{y}_1, \ldots, \overline{y}_m]$ is defined as the vector $\overline{\mathbf{y}}_{gS} = [\overline{y}_{1S}, \ldots, \overline{y}_{mS}]$, where:

$$\overline{y}_{iS} = \frac{1}{gM} \sum_{p \in S} \sum_{k \in \Omega_p} y_{ki} = \frac{1}{gM} \sum_{p \in S} z_{pi} \,. \tag{1}$$

The vector $\overline{\mathbf{y}}_S$ is the unbiased estimator of the mean vector $\overline{\mathbf{y}}$.

The covariance of the estimators $\overline{y}_{iS}$, $\overline{y}_{jS}$ $(i, j = 1, \ldots, m,\ i \neq j)$ can be derived similarly as the variance of the statistic $\overline{y}_{iS}$ $(i = 1, \ldots, m)$, see e.g. Särndal, Swenson,Wretman (1992), p.128-129. We obtain:

$$Cov\left(\overline{y}_{iS}, \overline{y}_{jS}\right) = \frac{G-g}{GgM^2} c(z_i, z_j) \,. \tag{2}$$

The variance-covariance matrix of the $\overline{\mathbf{y}}_S$ can be written down in the following way:

$$\mathbf{V}\left(\overline{\mathbf{y}}_S\right) = \frac{G-g}{GgM^2} C(z) \,, \tag{3}$$

where: $\mathbf{C}(z) = [c(z_i, z_j)]$. The unbiased estimator of the covariance is obtained by substitution of the following statistic for the parameter $c(z_i, z_j)$:

$$\tilde{c}(z_i, z_j) = \frac{1}{g-1} \sum_{p \in S} (z_{pi} - \overline{z}_{iS})(z_{pj} - \overline{z}_{jS}) \,, \quad \overline{z}_{iS} = \frac{1}{g} \sum_{p \in S} z_{pi}.$$

2 Homogeneity coefficient of a vector variable

Let $\mathbf{C}_w = [c_w(y_i, y_j)]$ be the within-cluster matrix of the variances and covariances where

$$c_w(y_i, y_j) = \frac{1}{G(M-1)} \sum_{p=1}^{G} \sum_{k \in \Omega_p} (y_{ki} - \overline{y}_{ip})(y_{kj} - \overline{y}_{jp}) \,.$$

Similarly to the one dimensional case (see e.g. Cochran (1963), p. 243) the *homogeneity coefficient of the vector variable* is defined in the following way:

$$\Delta = \mathbf{I} - \mathbf{C}^{-1}\mathbf{C}_w \,. \tag{4}$$

In the case of an one-dimensional variable y_i when $\mathbf{C}$ reduces to the variance v and $\mathbf{C}_w$ is the within-cluster variance v_w the matrix Δ reduces to the homogeneity coefficient (see Särndal, Swenson, Wretman (1992), p. 130):

$$\delta(y_i) = 1 - \frac{v_w(y_i)}{v(y_i)} \qquad with \qquad -\frac{G-1}{N-G} \leqslant \delta(y_i) \leqslant 1 \tag{5}$$

where $$v(y_i) = \frac{1}{N}\sum_{p=1}^{G}\sum_{k\in\Omega_p}(y_{ki}-\overline{y}_i)^2\,, \qquad \overline{y}_i = \frac{1}{N}\sum_{p=1}^{G}\sum_{k\in\Omega_p} y_{ki}\,,$$

$$v_w(y_i) = \frac{1}{G(M-1)}\sum_{p=1}^{G}\sum_{k\in\Omega_p}(y_{ki}-\overline{y}_{ip})^2\,, \qquad \overline{y}_{ip} = \frac{1}{M}\sum_{k\in\Omega_p} y_{ki}\,.$$

Insofar, the matrix Δ can be treated as a generalization of the homogeneity coefficient δ.

Theorem 1 (Wywiał 2001) *If the variance-covariance matrix* $\mathbf{C}$ *is non-singular then the eigenvalues* λ_i $(i = 1, \ldots, m)$ *of the matrix* Δ *fulfill the following inequalities:*

$$-\frac{G-1}{N-G} \leqslant \lambda_i \leqslant 1 \qquad \text{for } i = 1, \ldots, m\,. \tag{6}$$

The within-cluster spread of observations of a vector variable is less than their population spread if the matrix Δ is positive definite. When Δ is negative definite, the population spread of values of a vector variable is less than the within-cluster spread.

3 Accuracy of a cluster sample mean vector in relation to the simple sample mean vector

Let $\overline{\mathbf{y}}_U$ be the vector of means from a simple random sample U of size n, selected without replacement from a population of the size N according to the design $P(U) = \binom{N}{n}^{-1}$ for all possible U. Its variance-covariance matrix is of the following form:

$$\mathbf{V}\left(\overline{\mathbf{y}}_U\right) = \frac{N-n}{Nn}\mathbf{C} \tag{7}$$

Wywiał (2001) derived the following expression:

$$\mathbf{V}\left(\overline{\mathbf{y}}_S\right) = \frac{G-g}{GgM}\mathbf{C}\left(\mathbf{I} + \frac{N-G}{G-1}\Delta\right) \tag{8}$$

The expressions (7) and (8) lead to the following property.

Theorem 2 (Wywiał 2001) *If* $GM = N$, $gM = n$ *and the matrix* $\mathbf{C} - \mathbf{C}_w$ *is non-positive definite (non-negative definite) then the estimator* $\overline{\mathbf{y}}_S$ *is not worse (not better) than the estimator* $\overline{\mathbf{y}}_U$*. Particularly, if the matrix* $\mathbf{C}$ *is nonsingular and* Δ *is negative (positive) definite then the estimator* $\overline{\mathbf{y}}_S$ *is better (worse) than the estimator* $\overline{\mathbf{y}}_U$*.*

The estimator $\overline{\mathbf{y}}_S$ is better than the estimator $\overline{\mathbf{y}}_U$, if the within-cluster spread of a vector variable represented by the matrix $\mathbf{C}_w$ is larger than its population spread represented by the matrix $\mathbf{C}$.

Let us denote the variance, the determinant, the trace and the maximal eigenvalue of a variance-covariance matrix of a vector estimator by $D^2(\cdot)$, $\det(\cdot)$, $\mathrm{tr}(\cdot)$ and $\lambda_1(\cdot)$, respectively. The *relative efficiency coefficients* are defined as follows:

$$e_{0i} = \frac{D^2(\overline{y}_{iS})}{D^2(\overline{y}_{iU})} = 1 + \frac{N-G}{G-1}\delta(y_i), \qquad i = 1, \ldots, m, \tag{9}$$

$$e_1 = \frac{\det \mathbf{V}(\overline{\mathbf{y}}_S)}{\det \mathbf{V}(\overline{\mathbf{y}}_U)} = \det\left(\mathbf{I} + \frac{N-G}{G-1}\varDelta\right), \tag{10}$$

$$e_2 = \frac{\mathrm{tr}\mathbf{V}(\overline{\mathbf{y}}_S)}{\mathrm{tr}\mathbf{V}(\overline{\mathbf{y}}_U)} = 1 + \frac{N-G}{G-1}\overline{\delta}, \quad \overline{\delta} = \sum_{i=1}^{m}\delta(y_i)a_i, \quad a_i = \frac{v(y_i)}{\sum_{j=1}^{m} v(y_j)}, \tag{11}$$

$$e_3 = \frac{\lambda_1 \mathbf{V}(\overline{\mathbf{y}}_S)}{\lambda_1 \mathbf{V}(\overline{\mathbf{y}}_U)}. \tag{12}$$

On the basis of Theorem 2 and well known matrix properties we conclude that if the matrix $\varDelta$ is negative definite, then $e_a < 1$ for $a = 1, 2, 3$ and $e_{0i} \leqslant 1$ for $i = 1, \ldots, m$ and $e_{0j} < 1$ for at least one index $j = 1, \ldots, m$.

4 Clustering algorithm

The estimator $\overline{\mathbf{y}}_S$ can be better than the estimator $\overline{\mathbf{y}}_U$ if it is possible to cluster a population in such a way that the matrix $\varDelta$ is negative definite. It means that the within-cluster spread of values of the vector variable should be sufficiently bigger than the population spread of observations of those variables. So, a population should be divided into such clusters of the same size that all eigenvalues of the matrix $\varDelta$ take values as small as possible and each of them is less than zero.

Let $W = \{\Omega_1, \ldots \Omega_G\}$ be a set of G disjoint clusters of a fixed population such that $\bigcup_{p=1}^{G} \Omega_p = \Omega$ and each cluster consists of the same number M of elements. Let $\lambda^T(W) = [\lambda_1(W) \ldots \lambda_m(W)]$, where $\lambda_k(W) \geq \lambda_{k+1}(W)$, $k = 1, \ldots, m-1$, be the vector of the eigenvalues of the matrix $\varDelta = \varDelta(W)$ for some partition W. We say that a partition W is admissible if $\lambda(W) \prec 0$. It means that $\lambda_k(W) \leq 0$ for $k = 1, \ldots, m$ and $\lambda_h(W) < 0$ for at least one index $h = 1, \ldots, m$. The set of the *admissible partitions* is defined by $A = \{W : \lambda(W) \prec 0\}$. Let $B \subseteq A$ be defined as follows: for each $W_1, W_2 \in B \subseteq A$ it is not true that $\lambda(W_1) \prec \lambda(W_2)$. So, the set B consists of all *non-dominated partitions.*

Our purpose is to find such a partition $\underline{W} \in B$ where a criterion function $f(\underline{W}) = \min$. The function $f(W)$ depends on parameters of the matrix Δ. In particular:

$$f_1(W) = \mathrm{tr}\Delta = \sum_{k=1}^{m} \lambda_k\,, \quad f_2(W) = \det\Delta = (-1)^{m-1} \prod_{k=1}^{m} \lambda_k\,, \quad f_3(W) = \lambda_1\,.$$

The function f_a $(a = 1, 2, 3)$ can be treated as the *scalar homogenity coefficients of a vestor variable.* Let us note that the criterion function based on trace, determinants and maximal eigenvalue of some matrix measuring intra-cluster spread were considered by McQueen (1967) Friedman and Rubin (1967) and Wywiał (1995), respectively.

Hence, the criterion of clustering can be defined as follows. Finding such partition $\underline{W}$ that $f_i(\underline{W}) = min$ for a fixed $i = 1, 2, 3$ provided all eigenvalues of the homogenity coefficient $\Delta(\underline{W})$ are negative.

To determine the set $\underline{\mathrm{W}}$, we can construct the following iterative algorithm: Let $W_0 = \{\Omega_{10}, \ldots, \Omega_{G0}\}$ be an arbitrary initial partition of the population. Let $\underline{W}_t = \{\Omega_{1t}, \ldots, \Omega_{Gt}\}$ be the set of clusters resulting from the t-th iteration of the clustering algorithm. The partition $W_{k+1}(h, k)$ is obtained through moving the element h from the cluster $\Omega_{i,t}$ to the cluster $\Omega_{j,t}$ and moving the element k from the cluster $\Omega_{i,t}$ to the cluster $\Omega_{j,t}$. A cluster created during the $(t+1)$ iteration is denoted by $\Omega_{i,t+1}(h, k)$. It is obtained in the following way:

$$W_{k+1}(h, k) = \{W_t - \Omega_{i,t} - \Omega_{j,t}, \Omega_{i,t+1}(h, k), \Omega_{j,t+1}(h, k)\}\,, \tag{13}$$

where:

$$\Omega_{i,t+1}(h, k) = \Omega_{i,t} - \{h\} \cup \{k\}\,, \quad \Omega_{j,t+1}(h, k) = \Omega_{j,t} - \{k\} \cup \{h\}$$

At the end of the $(t + 1)$ iteration the set of admissible partitions

$$A_{t+1} = \{W : \lambda(W_{k+1}(h, k)) \prec 0 \quad \text{for} \quad h \neq k \quad \text{and} \quad h, k = 1, ..., N\}$$

is selected. If the set A_{t+1} is not empty then the final partition of the $(t+1)$-iteration, denoted by $\underline{W}_{k+1} = W_{k+1}(\underline{h}, \underline{k})$, is determined through the minimization of the scalar homogenity coefficient in the following way:

$$f_a(W_{k+1}(\underline{h}, \underline{k})) = \min_{f_a(W_{t+1}(h,k)) \in A_{t+1}} \{f_a\,(W_{t+1}(h, k))\} \tag{14}$$

If the set A_{t+1} is empty, and the final partition of the previous iteration $W_k(\underline{v}, \underline{u})$ is admissible than the clustering algorithm is stopped and $W_k(\underline{v}, \underline{u})$ is treated as final partition of a population. If the set A_{t+1} is empty, and the partition $W_k(\underline{v}, \underline{u})$ is not admissible than the clustering algorithm is continued. The partition $W_{k+1}(\underline{h}, \underline{k})$ is determined as follows:

$$f_a(W_{k+1}(\underline{h}, \underline{k})) = \min_{\substack{k=1,\ldots,N \\ k \neq h}} \min_{h=1,\ldots,N} \{f_a\left(W_{t+1}(h, k)\right)\} \tag{15}$$

The clustering algorithm is continued until a) iteration T when the set A_T is empty and the partition $W_{T-1}(\underline{v}, \underline{u})$ is admissible, a) no more elements of the population are moved from one cluster to another, c) until the number of iterations reaches the admissible level (which is usually chosen arbitrarily).

5 Conclusions

The algorithm can lead to the non-optimal partition. When at least one time the admissible partition will be obtained during the iteration process then the final partition of the algorithm is admissible, too. In this case the vector estimator from the cluster sample is more precise estimator of population means than the vector of order average from the simple sample. The obtained partition can be unadmissible in the sense of the definition of the set A_t. In this case the estimator $\overline{\mathbf{y}}_S$ is not better than the vector $\overline{\mathbf{y}}_U$. But in this case the estimator $\overline{\mathbf{y}}_S$ can be not less precise than the vector $\overline{\mathbf{y}}_U$ in the sense of the relative efficiency coefficient e_a which is a function of the scalar homogenity coefficient f_a. Moreover, it seems that it is possible that admissible partition of a population do not exist for some particular distribution of the vector variable in population. This can depend on size M of clusters.

Let us note that similar clustering algorithms can be based on the matrices $\mathbf{V}\left(\overline{\mathbf{y}}_S\right)$, $\mathbf{C} - \mathbf{C}_w$ or $\mathbf{C}(z)$, too.

In practice of surveys sampling the partition of a population can be processed on the basis of data from census. This could lead to improving one-stage or two-stage cluster sampling scheme in next surveys.

Acknowledgements

The research was supported by the grant number 1 H02B 015 10 from the Polish Scientific Research Committee. I am grateful to the reviewers for the valuable comments on the manuscript.

References

COCHRAN W.G. (1963), *Sampling Techniques*, John Wiley, New York.

FRIEDMAN J. H., RUBIN J. (1967), On some invariant criteria for grouping data. *Journal of the American Statistical Association* no. 62.

HUISMAN M. (2000), Post-stratification to correct fornonresponse: classification of ZIP code areas. In: *Proc. in Computational Statistics 14th Symp. held in Utrecht, The Netherlands*, Ed. J.G. Bethlehem and P.G.M. van der Huijden. Physica-Verlag, Heidelberg, New York.

MACQUEEN J. P. (1967), Some methods for classification and analysis of multivariate observations. In: *Proc. of the 5th Berkeley Symp. on Mathematical Statistics and Probability*, Berkeley.

SÄRNDAL C.E.,SWENSON B., WRETMAN J. (1992), *Model Assisted Survey Sampling*, Springer-Verlag, New York.

WYWIAŁ J. (1995), On optimal stratification of population on the basis of auxiliary variables. *Statistics in Transition* vol. 2, 831-837.

WYWIAŁ J. (1998), Estimation of population average on the basis of strata formed by means of discrimination functions. *Statistics in Transition* vol. 3, 903-912.

WYWIAŁ J. (2001), Estimation of population averages on the basis of a vector of cluster means. Paper presented at the Conf. Multivariate Statistical Analysis, Łódź, Department of Statistical Methods, University of Łódź.

Symbolic Data Analysis

Symbolic Regression Analysis

Lynne Billard[1] and Edwin Diday[2]

[1] Department of Statistics, University of Georgia,
Athens, GA 30602, USA
(e-mail: lynne@stat.uga.edu)

[2] CEREMADE, Universite de Paris 9 Dauphine
75775 PARIS Cedex 16 France
(e-mail: diday@ceremade.dauphine.fr)

Abstract. Billard and Diday (2000) developed procedures for fitting a regression equation to symbolic interval-valued data. The present paper compares that approach with several possible alternative models using classical techniques; the symbolic regression approach is preferred. Thence, a regression approach is provided for symbolic histogram-valued data. The results are illustrated with a medical data set.

1 Introduction

Billard and Diday (2000) developed a methodology to fit multiple linear regression equations to symbolic interval-valued data. This involved among other entities deriving formula for the calculation of a symbolic covariance function and hence a symbolic correlation function for such data. Predicted values were also calculated and compared with those predictions obtained by fitting the lower (upper) points only. It was observed that the prediction interval obtained from the symbolic regression equation was preferable.

The purpose of the present paper is two-fold. First, continuing a focus on interval-valued data, we extend the previous work to fit a total of ten possible regression models (of which the first only represents a true symbolic regression analysis). We shall see that the symbolic regression model seems to provide a better fit as far as prediction is concerned. This is done in Section 2.

Secondly, we introduce a methodology for fitting symbolic regression equations to histogram-valued symbolic data. This involves providing formula for calculating symbolic sample variances, covariances and hence correlation functions for such data. It follows that interval-valued data are a special case of this more general data format. The basic formula are given in Sections 3 and 4, and illustrated in Section 5.

2 Comparisons of Regressions for Interval-valued Data

We assume that we are interested in fitting a simple linear regression

$$Y_1 = \alpha + \beta Y_2 \tag{1}$$

to the data (Y_1, Y_2) which take values in the space $R \times R$. Extension to the general multiple regression model is reasonably straight-forward (see Billard and Diday, 2000, for p-dimensional interval-valued symbolic regression models).

Suppose $u \in E$ where E is a set of m objects with observations $Y(u) = \{Y_1(u), Y_2(u)\}$, $u = 1, \ldots, m$. The $Y(u)$ takes particular realizations $Z(u)$ over the rectangle $(\xi_1^u, \xi_2^u) = ([a_{1u}, b_{1u}], [a_{2u}, b_{2u}])$, $u = 1, \ldots, m$. An underlying assumption is that each description vector is uniformly distributed across the rectangle $Z(u)$. We also assume henceforth that the data set under consideration is such that all individual description vectors $x \in vir(d_u)$ are each uniformly distributed over the rectangle $Z(u)$ where $vir(d_u)$ is the virtual description of x defined as the set of all individual descriptions x which satisfy a set of rules ν; see Bertrand and Goupil (2000) and Billard and Diday (2001).

There are ten possible regression models that will be fitted and compared. This is executed through the fitting of the data of Table 1, consisting of hematocrit (Y_1) values recorded in intervals $\xi_1 = [a_{1u}, b_{1u}]$ and hemoglobin values (Y_2) recorded in intervals $\xi_2 = [a_{2u}, b_{2u}]$, which are then used to predict hematocrit values for a given hemoglobin interval $\xi_2 = (12, 13)$.

Table 1 - Data

Object	Hematocrit	Hemoglobin	Object	Hematocrit	Hemoglobin
u	a_{1u} b_{1u}	a_{2u} b_{2u}	u	a_{1u} b_{1u}	a_{2u} b_{2u}
1	33.296, 39.601	11.545, 12.806	9	28.831, 41.980	9.922, 13.801
2	36.694, 45.123	12.075, 14.177	10	44.481, 52.536	15.374, 15.755
3	36.699, 48.685	12.384, 16.169	11	27.713, 40.499	9.722, 12.712
4	36.386, 47,412	12.354, 15.298	12	34.405, 43.027	11.767, 13.936
5	39.190, 50.866	13.581, 16.242	13	30.919, 47.091	10.812, 15.142
6	39.701, 47.246	13.819, 15.203	14	39.351, 51.510	13.761, 16.562
7	41.560, 48.814	14.341, 15.554	15	41.710, 49.678	14.698, 15.769
8	38.404, 45.228	13.274, 14.601	16	35.674, 42.382	12.448, 13.519

The first regression model is the symbolic regression model, found from the methods of Billard and Diday (2000). This gives

$$\text{Model 1: } Y^{(1)} = \text{Hematocrit} = 0.497 + (2.978) \text{ Hemoglobin}. \tag{2}$$

The predicted interval for hematocrit over the hemoglobin interval (12, 13) is

$$Y_{12}^{(1)} = 0.497 + (2.978)(12) = 36.234, \quad Y_{13}^{(1)} = 0.497 + (2.978)(13) = 39.212;$$

that is, $Y^{(1)}$ (predicted) $= (Y_{12}^{(1)}, Y_{13}^{(1)}) = (36.234, 39.212)$.

The other nine models considered all take various specific combinations of the end-point(s) of the interval-valued data (i.e., specific apex points of the data rectangle) to which a standard linear regression model is fitted using classical methods.

Thus, we have the following models for the indicated data values.
Model 2 - Data $(Y_1, Y_2) = \{(b_{1u}, a_{2u}), u = 1, \ldots, m\}$:

$$Y^{(2)} = \text{Hematocrit} = 22.262 + (1.909) \text{ Hemoglobin}. \tag{3}$$

Model 3 - Data $(Y_1, Y_2) = \{(b_{1u}, b_{2u}), u = 1, \ldots, m\}$:

$$Y^{(3)} = \text{Hematocrit} = 0.922 + (3.051) \text{ Hemoglobin}. \tag{4}$$

Model 4 - Data $(Y_1, Y_2) = \{(a_{1u}, a_{2u}), u = 1, \ldots, m\}$:

$$Y^{(4)} = \text{Hematocrit} = 0.820 + (2.832) \text{ Hemoglobin}. \tag{5}$$

Model 5 - Data $(Y_1, Y_2) = \{(a_{1u}, b_{2u}), u = 1, \ldots, m\}$:

$$Y^{(5)} = \text{Hematocrit} = -3.832 + (2.713) \text{ Hemoglobin}. \tag{6}$$

Model 6 - Data $(Y_1, Y_2) = \{(b_{1u}, a_{2u})$ and $(b_{1u}, b_{2u}), u = 1, \ldots, m\}$:

$$Y^{(6)} = \text{Hematocrit} = 26.579 + (1.436) \text{ Hemoglobin}. \tag{7}$$

Model 7 - Data $(Y_1, Y_2) = \{(a_{1u}, a_{2u})$ and $(a_{1u}, b_{2u}), u = 1, \ldots, m\}$:

$$Y^{(7)} = \text{Hematocrit} = 13.096 + (1.706) \text{ Hemoglobin}. \tag{8}$$

Model 8 - Data $(Y_1, Y_2) = \{(b_{1u}, a_{2u}), (b_{1u}, b_{2u}), (a_{1u}, a_{2u})$ and $(a_{1u}, b_{2u}), u = 1, \ldots, m\}$:

$$Y^{(8)} = \text{Hematocrit} = 19.847 + (1.571) \text{ Hemoglobin}. \tag{9}$$

Model 9 - Data $(Y_1, Y_2) = \{(b_{1u}, a_{2u})$ and $(a_{1u}, b_{2u}), u = 1, \ldots, m\}$:

$$Y^{(9)} = \text{Hematocrit} = 45.787 + (-0.315) \text{ Hemoglobin}. \tag{10}$$

Model 10 - Data $(Y_1, Y_2) = \{(b_{1u}, b_{2u})$ and $(a_{1u}, a_{2u}), u = 1, \ldots, m\}$:

$$Y^{(10)} = \text{Hematocrit} = -6.094 + (3.457) \text{ Hemoglobin}. \tag{11}$$

Each model is then used to predict the hematocrit interval for the hemoglobin interval (12, 13). These predicted values are displayed in Table 2.

Table 2 - Predictions $(Y_{12}^{(k)}, Y_{13}^{(k)})$

k	Regression Model	Hematorcrit
1	Symbolic	(36.2336, 39.2117)
2	Upper left apex	(45.1733, 47.0825)
3	Upper right apex	(37.5364, 40.5864)
4	Lower left apex	(34.8095, 37.6420)
5	Lower right apex	(28.7223, 31.4351)
6	Top apexes	(43.8347, 45.2712)
7	Bottom apexes	(33.5695, 35.2757)
8	All apexes	(38.7021, 40.2734)
9	Upper left and lower right	(42.0114, 41.6968)
10	Upper right and lower left	(35.3927, 38.8499)

It is clear that some of these possible regressions are less valid than others. It is also apparent from Table 2, that the symbolic regression model gives a good fit. For example, if we compare the prediction obtained from the symbolic regression Model 1 with that from Models 9 and 10, we observe that at the lower interval endpoint, the symbolic prediction, 36.224 is framed by the other two, i.e., $35.323 < 36.224 < 42.011$; and likewise at the upper end, $38.850 < 39.212 < 41.697$. Using classical regression methods, we would not unreasonably choose the prediction interval (35.323, 41.697). The symbolic prediction interval at (36.234, 39.212) is clearly tighter. If instead we compare the symbolic prediction from Model 1 with the predictions from Models 3 and 4 (i.e., the set of "maximum" values and the set of "minimum" values), the symbolic prediction is framed by those from Models 3 and 4 according to $34.810 < 36.234 < 37.536$ at $Y_2 = 12$, and by $37.642 < 39.212 < 40.586$ at $Y_2 = 13$. Using classical methods directly, we would choose the prediction interval (34.810, 40.586), which is broader than is the symbolic prediction interval. Other reasonable comparisons also suggest the symbolic prediction is preferable. Note however no attempt has been made herein to calculate error bounds for the predictions; see also the discussion of Section 6.

3 Covariance Function for Histogram Data

Let us now suppose that the observations are recorded as histograms. That is, for each object $u = 1, \ldots, m$, each variable $Y_j(u), j = 1, 2$, takes values on the subintervals $\xi_{juk} = [a_{juk}, b_{juk})$ with probability p_{juk}, $k = 1, \ldots, s_{ju}$, and where $\sum_{k=1}^{s_{ju}} p_{juk} = 1$. An example of such data is that of Table 3, where for object $u = 3$ (say), the hemoglobin values fall in the interval $[12.384, 14.201)$ with probability 0.3 and in the interval $[14.201, 16.169]$ with probability 0.7. Notice in this data set, $s_{1u} = 1$ for all u, and $s_{2u} = 2$ for all u. Note that when $s_{ju} = 1$, and $p_{juk} = 1$ for all j, k, and u, we have the special case that the data are all interval-valued which was covered in Section 2.

By extending the results of Billard and Diday (2000, 2001), we can show that the symbolic covariance function for such histogram data is

$$Cov(Y_1, Y_2) = S_{12} = \int_{-\infty}^{\infty}\int_{-\infty}^{\infty}(\xi_1 - \bar{Y}_1)(\xi_2 - \bar{Y}_2)f(\xi_1, \xi_2)d\xi_1\xi_2 \tag{12}$$

where $\bar{Y}_j$ are the symbolic empirical means for histogram data given by Billard and Diday (2001) as

$$\bar{Y}_j = \frac{1}{2m}\sum_{u\in E}\{\sum_{k=1}^{s_{ju}} p_{juk}(b_{juk} + a_{juk})\} \tag{13}$$

and where $f(\xi_1, \xi_2)$ is the empirical joint density function for (Y_1, Y_2) given as

$$f(\xi_1, \xi_2) = \frac{1}{m}\sum_{u\in E}\{\sum_{k_1=1}^{s_{1u}}\sum_{k_2=1}^{s_{2u}}\frac{p_{1uk_1}p_{2uk_2}I_{uk_1k_2}(\xi_1, \xi_2)}{||Z_{k_1k_2}(u)||}\} \tag{14}$$

where $I_{uk_1k_2}(\xi_1, \xi_2)$ is the indicator function that (ξ_1, ξ_2) is or is not in the rectangle $Z_{k_1k_2}(u)$ and where $||Z_{k_1k_2}(u)||$ is the area of the rectangle $Z_{k_1k_2}(u) = [a_{1uk_1}, b_{1uk_1}) \times [a_{2uk_2}, b_{2uk_2})$. It follows, from (12) and (14), that

$$Cov(Y_1, Y_2) = \frac{1}{4m}\sum_{u\in E}\{\sum_{k=1}^{s_{1u}} p_{1uk}(b_{1uk} + a_{1uk})\}\{\sum_{k=1}^{s_{2u}} p_{2uk}(b_{2uk} + a_{2uk})\}$$

$$-\frac{1}{4m^2}[\sum_{u\in E}\{\sum_{k=1}^{s_{1u}} p_{1uk}(b_{1uk} + a_{1uk})\}][\sum_{u\in E}\{\sum_{k=1}^{s_{2u}} p_{2uk}(b_{2uk} + a_{2uk})\}]. \tag{15}$$

The symbolic empirical variance for Y_j, $j = 1, 2$, is

$$S_j^2 = \frac{1}{4m}\sum_{u\in E}\{\sum_{k=1}^{s_{ju}} p_{juk}(b_{juk} + a_{juk})\}^2$$

$$-\frac{1}{4m^2}[\sum_{u\in E}\{\sum_{k=1}^{s_{ju}} p_{juk}(b_{juk} + a_{juk})\}]^2. \tag{16}$$

Hence, the symbolic correlation coefficient between two variables realized as symbolic histogram data is given by

$$R \equiv R(Y_1, Y_2) = S_{12}/\sqrt{S_1^2 \times S_2^2} \tag{17}$$

where S_{12} is given by (15) and S_j^2 are given by (16).

4 Symbolic Regression Model

We wish to fit a symbolic linear regression equation (1) to the histogram-data. We can show that the regression coefficients are estimated by

$$\hat{\beta} = Cov(Y_1, Y_2)/S_1^2, \quad \hat{\alpha} = \bar{Y}_1 - \hat{\beta}\bar{Y}_2 \tag{18}$$

where now the $Cov(Y_1, Y_2)$ and S_1^2 terms are calculated from (15) and (16), respectively; and where $\bar{Y}_j$ are calculated from (13).

5 Illustration

Table 3 records hematocrit (Y_1) values as interval-valued data as a particular case of histogram-valued data, and hemoglobin (Y_2) values as histogram-valued data for each of $m = 16$ objects.

Table 3

Object	Hematocrit Y_1		Hemoglobin Y_2			
	a_{1u}, b_{1u}	p_1	a_{2u1}, b_{2u1}	p_{11}	a_{2u2}, b_{2u2}	p_{22}
1	33.296, 39.610	1.0	11.545, 12.195	.4	12.195, 12.806	.6
2	36.694, 45.123	1.0	12.075, 13.328	.5	13.328, 14.177	.5
3	36.699, 48.685	1.0	12.384, 14.201	.3	14.201, 16.169	.7
4	36.386, 47.412	1.0	12.384, 14.264	.5	14.264, 15.298	.5
5	39.190, 50.866	1.0	13.581, 14.289	.3	14.289, 16.242	.7
6	39.701, 47.246	1.0	13.819, 14.500	.4	14.500, 15.203	.6
7	41.560, 48.814	1.0	14.341, 14.812	.5	14.812, 15.554	.5
8	38.404, 45.228	1.0	13.274, 14.000	.6	14.000, 14.601	.4
9	28.831, 41.980	1.0	9.922, 11.989	.4	11.989, 13.801	.6
10	44.481, 52.536	1.0	15.374, 15.780	.3	15.780, 16.755	.7
11	27.713, 40.499	1.0	9.722, 10.887	.4	10.887, 12.712	.6
12	34.405, 43.027	1.0	11.767, 12.672	.4	12.672, 13.936	.6
13	30.919, 47.091	1.0	10.812, 13.501	.6	13.501, 15.142	.4
14	39.351, 51.510	1.0	13.761, 14.563	.5	14.563, 16.562	.5
15	41.710, 49.678	1.0	14.698, 15.143	.4	15.143, 15.769	.6
16	35.674, 42.382	1.0	12.448, 13.195	.7	13.195, 13.519	.3

The symbolic linear regression equation fitting these data is, by using (18) in (1),

$$\text{Hematocrit} = -0.648 + (3.052)\ \text{Hemoglobin}. \tag{19}$$

To predict the hematocrit value in the hemoglobin interval (12, 13) then, we calculate, from (19),

$$Y_{12} = -0.648 + (3.052)(12) = 35.979, \quad Y_{13} = -0.648 + (3.052)(13) = 39.031;$$

that is, when the hemoglobin values are in the interval (12, 13), the hematocrit values are predicted to be in the interval $\xi = (35.979, 39.031)$. Were the hemoglobin values in (12, 13) recorded as a histogram, then the corresponding predicted values for hematocrit can be found by appropriate adjustments along the lines used in Section 3. Comparisons (not done herein) like those for our interval-valued data in Section 2, would likely draw the same conclusions, that the symbolic predictions are preferable over classical predictions for these symbolic data.

6 Conclusion

The ad hoc empirical study of Section 2 for interval-valued data is somewhat limited in value even though enlightening. More generally, it is important to develop more rigorous measures of the quality of the symbolic regression methods. Specifically, basic questions such as calculating bounds of the parameter estimators and of the predictors, hypothesis testing of these values, robustness issues and so forth still remain to be addressed. One approach would be to use classical analogues but with the component terms replaced by their symbolic counterparts. For example, one measure of the quality of fit is provided by R^2 where R is the correlation coefficient. Thus, using (17), for the data of Table 1, we have the symbolic $R^2 = 0.990$; and for the data of Table 3, the symbolic $R^2 = 0.979$. Other measures would involve more direct analyses of the associated error or residual terms.

In a different direction, cross-validation methods could be used. For example, for each of $q = 1, \ldots, m$, the symbolic regression model can be fitted to the data set consisting of all but the data for object $u = q$. The resulting model is then used to predict values, $Y_1^*(q)$ say, for the Y_2 values associated with object q. The set of such predicted values $\{Y_1^*(q), q = 1, \ldots, m\}$ can then be compared with the set of actual values $\{Y_1(u), u = 1, \ldots, m\}$, with these comparisons taking various formats such as fits to white noise processes and the like. Alternatively, bootstrap, Gibbs sampling or other numerical techniques could be developed. Further details and results of such investigations will be reported elsewhere.

References

BERTRAND, P. and GOUPIL, F. (2000): Descriptive Statisitcs for Symbolic Data. In: *Analysis of Symbolic Data Sets* (eds. H. -H. Bock and E. Diday), Springer, 103-124.

BILLARD, L. and DIDAY, E. (2000): Regression Analysis for Interval-Valued Data. In: *Data Analysis, Classification, and Related Methods* (eds. H. A. L. Kiers, J. -P. Rasson, P. J. F. Groenen and M. Schader), Springer, 369-374.

BILLARD, L. and DIDAY, E. (2001): From the Statistics of Data to the Statistics of Knowledge: Symbolic Data Analysis, submitted.

RODRIGUEZ, O. (2001): Classification et Modéles Linéaires en Analyse des Données Symboliques. Doctoral Thesis, University of Paris, Dauphine.

Modelling Memory Requirement with Normal Symbolic Form

Marc Csernel[1] and Francisco de A. T. de Carvalho[2]

[1] INRIA - Rocquencourt, Domaine de Voluceau - Rocquencourt - B. P. 105 78153 Le Chesnay Cedex - France, Email: Marc.Csernel@inria.fr
[2] Centro de Informatica - CIn / UFPE, Av. Prof. Luiz Freire, s/n - Cidade Universitaria, CEP: 50740-540 - Recife - PE - Brasil, Email: fatc@cin.ufpe.br

Abstract. Symbolic objects can deal with domain knowledge expressed by dependency rules between variables. However taking into account this rules in order to analyze symbolic data can lead to exponential computation time. It's why we introduced an approach called Normal Symbolic Form (NSF) which lead to a polynomial computation time, but may sometimes bring about an exponential explosion of space memory requirement. In a previous paper we studied this possible memory requirement and we saw how it is bounded according to the nature of the variables and rules. The aim of this paper is to model this memory space requirement. The proposed modelling of this problem is related to the Maxwell-Boltzmann statistics currently used on thermodynamics.

1 Introduction

Symbolic Data Analysis (*SDA*) aims to extend standard data analysis to more complex data, called *symbolic data*, as they contain internal variation and they are structured (Bock and Diday (2000)). The *SDA* methods have as input a Symbolic Data Table. The columns of this Symbolic Table are the symbolic variables. A symbolic variable is *set-valued*, i.e., for an object, it takes a subset of categories of its domain. The rows of this data table are the symbolic descriptions of the objects, i.e., the symbolic data vectors whose columns are symbolic variables. A cell of such a data table does not necessarily contain, a single value, but a set of values.

Symbolic descriptions can be constrained by some domain knowledge expressed by dependency rules between variables. Taking this rules into account in order to analyze the symbolic data, requires usually an exponential computation time. In a previous work (Csernel and De Carvalho (1999)) we presented an approach called Normal Symbolic Form (NSF). The aim of this approach is to implement a decomposition of a symbolic description in such a way that only coherent descriptions (i.e., which do not contradict the rules) are represented. Once this decomposition is achieved, the computation can be done in a polynomial time (Csernel 1998), but in some cases, it can lead to a combinatorial explosion of space memory requirement.

In a previous paper (Csernel and De Carvalho (2002)) we studied this possible memory requirement and we showed how it is bounded according to the nature of the variables and nature of the rules, and that in most cases, we will obtain a reduction rather than a growth.

The aim of this paper is to model this memory space requirement. The proposed modelling is based on the classical combinatory occupancy problem of balls and boxes and is related to the Maxwell-Boltzmann statistics currently used on thermodynamics. In this model each description corresponds to a ball which has to be placed in a box (possible description).

2 Symbolic data and normal symbolic form

2.1 Symbolic data

In classical data analysis, the input is a data table where the rows are the descriptions of the individuals, and the columns are the variables. One cell of such data table contains a single quantitative or categorical value.

However, sometimes in the real world the information recorded is too complex to be described by usual data. That is why different kinds of symbolic variables and symbolic data have been introduced (Bock and Diday (2000)). A quantitative variable takes, for an object, an interval of its domain, whereas a categorical multi-valued variables takes, for an object, a subset of its domain.

2.2 Constraints on symbolic descriptions

Symbolic descriptions can be constrained by dependencies between couples of variables expressed by rules. We take into account two kinds of dependencies: hierarchical and logical. We will call premise variable and conclusion variable the variable associated respectively with the premise and the conclusion of each rule.

In the following $\mathcal{P}^*(D)$ will denote the power set of D without the empty set. Let y_1 and y_2 be two categorical multi-valued variables whose domains are respectively $\mathcal{D}_1$ and $\mathcal{D}_2$. A hierarchical dependence between the variables y_1 and y_2 is expressed by the following kind of rule:

$$y_1 \in \mathcal{P}^*(D_1) \Longrightarrow y_2 = NA$$

where $D_1 \subset \mathcal{D}_1$ and the term NA means *not applicable* hence the variable does not exist. With this kind of dependence, we sometimes speak of mother-daughter variables. In this paper, we will mostly deal with hierarchical rules.

A logical dependence between the variables y_1 and y_2 is expressed by the following kind of rule

$$y_1 \in \mathcal{P}^*(D_1) \Longrightarrow y_2 \in \mathcal{P}^*(D_2).$$

Both of these rules reduce the number of individual descriptions of a symbolic description, but the first kind of rule reduces the number of dimensions on a symbolic description, whereas the second does not. We have seen in De Carvalho (1998) that computation using rules leads to exponential computation time depending on the number of rules. To avoid this explosion we introduced the Normal Symbolic Form.

$$Wings \in \{absent\} \Longrightarrow Wings\_colour = NA \qquad (r_1).$$
$$Wings\_colour \in \{red\} \Longrightarrow Thorax\_colour \in (\{blue\}) \quad (r_2)$$

	Wings	Wings_colour	Thorax_colour	Thorax_size
d_1	{absent,present}	{red,blue}	{blue,yellow}	{big,small}
d_2	{absent,present}	{red,green}	{blue,red}	{small}

Table I: original table.

In the symbolic data table presented above there are two symbolic descriptions d_1, d_2, and three categorical multi-valued variables. The values are constrained by rules r_1 and r_2.

2.3 The normal symbolic form

The idea of the NSF is slightly related to Codd's normal form for relational databases (Codd (1972)). The aim is to represent the data in such a way that only coherent descriptions (i.e., which do not contradict the rules) are represented, and the rules will no longer be needed any more. As a consequence, any computation made using these data will not have to take the rules into account.

In order to achieve such a goal the initial table is decomposed into several tables (according to the number of different premise variables) as it is done in databases.

The variables not concerned by the rules remain in the original table. Each of these new tables contains variables to which the rules apply: a premise variable and all linked conclusion variables.

It is easy to check wether the premise and conclusion variables does not contradict the rules for each line of these tables. If a contradiction appears, it is easy to split the value into two parts: one where the premise is true, and one where the premise is false.

We will decompose the original data table according to the dependence graph between the variables, induced by the rules. We only carry out this decomposition if the graph is a tree. Then we will obtain a tree of data tables. We call the root of the data tree the *Main table*, we call the other tables *secondary tables.*

We say that a set of Boolean SO is conform to the NSF (Csernel and De Carvalho (1999)) if the following conditions hold:

First NSF condition: Either no dependence occurs between the variables belonging to the same table, or a dependence occurs between the first variable (premise variable) and all the other variables.

Second NSF condition: all the values taken by the premise variable for one table line lead to the same conclusion.

The following example shows the result of the NSF transformation of Table I. It can be seen that we now have now three tables instead of a single one, but only the valid parts of the objects are represented: now, the tables include the rules.

	wings_r	Thorax_size
d1	{ 1, 3}	{big,small}
d2	{2, 4}	{small}

main table

wings_t	wings	colour_r
1	absent	4
2	absent	5
3	present	{ 1, 2 }
4	present	{ 1, 3 }

secondary table 1

colour_t	wings_colour	Thorax_colour
1	{ red }	{blue }
2	{ blue }	{ blue, yellow }
3	{ green }	{blue, red }
4	NA	{ blue, yellow }
5	NA	{ blue, red }

secondary table 2

The data form a unique (degenerated) data tree where each node is a table. To refer from one table to another, we need to introduce some new variables, called reference variables, which introduce a small space overhead. In the example these variables are denoted by '_r' at the end and the corresponding table by a '_t' at the end. The corresponding values refer to a line number within the table. The first column of *secondary tables* contain the name of the table in the first line and the line number in the other lines. The initial symbolic description can be found from the *Main table.* All the tables, except the *main table*, are composed in the following way:

1) the first variable in a table is a premise variable, all the other variables are conclusion variables;

2) in each line the premise variable leads to a unique conclusion for each of the conclusion variables;

The second NSF condition has two consequences:

1) we have to decompose each individual in a table into two parts: one part where the premise is true, and one part where the premise is false. In order to have an easier notation we will denote this consequence $CQ1$.

2) if we want to represent different conclusions in one table, we need to represent each description by as many lines as we have conclusions.

The main advantage of the NSF is that, after this transformation, the rules are included in the data, there are no longer any rules to be taken into account, and so no more exponential growth in computation time. Instead, computation time required to analyze symbolic data is polynomial as if there were no rules to be taken into account.

The $CQ1$ can induce a memory growth. In the following, we will consider this growth more closely according hierarchical rules.

3 Memory requirement with hierarchical dependencies

In this section, we will consider the possible memory growth, using only categorical set-valued variables. We recall here the results we obtained in one of our previous article (Csernel and De Carvalho (2002)).

We define the *local growing factor* F_l as the ratio between the number of lines N_d of a daughter table and the number of lines of its mother table N_m: $F_l = N_d/N_m$. Its upper bound is given by:

$$F_l = \frac{N_d}{N_m} \leq 2 \tag{1}$$

We showed, as a consequence, that the size of the daughter table is at most twice as big as the initial table.

If the conclusions are determined by n different sets of premises values then we will have the size of a leaf table being at most $n+1$ times the size of its mother table

$$N_l = \frac{N_d}{N_m} \leq n+1 \tag{2}$$

As a consequence, the size of a leaf table is at most $n+1$ times the size of the initial table.

4 Modelling the table filling

The above description establish a limit to the size of a secondary table when we use hierarchical rules and set-valued variables. We can have a better idea of that size by modelling it and provide an estimation. We will see that the limit we defined above can hardly be reached.

This kind of problem already arises in the field of thermodynamic when one try to determinate the energy level of molecule in a gas, there is a correspondence between each energy level an our array cell and the problem is rather similar.

4.1 The problem without rules

We will start this study without taking the rules into account. As an hypothesis we consider an equidistribution of the descriptions provided by variables x_i, $i \in \{1, \ldots, p\}$, where x_i take its (finite) domain in X_i.

Let x_i be the variables taking its values in the domains X_i, where X_i is finite. If we consider only *set valued* variable, the number of different possible descriptions of an array line describing an object with p variables is:

$$m = |\mathcal{P}^*(X_1)| \ldots |\mathcal{P}^*(X_j)| \ldots |\mathcal{P}^*(X_p)| = \prod_{j=1}^{p} (2^{|X_j|} - 1)$$

This number if for sure a maximum.

The prior probability to observe one description among all the m possible description, assuming variables independence, is then:

$$P(x_1 = d_1 \dots x_p = d_p) =$$

$$P(x_1 = d_1) \dots P(x_p = d_p) = \prod_{i=1}^{p} \frac{1}{|\mathcal{P}^*(X_i)|} = \frac{1}{\prod_{j=1}^{p} (2^{|X_j|} - 1)} \quad (3)$$

This problem can be compared with the well known problem of boxes and balls. Boxes represent the different array lines (and the corresponding descriptions), balls represent the different objects that must be matched with a line. The problem is to evaluate the occupancy of the m boxes (possible descriptions) by the N balls (descriptions within the sample) i.e. to put the N balls within the m boxes.

For example, it may arise that all the sample balls has to be put in the first box. In this case we have the following repartition:

1	2	....	m-1	m
N	0	0	0	0

Table 2 : repartition of the balls in the boxes

Let be m random variables z_i taking their values in the domain $Z_i = \{0, \dots, N\}$, i $\in \{1, \dots, m\}$. The prior probability to affect the object to the line I is $1/m$. Let N_i be the number of ball affected to the i-th box, let $N = N_1 + \dots + N_i + \dots N_m$, we have :

$$P(z_1 = N_1, \dots, z_i = N_i, \dots, z_m = N_m) = \frac{N!}{N_1! \cdots N_i! \dots N_m!} \left(\frac{1}{m}\right)^N \quad (4)$$

Each of the z_i has a binomial distribution with N and $1/m$ as parameters:

$$P(z_i = N_i) = \binom{N}{N_i} \left(\frac{1}{m}\right)^{N_i} \left(1 - \frac{1}{m}\right)^{N-N_i} \text{ and } \mu_{z_i} = \frac{N}{m}, \sigma^2_{z_i} = \frac{N(m-1)}{m^2}$$

This is equivalent to the Maxwell-Boltzman statistic used in thermodynamic.

Let w be the random variable representing the number of busy lines among the m lines available, the domain of w is $W = \{1, \dots, m\}$. We have:

$$P(w = k) = \frac{\binom{m}{m-k} \sum_{j=1}^{m} (-1)^{k-j} \binom{k}{j} j^N}{m^N}, k \in \{1, \dots, m\} \quad (5)$$

We want evaluate the expected value of w, i.e. the average number of occupied lines in a table. We may show that:

$$\mu_w = m \left(1 - \left(\frac{m-1}{m}\right)^N\right) \quad (6)$$

4.2 The problem with the rules

Till now we only partially solve the occupancy problem. We can consider that the presence of rules involve the generation of two different tables, one where the premise is false T_1, one where the premise is true of T_2. The size of T_2 is more or less the same as when we do not take the rules into account.

The size of T_1 must be examined more carefully. To simplify, we will consider the case where the premise variable y is divided in two parts A and $\bar{A}$, the premises is true if the value belongs to A and in that case all the others variables have NA as value. This means that the number of cells occupied is at most $|A|$.

In general this number is small compared with N, the number of sampled descriptions, and, as a consequence, the table is generally full or nearly full. The second table will have a more important size, and the ball and boxes model will fully applied. Generally the maximum size of the secondary table is large compared with the number of objects.

The problem of the tables occupation in presence of rules will therefore depend on relation between $|A|$ and $|\bar{A}|$, and it will be slightly complicated because the membership of a value as coming from A or from $|\bar{A}|$ is not an independent phenomenon. The previous modelling will only furnish an upper limit to the average number of cells occupied.

Roughly speaking, as the relation $|A|$ / $|\bar{A}|$ is large the factorization of the data will be strong. Indeed, if $|A|$ is large a lot of objects will take place in table T_1 which is small, and a lot of objects will therefore occupy the same cell. If $|\bar{A}|$ is small, few objects will go in the table T_2 which is large and little lines will be occupied.

5 Application and conclusion

As an example we took a knowledge base describing the 73 different male of all Phlebotom sandfly species in French Guiana (Vignes (1991)), each corresponding to a symbolic description described by 53 categorical set-valued variables. There are 5 hierarchical rules. These rules are represented by 3 different connected graphs involving 8 variables. In the first and second graphs two daughter variables become inapplicable if the mother variable takes its values in $\mathcal{P}^*(4)$. In the third graph a daughter variable becomes inapplicable if the mother variable takes its values in $\mathcal{P}^*(2)$.

The size of the secondary tables was 32, 18 and 16 lines.

As we saw in section 4.2 the resulting table will be devided in two parts: T1(premise true) and T2 (premise false). Concerning the first 2 graphs:

- T1 is bouded by P*(4) = 15. The estimation of the average number of occupied lines according to our model is 14.9.

- T2 is bounded by the number of initial description: 73. The estimation of the average number of occupied lines according to our model gives 46.

Our estimation for the size of T1 and T2 is S = 15 + 46 = 61. We observed 32 and 19, the model estimates 61 and 61. For the third table, we observed 16, the model estimates 49.

The complete commented results can be seen on:

http://www-rocq.inria.fr/sodas/NSF/.

Our method provide a upper limit of the average number of cells occupied, the difference we observed is not totally surprising. Obvisously the independant equirepartion hypothesis is, in most of the cases, not relevant.

Acknowledgements. This paper is supported by grants from the joint project INRIA-France CNPq-Brasil CLADIS (Proc. 480190/00-1) and by grants from CNPq (Proc. 301387-92-3)

References

BOCK, H.-H. and DIDAY, E. (2000): *Analysis of Symbolic Data.* Springer, Berlin.

CODD, E.E. (1972): Further Normalization of the Data Relational Model. In: R. Rustin (Ed.): *Data Base Systems.* Prentice-Hall, Englewood Cliffs, N.J., 33–64.

CSERNEL, M. (1998): On the Complexity of Computation with Symbolic Objects using Domain Knowledge. In: A. Rizzi, M. Vichi, and H.-H. Bock (Eds.): *Advances in Data Science and Classification.* Springer, Heidelberg, 403–408.

CSERNEL, M. and DE CARVALHO, F. A. T. (2002): On memory requirement with Normal Symbolic Form. In: O. Opitz and M. Schwaiger (Eds.): *Exploratory Data Analysis in Empirical Research.* Springer, Heldelberg (accepted to be published)

CSERNEL, M. and DE CARVALHO, F. A. T. (1999): Usual Operations with Symbolic Data under Normal Symbolic Form. *Applied Stochastic Models in Business and Industry, 15*, 241–257.

DE CARVALHO, F. A. T. (1998): Extension based Proximities between Constrained Boolean Symbolic Objects. In: Hayashi, Ohsumi, Yajima, Tanaka, Bock, Baba (Eds.): *Data Science, Classification and Related Methods.* Springer, Berlin, 370–378.

VIGNES, R. (1991): Caractérisation Automatique de Groupes Biologiques. Thèse de Doctorat. Université Paris VI, Paris.

Mixture Decomposition of Distributions by Copulas

Edwin Diday

CEREMADE, Paris IX Dauphine University

Abstract. Each unit is described by a vector of p distributions each one associated to one of p variables. Our aim is to find simultaneously a "good" partition of these units and a model using "copulas" associated to each class of this partition. Different copulas models are recalled. The mixture decomposition problem is settled in this general case. It extends the standard mixture decomposition problem to the case where each unit is described by a vector of distributions instead as usual, by a vector of a single (categorical or numerical) values. Several generalization of standard algorithms are suggested. All these results are first considered in the case of a single variable and then extended to the case of a vector of p variables by using a top-down binary tree approach.

1 Introduction

In a symbolic data table, a cell can contain, a distribution (Schweitzer (1984) says that "distributions are the number of the future"!), or several values linked by a taxonomy and logical rules, etc.. The need to extend standard data analysis methods (exploratory, clustering, factorial analysis, discrimination,...) to symbolic data table is increasing in order to get more accurate information and summarise extensive data sets. For more details on symbolic data analysis see for example Diday (1998), Bock, Diday (2000). The idea of operating with distribution functions as data values has been applied in clustering by Janowitz and Schweitzer (1989). We are here interested to extend the mixture decomposition problem (as defined for instance in Dempster and al (1977)) to the case where the units are described by distributions. Here a "variable" Y_j is considered to be a random variable from a set of units Ω in an infinite set of distributions $\mathbf{F}_j$. We consider a data table of N rows and p columns, where the ith row is associated to the unit (or "individual") $\omega_i \in \Omega$, the set of units of this data table is denoted $\mathbf{E}$ and each column is defined by a variable $Y_j \in \{Y_1, ..., Y_p\}$. Each cell (i, j) of this data table contains a distribution $Y_j(\omega_i) \in \mathbf{F}_j$. This sample of the N x p distributions of this data table is called the "distribution base".

We consider first, the case of a single variable $Y \in \{Y_1, ..., Y_p\}$ such that $Y(\omega) \in \mathbf{F}$ an infinite set of distributions. The case of several variables will be only considered in the section 6. The cell associated to a given unit ω_i and for the column associated to the variable Y of the data table, is a distribution denoted $F_i = Y(\omega_i)$. We denote by X_i the random variable associated with

F_i such that $F_i(t) = Pr(X_i \leqslant t)$ for $t \in \overline{\mathbb{R}}$ where $\overline{\mathbb{R}} = [-\infty, +\infty]$. The distribution base is here reduced to the sample set $\mathcal{F} = \{F_i / \ i = 1, N\}$ of all the distributions contained in the column associated to the variable Y in the data table.

First, we define a single "point distribution of distributions" associated to Y at a given point t by: $G_t(x) = Pr([Y = F / F(t) \leqslant x])$ where $x \in \overline{\mathbb{R}}$. Notice that $G_t(x) = 1$ if $x \geq 1$ and $G_t(x) = 0$ if A point distribution of distributions can also be interpreted as the distribution function of the random variable y_t from Ω in $\overline{\mathbb{R}}$ such that $y_t(\omega) = F(t)$ if $Y(\omega) = F$ as we have: $Pr([y_t(\omega) \leqslant x])$=$Pr([Y = F / F(t) \leqslant x])$.

We get the empirical frequency $\overline{G}_{t_n}$ by: $\overline{G}_{t_n}(x) = card(\{F_i \in \mathcal{F} / F_i(t_n) \leq x\}) / N$. Hence, $\overline{G}_{t_n}(x)$ is the frequence of units whose probability of taking a value less than t_n is equal or less than x. For instance, if the units are pupils and the variable Y is their results in mathematics, then, $\overline{G}_{10}(1/2)$ is the frequence of pupils whose, probability of having a mark less than 10 in mathematics is equal or less than 1/2. We define a "k-point joint distribution of distributions" by: $H_{t_1,\dots,t_k}(x_1, \dots, x_k) = Pr([Y = F / F(t_1) \leqslant x_1] \cap \dots \cap [Y = F / F(t_k) \leqslant x_k])$. A k-point joint distribution of distributions can be interpreted as the distribution of the random variable: $y = (y_{t_1}, \dots, y_{t_k})$. Therefore, we can also write: $H_{t_1,\dots,t_k}(x_1, \dots, x_k) = Pr([y_{t_1} \leqslant x_1] \cap \dots \cap [y_{t_k} \leqslant x_{t_k}])$. The "k-point joint density of distributions" is defined by $h_{t_1,\dots,t_k}(x_1, \dots, x_k) = \partial^k H_{t_1,\dots,t_k}(x_1, \dots, x_k) / \partial x_1 \dots \partial x_k$. It is the probability density associated to the random variable $y = (y_{t_1}, \dots, y_{t_k})$. In the following, we suppose that $t_1 < \dots < t_k$ and $Min \{F_i(t_1) / F_i \in F\} > 0$.

2 Link with copulas

Proposition 1
G_{t_n} is a distribution function. $H_{t_1,\dots,t_k}$ is a k-dimensional joint distribution function with margin $G_{t_1}, \dots, G_{t_k}$
Proof: this results from the definition of a distribution function and a n-dimensional distribution function and from the assumption: $Min\{F_i(t_1)/F_i \in F\} > 0$.

Before giving the definition of a k-copula, we need to define a "H-volume".
Definition of a H-volume (Nelsen (1998)):
Let $S_1, \dots S_k$ be non empty subsets of $\overline{\mathbb{R}}$, and let H be a mapping from $S = S_1 \times \dots \times S_k$ in $\overline{\mathbb{R}}$. Let $\mathbf{a} = (a_1, \dots, a_k)$ and $\mathbf{b} = (b_1, \dots, b_k)$ two points of $\overline{\mathbb{R}}^k$. Let $B = [\mathbf{a}, \mathbf{b}]$ denotes the n-box $[a_1, b_1] \times \dots \times [a_k, b_k]$ where all vertices B' are in S.Then, the H-volume of B is given by $V_H(B) = \sum_{c \in B'} sign(c) H(c)$ where $sign(c)$ is given by : $sign(c) = 1$ if $c_i = a_i$ for an even number of i' s and $sign(c_i) = -1$ for an odd number of i's.
Definition of a k-copula (Schweizer and Sklar (1983), Nelsen (1998)):
A k-copula is a function C from $[0, 1]^k$ to $[0, 1]$ with the following properties:

1. For every $u \in [0, 1]^n$, $C(u) = 0$ if at least one coordinate of u is 0

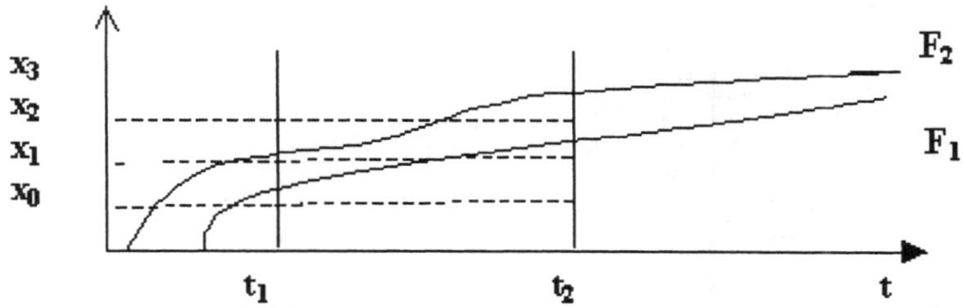

Fig. 1. The distribution base is reduced to F_1 and F_2

2. If all coordinates of u are 1 except one which is $u*$, then $C(u) = u*$
3. For every $\mathbf{a}$ and $\mathbf{b}$ in $[0,1]^n$ such that $\mathbf{a} \leqslant \mathbf{b}$, $V_C([\mathbf{a}, \mathbf{b}]) \geq 0$.

For example, in two dimensions ($k = 2$), the third condition gives: $C(b_1, b_2) - C(b_1, a_2) - C(a_1, b_2) + C(a_1, a_2) \geq 0$. In the following, we denote $RanG$ the range of the mapping G.

Proposition 2
There exists a k-copula C such that for all $X = (x_1, ..., x_k) \in [-\infty, +\infty]^k$: $H_{t_1,...,t_k}(x_1, ..., x_k) = C(G_{t_1}(x_1), ..., G_{t_k}(x_k))$. Moreover, C is uniquely determined on $Ran\ G_{T_1} \times ... \times Ran\ G_{T_k}$.

Proof: this results directly from Sklar theorem (Schweizer and Sklar (1983), Nelsen (1998)) and proposition 1, as this theorem says that when H is a k-dimensional distribution function with margins $h_1, h_2, ..., h_k$ then, there exists a copula C such that for all $\mathbf{x} = (x_1, ..., x_k)$ in $\overline{\mathbb{R}}^k, H(x_1, ..., x_k) = C(h(x_1), ..., h(x_k))$, moreover it says that C is uniquely determined on $Ranh_1 \times ... \times Ran\, h_k$.

Parametric families of copulas:
The most simple copulas denoted M, Π and W are $M(u,v) = min(u,v)$, $\Pi(u,v) = uv$ and $W(u,v) = max(u + v - 1, 0)$. These copulas are special cases of some parametric families of copulas as the followings: $C_b(u,v) = max([u^{-b} + v^{-b} - 1]^{1/b}, 0)$ discussed by Clayton (1978) and where $b \in [-1, \infty) \setminus \{0\}$ has the following special cases: $C_{-1} = W, C_0 = \Pi, C_\infty = M$.
Frank (1979) has defined $F_b(u,v) = -1/b\ \ell n(1 + (e^{-bu} - 1)(e^{-bv} - 1)/(e^{-b} - 1))$ where $b \in \overline{\mathbb{R}} \setminus \{0\}$ has the following special cases: $F_{-\infty} = W, F_0 = \Pi, F_\infty = M$. Many other families are defined in Nelsen (1998).

Example: the distribution base $\mathcal{F}$ (see figure 1) consists of two distributions F_1 and F_2 and the copula C is defined by: $C(u,v) = \overline{H}_{t_1,t_2}(x_1, x_2)$ with $u = \overline{G}_{t_\ell}(x_\ell) = card(\{F_i \in \mathcal{F}/F_i(t_\ell) \leqslant x_\ell\})/N$, $v = G_{t_2}(x_2) = card(\{F_i \in \mathcal{F}/F_i(t_2) \leqslant x_2\})\,/\,N$ and $\overline{H}_{t_1,t_2}(x_1, x_2) = card(\{F_i \in \mathcal{F}/F_i(t_1) \leqslant x_1\} \cap \{F_i \in \mathcal{F}/F_i(t_2) \leqslant x_2\})\,/\,N$.

The possible values of $\overline{G}_{t_1}(x_i)$and $\overline{G}_{t_2}(x_j)$ are only 0, 1/2, 1. For instance, from: $u = \overline{G}_{t_1}(x_0) = 0$, $v = \overline{G}_{t_2}(x_2) = 1/2$ and $\overline{H}_{t_1,t_2}(x_1, x_2) = 0$, it results

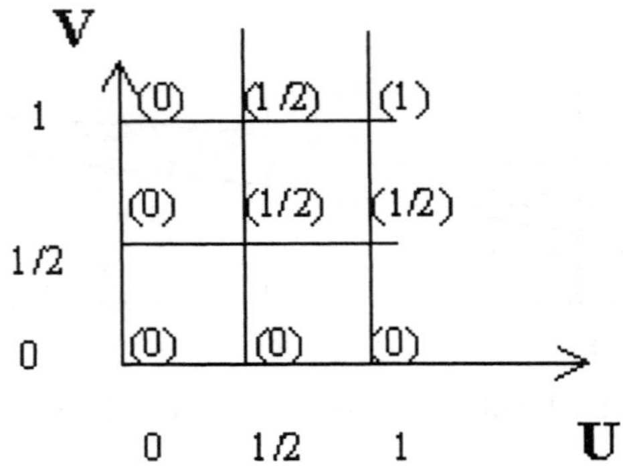

Fig. 2. The copulas $C(u,v)$, for instance $C(0,1) = 0$, $C(1,1/2) = 1/2$

that $C(0,1/2) = 0$. In the same way, we get: $C(0,0) = C(0,1/2) = C(0,1) = 0$, $C(1,0) = 0, C(1,1/2) = 1/2$, $C(1,1) = 1$.

From all calculation of $C(u,v)$ (see figure 2), it results that the model $C \equiv Min$ is always satisfied. This can be proved more generally for any pair of such kind of distribution functions without crossing.

3 Fit between a unit and a copula by using an approximation of the density

We define a mapping a which measures the "fit" between a distribution $Y(\omega) = L$ and a given copula C(G_{t_1},G_{t_2})=H_{t_1,t_2} associated to the distribution base, such that $a : \Omega \times [0,1]^2 \longrightarrow \mathbb{R}^+$:
$a(\omega,\varepsilon) = [Y(\omega)R(\varepsilon)C(G_{t_1},G_{t_2})] \in \mathbb{R}^+$ where $\varepsilon = (\varepsilon_1,\varepsilon_2)$ is a threshold. $x_i = L(t_i)$ and $R(\varepsilon)$ is defined by: $LR(\varepsilon)C(G_{t_1},G_{t_2}) = (C(G_{t_1}(x_1 + \varepsilon_1), G_{t_2}(x_2+\varepsilon_2)) - C(G_{t_1}(x_1+\varepsilon_1), G_{t_2}(x_2-\varepsilon_2)) - C(G_{t_1}(x_1-\varepsilon_1)), G_{t_2}(x_2 + \varepsilon_2)) + C(G_{t_1}(x_1-\varepsilon_1)), G_{t_2}(x_2-\varepsilon_2)))$.
Notice that due to proposition 1, G_{t_1} and G_{t_2} are increasing, so we have $G_{t_i}(x_i+\varepsilon_i) \geq G_{t_i}(x_i - \varepsilon_i)$.
If h is the density function of the random variable $y = (\, y_{t_1}, y_{t_2})$ defined in section 2, we have:
Proposition 3
$a(\omega,\varepsilon) \in [0,1]$ and if $\partial\varepsilon = \varepsilon_1\varepsilon_2$ then, $h_\varepsilon(x) = a(\omega,\varepsilon)/\partial\varepsilon \to h(x)$ when $\partial\varepsilon \to 0$.
Proof: This can be proved easily from:
$a(\omega,\varepsilon) = (H_{t_1,t_2}(x_1+\varepsilon_1, x_2+\varepsilon_2) - H_{t_1,t_2}(x_1+\varepsilon_1, x_2-\varepsilon_2) - H_{t_1,t_2}(x_1 - \varepsilon_1, x_2+\varepsilon_2) + H_{t_1,t_2}(x_1-\varepsilon_1, x_2-\varepsilon_2))$. From which it results: $a(\omega,\varepsilon) = Pr([Y = F \,/ x_1 - \varepsilon_1 \leqslant F(t_1) \leqslant x_1+\varepsilon_1] \cap [Y = F \,/ x_2 - \varepsilon_2 \leqslant F_i(t_2)) \leqslant x_2 + \varepsilon_2\}) \in [0,1]$.

If we set: $h(x;\varepsilon) = a(\omega,\varepsilon)/\partial\varepsilon$, then by definition the limit of $a(\omega,\varepsilon)/\partial\varepsilon$ when $\varepsilon \to 0$ is the density $h(x)$ of the random variable y.

4 The mixture decomposition problem of distributions and three algorithms for solving it

The set of units $E = \{\omega_1, ..., \omega_N\} \subset \Omega$ is considered as a sample of a random variable Y such that $Y(\omega) \in \mathbf{F}$ and (see section 2) its associated k-point distribution of distributions $H_{t_1,...,t_k}(x_1, ..., x_k)$ is denoted $H(X)$. We denote $P = (P_1, ..., P_\ell)$ a partition of E such that each class P_i be considered as a sample of a random variable $\mathbb{Y}_i : \Omega \to \mathbf{F}$ such that $\mathbb{Y}_i(\omega) \in \mathbf{F}$ and its associated k-point joint distribution of distributions depending on a parameter α_i is $H^i_{t_1,...,t_k}(x_1, ..., x_k; \alpha_i)$ and denoted.$H_i\ (X; \alpha_i)$.

The mixture decomposition problem of a distribution of distributions can be settled in the following way: find a partition $P = (P_1, ..., P_\ell)$ of E, the proportions $(p_1, ..., p_\ell)$ of the mixture and the parameters $(\alpha_1, ..., \alpha_\ell)$. such that: $H(X) = \sum_{i=1,\ell} p_i\ H_i\ (X; \alpha_i) \quad (1)$.

It can be shown (see Diday (2001)), that the standard mixture decomposition problem with standard data, is a special case of this general problem.

In order to solve this problem we can settle it in term of mixture decomposition of density functions, by setting in case of H derivability : $h(X) = \partial H(X)/\partial x_1 ... \partial x_k$ and $h_i(X, \alpha_i) = \partial H_i\ (X; \alpha_i)/\partial x_1 ... \partial x_k$. Then, the mixture decomposition equation (1) becomes: $h(X) = \sum_{i=1,\ell} p_i h_i(X; \alpha_i)$.

Notice that in the case of k= 2 the link between h and H and the copulas model is given by:

$h(X) = \partial^2 H_{t_1,t_2}(x_1, x_2)/\partial x_1 \partial x_2 = \partial^2 C(G_{t_1}(x_1).G_{t_2}(x_2))/\partial x_1 \partial x_2$

$h(X) = (\partial G_{t_1}(x_1)/\partial x_1).(\partial G_{t_2}(x_2)/\partial x_2).(\partial^2 C(G_t(x_1), G_t(x_2))/\partial u_1 \partial u_2)$

$h(X) = G'_t(X) C'(G_t(X))$ where $C'(u_1, u_2) = \partial^2 C(u_1, u_2)/\partial u_1 \partial u_2$

and $G'_t(x_1, x_2) = \partial G_{t_1}(x_1)/\partial x_1 . \partial G_{t_2}(x_2)/\partial x_2$

The parameters $\alpha_i = (d_i, b_i)$ depend on the parameters of a chosen copulas family model (for instance, the Frank family, see section 3) denoted b_i and on the parameters of a chosen distribution family model denoted d_i. In order to approximate the density function $h_i(X, \alpha_i)$ or to calculate it, we can use the way presented in section 3: $h_i(X; \alpha_i, \varepsilon_i)\partial\varepsilon_i = a_i(\omega; \alpha_i, \varepsilon_i) = [Y(\omega)\ R(\varepsilon_i)\ C_i((G^i_{t_1}(.; d_{i_1}) G^i_{t_2}(.; d_{i_2})), b_i)] \quad (2)$, where $X = (L(t_1), L(t_2))$, $L = Y(\omega)$, $\alpha_i = (d_i, b_i)$ with $d_i = (d_{i_1}, d_{i_2})$ and $\varepsilon_i = (\varepsilon_{1i}, \varepsilon_{2i})$. Hence, the mixture decomposition model can be settled in the following way: $h(X) = \Sigma_{i=1,k}\, p_i h_i(X, \alpha_i, \varepsilon_i)$.

Given the models associated to G and C, the decomposition can be obtained by maximizing alternatively (Diday et al. 1974) a criterion based on the likelihood. With $L = Y(\omega), x_i = L(t_i)$ and $X = (x_1, ..., x_k)$ this criterion can be, the likelihood: $v(X,\beta) = \Pi_{i=1,\ell} p_i \Pi_{\omega \in P_i} h_i(X,\beta_i))$ where $\beta = (\beta_1, ..., \beta_\ell)$ are the parameters of the densities h_i, or the log likelihood:

$V(X,\beta) = \Sigma_{i=1,\ell} \Sigma_{\omega \in P_i} Log(p_i h_i(X,\beta_i))$. We suggest the two following algorithms with:

Input: a set of units. E described by distributions, a given partition $(P_1, ..., P_\ell)$ a parametric copula family $\mathbb{C}$, optionally a parametric distribution familly law $\mathbb{G}$.

Output: a partition and a copula C_i for each class , optionally a distribution law for each G_i at each t_i.

Algorithm 1:

This algorithm is defined in two steps of representation and allocation (Diday, Ok, Schroeder (1974)):

Representation: $g(P_1, ..., P_\ell) = (\beta_1, ..., \beta_\ell)$ by finding the parameters $(\beta_1, ..., \beta_\ell)$ which maximise the chosen criterion (v, or V).

Maximisation: $f(\beta_1, ..., \beta_\ell) = (P_1, ..., P_k)$ by finding the partition $(P_1, ..., P_\ell)$ where P_i is the set of distributions: $P_i = \{X \,/\, p_i h_i(X,\beta_i) \geq p_m h_m(X,\beta_m))$, with $i < m$ in case of equality$\}$.

When the criterion are bounded (which is the case of v, V), it is easy to show that this algorithm converges as it increases the criterion at each step.There are several variants for the choice of $p_l : p_i = card(\ P_i\ ) / N$ at the last step, or at each step. In the case where an approximation of the derivative is needed (see (2)), $\varepsilon_i = Min(\varepsilon_{1i}, \varepsilon_{2i})$ is increased until $Max_i\ a_i(\omega;\alpha_i)$ becomes different from all (or a given number) of $a_j(\omega;\alpha_i)$.

Algorithm 2:

This algorithm is based on the two steps of the EM algorithm (Dempster, Laird, Rubin (1977)): we start from an initial solution at step $n = 0 : (p_l^0, \beta_l^0)$ for $l = 1, \ell$ and then at step n we have:

Estimation: for $l = 1, \ell$ and any $\omega \in \Omega$, $t_l^n(X) = p_l^n h_l(X, \beta_l^n) / \Sigma_{\omega \in \Omega} p_l^n h_i$ (X, β_l^n) which is the posterior conditional probability of an individual ω to belong to the class l at the nth iteration.

Maximisation: with $p_l^{n+1} = 1/N, \sum_{\omega \in \Omega} t_l^n(Y(\omega))$ maximise:
$\sum_{\omega \in \Omega} t_l^n(Y(\omega)) \partial Log\ h(Y(\omega), \beta_l^{n+1}) / \partial \beta_l^{n+1}$

A simple example of mixture decomposition:

The data table is defined by five units w_i, each one described by a distribution F_i. The five distributions are given on figure 3.

The family of copulas parametric model is simple : $b_i \in \{0,1\}$, $C(u,v) = M$ if $b_i = 1$ and $C(u,v) = W$ (see 2.4) if $b_i = 0$.

There is no parametric model for $G_{T_1}^i$ and G_{T2}^i as we use only their empirical distribution. Hence, $G_T^i(x) = Pr(L \in P_i / L(T_j) \leq x)$ and there are no parameters d_i.

In order to simplify calculations, we use the following criterion: $Z(X,\beta) = \Sigma_{i=1,2} p_i \Sigma_{w \in P_i} h_i(X,\beta_i))$ where $\beta_i = (\alpha_i, \varepsilon_i) = ((d_i, b_i), (\varepsilon_{1i}, \varepsilon_{2i})) = (b_i, (0.1, 0.1))$. In order to maximise Z, it suffices to look for $b_i \in \{0,1\}$ which

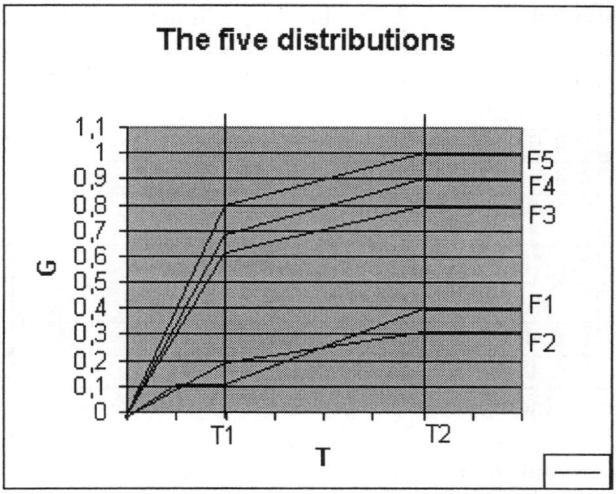

Fig. 3. The five distributions

maximizes : $Z(X,\beta) = \sum_{i=1,2} p_i \Sigma_{w\in P_i} h_i(X,\beta_i) = \sum_{i=1,2} p_i \sum_{w\in P_i} C((G^i_{T_1}(x_1 + \varepsilon_1), G^i_{T2}(x_2 + \varepsilon_2), b_i) - C((G^i_{T_1}(x_1 + \varepsilon_1), G^i_{T_2}(x_2 - \varepsilon_2)), b_i)$- $C((G^i_{T_1}(x_1 - \varepsilon_1)), G^i_{T_2}(x_2 + \varepsilon_2)), b_i) + C((G^i_{T_1}(x_1 - \varepsilon_1), G^i_{T_2}(x_2 - \varepsilon_2)), b_i)$. The initial given partition is $P = (P1, P2)$ with $P^0_1 = \{F1, F2, F3\}$ and $P^0_2 = \{F4, F5\}$.

We choose $\varepsilon_i = (Max_{L\in F} G^i_{T_i}(L(T_i)) - Min_{L\in F} G^i_{T_i}(L(Ti)))/N$. So, we find $\varepsilon_1 = \varepsilon_2 \approx 0.1$.

Step 1: induction of parameters from classes
$g(P_1, ..., P_k) = (\beta_1, ...\beta_k)$ by maximisation of the criterion Z where $X_i = (x_{i1}, x_{i2}), x_i = L_i(T_i)$ and $L_i = Y(w_i)$. Here, we have : $\beta_i = (\alpha_i, \varepsilon_i)$ where α_i generally equal to (d_i, b_i) is reduced to $b_i = (bi1, bi2)$ and $\varepsilon_i = (\varepsilon_{1i}, \varepsilon_{2i}) = (0.1, 0.1)$. For each $w \in \{w_1, ..., w_5\}$ and for $b_{ij} \in \{0,1\}$ (i.e. $C \in \{M, W\}$), we compute: $e^i_1 = C^s((G^i_{T_1}(x_1 + \varepsilon_1), G^i_{T_2}(x_2 + \varepsilon_2)), bij)$, $e^i_2 = C^s((G^i_{T_1}(x_1 + \varepsilon_1), G^i_{T_2}(x_2 - \varepsilon_2)), b_{ij})$, $e^i_3 = -C^s((G^i_{T_1}(x_1 - \varepsilon_1)), G^i_{T_2}(x_2 + \varepsilon_2)), bij)$, $e^i_4 = C^s((G^i_{T_1}(x_1 - \varepsilon_1))G^i_{T_2}(x_2 - \varepsilon_2), bij)$

Hence, we obtain the following table:

Class	Copula	Individual: ω_1	Individual : ω_2	Individual: ω_3	$\sum_{\omega \in P_i} h_i(X, \beta_i)$
P_1^0		$e_1^1, e_2^1, e_3^1, e_4^1(\omega_1)$	$e_1^1, e_2^1, e_3^1, e_4^1(\omega_2)$	$e_1^1, e_2^1, e_3^1, e_4^1(\omega_3)$	$\sum_{i=1,4 \atop j=1,3} e_i^1(\omega_j)$
	W	$1/3, 0, 0, 0$	1/3, 0, 0, 0	1, -2/3, -2/3, 1/3	2/3
	Min	$2/3, -1/3, 0, 0$	$2/3, 0, -1/3, 0$	1, -2/3, -2/3, 2/3	1
Class	Copula	Individual: $\omega 4$	Individual: ω_5		$\sum_{\omega \in P_i} h_i(X, \beta_i)$
P_2^0		$e_1^2, e_2^2, e_3^2, e_4^2(\omega_1)$	$e_1^2, e_2^2, e_3^2, e_4^2(\omega_2)$		$\sum_{i=1,4 \atop j=1,3} e_i^2(\omega_j)$
	W	1, 0, 0, 0	1, -1/2, -1/2, 0		1
	Min	1, 0, 0, 0	1, -1/2, -1/2, 1/2		3/2

Table 1 Induction of the copulas from the classes P_1^0 and P_2^0.

As $p_1 = 3/5$ and $p_2 = 2/5$, it results from the Table 1, that for $b_1 = b_2 = Min, Z(X, \beta) = 6/5$, for $b_1 = b_2 = W, Z(X, \beta) = 4/5$, for $b_2 = W$, or $b_1 = W$, $b_2 = Min, Z(X, \beta) = 1$. Therefore $b_1 = b_2 = Min$ gives the best solution. In other words the best parameters are defined by $b_2 = 1$ for the class P_1^0 and $b_2 = 1$ for the class P_2^0.

Step 2: for each individual ω we compute
$h_i(X, \beta_i) = a_i(w, \varepsilon) = [Y(w)R(\varepsilon)C_i^s\ (G_{T_1}^i, G_{T_2}^i)]$
$= C^s((G_{T_1}^i(x_1 + \varepsilon_1))G_{T_2}^i(x_2 + \varepsilon_2)), b_i) - C^s G_{T_1}^i(x_2 + \varepsilon_1), G_{T_2}^i(x_2 - \varepsilon_2)), b_i) -$
$C^s((G_{T_1}^i(x_1 - \varepsilon_1)), G_{T_2}^i(x_2 + \varepsilon_2)), b_i) + C^s((G_{T_1}^i(x_1 - \varepsilon_1)), G_{T_2}^i(x_2 - \varepsilon_2)), b_i) =$
$C^s((G_{T_1}^i(x_1 + \varepsilon_1)), G_{T_2}^i(x_2 + \varepsilon_2))), 1) - C^s((G_{T_1}^i(x_1 + \varepsilon_1), G_{T_2}^i(x_2 - \varepsilon_2)), 1) -$
$C^s((G_{T_1}^i)(x_1 - \varepsilon_1)), G_{T_2}^i(x_2 + \varepsilon_2)), 1) + C^s((G_{T_1}^i(x_1 - \varepsilon_1)), G_{T_2}^i(x_2 - \varepsilon_2)), 1) =$
$Min((G_{T_1}^i(x_1 + \varepsilon_1), G_{T_2}^i(x_2 + \varepsilon_2)) - \min G_{T_1}^i(x_1 + \varepsilon_1), G_{T_2}^i(x_2 - \varepsilon_2))) - Min((G_{T_1}^i (x_1 - \varepsilon_1)), G_{T_2}^i(x_2 + \varepsilon_2))) + \min((G_{T_1}^i(x_1 - \varepsilon_1)G_{T_2}^i(x_2 - \varepsilon_2)))$. Hence, we obtain the following table where $a_i(w, \varepsilon)$ measure the fit of w to the class P_i^0 and the class membership is i when: $a_i(w, \varepsilon)/card(P_i^0) > a_j(w, \varepsilon)/card(P_j^0)$.

Individual	$e_1^1, e_2^1, e_3^1, e_4^1$	$a_1(\omega, \varepsilon)$	$e_1^2, e_2^2, e_3^2, e_4^2$	$a_2(\omega, \varepsilon)$	Class
ω_1		1/3	0, 0, 0, 0	0	1
ω_2		1/3	0, 0, 0, 0	0	1
ω_3		1/3	1/2, 0, 0, 0	1/2	2
ω_4	1, 1, 1, 1	0		1	2
ω_5	1, 1, 1, 1	0		1/2	2

Table 2: Allocation of each individual to the best fitting class. Some e_j^i are already computed in table 1.

It results that the new classes are $P_1^1 = \{F1, F2\}$ and $P_2^1 = \{F3, F4, F5\}$.

Step 3: induction of parameters from classes. We have $P_1^0 = \{\omega_1, \omega_2, \omega_3\}$ and $P_2^0 = \{\omega_4, \omega_5)$

Class	Copula	Individual: ω_1	Individual: ω_2	g	$\sum_{\omega \in P_i} \text{h}_i(\text{X},\beta_i)$
P_1^1		$e_1^1, e_2^1, e_3^1, e_4^1(\omega_1)$	$e_1^1, e_2^1, e_3^1, e_4^1(\omega_2)$		$\sum_{\substack{i=1,4\\j=1,3}} \text{e}_i^1(\omega_j)$
	W	$1, 0, 1/2, 0$	$1, 1/2, 0, 0$		2
	Min	$1, -1/2, 0, 0$	$1, 0, -1/2, 01$		1
Class	Copula	Individual: ω_3	Individual: ω_4	Individual: ω_5	$\sum_{\omega \in P_i} \text{h}_i(\text{X}, \beta_i)$
P_2^1		$e_1^1, e_2^1, e_3^1, e_4^1(\omega_3)$	$e_1^2, e_2^2, e_3^2, e_4^2(\omega_4)$	$e_1^2, e_2^2, e_3^2, e_4^2(\omega_5)$	$\sum_{\substack{i=1,4\\j=1,3}} \text{e}_i^2(\omega_j)$
	W	1/3, 0, 0, 0	1, -1/3, -1/3, 0	1, -2/3, -2/3, 1/3	2/3
	Min	2/3, 0, 0, 0	1, -1/3, -1/3, 1/3	1, -2/3, -2/3, 1/3	4/3

Table 3: Induction of copulas from the classes P_1^1, P_2^1. It results that W for the first class and Min for the second class are the best.

Step 4: Building new classes

Individual	$e_1^1, e_2^1, e_3^1, e_4^1$	$a_1(\omega, \varepsilon)$	$e_1^2, e_2^2, e_3^2, e_4^2$	$a_2(\omega, \varepsilon)$	Class
ω_1		1/2		0	1
ω_2		1/2		0	1
ω_3		1/2		2/3	2
ω_4		1/2		2/3	2
ω_5		1/2		2/3	2

Table 4: Allocation of each individual to the best fit class. The new partition (P_1^2, P_2^2) is the same as the preceeding (P_1^1, P_2^1).

The process has converged towards the partition: $(P_1^2, P_2^2) = \{\{F_1, F_2\}, \{F_3, F_4, F_5\}\}$ and the mixture decomposition $H(X) = \sum_{i=1,2} p_i \, H_i\,(X;\alpha_i) = 2/5W(G_{t_1}^1(x_1), G_{t_2}^1(x_2)) + 3/5M(G_{t_1}^2(x_1), G_{t_2}^2(x_2))$ where $X = (x_1, x_2)$ and $G_{t_j}^i$ is defined by the empirical distribution of the class P_i^2 at t_j.

5 The special case of the standard mixture decomposition problem

5.1 Properties of a distribution base of unit mass distributions

Our aim in this section, is to imbed the standard mixture decomposition problem in the mixture decomposition of distributions problem, in the case of a single quantitative random variable Z: $\Omega \longrightarrow \overline{\mathbb{R}}$. Each value taken by an individual "ω" can be transformed in a single way in a distribution which takes the value 0 until $Z(\omega)$ not included and the value 1 after. Such distribution is called "unit mass". More formally, if $Z(\omega_i) = z_i$, the distribution associated to ω_i is defined by $F_i(t) = Pr(X_i \leq t)$ where the random variable X_i associated to ω_i is such that its distribution F_i satisfies: $F_i(t) = 0$ if $t < z_i$ and $F_i(t) = 1$ if $t \geq z_i$.

Proposition 4
If the distribution base F contains only such distributions F_i,F_z is the distribution associated to the random variable Z and the t_i increases with i, we have the following results:

1) If $H(x_1, ..., x_p) = C(G_{t_1}(x_1), ..., G_{t_p}(x_p))$ then C is the M copulas.
2) If $x_p < 1$, then $Min\ (G_{t_1}(x_1), ..., G_{t_p}(x_p)) = G_{t_p}(x_p)$.
3) If $x < 1$ then $G_t(x) = Prob(Z > t) = 1 - F_z(t)$.
4) If $x_p < 1$ then $F_z(t_p) = 1 - H(x_1, ..., x_p)$.

Proof

Lemma

If F is a set of unit mass distributions , $x_i \in [0, 1[$ for $i = 1, j$, $A_j = \{f \in F\ /\ f(t_j) = 0\}$ and

$B_j = \{f \in F\ /\ f(t_i) \leq x_i, 1 \leq i \leq j\}$, then we have: A j = B j and $|A_j| = Min_{i=1,j}\ |A_i|$.

Proof of the Lemma

We have $B_j \subseteq A_j$ as $f \in B_j$ implies $f \in F$ and $f(t_j) < x_j$ by definition of B_j. As F is a set of unit mass distributions and x j ∈ [0, 1[, we have necessarily $f(t_j) = 0$. Therefore $f \in A_j$.

$A_j \subseteq B_j$ as $f \in A_j$ implies $f(t_j) = 0$ which implies $f(t_i) = 0$ for $i = 1, j$ as f is increasing as it is a distribution. So $f \in B_j$ and therefore $A_j = B_j$. As by definition of B_j we have $B_j = \cap_{i=1,j} A_i$ and moreover we have proved that $A_j \subseteq B_j$, it results that $|A_j| = Min_{i=1,j}\ |A_i|$.

With this lemma, we can now prove the proposition 4.

1) If $H(x_1, ..., x_p) = C(G_{t_1}(x_1), ..., G_{t_p}(x_p))$ then C is the M copulas.

This can be proved in the following way: if all x_i are equal to 1, as all the elements of a distribution base take a value smaller than 1 everywhere, we have by definition of a point distribution of distributions: $G_{t_i}(x_i) = 1$ and also, by definition of a k-point joint distribution of distribution $H(x_1, ..., x_p) = 1$. So, in that case 1) is true. Suppose now that some x_i are smaller than 1 and suppose by denoting them $x'_1, ..., x'_j$ such that their corresponding t denoted $t'_1, , t'_j$are increasing, then we have $H(x'_1, ..., x'_p) = H(x_1, ..., x_p)$. This comes from the fact that the set of distributions included in the distribution base, which are lower than $x'_1, ..., x'_j$ are the same as the ones which are also lower than $x_1, ..., x_p$. We can now apply the lemma 1 by denoting $Aj = \{f \in F\ /\ f(t'_j) = 0\}$ and $B_j = \{f \in F\ /\ f(t_i) \leq x_i, 1 \leq i \leq j\}$. As $G_{t'_j}(x_j) = 1 = \left|\{f \in F/f(t'_j) = 0\}\right|\ /\ |F|$ and $H(x'_1, ..., x'_j) = |\{f \in F/f(t'_i) \leq x_i, 1 \leq i \leq j\}|\ /\ |F|$ it results that $G_{t'_j}(x'_j) = |A_j|\ /\ |F|$, and $H(x'_1, ..., x'_j) = |B_j|\ /\ |F|$. As from the lemma we have $A_j = B_j$ it results that $G_{t'_j}(x'_j) = H(x'_1, ..., x'_j)$ and so, $G_{t'_j}(x'_j)\ H(x_1, ..., x_p)$. From the lemma, we have also $|A_j| = Min_{i=1,j}\ |A_i|$ which implies , $G_{t'_j}(x'_j) = Min\ _{i=1,j}G_{t'_i}(x'_i) = 1$. As $Min\ _{i=1,j}G_{t'_i}(x'_i) = Min_{i=1,p}\ G_{t_i}(x_i)$ (as for the i such that $x_i = 1$ we have $G_{t_i}(x_i) = 1$ and for the i such that $x_i < 1$ we have $G_{t_i}(x_i) \leq 1$), we get $H(x_1, ..., x_p) = Min_{i=1,p}\ G_{t_i}(x_i)$ which shows that $H(x_1, ..., x_p) = C(G_{t_1}(x_1), ..., G_{t_p}(x_p))$ where C is the Min copulas.

2) If $x_p < 1$, then $Min\ (G_{t_1}(x_1), ..., G_{t_p}(x_p)) = G_{t_p}(x_p)$.

As in the proof of 1) we denote $x'_1, ..., x'_j$ (associated to increasing $t'_1, , t'_j$), the x_i among $x_1, ..., x_p$ which are strictly lower than 1. It results that $x'_j = x_p$ and so from lemma 1 that $Min\ (G_{t_1}(x_1), ..., G_{t_p}(x_p)) = G_{t_p}(x_p)$. We have

$Min\ (G_{t_1}(x_1), ..., G_{t_p}(x_p)) = Min(G_{t'_1}(x'_1), ..., G_{t'_p}(x'_p))$ as shown in the preceding proof .Therefore we get finally: $Min\ (G_{t_1}(x_1), ..., G_{t_p}(x_p)) = G_{t_p}(x_p)$.

3) If $x < 1$ then $G_t(x) = Prob(Z > t) = 1 - F_z(t)$ as by definition $F_z(t) = Pr(\{Z(\omega) \le t\}\ )$ and $G_t(x) = Prob(\{F_i \in F / F_i(t) \le x\})$ is exactly the proportion of unit mass distributions F_i which value is $F_i(t) = 1$ only strictly after t (i.e. $t' > t$), as $F_i(t) = 0$ if $t < Z(\omega_i)$ and $F_i(t) = 1$ if $t < Z(\omega i)$. In other words, this means that $G_t(x)$ is the proportion of individuals ω such that $Z(\omega) > t$.

4) If $x_p < 1$ then $F_z(t) = 1 - H(x_1, ..., x_p)$ as from 1), $H(x_1, ..., x_p) = Min\ (G_{t_1}(x_1), ..., G_{t_p}(x_p))$ and from 2), $Min\ (G_{t_1}(x_1), ..., G_{t_p}(x_p)) = G_{t_p}(x_p)$ and from 3) $F_z(t_p) = 1 - G_{t_p}(x_p)$.

5.2 The standard mixture decomposition is a special case

Here we need to introduce the following notations: F_{z_i} is the distribution associated to a quantitative random variable Z^i defined on Ω. F^i is a distribution base whose elements are the units mass distributions associated to each value $Z^i(\omega_i)$ (i.e. they take the value 0 for $t \le Z^i(\omega_i)$ and 1 for $t > Z^i(\omega_i)$). G^i_t is a point-distribution of distributions at point t associated to the distribution base F_i. $H^i\ _{t_1,...,t_k}$ is a k-point joint distribution of distributions associated to the same distribution base.

Proposition 5

If $H_{t_1,...,t_k} = \sum_{i=1,\ell} p_i\ H^i\ _{t_1,...,t_k}$ with $\sum_{i=1,\ell} p_i = 1$, then $F_z = \sum_{i=1,\ell} p_i\ F_{z_i}$.

Proof

From the proposition 2, we have $H^i_{t_1,...,t_k}(x_1, ..., x_k) = C^i\ (G^i_{t_1}(x_1), ..., G^i_{t_k}(x_k))$ where C_i is a k-copula. Therefore $H_{t_1,...,t_k}(x_1, ..., x_k) = \sum_{i=1,p} p_i\ C^i\ (G^i_{t_1}(x_1)\ , ..., G^i_{t_k}(x_k))$

We choose $x_p < 1$ and we use 1), 2), 3),4) in proposition 4.

From 1) we get: $H_{t_1,...,t_k}(x_1, ..., x_k) = \sum_{i=1,\ell} p_i Min(G^i_{t_1}(x_1), ..., G^i_{t_k}(x_k))$.

From 2) we get: $H_{t_1,...,t_k}(x_1, ..., x_k) = \sum_{i=1,\ell} p_i G^i_{t_k}(x_k)$.

From 3) we get: $H_{t_1,...,t_k}(x_1, ..., x_k) = \sum_{i=1,\ell} p_i\ (1 - F_{z_i}(t_k)) =$ 1-$\sum_{i=1,p} p_i\ F_{z_i}(t_k)$.

From 4) we have $F_z(t_k) = 1 - H_{t_1,...,t_k}(x_1, ..., x_k)$ and therefore $F_z(t_k) = \sum_{i=1,p} p_i\ F_{z_i}(t_k)$.

As the same reasoning can be done for any sequence t1,,t p, it results finally that: $F_z = \sum_{i=1,p} p_i\ F_{z_i}$.

5.3 Links between the generalised mixture decomposition problem and the standard one

It results from the proposition 5 that by solving the mixture decomposition of distribution of distributions problem we solve the standard mixture decomposition problem. This results from the fact that it is possible to induce F_{z_i} (t_1),..., F_{z_i} (t_k), from $G^i_{t_1}(x_1), ..., G^i_{t_k}(x_k)$ and therefore, the parameters of the chosen model of the density law associated to each Z_i. Moreover, by choosing the "best model" among a given family of possible models (Gaussian, Gamma, Poisson,) for each Z_i, we can obtain a different model for each law of the mixture. "Best model" means model which fits the best with F_{z_i} (t_1),..., F_{z_i} (t_k) for each i. It would be interesting to compare the result of both approach: the mixture decomposition of a distribution of distributions algorithms, and the standard mixture distribution algorithms in the standard framework, when the same model is used for each class or more generally when each law of the mixture follows a different family model.

6 Mixture decomposition with copula model in the case of more than one variable

We have considered the mixture decomposition problem in the case of a single variable. In order to extend it to the case of several variables, we can proceed as follows: we look for the variable which gives the best mixture decomposition criteria value in two classes and we repeat the process to each class thus obtained until the size of the classes becomes too small.In order to select the best variable the choice of the t_j is important. As we are looking for a partition of the set of distributions, it is sure that a given t_j is bad if all the distributions F_i of the base F take the same value at t_j. Also t_i is a bad choice if all the $F_i(t_j)$ are uniformly distributed in $[0,1]$. In fact we can say that a t_j is good if distinct clusters of values exist among the set of values: $\{F_i(t_j) \,/\, i = 1, N\}$. For instance, in Jain and Dubes (1988) several methods are proposed in order to reveal clustering tendency. Here, we are in the special case where we look for such tendency among a set of points belonging in the interval $[0,1]$.

We suggest a method based on the number of triangles whose vertices are points of $[0,1]$ and where the two sides of closest size, are larger (resp. smaller) than the remaining side. These sets of triangle are denoted A (resp.B). For instance, let $(a_1, a_2, a_3) \in [0,1]^3$ be the vertices of a triangle a. The length of the sizes of this triangle are: $\mid a_1 - a_2 \mid, \mid a_1 - a_3 \mid, \mid a_2 - a_3 \mid$. If the two closest, are larger than the third one, then $a \in A$,if not $a \in B$. We define the hypotheses H^0 that there is no clustering tendency, by the distribution of a random variable X^0 which associates to $u = \{u_1, ..., u_N\}, N$ points randomly distributed in the interval $[0,1]$, to the value $X^0(u) = (\mid A \mid - \mid B \mid)/C^3_N = 6(\mid A \mid - \mid B \mid) \,/\, n(n-1)(n-2)$ which belongs to $[-1,1]$. The greater is

$X^0(u)$ the higher is the clustering tendency of the N points. We calculate the number of triangles whose vertices are points of $U = \{F_i(t_j) \,/\, i = 1, N\}$ for which the two closest sides are larger (resp. smaller) than the remaining side and denoted $A(U)$ (resp. $B(U)$). Having the distribution of X^0, the value $(A(U) - B(U))/C_N^3 = 6(A(U) - B(U))/n(n-1)(n-2)$ can reject or accept the null hypotheses at a given threshold.

When t_1 and t_2 have been found, for instance, by the preceding way, the mapping a defined in section 4, can be extended in the following way: a^* : $\Omega \times [0,1]^2 \longrightarrow \mathbb{R}^+$: $a^*(\omega, \varepsilon) = \int_{t_1}^{t_2} [Y(\omega) R(\varepsilon) C(G_{t_1}, G_t)]\, dt \in \mathbb{R}^+$.

7 Conclusion

Many things remain to be done, for instance, comparing the results obtained by the general methods and the standard methods of mixture decomposition on standard data (as they are a special case of distributions) and studying the case in which each class may be modelled by a different copula family. Also, the G_t can be modelled at each t by a different distribution family. We can also add other criteria taking care of a class variable and a learning set. Notice that the same kind of approach can be used in the case where, instead of having distributions, we have any kind of mapping.

References

BOCK H.H. and DIDAY E. (eds.) (2000): *Analysis of Symbolic Data.Exploratory methods for extracting statistical information from complex data.* Springer Verlag, Heidelberg, 425.

CLAYTON D.G. (1978): A model for association in bivariate life tables. *Biometrika, 65*, 141-151.

DEMPSTER A.P., LAIRD N.M. and RUBIN D.B. (1977): Mixture densities, maximum likelihood from incomplete data via the EM algorithm. *Journal of the Royal Stat. Society, 39, 1*, 1-38.

DIDAY E. (2001): A generalisation of the mixture decomposition problem in the symbolic data analysis framework. Ceremade report $n°$0112. May 2001, 14 pages University Paris IX Dauphine.

DIDAY E. (1998): Extracting Information from Multivalued Surveys or from Very Extensive Data Sets by Symbolic Data Analysis Advances in Methodology, Data Analysis and Statistics. Anuska Ferligoj (Editor), Metodoloski zveski, 14. ISBN 86-80227-85-4.

DIDAY E., OK Y. and SCHROEDER A. (1974): The dynamic clusters method in Pattern Recognition. Proceedings of IFIP Congress, Stockolm.

JAIN A.K. and DUBES R.C. (1988): *Algorithms for Clustering Data.* Prentice Hall Advanced Reference Series.

JANOWITZ M. and SCHWEITZER B. (1989): Ordinal and percentile clustering. *Math. Social Sciences, 18.*

NELSEN R.B. (1998): *An introduction to Copulas in Lecture Notes in Statistics.* Springer Verlag.

SCHWEITZER B. and SKLAR A. (1983): *Probabilist Metric Spaces*. Elsever North-Holland, New-York.

Determination of the Number of Clusters for Symbolic Objects Described by Interval Variables

André Hardy and Pascale Lallemand

Department of Mathematics, University of Namur,
Rempart de la Vierge 8, B - 5000 Namur, Belgium

Abstract. One of the important problems in cluster analysis is the objective assessment of the validity of the clusters found by a clustering algorithm. The problem of the determination of the "best "number of clusters has often been called the central problem of cluster validation. Numerous methods for the determination of the number of clusters have been proposed, but most of them are applicable only to classical data (qualitative, quantitative). In this paper we investigate the problem of the determination of the number of clusters for symbolic objects described by interval variables. We define a notion of convex hull for a set of symbolic objects of interval type. We obtain classical quantitative data, and consequently the classical rules for the determination of the number of clusters can be used. We consider the Hypervolumes test and the best stopping rules from the Milligan and Cooper (1985) study.

Two symbolic clustering methods are also used: Scluster (Bock et al. (2001)), a dynamic partitioning procedure, and a monothetic divisive clustering algorithm (Chavent (1997)). Two data sets illustrate the methods. The first one is an artificially generated data set. The second one is a real data set.

1 Introduction

The aim of cluster analysis is to identify a structure within a data set. Optimization methods for cluster analysis assume usually that the number of groups has been fixed a priori by the user. When hierarchical techniques are used, an important problem is then to select one solution in the nested sequence of partitions of the hierarchy. So most clustering procedures require the user to specify the number of clusters, or to determine it in the final solution.

Symbolic data analysis (Bock and Diday (2000)) is concerned with the extension of classical data analysis and statistical methods to more complex data called symbolic data. In this paper we are interested in the determination of the number of clusters for one particular type of symbolic data: interval data.

Let $E = \{x_1, ..., x_n\}$ denote the set of n individuals. Each individual is characterized by p interval variables $Y_1, ..., Y_p$ i.e.

$$Y_j : E \longrightarrow I : x_k \mapsto Y_j(x_k)$$

where I is the set of all closed intervals of the observation space R. So the value of a variable Y_j on the individual x_k is an interval.

2 Symbolic algorithms for cluster analysis

We will consider in this paper two symbolic clustering methods. The first one, Scluster (Bock et al. (2001)), is close to the Dynamic clouds clustering method (Diday (1972)). That dynamic programming procedure determines iteratively a series of partitions that improve at each step a mathematical criterion. The second symbolic method (Chavent (1997)) is a monothetic divisive clustering procedure, based on a generalization of the classical within clusters variance criterion. One of its characteristics is to give a symbolic interpretation of the clusters.

3 Methods for the determination of the number of clusters

3.1 The Hypervolumes test

The Hypervolumes clustering method (Hardy and Rasson (1982)) assumes that the n p-dimensional observation points x_1, x_2, ..., x_n are generated by a homogeneous Poisson process in a set D included in the Euclidean space R^p. The set D is supposed to be the union of k disjoint convex compact domains D_1, D_2, ..., D_k. We denote by $C_i \subset \{x_1, x_2, .., x_n\}$ the subset of the points belonging to D_i $(1 \leq i \leq k)$. The Hypervolumes clustering criterion is deduced from that statistical model, using maximum likelihood estimation. It is defined by

$$W(P,k) := \sum_{i=1}^{k} m(H(C_i))$$

where $H(C_i)$ is the convex hull of the points belonging to C_i and $m(H(C_i))$ is the multidimensional measure of that convex hull. That clustering criterion has to be minimised over the set of all the partitions of the observed sample into k clusters.

That model allows us to define a likelihood ratio test for the number of clusters (Hardy (1996)). Let us denote by $C = \{C_1, C_2, ..., C_l\}$ the optimal partition of the sample into l clusters and $B = \{B_1, B_2, ..., B_{l-1}\}$ the optimal partition into $l-1$ clusters.

We test the hypothesis H_0: $t = l$ against the alternative H_A: $t = l - 1$, where t represents the number of "natural "clusters ($l \geq 2$).

The test statistics is defined by

$$S(x) := \frac{W(P,l)}{W(P,l-1)}.$$

Unfortunately the sampling distribution of the statistics S is not known. But $S(x)$ belongs to $[0,1[$. Consequently, for practical purposes, we can use the following decision rule: reject H_0 if S is close to 1. So we apply the test in a sequential way: if l_0 is the smallest value of $l \geq 2$ for which we reject H_0, we take $l_0 - 1$ as the appropriate number of "natural "clusters. The easy determination of the best value for l_0 is due to the geometrical nature of the Hypervolumes criterion. The originality of the Hypervolumes test comes from the use of the Lebesgue measure of R^p, the convex hulls of the classes and the Poisson point process model. Furthermore that test performs at a competitive rate for the detection of "natural "convex clusters present in classical quantitative data sets (Hardy and Deschamps (1999)).

3.2 A notion of convex hull for symbolic objects described by interval variables

We consider a set E of n individuals. We measure on each of them the value of p interval variables. So each individual can be represented by a hyperrectangle in the p-dimensional Euclidean space R^p.

In order to apply the Hypervolumes test we define a notion of convex hull for symbolic objects described by interval variables. We simulate a homogeneous Poisson process into the hyperrectangle representing that symbolic object and we define the convex hull of a cluster as the convex hull of all the points generated into the hyperrectangles representing the individuals of that cluster.

Thanks to that simulation process, we obtain classical quantitative data, and so it's possible to apply classical clustering algorithms and classical methods for the determination of the number of clusters, and then to interpret the clusters in terms of symbolic data.

3.3 Other methods for the determination of the number of clusters

Many different stopping rules have been published. The most detailed and complete comparative study that has been carried out appears to be that undertaken by Milligan and Cooper (1985). They conducted a Monte Carlo evaluation of thirty indices for the determination of the number of clusters, and they investigate the extent to which these indices were able to detect the correct number of clusters in a series of simulated data sets containing a known structure.

In order to provide comparative performance information, we consider the five best stopping rules from the Milligan and Cooper (1985) study: the Caliński and Harabasz (1974) index, the Duda and Hart (1973) rule, the C index (Hubert and Levin (1976)), the γ index (Baker and Hubert (1976)) and the Beale (1975) test. For example, the Caliński and Harabasz, Duda and Hart, and Beale indices use various forms of sum of squares within and between clusters. The Duda and Hart rule and the Beale test are statistical hypothesis tests.

4 Examples

4.1 Symbolic artificial data set

The data set comprises 16 individuals described in terms of two interval variables (Table 1). So each individual can be represented by a rectangle in a two-dimensional Euclidean space (Fig. 1). The data were generated in order to have a well-defined structure into three clusters.

objects	variable 1	variable 2
1	[0.5;2]	[3;4]
2	[3;4]	[2.5;5.5]
3	[5;7.5]	[2.5;3.5]
4	[5.5;7]	[1;2]
5	[2;4]	[1;2]
6	[11.5;12.5]	[5.5;7.5]
7	[12.5;16]	[8.5;9.5]
8	[13.5;15]	[6;7.5]
9	[16;16.5]	[5;7]
10	[17;18]	[6;8.5]
11	[14;16.5]	[1;2]
12	[15.5;16.5]	[-1.5;0.5]
13	[14;15]	[-1;0]
14	[12.5;13.5]	[-1.5;0.5]
15	[13.5;15.5]	[-3;-2]
16	[17;18]	[-3;-1]

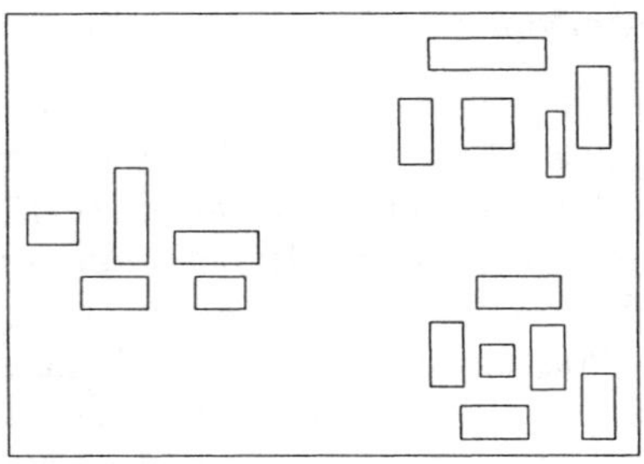

Fig. 1. Artificial data

Inside each rectangle, we've generated a homogeneous Poisson process. We obtain a set of classical quantitative data to which we can apply classical cluster analysis algorithms.

In order to generate sets of partitions we use four well-known hierarchical clustering methods: the single link, complete link, group average and Ward's minimum variance procedures. We apply the Hypervolumes test, and the five best stopping rules from the Milligan and Cooper's study, to determine the best number of natural clusters. The results are tabulated in Table 1.

In the first column we find the names of the clustering methods. When a clustering procedure reveals the "best" classification, a "+ " appears in the second column of Table 1. In this example, the four clustering methods retrieve the true classification if we fix the number of clusters to three. Furthermore,

		Hypervolumes test	Caliński and Harabasz	J index	C index	γ index	Beale test
single link	+	3	3	3	3	3	11
complete link	+	3	3	3	3	3	14
group average	+	3	3	3	3	3	13
Ward	+	3	3	3	3	3	14

Table 1. Number of clusters: artificial data

all the methods for the determination of the number of clusters (except the Beale test) give the correct number of clusters of symbolic objects.

The symbolic clustering method Scluster (Bock et al. (2001)) has been applied to the original symbolic objects. The optimal solution into three natural clusters is also given by that algorithm.

4.2 Symbolic real data set

The data set contains eight fats and oils described by four quantitative features of interval type: Specific gravity, Freezing point, Iodine value, Saponification (Ichino (1994), Gowda and Diday (1994)) (Table 2).

	Specific gravity	Freezing point	Iodine value	Saponification value
linseed oil	[0.930;0.935]	[-27;-18]	[170;204]	[118;196]
perilla oil	[0.930;0.937]	[-5;-4]	[192;208]	[188;197]
cottonseed oil	[0.916;0.918]	[-6;-1]	[99;113]	[189;198]
sesame oil	[0.920;0.926]	[-6;-4]	[104;116]	[187;193]
camelia oil	[0.916;0.917]	[-21;-15]	[80;82]	[189;193]
olive oil	[0.914;0.919]	[0;6]	[79;90]	[187;196]
beef tallow	[0.860;0.870]	[30;38]	[40;48]	[190;199]
hog fat	[0.858;0.864]	[22;32]	[53;77]	[190;202]

Table 2. Ichino's data

Let us simulate a homogeneous Poisson process into the eight hyperrectangles representing the symbolic objects. In order to generate partitions, we apply the four classical clustering methods already considered to the quantitative data set obtained, and then the six stopping rules for the determination of the number of clusters. The results are presented in Table 3.

The Hypervolumes test detects clearly four clusters:

- C_1 = {beef tallow, hog fat}
- C_2 = {linseed}
- C_3 = {camelia, olive, cottonseed, sesam}
- C_4 = {perilla}

	Hypervolumes test	Caliński and Harabasz	J index	C index	γ index	Beale test
single link	4	4	2	4	4	2
complete link	4	3	4	2	3	3
group average	4	3	3	4	3	3
Ward	4	3	4	2	3	3

Table 3. Number of clusters : Ichino's data

It is also the case, for example, of the Caliński and Harabasz index, the C index and the γ index, applied to the hierarchy of partitions produced by the single link clustering method.

The application of the symbolic clustering algorithm Scluster leads to the same structure into four clusters.

On the other hand, for example, the Caliński and Harabasz index, the γ index, and the Beale test, applied to the partitions produced by the complete link, the group average and the Ward clustering procedures, tend to validate the following partition into three clusters:

- $C'_1 = \{$beef tallow, hog fat$\}$
- $C'_2 = \{$linseed, perilla$\}$
- $C'_3 = \{$camelia, olive, cottonseed, sesam$\}$

Let us remark that the partition into 4 clusters in obtained by dividing C'_2 into two clusters.

Chouakria et al. (2000) have proposed two extensions of the principal components analysis to interval data: the vertices and the centers methods. They've applied these two methods to Ichino's data. The results given by both methods are similar, and Ichino's data projection on the factorial plane of the first two principal components of interval type shows the presence of the three well-separated clusters C'_1, C'_2 and C'_3.

Chavent (1997) has also applied her monothetic divisive clustering method to Ichino's data. The dendogram displayed in Fig. 2 is the dendogram of the hierarchy obtained with the Hausdorff distance, normalized with the inverse of the length of the domains. The resulting four-cluster partition is the same as the one validated by the Hypervolumes test. That monothetic procedure allows a symbolic description of these four clusters (Chavent (2000)).

5 Conclusion

In this paper we were interested in the determination of the number of clusters for symbolic objects of interval type. A first way to handle that problem was to generate a homogeneous Poisson process inside the hyperrectangles representing the symbolic objects, and to apply the classical clustering methods and stopping rules for the determination of the number of clusters to the classical quantitative data set obtained. Another solution was to apply

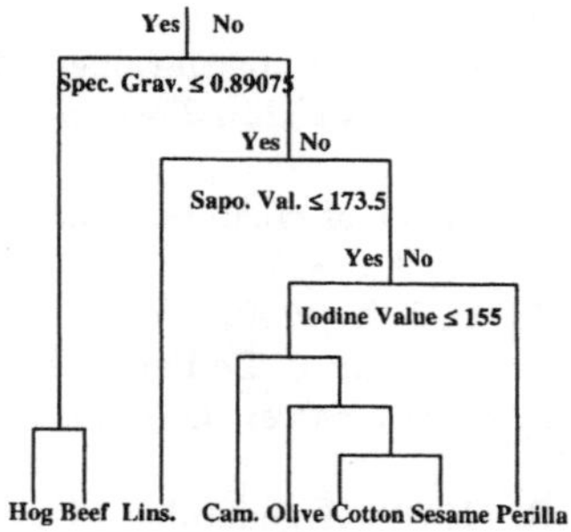

Fig. 2. Hierarchical tree

directly symbolic clustering methods to the original symbolic objects. Both approaches give interesting and similar results on the two examples.

References

BAKER, F. B. and HUBERT, L. J. (1976): Measuring the power of hierarchical cluster analysis. *Journal of the American Statistical Association 70*, 31–38.

BEALE, E. M. L. (1969): Euclidean cluster analysis. *Bulletin of the International Statistical Institute 43 (2)*, 92–94.

BOCK, H.-H. and DIDAY, E. (eds) (2000): Analysis of Symbolic Data, Exploratory methods for extracting statistical information from complex data. Springer Verlag.

BOCK, H.-H. et al. (2001): Report of the Meeting ASSO - WP6.2 Classification group (Munich). Technical report.

CALINSKI, T. and HARABASZ, J. (1974): A dendrite method for cluster analysis. *Communications in Statistics, 3*, 1–27.

CHAVENT, M. (1997): Analyse des données symboliques - Une méthode divisive de classification, Thèse, Université Paris Dauphine.

CHAVENT, M. (2000): Criterion-Based Divisive Clustering for Symbolic data. In: Analysis of Symbolic Data, H.-H. Bock, E. Diday (eds.): *Analysis of Symbolic Data.* Springer Verlag, 299–311.

CHOUAKRIA, A., CAZES, P., and DIDAY, E. (2000): Symbolic Principal Component Analysis. In: H.-H. Bock, E. Diday (eds.): *Analysis of Symbolic Data.* Springer Verlag, 200–212.

DIDAY, E. (1971): La méthode des Nuées Dynamiques. *Revue de Statistique Appliquée, 19, 2* , 19–34.

DUDA, R. O. and HART, P. E. (1973): *Pattern Classification and Scene Analysis.* Wiley, New York.

GOWDA, K. C. and DIDAY, E. (1994): Symbolic clustering algorithms using similarity and dissimilarity measures. In: E. Diday, Y. Lechevallier, M. Schader, P. Bertrand, B. Burtschy (eds): *New approaches in classification and data analysis.* Springer, Berlin, 414–422.

HARDY, A., and RASSON, J.-P. (1982): Une nouvelle approche des problèmes de classification automatique. *Statistique et Analyse des Données, 7*, 41–56.

HARDY, A. (1994): An examination of procedures for determining the number of clusters in a data set. In: E. Diday, Y. Lechevallier, M. Schader, P. Bertrand, B. Burtschy (eds): *New approaches in classification and data analysis.* Springer, Berlin, 178–185.

HARDY, A. (1996): On the number of clusters. *Computational Statistics & Data Analysis, 23*, 83–96.

HARDY, A., and DESCHAMPS, J.F. (1999): Apport du critère des Hypervolumes à la validation en classification. In: F. Le Ber, J.-F. Mari, A. Napoli, A. Simon (eds.): *Actes des Septièmes Rencontres de la Société Francophone de Classification.* Nancy, 201–207.

HUBER, L. J., LEVIN, J. R. (1976): A general statistical framework for assessing categorical clustering in free recall. *Psychological Bulletin, 83*, 1072–1080.

ICHINO, M. and YAGUCHI, H. (1994) : Generalized Minkowsky Metrics for Mixed Feature Type Data Analysis. *IEEE Transactions System, Man and Cybernetics, 24*, 698–708

MILLIGAN, G.W. and COOPER, M.C. (1985): An examination of procedures for determining the number of clusters in a data set. *Psychometrika, 50*, 159–179.

Symbolic Data Analysis Approach to Clustering Large Datasets

Simona Korenjak-Černe[1] and Vladimir Batagelj[2]

[1] University of Ljubljana, Faculty of Economics,
Kardeljeva ploščad 17, 1 101 Ljubljana, Slovenia,
and IMFM Ljubljana, Department of TCS,
Jadranska ulica 19, 1 000 Ljubljana, Slovenia
(e-mail: simona.cerne@uni-lj.si)

[2] University of Ljubljana, FMF, Department of Mathematics,
and IMFM Ljubljana, Department of TCS,
Jadranska ulica 19, 1 000 Ljubljana, Slovenia
(e-mail: vladimir.batagelj@uni-lj.si)

Abstract. The paper builds on the representation of units/clusters with a special type of symbolic objects that consist of distributions of variables. Two compatible clustering methods are developed: the *leaders method*, that reduces a large dataset to a smaller set of symbolic objects (clusters) on which a *hierarchical clustering method* is applied to reveal its internal structure. The proposed approach is illustrated on *USDA Nutrient Database*.

1 Introduction

Nowadays lots of large datasets are available in databases. One of possible ways how to extract information from these datasets is to find homogeneous clusters of similar units. For the description of the data vector descriptions are usually used. Each its component corresponds to a variable which can be measured in different scales (nominal, ordinal, or numeric). Most of the well known clustering methods are implemented only for numerical data (e.g., k-means method) or are too complex for clustering large datasets (such as hierarchical methods based on dissimilarity matrices). For these reasons we propose to use for clustering large datasets a combination of the adapted leaders and hierarchical clustering methods based on special descriptions of units and clusters. This description is based on a special kind of symbolic objects (Bock and Diday (2000)), formed by the distributions of partitioned variables over a cluster – histograms.

2 The descriptions of units and clusters

Let E be a finite set of units X, which are described by frequency/probability distributions of their descriptors $\{V_1, \cdots, V_m\}$ (Korenjak-Černe and Batagelj

(1998)). The domain of each variable V is partitioned into k_V sub-sets $\{V_i,\ i = 1, \ldots k_V\}$. For a cluster C we denote

$$Q(i, C; V) = \{X \in C : V(X) \in V_i\},\ i = 1, \ldots, k_V,$$

$$q(i, C; V) = \text{card}(Q(i, C; V)),\ \text{(frequency)}$$

$$f(i, C; V) = \frac{q(i, C; V)}{\text{card}(C)},\ \text{(relative frequency)}$$

where $V(X)$ is the value of variable V on unit X, and card(C) is the number of units in the cluster C. It holds

$$\sum_{i=1}^{k_V} f(i, C; V) = 1$$

The description of the cluster C by the variable V is the vector of the frequencies of V_i $(i = 1, \cdots, k_V)$. A unit is considered as a special cluster with only one element and can be in our approach represented either with a single value or by the distributions of the partitioned variables.

Such a description has the following important properties:

- it requires a *fixed space* per variable;
- it is *compatible* with merging of disjoint clusters – knowing the description of clusters C_1 and C_2, $C_1 \cap C_2 = \emptyset$, we can, without additional information, produce the description of their union

$$f(i, C_1 \cup C_2; V) = \frac{\text{card}(C_1)\, f(i, C_1; V) + \text{card}(C_2)\, f(i, C_2; V)}{\text{card}(C_1 \cup C_2)};$$

- it produces an *uniform description* for all the types of descriptors.

3 Dissimilarity

In the following we shall use two dissimilarities, both defined as a weighted sum of the dissimilarities on each variable:

$$d(C_1, C_2) = \sum_{j=1}^{m} \alpha_j\, d(C_1, C_2; V_j), \qquad \sum_{j=1}^{m} \alpha_j = 1 \tag{1}$$

where

$$d_{abs}(C_1, C_2; V_j) = \frac{1}{2} \sum_{i=1}^{k_j} |f(i, C_1; V_j) - f(i, C_2; V_j)| \tag{2}$$

or

$$d_{sqr}(C_1, C_2; V_j) = \frac{1}{2} \sum_{i=1}^{k_j} (f(i, C_1; V_j) - f(i, C_2; V_j))^2, \tag{3}$$

$k_j = k_{V_j}$. Here, $\alpha_j \geq 0$ $(j = 1, \ldots, m)$ denote weights, which could be equal for all variables or different if we have same information about the importance of the variables.

For the dissimilarity d_{abs} the triangle inequality also holds. Therefore it is also a semidistance.

4 The adapted leaders method

For clustering large datasets the clustering procedures based on dissimilarity matrix are too time consuming. A more appropriate approach is the adapted leaders method – a variant of the dynamic clustering method (Diday (1979), Korenjak-Černe and Batagelj (1998), Verde et al. (2000)). This method can be shortly described with the following procedure:

determine an initial clustering
repeat
  determine leaders of the clusters in the current clustering;
  assign each unit to the nearest new leader – producing a new clustering
until the leaders do not change more.

The leaders method is solving the following optimization problem:

Find a clustering $\mathbf{C}^*$ in a set of feasible clusterings Φ for which

$$P(\mathbf{C}^*) = \min_{\mathbf{C} \in \Phi} P(\mathbf{C}) \tag{4}$$

with the criterion function

$$P(\mathbf{C}) = \sum_{C \in \mathbf{C}} p(C) \qquad \text{and} \qquad p(C) = \sum_{X \in C} d(X, L_C), \tag{5}$$

where L_C represents the leader (a representative element) of the cluster C. In our case the set of feasible clusterings Φ is a set of partitions of the set E. The number of the clusters could be fixed a priori or could be determined with the selection of the maximal allowed dissimilarity between the unit and the nearest leader.

In the elaboration of the proposed approach we assume that the descriptions of the leaders have the same form as the descriptions of the units and clusters:

$$L = [L(V_1), \ldots, L(V_m)],$$
$$L(V) = [s(1, L; V), \ldots, s(k_V, L; V)],$$

where $\sum_{j=1}^{k_V} s(j, L; V) = 1$

It can be proved that for the first criterion function P_{abs}, where in the definition (5) the dissimilarity d_{abs} is used, the optimal leaders are determined with maximal frequencies

$$s(i, L; V) = \begin{cases} \frac{1}{t} & \text{if } j \in M \\ 0 & \text{otherwise} \end{cases}$$

where $M = \{j : q(j, C; V) = \max_i q(i, C; V)\}$ and $t = \text{card}(M)$. The precondition for this result is that all units should be represented with a single value for each variable (this is usually the case).

For the second criterion function P_{sqr} with dissimilarity d_{sqr} the optimal leaders are uniquely determined with the averages of relative frequencies

$$s(i, L; V) = \frac{1}{\text{card}(C)} \sum_{X \in C} f(i, X; V).$$

This is an extended version of the well-known k-means method, which is appropriate only for numerical variables (Hartigan (1975)). The main advantages of the second method are:

- the input unit can be represented also with distributions, and not only with a single value for each variable,
- the optimal leaders are uniquely determined.

5 Building a hierarchy

To produce a hierarchical clustering on the clusters represented with their leaders the standard agglomerative hierarchical clustering method is used:

each unit is a cluster: $\mathbf{C}_1 = \{\{X\} : X \in E\}$;
they are at level 0: $h(\{X\}) = 0, \ X \in E$;
for $k := 1$ **to** $n - 1$ **do**
  determine the closest pair of clusters
    $(p, q) = \text{argmin}_{i,j:\ i \neq j}\{D(C_i, C_j) : C_i, C_j \in \mathbf{C}_k\}$;
  join them
    $\mathbf{C}_{k+1} = (\mathbf{C}_k \setminus \{C_p, C_q\}) \cup \{C_p \cup C_q\}$;
    $h(C_p \cup C_q) = D(C_p, C_q)$
endfor

The level $h(C)$ of the cluster $C = C_p \cup C_q$ is determined by the dissimilarity between the joint clusters C_p and C_q by $h(C_p \cup C_q) = D(C_p, C_q)$. The units X are the clusters from the initial clustering, represented with their leaders. $h(C) = 0$ for C from the initial clustering.

The dissimilarity between clusters $D(C_p, C_q)$ measures the change of the value of the criterion function produced by the merging of the clusters C_p and C_q

$$D(C_p, C_q) = p(C_p \cup C_q) - p(C_p) - p(C_q) \tag{6}$$

For the second criterion function P_{sqr} the dissimilarity $D(C_p, C_q)$ can be determined using the analogue of the Ward's relation (Batagelj (1988)):

$$D(C_p, C_q) = \frac{\mathrm{card}(C_p) \cdot \mathrm{card}(C_q)}{\mathrm{card}(C_p) + \mathrm{card}(C_q)} \, d(L_p, L_q).$$

6 Example

The proposed approach was successfully applied to some large datasets (for example, the dataset on the topic *Family and Changing Gender Roles I* and *II* with 45 785 units and 33 selected variables from ISSP datasets). We are presenting here the results on the nutrient database from U.S. Department of Agriculture. The dataset contains data on 6039 foods – units. We considered in this study 31 nutrients – numerical variables describing each food. This dataset was selected because the results can be interpreted in easy-to-understand way.

6.1 Partition of domains of variables

The domain of each variable is divided into 10 sub-sets: one with the value, that indicates missing value, one with the value zero, and one special sub-set with outlying (extremely large) values. The rest of the values are divided into 7 sub-sets with equal number of values (Dougherty et al. (1995)).

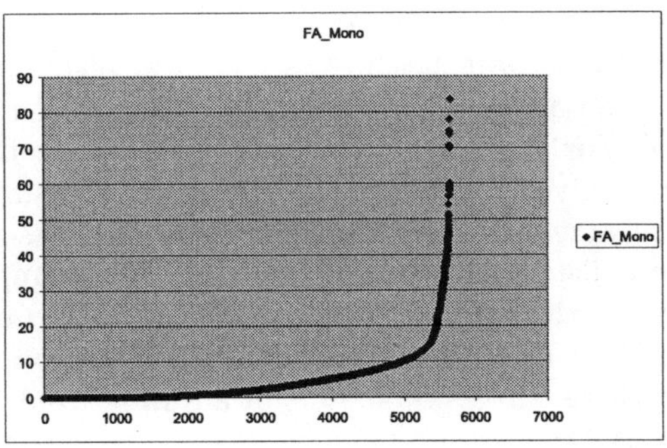


```
var=fa_mono
MAP
-1 (missing)
1={0}
2=(0,0.035]
3=(0.035,0.3]
4=(0.3,1.25)
5=[1.25,3]
6=(3,5.6)
7=[5.6,9.5)
8=[9.5,65)
9=[65,85] (outliers)
```

Fig. 1. The graph of the distribution of the variable *fa-mono*.

For example, the variable *fa-mono* (total monounsaturated fatty acids) has 395 missing values, 128 units have value 0 and 7 units have extremely large values. The distribution of the values for this variable is presented in the Figure 1 and next to it is our partition of it's domain.

6.2 Transformed data

Each unit is represented with the vector of indices of sub-sets in which ly its real values. For example: the food BUTTER,WITH SALT with the ID = 1001 has for the first five variables and their indices the following values:

ID	water	food energy	protein	total lipid (fat)	carbohydrate
1001	15.87	717	0.85	81.11	0.06
1001	3	8	2	8	2

because for the *water* (g/100g) the third sub-set is $3 = (5.65, 29.5]$, the eighth sub-set for *energy* (kcal/100g) is $8 = (386, 800)$, the second sub-set for *protein* (g/100g) is $2 = (0, 1.5]$, the eighth sub-set for *fat* (g/100g) is $8 = [23.5, 85)$ and the second sub-set for *carbohydrate* (g/100g) is $2 = (0, 3.5]$.

V	modus sub-set	% of units	extanded sub-set	% of units
$fiber - td$	$\{0\}$	99.49	$\{0\} \cup \{missing\}$	100.00
$vit - A$	$\{0\}$	99.49	$\{0\} \cup \{missing\}$	100.00
$vit - C$	$\{0\}$	99.32	$\{0\} \cup \{missing\}$	100.00
$carbohyd$	$\{0\}$	99.32	$[0, 6.85]$	100.00
$zinc$	$[4.12, 20)$	72.93	$[2.23, 20)$	100.00
$sodium$	$[50, 66)$	69.88	$[10, 121)$	99.32

6.3 Clustering results

In the leaders program the initial clustering with 30 clusters was randomly selected. For the selected dissimilarity d_{sqr} the 30 leaders stabilized after 29 iterations. On these leaders the hierarchy based on the same dissimilarity was built. The dendrogram displayed in Figure 2 was obtained.

The hierarchy we got has three main branches: meats, (mainly) vegetables, and (mainly) cereals. For each node of the dendrogram, the distribution for each variable is also determined. For example, the cluster *Beefs* = *Beef1* ∪ *Beef3* ∪ *Beef2* has the description given in Table 1. It consists of 591 units. From this table the following characteristics of the cluster *Beefs* can be seen:

For each variable the complete distribution can be observed. For example, from the Table 1 we can see that for the variable *fa-mono* 184 (31.13 %) units from this cluster have values in the 6th sub-set $(3, 5.6)$. But if we extend the interval to $(3, 65)$ (union of three sub-sets) 85.96 % of all units from the cluster *Beefs* are included in it. Detailed results and programs are available at `http://www.educa.fmf.uni-lj.si/datana/` .

Acknowledgment

This work was supported by the Ministry of Science and Technology of Slovenia, Project J1-8532.

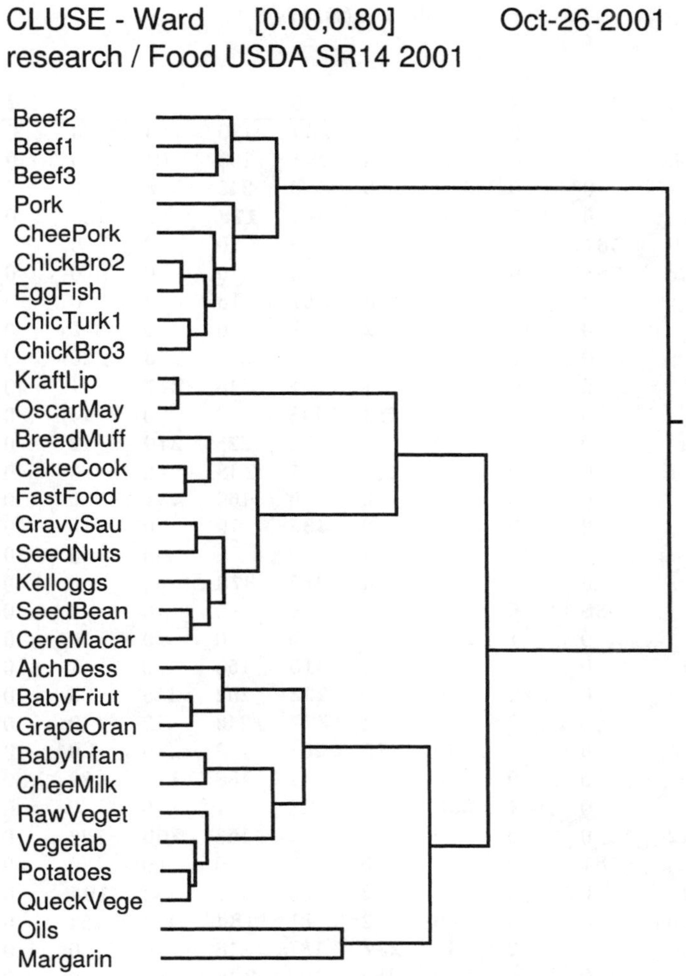

Fig. 2. The dendrogram on 30 leaders of food's clusters.

Table 1. $q(Beefs, V)$

V	{0}	2	3	4	5	6	7	8	*out*	*mis*
water	0	0	1	167	**264**	130	29	0	0	0
energ − kc	0	0	0	54	**231**	183	103	20	0	0
protein	0	0	0	0	8	245	**338**	0	0	0
tot − lipi	0	0	0	10	101	**178**	169	133	0	0
carbohyd	**587**	3	1	0	0	0	0	0	0	0
fiber − td	**588**	0	0	0	0	0	0	0	0	3
ash	0	0	195	**278**	101	13	4	0	0	0
calcium	0	**294**	241	52	2	0	0	2	0	0
phosphor	0	0	0	0	71	227	**273**	20	0	0
iron	0	0	0	1	61	248	**277**	4	0	0
sodium	0	0	26	**413**	148	0	0	4	0	0
potassiu	0	0	0	0	75	225	**279**	12	0	0
magnesiu	0	0	11	166	197	**208**	9	0	0	0
zinc	0	0	0	0	0	160	**431**	0	0	0
copper	0	0	76	0	**436**	79	0	0	0	0
manganes	0	**346**	242	3	0	0	0	0	0	0
selenium	0	0	0	4	168	**378**	41	0	0	0
vit − A	**588**	0	0	0	0	0	0	0	0	3
vit − E	0	1	**218**	65	0	0	0	0	0	307
thiamin	0	0	3	122	**315**	150	0	1	0	0
ribolfla	0	0	0	41	**233**	202	115	0	0	0
niacin	0	0	0	5	**327**	240	19	0	0	0
panto − ac	0	0	11	**375**	201	2	0	0	0	2
vit − B6	0	0	0	1	2	188	**311**	89	0	0
folate	0	1	**335**	228	23	1	0	0	0	3
vit − B12	0	0	0	0	1	163	**300**	127	0	0
vit − C	**587**	0	0	0	0	0	0	0	0	4
fa − sat	0	0	0	2	65	147	183	**194**	0	0
fa − mono	0	0	0	2	81	**184**	170	154	0	0
fa − poly	0	2	147	**237**	187	18	0	0	0	0
cholestr	0	0	3	150	169	**205**	64	0	0	0

References

BATAGELJ, V.: *Generalized Ward and related clustering problems.* (H.H. Bock, ed.: Classification and related methods of data analysis), North-Holland, Amsterdam, 1988, 67-74.

BOCK, H.-H. (2000): Symbolic Data. In: H.-H. Bock and E. Diday (Eds.): *Analysis of Symbolic Data.* Exploratory methods for extracting statistical information from complex data. Springer, Heidelberg.

BOCK, H.-H. and DIDAY, E. (2000): Symbolic Objects. In: H.-H. Bock and E. Diday (Eds.): *Analysis of Symbolic Data.* Exploratory methods for extracting statistical information from complex data. Springer, Heidelberg.

DIDAY, E. (1979): *Optimisation en classification automatique*, Tome 1.,2.. INRIA, Rocquencourt (in French).

DOUGHERTY, J., KOHAVI, R., and SAHAMI, M. (1995): *Supervised and unsupervised discretization of continuous features.* Proceedings of the Twelfth International Conference on Machine Learning (pp. 194–202). Tahoe City, CA: Morgan Kaufmann. `http://citeseer.nj.nec.com/dougherty95supervised.html`

HARTIGAN, J.A. (1975): *Clustering Algorithms.* Wiley, New York.

KORENJAK-ČERNE, S. and BATAGELJ, V. (1998): Clustering large datasets of mixed units. In: Rizzi, A., Vichi, M., Bock, H.-H. (Eds.): *Advances in Data Science and Classification.* Springer.

VERDE, R., DE CARVALHO, F.A.T. and LECHEVALLIER, Y. (2000): A Dynamic Clustering Algorithm for Multi-nominal Data. In: Kiers, H.A.L., Rasson, J.-P., Groenen, P.J.F., Schader, M. (Eds.): *Data Analysis, Classification, and Related Methods.* Springer.

USDA Nutrient Database for Standard Reference, Release 14. U.S. Department of Agriculture, Agricultural Research Service. 2001: Nutrient Data Laboratory Home Page, `http://www.nal.usda.gov/fnic/foodcomp`.

Symbolic Class Descriptions

Mathieu Vrac[1], Edwin Diday[1], Suzanne Winsberg[2], and Mohamed Mehdi Limam[1]

[1] LISE-CEREMADE, Université Paris IX Dauphine,
Place du Maréchal de Lattre-de-Tassigny,75775, Paris, France
[2] IRCAM
1 Place Igor Stravinsky, Paris 75004, France

Abstract. Our aim is to describe a partition of a class by a conjunction of characteristic properties. We use a stepwise top-down binary tree method. At each step we select the best variable and its optimal splitting to optimize simultaneously a discrimination criterion given by a prior partition and a homogeneity criterion. Moreover, this method deals with a data table in which each cell contains a histogram of nominal categories but the method can be extended or reduced to other types of data. The method is illustrated on both simulated data and real data.

1 Introduction

Classification methods are often designed to split a population of statistical individuals to obtain a partition into L classes. Generally, a partition is designed to optimize an intra-class homogeneity criterion as in classical clustering, or equivalently an inter-class criterion, as in classical decision or regression trees. In practice, when the aim is class description, it may be desireable to consider both types of criteria simultaneously. Here, our aim is to produce a description of a class which induces a partition of the class, both satisfying an intra-class homogeneity criterion and a discrimination criterion with respect to a prior partition. So, our approach has both unsupervised and supervised aspects. For example, the class to describe, C, could be young people of ages between 15 and 25, and the discriminatory categorical variable or prior partition could be smokers and nonsmokers. We want to obtain a description of C which induces a homogeneous partition of C which is well discriminated for the prior partition. The context here differs from that considered in Huygen's theorem, in which the intra-class and inter-class inertias are based on the same initial set. Here, we calculate the homogeneity criterion for the class we want to describe, but the discrimination criterion is based on the prior partition.

Our approach is based on divisive top-down methods, which successively divide the population into two classes, until a suitable stopping rule prevents further divisions. We use a monothetic approach such that each split is carried out using only one variable, as it provides a clearer interpretation of the obtained clusters. Divisive methods of this type are often referred to as tree structured classifiers with acronyms such as CART and ID3 (see Breiman et al.(1984), Quinlan (1986)).

Not only does our paper combine the two approaches: supervised and non-supervised learning, to obtain a description induced by the synthesis of the two methods, which is in itself an innovation, but it can deal with histogram data. We call histogram data, data in which the entries of the data table are weighted categorical or ordinal variables. These data are inherently richer, possesing potentially more information than the data previously considered in the classical algorithms mentioned above. This type of data is encountered when dealing with more complex, aggregated statistical units found when analyzing very large data sets. It may be more interesting to deal with aggregated units such as towns rather than with the individual inhabitants of the towns. Then the resulting data set, after the aggregation will most likely contain symbolic data rather than classical data values. By symbolic data we mean that rather than having a specific single value for an observed variable, an observed value for an aggregated statistical unit may be multivalued. For example, as in the case under consideration, the observed value may be a multivalued weighted categorical variable. For a detailed description of symbolic data analysis see Bock and Diday (2000). Naturally, classical data are a special case of the histogram type of data considered here. This procedure thus works for classical numerical or nominal data. It can also be applied when dealing with other types of symbolic data such as interval data. Others have developped divisive algorithms for data types encountered when dealing with symbolic data, considering either a homogeneity criterion or a discrimination criterion based on an a priori partition, but not both simultaneously. Chavent (1997) has proposed a method for unsupervised learning, while Périnel (1999), and Gettler-Summa (1999) have proposed ones for supervised learning. This method is an extension of that proposed by Vrac and Diday (2001).

First we describe our method, including some practical details necessary to implement it. For example we define a cutting or split for weighted categorical variables and we define a cutting value for this type of data. Then we outline the approach used to combine the two criteria. We present some examples of simulated data of histogram type to test the behavior of our new method. Finally we illustrate the algorithm with a real example dealing with unemployment data.

2 The method

Four inputs are required for this method: 1) the data, consisting of n statistical units, each described by K histogram variables; 2) the prior partition into classes; 3) the class, C, the user aims to describe; and 4) a coefficient which gives more or less importance to the discriminatory power of the prior partition or to the homogeneity of the description of the given class, C. Alternatively, instead of specifying this last coefficient, the user may choose to determine an optimum value of this coefficient, using this algorithm.

The method uses a monothetic hierarchical descending approach working by division of a set into two nodes, that is sons. At each step l (l nodes corresponding to a partition into l classes), one of the nodes (or leaves) of the tree is cut into two nodes in order to optimize a quality criterion Q for the constructed partition into $l+1$ classes. The division of a node N into two nodes N_1 and N_2 is done by "cutting", where y is called the cutting variable and c the cutting value. We denote as N_1 and N_2, respectively, the left and right node of N.

The algorithm always generates two kinds of output. The first is a graphical representation, in which the class to describe, C, is represented by a binary tree. The final leaves are the clusters constituting the class and each branch represents a cutting (y, c). The second is a description: each final leaf is described by the conjunction of the cutting values from the top of the tree to this final leaf. The class, C, is then described by a disjunction of these conjunctions. If the user wishes to choose an optimal value of α using our data driven method, a graphical representation enabling this choice is also generated as output.

Let $H(N)$ and $h(N_1; N_2)$ be respectively the homogeneity criterion of a node N and of a couple of nodes $(N_1; N_2)$. then we define $\Delta H(N) = H(N) - h(N_1; N_2)$. Similarly we define $\Delta D(N) = D(N) - d(N_1; N_2)$ for the discrimination criterion. The quality Q of a node N (respectively q of a couple of nodes $(N_1; N_2)$) is the weighted sum of the two criteria, namely $Q(N) = \alpha H(N) + \beta D(N)$ (respectively $q(N_1; N_2) = \alpha h(N_1; N_2) + \beta d(N_1; N_2)$) where $\alpha + \beta = 1$. So the quality variation induced by the splitting of N into $(N_1; N_2)$ is $\Delta Q(N) = Q(N) - q(N_1; N_2)$. We maximize $\Delta Q(N)$. Note that since we are optimizing two criteria the criteria must be normalized. The user can modulate the values of α and β so as to weight the importance that he gives to each criterion.

To determine the cutting $(y; c)$ and the node to cut: first, for each node N select the cutting variable and its cutting value minimizing $q(N_1; N_2)$; second, select and split the node N which maximizes the difference between the quality before the cutting and the quality after the cutting, $max \Delta Q(N) = max[\alpha \Delta H(N) + \beta D(N)]$.

We recall that we are working with multivalued weighted categorical variables (histograms). So we must define what constitutes a cutting for this type of data and what constitutes a cutting value. The main idea is that the cutting value of a histogram variable is defined on the value of the frequency of just one category, or on the value of the sum of the frequencies of several categories.

To illustrate consider the following example: we have n statistical units in the class N (take $n = 3$), and consider variable Y_k, (say the variable color with categories red(r), blue(b), green(g), yellow(y)). Say, $unitA$ has values

$r = 0.2, b = 0.1, g = 0.2, y = 0.5$; *unitB* has values $r = 0.5, b = 0.2, g = 0.1, y = 0.2$; and *unitC* has values $r = 0.1, b = 0.4, g = 0.1, y = 0.4$. For this variable Y_k we first order the units in increasing order of the frequency of just one category, eg red. We obtain $C(r = 0.1) < A(r = 0.2) < B(r = 0.5)$. So we can determine $n - 1 = 2$ cutting values by taking the mean of two different consecutive values (here cutting value $1 = (0.1 + 0.2)/2 = 0.15$ and cutting value $2 = (0.2 + 0.5)/2 = 0.35$). Therefore we can also determine $n - 1$ partitions into two classes (here partition $1 = \{N_1 = \{unitC\}; N_2 = \{unitA; unitB\}\}$ and partition $2 = \{N_1 = \{unitC; unitA\}; N_2 = \{unitB\}\}$ and so we have $n - 1$ quality criterion values $q(N_1; N_2)$. We do it in turn for each single category (just red, just blue, just green, just yellow). Then we sort units in increasing order of the sum of frequencies of two categories. For example, with " (red + blue)" we obtain $unitA(r+b = 0.3) < unitC(r+b = 0.5) < unitB(r + b = 0.7)$. We thus get $n - 1$ new cutting values, ($n - 1$ new partitions into two classes and $n - 1$ new quality criterion values $q(N_1; N_2)$. In general if a histogram allows m categories we can look at the sorting on the sum of at most $m/2$ categories for even m and on the sum of at most $[m/2] = (m - 1)/2$ categories for odd m. Indeed, in our little example we can see that the partitions obtained with "red + blue" are the same as those obtained with "green + yellow". We remark that if a multivalued weighted categorical variable Y has m categories, we have 2^{m-1} ways to sort the units in increasing order. Indeed, $2^{(m-1)} - 1$ is the number of partitions in two non-empty classes from a set with m categories. Moreover, for each way of sorting we can have $(n - 1)$ partitions of the units, so for a variable with m categories we have at most $(2^{(m-1)} - 1)(n - 1)$ partitions of the units.

The clustering or homogeneity criterion we use is an inertia criterion. This criterion is used in Chavent (1997). The inertia of a class N is

$$H(N) = \sum_{w_i \in N} \sum_{w_j \in N} \frac{p_i p_j}{2\mu} \delta^2(w_i, w_j),$$

and

$$h(N_1, N_2) = H(N_1) + H(N_2);$$

where $p_i =$ the weight of individual w_i, and $\mu = \sum_{w_i \in N} p_i =$ the weight of class N, and δ is a distance between individuals. For histograms with weighted categorical variables, we can imagine many distances. We choose δ, defined as,

$$\delta(w_i; w_j) = \sum_{k=1}^{K} \sum_{m=1}^{mod[k]} |y_k^m(w_i) - y_k^m(w_j)|^2,$$

where $y_k^m(w)$ is the value of the category m of the variable k for the individual w, $mod[k]$ is the number of categories of the variable k, and K is the number of variables.

This distance must be normalized. We normalize it to fall in the interval [0,1]. So δ must be divided by K to make it fall in the interval [0,1].

Let us turn to the discrimination criterion. The discrimination criterion we choose is an impurity criterion, Gini's index. Gini's index, which we denote as D, was introduced by Breiman et al (1984) and measures the impurity of a node N with respect to the prior partition $G_1, G_2, ..., G_J$ by

$$D(N) = \sum_{l \neq j} p_l p_j = 1 - \sum_{j=1,\ldots,J} p_j^2,$$

with $p_j = n_j/n$, $n_j = card(N \cap G_j)$ and $n = card(N)$ in the classical case. In our case n_j = the number of individuals from G_j such that their characteristics verify the current description of the node N. To normalize $D(N)$ we multiply it by $J/(J-1)$; where J is the number of prior classes; it then lies in the interval [0,1].

Let us now discuss the robustness of the results. In many situations we obtain trees yielding unstable predicting models. Then it is necessary to prune the tree by removing the less significant branches. Consider a fixed value of α. The main idea is to estimate the inertia and discrimination rate for every subtree. The inertia and discrimination rate R associated with tree A is $R(A) = \sum_{t \in A} \frac{n_t}{n} Q(t)$ where n_t is the number of individuals in terminal node t and n is the total number of individuals. The optimal tree is the subtree minimizing this rate. We use a bootstrap method to estimate these rates. Here, pruning consists of selecting the best subtree from all subtrees obtained by removing branches from the main or starting tree. The tree with lowest value of R is the "best" subtree. Starting from the set of individuals we construct the main tree A_{max}. For each subtree A_h we estimate R using the bootstrap, so we have B samples, say 100, from the initial set of individuals. Then for each bootstrap sample we calculate R for each subtree A_h, and for each subtree A_h we calculate the mean $\overline{R}(A_h)$ of all the samples. Finally we choose the best subtree A_h^*, that is the one that has the minimum $\overline{R(A_h)}$.

The user may choose to optimize the value of the coefficient α. To do so, one must fix the number of terminal nodes. The influence of the coefficient α can be determinant both in the construction of the tree and in its prediction qualities. The variation of α (or of β since $\alpha + \beta = 1$) from 0 to 1 increases the importance of the homogeneity and decreases the importance of discrimination. This variation influences splitting and consequently results in different terminal nodes. We need to find the inertia of the terminal nodes and the rate of misclassification as we vary α. Then we can determine the value of α which optimizes both the inertia and the rate of misclassification ie the homogeneity and discrimination simultaneously. If on the contrary the user fixes the value of $\alpha = 0$, considering only the discrimination criterion, and in addition the data are classical, the algorithm functions just as CART. So CART is a special case of this algorithm.

For each terminal node t of the tree T associated with class c_s we can calculate the corresponding misclassification rate $R(s/t) = \sum_{r=1}^{L} P(r/t)$ where $r \neq s$ and $P(r/t) = \frac{n_r(t)}{n_t}$ is the proportion of the individuals of the node t allocated to the class c_s but belonging to the class c_r. The misclassification MR of the tree T is the sum over all terminal nodes ie $MR(A) = \sum_{t \in A} \frac{n_i}{n} R(s/t) = \sum_{t \in A} \sum_{r=1}^{L} \frac{n_r(t)}{n}$, where $r \neq s$. For each terminal node of the tree T we can calculate the corresponding inertia, $H(t)$ and we can calculate the total inertia by summing over all the terminal nodes. So, $H(t) = \frac{1}{2n|t|} \sum_{w_i \in t} \sum_{w_j \in t} \delta(W_i, W_j)$ with $|t| = card(t)$, and the total inertia of T, $I(A) = \sum_{t \in T} H(t)$.

The idea is to build for each value of α several trees from many samples and then to calculate the inertia and misclassification rate for each tree. Starting from our initial set of n individuals we extract B bootstrap samples of size n (by randomly sampling n individuals with replacement). For each sample and for each value of α between 0 and 1, we build a tree and calculate our two parameters (inertia and misclassification rate). Varying α from 0 to 1 (say with a stepsize of 0.1) gives us 11 couples of values of inertia and misclassification rate corresponding to the mean values of these parameters for the B bootstrap samples.

In order to visualize the variation of the two parameters, we display a curve showing the inertia and a curve showing the misclassification rate as a function of α. The optimal value of α is the one which minimizes the sum of the two parameters.

3 Examples

First we consider three sets of simulated data. These simulated data examples are presented to give a clear picture under controlled conditions, of how the algorithm works and how it permits an optimal choice of α. We also consider a real data set. The first example consists of 90 individuals with a prior partition. The class to describe, $C = C1 \cup C2$. Each individual is described by two variables of histogram type. This first simulation is constructed so as to make the class to describe have two subclasses, $C1$ and $C2$ with optimal homogeneity. Then we added a prior partition which perfectly distinguishes C from the rest of the population. In this very special case the inertia and the misclassification rate should not vary with the choice of α. Any value of α from 0 to 1 should give a result with the same inertia and misclassification rate. In fact for this example when we graphically display the results we obtain for these two parameters as a function of α we obtain a constant value.

The second simulated example is a modification of the first example. We modify the data from example 1 such that we deteriorate only the discrimination by changing the value of the discriminatory variable for some individuals, while keeping the homogeneity of the class to describe identical to that in example 1. This change should make it necessary to increase the importance of discrimination and thus the optimum value of β should be close to one (that is, α close to zero). In fact our results show that the inertia remains constant as a function of α while the misclassification rate increases as α increases from 0 to 1 indicating a choice of $\alpha = 0$ as expected.

In the third simulated example we modify the data from example one by deteriorating both the discrimination and the inertia. We find as expected that the optimal level of α depends upon the degree of deterioration of these two factors. For example, the optimal value of α decreases as the number of individuals whose discriminatory variable is changed increases.

Finally the fourth example deals with real unemployment data from 35 towns (districts) in Great Britain. The aim is to explain the factors which discriminate towns with high unemployment from those with low unemployment. But we also wish to have good descriptors of the resultant clusters due to their homogeneity. Because we have aggregated data for the inhabitants of each town, we are not dealing with classical data with a single value for each variable for each statistical unit, here the town. The class to describe is the 35 towns, and the prior partition is low versus high employment. Here, each variable for each town is a histogram. There are $K = 4$ variables. The first is age with 6 categories: 0-4 years; 5-14 years; 15-24 years; 25-44 years; 45-64 years; greater than or equal to 65. The second is racial origin with 4 categories: Whites; Blacks; Asians; Others. The third is type of dwelling with 4 categories: owner occupied; public sector accomation; private sector accomadation; other. The fourth is social class with 4 categories: household head in social class 1 or social class 2; household head in social class 3; household head in social class 4 or social class 5; other. The discriminatory variable is unemployment rate for men and women. For these districts the rate varies between 0% and 18%. Two prior classes are defined: class1 for unemployment rate $\leq 9\%$ and class2 for unemployment rate $> 9\%$.

We stopped the algorithm with four terminal nodes of description. We obtain four symbolic descriptions of each node. An example of such a description is: [the proportion of people in social classes 1 and 2 is less than 33.9%], and [the proportion of people between 45 and 64 years of age is less than 21.8%]. When we use only a homogeneity criterion, that is we fix $\alpha = 1$, ($\beta = 0$), each description gathers homogenous groups of towns. The total inertia of the terminal nodes is minimized and equals 0.135. However, the misclassification rate is high, equal to 20.4%. Next we use only a discrimination criterion, (that is we fix $\alpha = 0, \beta = 1$). We choose an intial partition with P1 = towns where the unemployment rate is high and P2 = towns where the unemployment rate is low. We have the same set of towns to describe. In this case we have a misclassification rate of 6.25% considerably reduced from 20.4%. However the inertia is equal to 0.26 which is higher than above. So we have good discrimination but inferior homogeneity. Finally, we use our method and choose a value of α based on the data which optimizes both the inertia and the misclassification rate simultaneously. The inertia decreases when we increase α. But, there is a gradient from $0.4 \leq \alpha \leq 0.6$ showing that the inertia decreases sharply in this region. The misclassification rate increases when we increase α. However, there is a gradient between $0.5 \leq \alpha \leq 0.7$; showing that the rate increases sharply between these values. However, at $\alpha = 0.6$ the rate of misclassification is only slightly increased over that for $\alpha = 0$, which is the best rate. If we choose $\alpha = 0.6$ the inertia is 0.233 and the misclassification rate is 6.52%. So we have an almost optimal misclassification rate and a better class description, than that which we obtain when considering only a discrimination criterion; and we have a much better misclassification rate than that which we obtain when considering only a homogeneity criterion.

4 Conclusion

In this paper we present a new approach to get a description of a set. This method applicable to histogram data is new for the classical case as well. The main idea is to mix a homogeneity criterion and a discrimination criterion to describe a set according to an initial partition. The set to describe can be a class from a prior partition, the whole population or any class from the population. Having chosen this class, the interest of the method is that the user can choose the weights α and $\beta = 1 - \alpha$ he/she wants to put on the homogeneity and discrimination criteria respectively, depending on the importance of these criteria to reach a desired goal. Alternatively, the user can optimize both criteria simultaneously, choosing a data driven value of α. We show on a real data set that a data driven choice can yield an almost optimal discrimination, while improving homogeneity, leading to improved class description. One of the future evolutions of this algorithm will be the treatment of other types of symbolic data, such as symbolic data dealing with taxonomies and rules.

References

BREIMAN, L., FRIEDMAN, J.H., OLSHEN, R.A., and STONE, C.J. (1984): *Classification and Regression Trees.* Wadsworth, Belmont, California.

CHAVENT, M. (1997): *Analyse de Données Symboliques, Une Méthode Divisive de Classification.* Thèse de Doctorat, Université Paris IX Dauphine.

DIDAY, E. (1999): Symbolic Data Analysis and the SODAS Project: Purpose, History, and Perspective In: H.H. Bock, and E. Diday (Eds.): *Analysis of Symbolic Data.* Springer, Heidelberg, 1–23.

GETTLER-SUMMA, M. (1999): *MGS in SODAS : Marking and Generalization by Symbolic Objects in the Symbolic Official Data Analysis Software.* Cahiers du CEREMADE, Paris, France.

PÉRINEL, E. (1999): Construire un arbre de discrimination binaire à partir de données imprécises. *Revue de Statistique Appliquée, 47*, 5–30.

QUINLAN, J.R. (1986): Induction of Decision Trees. *Machine Learning, 1*, 81–106.

VRAC, M. and DIDAY, E. (2001): *Description Symbolique de Classes*, Cahiers de CEREMADE, Paris,France.

Consensus Trees and Phylogenetics

A Comparison of Alternative Methods for Detecting Reticulation Events in Phylogenetic Analysis

Olivier Gauthier and François-Joseph Lapointe

Département de sciences biologiques, Université de Montréal,
C.P. 6128, Succ. Centre-Ville, Montréal (Québec), Canada, H3C 3J7
(e-mail: Olivier.Gauthier@Umontreal.CA)
(e-mail: Francois-Joseph.Lapointe@Umontreal.CA)

Abstract. A growing concern in phylogenetic analysis is our ability to detect events of reticulate evolution (e.g. hybridization) that deviate from the strictly branching pattern depicted by phylogenetic trees. Although algorithms for estimating networks rather than trees are available, no formal evaluation of their ability to detect actual reticulations has been performed. In this paper, we evaluate the performance of reticulograms and split decomposition graphs (or splitsgraphs) in the identification of known hybridization events in a phylogeny. Our results show that neither technique permits unambiguous identification of hybrids. We thus introduce a quartet-based approach used in combination with these two methods and show that quartet analysis of splitsgraphs lead to a near perfect identification of hybrids. We also suggest ways in which the reticulogram reconstruction algorithm could be improved.

1 Introduction

The problem of reticulate evolution represents a growing concern in phylogenetic analysis (see Legendre, 2000). Reticulate events of evolution are brought upon by lateral gene transfer, introgression, and hybridization, among other phenomena. They do not result in a strictly bifurcating branching pattern such as those depicted by phylogenetic trees, but rather in a graph with multiple routes between some of the nodes, which cannot be elucidated by classical phylogenetic reconstruction methods. McDade (1990, 1992, 1997) has studied the impact of hybrids in phylogenetic analysis and compared the behavior of parsimony and distance-based tree reconstruction methods. Her results illustrated the poor performance of these techniques and showed that new methods were badly needed to solve the so-called 'hybrid problem'. Although some algorithms of network estimation are currently available (for reviews see Lapointe, 2000; Posada and Crandall, 2001) their relative performance has never been determined for detecting hybridization events in a phylogeny. Using McDade's data we compared the behavior of two distance-based methods of reticulate analysis. Based on our results, a new quartet-based approach for hybrid detection is introduced.

2 Distance-based methods of reticulate analysis

The reticulogram reconstruction method of Makarenkov and Legendre (2000) was designed specifically to detect events of reticulate evolution. It starts from an additive tree and adds additional edges, or reticulations, until a goodness-of-fit criterion is minimized, or a fixed number of reticulations specified by the user is added. Two criteria have been proposed by Makarenkov and Legendre (2000) as different stopping rules:

$$Q_1 = \frac{\sqrt{Q(N)}}{n(n-1)/2 - N} \tag{1}$$

and

$$Q_2 = \frac{Q(N)}{n(n-1)/2 - N} \tag{2}$$

where $Q(N) = \Sigma\Sigma(d_{ij} - \delta_{ij})^2$, d_{ij} and δ_{ij} are the original dissimilarities and the reticulogram (or tree) distances respectively, n is the number of objects, and N the number of edges in the reticulogram (or tree). The result is presented in the form of a tree with extra edges superimposed onto it to depict possible reticulation events. Notice that this method will return a tree if the input data satisfies the four-point condition. Reticulogram reconstruction is implemented in the T-Rex program (Makarenkov, 2001) available from the WWWeb at <http://www.fas.umontreal.ca/biol/casgrain/en/labo/t-rex/>.

Bandelt and Dress (1992) developed split decomposition in a totally different perspective. Their method aims at representing the conflicting signals in a phylogenetic data set. It uses the four-point condition on quartets to reject the most improbable tree among the three distinguishable topologies involving four objects. If the two remaining trees are both supported by the data, a pair of weakly compatible splits is shown to represent the conflict. The full representation on all quartets is a set of weakly compatible splits, called a splitsgraph. Notice that if no conflict is present in the data set, the split decomposition method will output a tree. The SplitsTree program (Huson, 1998) implements split decomposition and is available by anonymous ftp at <ftp://ftp.uni-bielefeld.de/pub/math/splits/>.

3 Hybrid detection analysis

To assess the relative performance of the reticulograms and splitsgraphs to detect actual hybridization events, we applied both techniques to a data set containing known hybrids. The morphological data collected by

McDade (1984, 1990) included 12 species of the plant genus *Aphelandra* and 17 hybrids produced in the lab by crossing parents representing 9 of the 12 species (see Table 1). A species of this genus, the zebra plant (*Aphelandra squarrosa*), is a common house plant. For simplicity, our analyses were conducted on 17 different data sets, each containing the 12 species and one single hybrid. The 17 distance matrices computed from the raw data were submitted to T-Rex and SplitsTree and the position of the hybrids with respect to their parents were noted to determine the hybrid detection rate of the competing approaches. An hybrid was detected in a reticulogram if it was the sister taxa of one of its parents and had a reticulation to the other. In splitsgraphs, it had to form a pair of weakly compatible splits with its parents.

Our results indicated that a direct application of reticulate methods did not enable hybrid detection. In reticulograms, an average of 8.4 reticulations were added to the tree by the algorithm (7.0 and 9.7 for Q_1 and Q_2 respectively), making the interpretation of the resulting graph difficult, if at all possible. Hybrids were unambiguously detected in one single case with Q_1 and three times with Q_2 (these reticulations were always among the last ones added to minimize the criteria). Direct application of split decomposition did not provide better results, with only two hybrids detected in total. These were hybrids between closely related parents that grouped together in all of the analyses.

Table 1. Hybrid detection rate for the 17 hybrids and the two methods used with quartet analysis

Parents			
Female	**Male**	**Reticulogram**	**Split Decomposition**
	A. panamensis	5	10
A. deppeana	*A. sinclairiana*	5	10
	A. storkii	4	9
	A. deppeana	3	10
A. golfodulcensis	*A. leonardii*	3	9
	A. sinclairiana	7	10
	A. campanensis	4	9
A. leonardii	*A. golfodulcensis*	1	7
	A. sinclairiana	6	10
	A. deppeana	6	10
A. panamensis	*A. golfodulcensis*	6	9
	A. leonardii	4	9
	A. sinclairiana	4	6
	A. deppeana	5	9
A. sinclairiana	*A. golfodulcensis*	2	10
	A. gracilis	1	10
	A. terryae	1	10

4 Hybrid detection through quartet analysis

Given the poor performance of the reticulate analysis methods in our comparisons, we have developed a different approach for detecting hybridization events in a data set. Based on morphological character patterns, it has long been known that hybrids should be placed between their parents in a phylogenetic tree (Wagner, 1969), but the presence of additional species may obscure these relationships (McDade, 1990, 1992, 1997). This led us to turn towards quartet analysis using a Hybrid Detection Criterion (HDC), a technique that can be applied in combination with any method of phylogenetic analysis. HDC is defined in the following way: in quartets made up of one hybrid (AB), its two parents (A and B) and any other species (X), the hybrid should never group with X but should be positioned next to A or B with an equal frequency over all quartets, or, if the method allows for hybridization events, with both A and B in every quartet. For each of the 17 data sets, we looked at all possible quartets {A, B, AB, X}, for a total of ten quartets per analysis. For each data set and each method, the number of quartets satisfying HDC was used as a measure of hybrid detection, yielding values ranging from zero (no quartet meet HDC) to ten (all quartets meet HDC). In order to detect hybrids through quartet analysis using the reticulogram algorithm, we fixed the number of reticulations to one; for a given quartet, HDC was met if the hybrid grouped with one parent and had a reticulation to the other (Figure 1). For split decomposition quartets, HDC was satisfied if the hybrid was placed between its two parents (Figure 2).

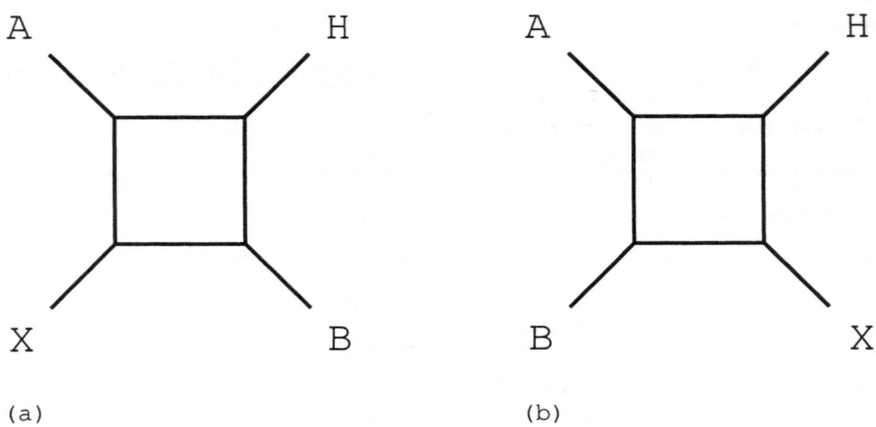

Fig. 1. Illustration of the Hybrid Detection Criterion (HDC) for splitsgraph quartets. (a) HDC is met if the hybrid (H) forms a pair of weakly compatible splits with its parents (A and B). (b) HDC is not met if the hybrid forms a weakly compatible split with the other species (X). The positions of parents A and B are interchangeable.

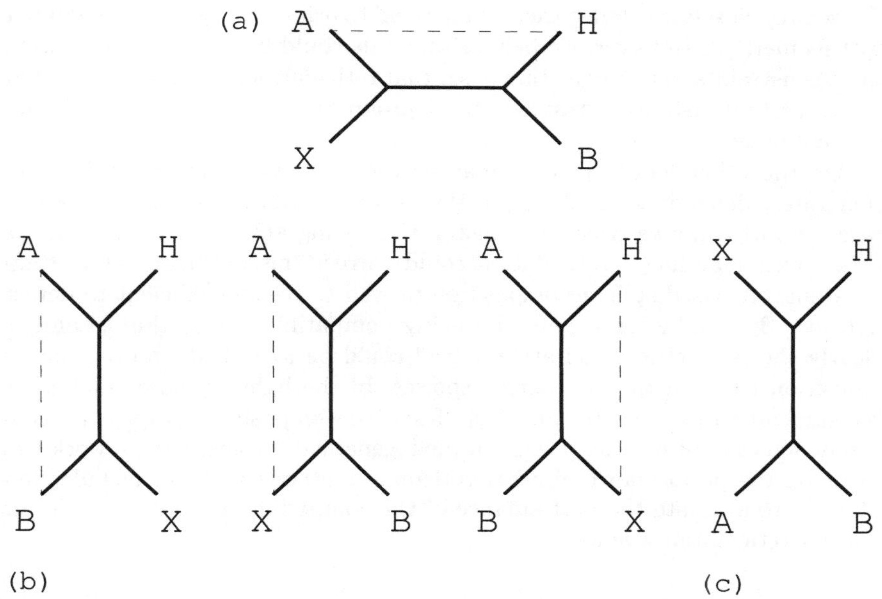

Fig. 2. Illustration of the Hybrid Detection Criterion (HDC) for reticulogram quartets. (a) HDC is met if the hybrid (H) is the sister taxa of either one of its parents (A or B) and has a reticulation to its other parent. HDC is not met if (b) the hybrid is the sister taxa of one of its parents, but does not have a reticulation to the other, or, (c) the hybrid is the sister taxa of the other species (X). The positions of parents A and B are interchangeable.

Overall, quartet analysis increased the hybrid detection rate compared to the direct application of reticulate methods (Table 1). While split decomposition permitted unambiguous identification of the majority of hybrids, reticulograms produced spurious results. Closer examination of individual reticulograms indicated that the single reticulation was added between one of the parents and the more distant species (X) in the vast majority of quartets; this situation violates HDC.

5 Discussion

Our results showed that neither reticulograms nor splitsgraphs allow for efficient hybrid detection when these methods are blindly applied. Whereas split decomposition only identified hybrids between closely-related parents, the main problem with the reticulogram approach appears to be caused by the goodness-of-fit criterion ($Q1$ or $Q2$) for adding reticulations. In trying to maximize the fit of a reticulogram to a distance matrix, it follows that the first reticulations usually connect the most distant pairs of nodes in the tree.

Therefore, closely-related parents and their hybrids will rarely be detected by this method. However, we believe that this could be corrected by adding some constraints to the algorithm, such that reticulations are never generated over a certain distance threshold, or by prohibiting reticulations to or from internal nodes.

On the other hand, quartet analysis provided very good hybrid detection rates, depending on the algorithm selected. Reticulograms performed rather poorly and we trust, here again, that being able to define a distance threshold for adding a reticulation could correct this problem. Interestingly, a quartet-based split decomposition proved to be very efficient to detect hybrids. By producing a pair of weakly compatible splits, this technique clearly shows whether a putative hybrid could be identified simply from its position between the two parent species. In the light of those results, we recommend to use quartet analysis of split decomposition graphs to accurately detect hybridization events in phylogenetic data sets. Future work will focus on the performance of other reticulate methods and will include simulations to evaluate the performance of the competing approaches to detect ancient reticulation events.

Acknowledgments

The authors are grateful to L. A. McDade for providing the data set used in the present study. This work was made possible by a NSERC scholarship to O. Gauthier and by NSERC grant no. OGP0155251 to F.-J. Lapointe.

References

BANDELT, H.-J. and DRESS, A.W.M. (1992): Split Decomposition: A New and Useful Approach to Phylogenetic Analysis of Distance Data. *Molecular Phylogenetics and Evolution, 1*, 242–252.

HUSON, D.H. (1998): SplitsTree: A Program for Analyzing and Visualizing Evolutionary Data. *Bioinformatics, 14*, 68–73.

LAPOINTE, F.-J. (2000): How to Account for Reticulation Events in Phylogenetic Analysis: A Comparison of Distance-Based Methods. *Journal of classification, 17*, 175–184.

LEGENDRE, P. (2000): Special Section on Reticulate Evolution. *Journal of Classification, 17*, 153–195. Including papers from F.-J. Lapointe, P. Legendre, F.J. Rohlf, P.E. Smouse, and, P.H.A. Sneath.

MAKARENKOV, V. (2001): T-Rex: Reconstructing and Visualizing Phylogenetic Trees and Reticulation Networks, *Bioinformatics, 17*, 664–668.

MAKARENKOV, V. and LEGENDRE, P. (2000): Improving the Additive Tree Representation of a Given Dissimilarity Matrix Using Reticulations. In: H.A.L. Kiers, J.-P. Rasson, P.J.F. Groenen, and M. Schader (Eds.), *Data Analysis, Classification, and Related Methods*, Berlin: Springer, 35–46.

McDADE, L.A. (1984): Systematics and Reproductive Biology of the Central American species of the *Aphelandra pulcherrima* complex (Acanthaceae). *Annals of the Missouri Botanical Garden*, *71*, 104–165.

McDADE, L.A. (1990): Hybrids and Phylogenetic Systematics I. Patterns of Character Expression in Hybrids and their Implication for Cladistic Analysis. *Evolution*, *44*, 1685–1700.

McDADE, L.A. (1992): Hybrids and Phylogenetic Systematics II. The Impact of Hybrids on Cladistic Analysis. *Evolution*, *46*, 1329–1346.

McDADE, L.A. (1997): Hybrids and Phylogenetic Systematics III. Comparison with Distance Methods. *Systematic Botany*, *22*, 669–683.

POSADA, D. and CRANDALL, K.A. (2001): Intraspecific Gene Genealogies: Trees Grafting Into Networks. *Trends in Ecology and Evolution*, *16*, 37–45.

WAGNER, W.H.Jr. (1969): The Role and Taxonomic Treatment of Hybrids. *Bioscience*, *19*, 785–789.

Hierarchical Clustering of Multiple Decision Trees

Branko Kavšek[1], Nada Lavrač[1], and Anuška Ferligoj[2]

[1] Institute Jožef Stefan, Jamova 39, 1000 Ljubljana, Slovenia
branko.kavsek@ijs.si, nada.lavrac@ijs.si

[2] University of Ljubljana, 1000 Ljubljana, Slovenia
anuska.ferligoj@uni-lj.si

Abstract. Decision tree learning is relatively non-robust: a small change in the training set may significantly change the structure of the induced decision tree. This paper presents a decision tree construction method in which the domain model is constructed by consensus clustering of N decision trees induced in N-fold cross-validation. Experimental results show that consensus decision trees are simpler than C4.5 decision trees, indicating that they may be a more stable approximation of the intended domain model than decision trees, constructed from the entire set of training instances.

1 Introduction

Decision tree induction (Breiman et al. 1984, Quinlan, 1986) has been recognized as one of the standard data analysis methods. In particular, variants of Quinlan's C4.5 (Quinlan, 1993) can be found in virtually all commercial and academic data mining packages.

The main advantages of decision tree learning are reasonable accuracy and simplicity of explanations and computational efficiency. It is well known, however, that decision tree learning is a rather non-robust method: a small change in the training set may significantly change the structure of the induced decision tree, which may result in experts' distrust in induced domain models. Improved robustness and improved accuracy results can be achieved e.g., by bagging/boosting (Breiman, 1996) at a cost of increased model complexity and decreased explanatory potential.

To improve robustness, this paper presents a novel decision tree construction method in which the domain model in the form of a decision tree is constructed by consensus clustering of N decision trees induced in N-fold cross-validation. Experimental results show that consensus decision trees are simpler than C4.5 decision trees, indicating that they may be a more stable approximation of the intended domain model than decision trees constructed from the entire set of training instances.

The paper is organized as follows. Section 2 presents the basic methodology of decision tree induction and hierarchical clustering, Section 3 outlines the novel approach of consensus decision tree construction, and Section 4 provides the experimental evaluation of the proposed approach. We conclude by a summary and plans for further work.

2 Background methodology

2.1 Decision trees

Induction of decision trees is one of the most popular machine learning methods for learning of attribute-value descriptions (Breiman et al., 1984, Quinlan, 1986). The basic decision tree learning algorithm builds a tree in a top-down greedy fashion by recursively selecting the 'best' attribute on the basis of an information measure, and splitting the data set accordingly. Various modifications of the basic algorithm can be found in the literature, the most popular being Quinlan's C4.5 (Quinlan, 1993). In our work we used the WEKA (Witten and Frank, 1999) implementation of C4.5.

2.2 Hierarchical clustering

Clustering methods in general aim at building clusters (groups) of objects so that similar objects fall into the same cluster (internal cohesivity) while dissimilar objects fall into separate clusters (external isolation). A particular class of clustering methods, studied and widely used in statistical data analysis (e.g., Sokal and Sneath, 1963; Gordon, 1981; Hartigan, 1975) are *hierarchical clustering* methods.

The purpose of (agglomerative) hierarchical clustering is to fuse objects (instances) into successively larger clusters, using some measure of (dis)similarity and an agglomeration or linkage rule (*complete linkage* in our case). A typical result of this type of clustering is a hierarchical tree or dendrogram.

Consensus clustering Consensus hierarchical clustering deals with the following problem: given a set of concept hierarchies (represented by dendrograms), find a *consensus concept hierarchy* by merging the given concept hierarchies in such a way that similar instances (those that belong to the same concept/cluster) will remain similar also in the merged concept hierarchy. In the last thirty years many consensus clustering methods have been proposed (e.g., Regnier, 1965; Adams, 1972; McMorris and Neuman, 1983; Day, 1983). In 1986, a special issue of the Journal of Classification was devoted to consensus classifications. Excellent reviews of this topic are also available (Faith, 1988; Leclerc, 1988).

3 Consensus decision tree construction

3.1 Motivation

As pointed out by Langley (Langley, 1996), decision tree induction can be seen as a special case of induction of concept hierarchies. A concept is associated with each node of the decision tree, and as such a tree represents a kind of taxonomy, a hierarchy of many concepts. In this case, a concept is identified by the set of instances in a node of the decision tree. Hierarchical clustering also results in a taxonomy of concepts, equally identified by the set of instances in a 'node' of a dendrogram representing the concept hierarchy. Concept hierarchies can be induced in a supervised or unsupervised manner: decision tree induction algorithms perform supervised learning, whereas induction by hierarchical clustering is unsupervised.

Our idea of building consensus decision trees is inspired by the idea of consensus hierarchical clustering. A consensus decision tree should be constructed in such a way that instances that are similar in the original decision trees should remain being similar also in the consensus decision tree. To this end, it is crucial to define an appropriate measure of similarity between instances.

3.2 CDT: An algorithm for consensus decision tree construction

The consensus tree building procedure consists of the following four steps:

1) perform N-fold cross-validation resulting in N decision trees induced by a decision tree learning algorithm (e.g., C4.5),
2) use these decision trees for computing a dissimilarity matrix that measures the dissimilarity of pairs of instances,
3) construct a concept hierarchy using the dissimilarity matrix of step 2 and define concepts by 'cutting' the dendrogram w.r.t. the maximal difference in cluster levels,
4) induce a consensus decision tree using the same decision tree algorithm as in step 1.

Decision tree construction First, N-fold cross-validation is performed resulting in N decision trees induced by the C4.5 learning algorithm (the WEKA implementation of C4.5 with default parameters[1] is used for this purpose). The decision trees are then stored and used to compute the dissimilarity between pairs of instances.

Dissimilarity between instances The dissimilarities between pairs of instances are computed from the N stored decision trees in the following way:

- first we measure the similarity s between instances i and j by counting (for all N decision trees) how many times the two instances belong to the same leaf (i.e., are described by the same path of attribute-value tests

[1] The default parameters were: *NO binary splits, confidence factor for pruning 0.25, minimum number of objects in a leaf 2.*

leading from the root to the leaf of the decision tree). Therefore $s(i,j)$ is defined as follows:

$$s(i,j) = \sum_{l=1}^{N} T_l(i,j)$$

where

$$T_l(i,j) = \begin{cases} 1, & \text{if } i \text{ and } j \text{ belong to the same leaf (are described by the same attribute values) in the } l\text{-th decision tree} \\ 0, & \text{otherwise} \end{cases}$$

- then we compute the dissimilarity measure $d(i,j)$ by simply subtracting the similarity $s(i,j)$ from the number of trees N, i.e., $d(i,j) = N - s(i,j)$.

Concept hierarchy construction A concept hierarchy is constructed using the well known agglomerative hierarchical clustering algorithm mentioned in Section 2.2. However, if the number of clusters produced by the algorithm is smaller than the number of classes of the given classification problem, we stop the "agglomeration" earlier. Consequently, we sometimes force the algorithm to produce more clusters than the optimal number of clusters according to the algorithm.

Induction of consensus decision trees Within each cluster of instances, we select the majority class and remove from the cluster all instances not belonging to this majority class. We then use the C4.5 learning algorithm to induce the consensus decision tree from the remaining subset of instances. In all runs of the C4.5 algorithm the same (default) parameter setting is used (as in the first step of this algorithm). Notice that in the case of a tie (two or more classes being the majority class), a random choice between these class assignments is made.

4 Experimental evaluation

4.1 Experimental design

In standard 10-fold cross-validation, the original data set is partitioned into 10 folds with (approximately) the same number of examples. Training sets are built from 9 folds, leaving one fold as a test set. Let G denote the entire data set, T_i an individual test set (consisting of one fold), and G_i the corresponding training set ($G_i \leftarrow G \setminus T_i$, composed of nine folds). In this way, 10 training sets G_1– G_{10}, and 10 corresponding test sets T_1–T_{10} are constructed. Every example occurs exactly once in a test set, and 9 times in training sets.

In the first experiment we used C4.5 (WEKA implementation, default parameter setting) to induce decision trees on training sets G_1–G_{10}. We measured the average accuracy $Acc(C4.5(G))$ (and standard deviation) and the information score[2] $Info(C4.5(G))$ of ten hypotheses $C4.5(G_i)$ constructed by C4.5 on training sets G_i, $i \in [1, 10]$, where $Acc(C4.5(G)) = \frac{1}{10}\sum_1^{10} Acc(C4.5(G_i))$, and $Info(C4.5(G)) = \frac{1}{10}\sum_1^{10} Info(C4.5(G_i))$. The average size of decision trees $Leaves/Nodes(C4.5(G))$ was measured by the number of leaves and the number of all decision tree nodes (number of leaves + number of internal nodes), averaged over 10 folds.

The above result presents the baseline for comparing the quality of our consensus tree building algorithm CDT, measured by the average accuracy $Acc(CDT(G))$, $Leaves/Nodes(CDT(G))$, and information score $Info(CDT(G))$ over ten consensus decision trees $CDT(G_i)$.

As described in Section 3.2, a consensus decision tree is constructed from ten C4.5 decision trees. Building of consensus decision trees was performed in a nested 10-fold cross-validation loop: for each G_i, $i \in [1, 10]$, training sets G_{ij}, $j \in [1, 10]$ were used to construct decision trees $C4.5(G_{ij})$ by the C4.5 algorithm. Training sets G_{ij} were obtained by splitting each G_i into ten test sets T_{ij} (consisting of one sub-fold), and ten training sets $G_{ij} \leftarrow G_i \setminus T_{ij}$ (composed of nine sub-folds).

Ten decision trees $C4.5(G_{ij})$ were merged into a single consensus decision tree $CDT(G_i)$. Let $Acc(CDT(G_i))$ denote its accuracy tested on T_i. Accordingly, $Acc(CDT(G)) = \frac{1}{10}\sum_1^{10} Acc(CDT(G_i))$, and information contents $Info(CDT(G)) = \frac{1}{10}\sum_1^{10} Info(CDT(G_i))$. $Leaves/Nodes(CDT(G))$ is also the average of $Leaves/Nodes(CDT(G_i))$.

In order to compare accuracy, tree size and information score of consensus trees and C4.5 trees, we calculate their *relative improvements* as follows: $Rel(Acc(G)) = \frac{Acc(CDT(G))}{Acc(C4.5(G))} - 1$, $Rel(Leaves/Nodes(G)) = 1 - \frac{Leaves/Nodes(CDT(G))}{Leaves/Nodes(C4.5(G))}$, $Rel(Info(G)) = \frac{Info(CDT(G))}{Info(C4.5(G))} - 1$.

4.2 Results of experiments

Experiments were performed on 28 UCI data sets whose characteristics are outlined in Table 1 (boldface denoting the majority class). Results of experiments are shown in Table 2 (boldface meaning that CDT performed equally well or better than C4.5).

Results of experiments show that there is no significant difference in average accuracy between the consensus decision trees and the decision trees induced by C4.5 ($t = 1.8664$, $df = 27$, $p = 0.0729$, using two-tailed t-test for

[2] Whereas accuracy computes the relative frequency of correctly classified instances, the information score takes into the account also the improvement of accuracy compared to the prior probability of classes, see (Kononenko and Bratko, 1991).

Table 1. Characteristics of data sets

Data set	#Attributes	#Classes	#Instances	Class distribution (%)
Anneal	38	5	898	1:11:**76**:8:4
Audiology	69	24	226	1:1:**25**:9:1:1:8:21:1:2:1:2:1:1:3:1:1:1:9:2:1:2:4:1
Australian	14	2	690	**56**:44
Autos	25	6	205	1:11:**33**:26:16:13
Balance	4	3	625	**46**:8:**46**
Breast	9	2	286	**70**:30
Breast-w	9	2	699	**66**:34
Car	6	4	1728	**70**:22:4:4
Colic	22	2	368	**63**:37
Credit-a	15	2	690	44:**56**
Credit-g	20	2	1000	**70**:30
Diabetes	8	2	768	**65**:35
Glass	9	6	214	33:**36**:8:6:4:14
Heart-c	13	2	303	**54**:46
Heart-stat	13	2	270	**56**:44
Hepatitis	19	2	155	21:**79**
Ionosphere	34	2	351	36:**64**
Iris	4	3	150	**33**:**33**:**33**
Labor	16	2	57	35:**65**
Lymph	18	4	148	1:**55**:41:3
Prim. tumor	17	21	339	**25**:5:3:4:11:1:4:2:1:8:4:2:7:1:1:3:8:2:1:1:7
Segment	19	7	2310	**14**:**14**:**14**:**14**:**14**:**14**:**14**
Sonar	60	2	208	47:**53**
Tic-tac-toe	9	2	958	**65**:35
Vehicle	18	4	846	25:**26**:**26**:24
Vote	16	2	435	**61**:39
Wine	13	3	178	33:**40**:27
Zoo	17	7	101	**41**:20:5:13:4:8:10

Table 2. Results of the experiments

Data set	C4.5 Acc(Sd)	C4.5 leaves /nodes	C4.5 info.	CDT Acc(Sd)	CDT leaves /nodes	CDT info.	Relative improvement Acc	Relative improvement leaves /nodes	Relative improvement info.
Anneal	97,14 (0,0893)	39/77	2,7130	**99,13** (0,0497)	52/103	**2,7811**	**0,0205**	-0,333/-0,338	**0,0251**
Audiol.	77,88 (0,1193)	32/54	2,6579	**80,09** (0,1288)	**29/49**	2,5316	**0,0284**	**0,094/0,093**	-0,0475
Austral.	85,51 (0,3455)	31/45	0,6183	84,64 (0,3919)	**17/24**	**0,6813**	-0,0102	**0,452/0,467**	**0,1019**
Autos	82,44 (0,2022)	49/69	1,8198	80,49 (0,2361)	**39/57**	1,7598	-0,0237	**0,204/0,174**	-0,0330
Balance	77,76 (0,3567)	58/115	0,6778	69,12 (0,4537)	**8/15**	0,5598	-0,1111	**0,862/0,870**	-0,1741
Breast	75,17 (0,4423)	4/6	0,1005	72,73 (0,5222)	13/17	**0,2630**	-0,0326	-2,250/-1,833	**1,6169**
Brst-w	95,28 (0,2116)	16/31	0,8020	94,85 (0,2269)	**9/17**	**0,8189**	-0,0045	**0,438/0,452**	**0,0211**
Car	92,48 (0,1628)	131/182	1,0218	68,66 (0,3946)	**29/41**	0,2687	-0,2576	**0,779/0,775**	-0,7370
Colic	85,87 (0,3518)	4/6	0,4993	83,42 (0,4071)	11/17	**0,6023**	-0,0285	-1,750/-1,833	**0,2063**
Crdt-a	85,94 (0,3402)	30/42	0,6221	85,07 (0,3864)	**18/25**	**0,6901**	-0,0101	**0,400/0,405**	**0,1093**
Crdt-g	69,70 (0,4922)	103/140	0,1193	69,50 (0,5523)	**82/114**	**0,1949**	-0,0029	**0,204/0,186**	**0,6337**
Diab.	74,09 (0,4356)	22/43	0,2993	73,96 (0,5103)	**21/41**	**0,3766**	-0,0018	**0,045/0,047**	**0,2583**
Glass	67,29 (0,2837)	30/59	1,3307	**68,69** (0,2991)	**23/45**	**1,3524**	**0,0208**	**0,233/0,237**	**0,0163**
Heart-c	79,21 (0,2619)	30/51	0,5354	**79,21** (0,2884)	**18/29**	**0,5917**	**0,0000**	**0,400/0,431**	**0,1052**
Heart-s	77,78 (0,4322)	18/35	0,4847	75,93 (0,4907)	21/41	**0,5054**	-0,0238	-0,167/-0,171	**0,0427**
Hepat.	79,35 (0,4200)	11/21	0,1445	77,42 (0,4752)	**8/15**	**0,1510**	-0,0244	**0,273/0,286**	**0,0450**
Ionos.	90,88 (0,2887)	18/35	0,7416	89,74 (0,3203)	**16/31**	0,7247	-0,0125	**0,111/0,114**	-0,0228
Iris	95,33 (0,1707)	5/9	1,4663	94,67 (0,1886)	**3/5**	**1,4692**	-0,0070	**0,400/0,444**	**0,0020**
Labor	78,95 (0,4285)	3/5	0,3496	**80,70** (0,4393)	3/5	**0,5257**	**0,0222**	**0,000/0,000**	**0,5037**
Lymph	77,03 (0,3274)	21/34	0,6074	**77,70** (0,3339)	**18/28**	**0,6712**	**0,0088**	**0,143/0,176**	**0,1050**
Prim.t.	40,71 (0,1961)	47/88	1,2050	**41,30** (0,2310)	**29/54**	**1,2396**	**0,0145**	**0,383/0,386**	**0,0287**
Segment	97,14 (0,0893)	39/77	2,7130	96,02 (0,1067)	52/103	2,6867	-0,0116	-0,333/-0,338	-0,0097
Sonar	74,04 (0,4986)	18/35	0,4662	**75,81** (0,4952)	**15/29**	**0,5050**	**0,0239**	**0,167/0,171**	**0,0832**
T-tac-t	84,76 (0,3485)	95/142	0,5613	78,91 (0,4592)	**71/106**	0,4793	-0,0690	**0,253/0,254**	-0,1461
Vehicle	73,40 (0,3272)	98/195	1,3607	70,92 (0,3813)	**56/111**	1,2996	-0,0338	**0,429/0,431**	-0,0449
Vote	96,78 (0,1650)	6/11	0,8580	96,55 (0,1857)	6/11	**0,8908**	-0,0024	**0,000/0,000**	**0,0382**
Wine	94,94 (0,1776)	5/9	1,4476	93,26 (0,2120)	7/13	1,4192	-0,0178	-0,400/-0,444	-0,0196
Zoo	92,08 (0,1359)	9/17	2,1435	**92,08** (0,1504)	9/17	2,1014	**0,0000**	**0,000/0,000**	-0,0196
Average	82,10 (0,2900)	34,71/ 58,32	1,01	80,38 (0,3300)	**24,39/ 41,54**	1,01	-0,0195	**0,037/ 0,051**	**0,0960**

dependent samples, where t, df and p stand for *t-statistics*, *degrees of freedom* and *significance level*, respectively), using a 95% significance level (the bound used throughout this paper). Notice, however, that the CDT algorithm improves the information score (compared to C4.5) in 18 domains (9.6% improvement on the average).

Our hypothesis that the structure of CDT is simpler than the structure of the induced C4.5 decision trees was confirmed: indeed, the average number of leaves of CDT is significantly smaller than the average number of leaves of the C4.5 decision trees ($t = 2.3787$, $df = 27$, $p = 0.0247$, using two-tailed t-test for dependent samples). Moreover, the average tree size (measured by the number of all decision tree nodes) of CDT is also significantly smaller than that of C4.5 ($t = 2.4413$, $df = 27$, $p = 0.0215$, using two-tailed t-test for dependent samples).

The relative improvement in tree size also shows that in 19 domains the CDT algorithm learned smaller decision trees than C4.5 yielding, on the average, 3.7% smaller trees according to the number of leaves (5.1% according to the number of all nodes).

5 Summary and conclusions

Results show that consensus decision trees are, on the average, as accurate as C4.5 decision trees, but simpler (smaller w.r.t. the number of leaves and nodes). Moreover, consensus decision trees improve the information score compared to C4.5 decision trees. In fact there are two data sets in which a C4.5 decision tree outperforms CDT in accuracy, simplicity and information score; on the other hand, in five domains CDT is better in all the three characteristics.

We also tested alternative ways of constructing consensus decision trees. First, the similarity measure that only considers the distances between instances was replaced by a measure that takes into the account that instances are labelled by class labels; the similarity value of two instances labelled by the same class label was appropriately increased. Opposed to our expectations, this way of measuring similarities between instances has not proved to be better than the one described in this paper.

Second, instead of labelling instances by class labels, instances may be labelled by cluster labels, considering clusters generated by consensus clustering as classes for learning by C4.5. This approach has turned out to be inferior compared to the approaches described in the paper.

Third, similarities between instances can be measured not only by counting how many times two instances belong to the same leaf (have the same attribute-value representation), but also by putting different weights on the segments of this path (higher weights assigned to segments closer to the root). Contrary to our expectations this approach produced worse results than the original approach described in this paper. Although CDTs (induced in such a way) were much smaller, they were also significantly less accurate than C4.5 decision trees.

In further work we are planning to test the hypothesis that consensus decision trees are more robust with respect to adding of new instances, i.e., that the structure of the consensus decision tree would change less than the structure of C4.5 decision trees. To this end we need to propose new measures of tree structure variability, and measure the robustness accordingly.

The current results indicate that a step in the direction of improving the robustness has been achieved, assuming that simpler tree structures are more robust.

There is however a performance drawback that we should take into account when using the CDT method for building decision trees. Since the CDT algorithm builds 11 decision trees (10 in the cross-validation process and the final one) and does also the hierarchical clustering, it is much slower than the traditional decision tree building algorithm.

Acknowledgements

Thanks to Saso Džeroski, Ljupčo Todorovski and Marko Grobelnik for useful comments on the draft of this paper. Thanks also to Bernard Ženko for his help when using WEKA. The work reported in this paper was supported by the Slovenian Ministry of Education, Science and Sport, and the IST-1999-11495 project Data Mining and Decision Support for Business Competitiveness: A European Virtual Enterprise.

References

ADAMS, E.N. (1972). Consensus techniques and the comparison of taxonomic trees. *Systematic Zoology*, 21, 390–397.

BREIMAN, L., FRIEDMAN, J., OLSHEN, R., and STONE, C. (1984). *Classification and Regression Trees.* Wadsworth International Group, Belmont, CA.

BREIMAN, L. (1996). Bagging predictors. *Machine Learning*, 24:123–140.

DAY, W.H.E. (1983). The role of complexity in comparing classifications. *Mathematical Biosciences*, 66, 97–114.

FAITH, D.P. (1988). Consensus applications in the biological sciences. In: Bock, H.H. (Ed.) *Classification and Related Methods of Data Analysis*, Amsterdam: North-Holland, 325–332.

FISHER, D.H. (1989). Noise-tolerant conceptual clustering. *Proceedings of the Eleventh International Joint Conference on Artificial Intelligence* 825–830. San Francisco: Morgan Kaufmann.

GORDON, A.D. (1981). *Classification.* London: Chapman and Hall.

HARTIGAN, J.A. (1975). *Cluster Algorithms.* New York: Wiley.

HUNT, E., MARTIN, J., and STONE, P. (1966). *Experiments in Induction.* New York: Academic Press.

KONONENKO, I., and BRATKO, I. (1991). Information based evaluation criterion for classifier's performance. *Machine Learning*, 6, (1), 67–80.

LANGLEY, P. (1996). *Elements of Machine Learning.* Morgan Kaufmann.

LECLERC, B. (1988). Consensus applications in the social sciences. In: Bock, H.H. (Ed.) *Classification and Related Methods of Data Analysis*, Amsterdam: North-Holland, 333–340.

McMORRIS, F.R. and NEUMAN, D. (1983). Consensus functions defined on trees. *Mathematical Social Sciences*, 4, 131–136.

QUINLAN, J.R. (1986). Induction of decision trees. *Machine Learning*, 1(1): 81–106.

QUINLAN, J.R. (1993). *C4.5: Programs for Machine Learning.* California: Morgan Kaufmann.

REGNIER, S. (1965). Sur quelques aspects mathematiques des problems de classification automatique. *I.I.C. Bulletin*, 4, 175–191.

SOKAL, R.R. and SNEATH, P.H.A. (1963). *Principles of Numerical Taxonomy.* San Francisco: Freeman.

WITTEN, I.H. and FRANK, E. (1999). *Data Mining: Practical Machine Learning Tools and Techniques with Java Implementations.* Morgan Kaufmann, San Francisco.

Multiple Consensus Trees

François-Joseph Lapointe[1] and Guy Cucumel[2]

[1] Département de sciences biologiques, Université de Montréal, C.P. 6128, Succ. Centre-Ville, Montréal (Québec), Canada, H3C 3J7 (e-mail: Francois-Joseph.Lapointe@Umontreal.CA)

[2] École des sciences de la gestion, Université du Québec à Montréal, C.P. 8888, Succ. Centre-Ville, Montréal (Québec), Canada, H3C 3P8 (e-mail: cucumel.guy@uqam.ca)

Abstract. A consensus function takes as input a profile of trees representing the relationships among overlapping or identical sets of objects and returns a tree that is in some sense closest to the entire profile. A single tree is obtained by most consensus functions, although this solution may not always be representative of the profile of input trees. If a unique consensus tree is not suitable, how many consensus trees are needed? In this paper, we propose a procedure for the computation of multiple consensus trees to summarize a profile of input trees. Different stopping rules are discussed to determine the optimal number of consensus trees. This procedure is developed in the special case of the average consensus method for weighted trees. An application to phylogenetic trees is presented.

1 Computation of the average consensus

The average consensus, originally proposed by Cucumel (1990) to combine dendrograms, is a method specifically designed to compute the consensus of trees with branch lengths; the solution is returned as a weighted tree of the same kind as the trees combined. The average procedure takes as input a profile of m weighted trees and returns a weighted tree that is in some sense 'closest' to the entire input set. Given that there exists a one-to-one correspondence between any weighted tree $\mathbf{T}$ and its associated path-length distance matrix $\mathbf{D}$ (Hartigan, 1967; Buneman, 1971), it is equivalent to deal with trees or with their matrix representations. Accordingly, the average procedure thus returns a path-length matrix associated with a consensus weighted tree that is closest to the entire input set. Let $\mathrm{S} = \{1, \ldots, i, \ldots, j, \ldots, n\}$ be a set of n objects. Let $\mathbf{T}_1$ and $\mathbf{T}_2$ be two weighted trees defined on S, and $\mathbf{D}_1$ and $\mathbf{D}_2$ their associated path-length distance matrices. The distance Δ between the trees $\mathbf{T}_1$ and $\mathbf{T}_2$ is calculated as follows (Hartigan, 1967):

$$\Delta(T_1, T_2) = \sum_{i=1}^{n} \sum_{j=1}^{n} [d_1(i,j) - d_2(i,j)]^2 \tag{1}$$

where $d_1(i,j)$ and $d_2(i,j)$ are the path length distances in $\mathbf{D}_1$ and $\mathbf{D}_2$ respectively.

Let us now consider a profile P = ($\mathbf{T}_1, \ldots, \mathbf{T}_k, \ldots, \mathbf{T}_m$) of m weighted trees defined on a common set of objects S. The average consensus tree $\mathbf{T}_c$ is defined as the solution that minimizes the following loss function Q (Lapointe and Cucumel, 1997):

$$Q = \sum_{k=1}^{m} \Delta(T_c, T_k) = \sum_{k=1}^{m} \sum_{i=1}^{n} \sum_{j=1}^{n} [d_c(i,j) - d_k(i,j)]^2 \tag{2}$$

where the $d_c(i,j)$ and $d_k(i,j)$ are the path-length distances in the matrices $\mathbf{D}_c$ and $\mathbf{D}_k$ respectively. With no loss of generality, one could consider this consensus procedure as one involving a profile of scaled trees $\mathbf{T}_k$, represented by their matrices $\mathbf{D}_k$ taking their values in the interval [0,1].

Practically, the average consensus is obtained in two steps. Let $\overline{\mathbf{D}}$ represent the average distance matrix containing distances $\bar{d}(i,j)$, such that,

$$\bar{d}(i,j) = \frac{1}{m} \sum_{k=1}^{m} d_k(i,j) \tag{3}$$

for every i and j. Because the average matrix $\overline{\mathbf{D}}$ of path-length distances does not necessarily represent a path-length distance matrix itself, the average consensus $\mathbf{T}_c$ (see eq. 2) is thus obtained as the solution that minimizes the following function Q':

$$Q' = \sum_{i=1}^{n} \sum_{j=1}^{n} [d_c(i,j) - \bar{d}(i,j)]^2 \tag{4}$$

where the $d_c(i,j)$ are the path-length distances in the $\mathbf{D}_c$ matrix associated with the consensus tree $\mathbf{T}_c$, and the $\bar{d}(i,j)$ are the distances in the average matrix $\overline{\mathbf{D}}$.

Minimizing Q' (eq. 4) is equivalent to minimizing Q (eq. 2) (see proof in Lapointe and Cucumel, 1997), so that applying any least-squares algorithm (Cavalli-Sforza and Edwards, 1967; Fitch and Margoliash, 1967; De Soete, 1983; Makarenkov and Leclerc, 1999) to the average matrix $\overline{\mathbf{D}}$ returns a weighted tree $\mathbf{T}_c$ representing the average consensus solution. Because this minimization problem is NP-hard, however, the solution may not be unique and the different heuristics may produce different results.

2 Partition of a profile of trees

Any profile of trees is defined in a tree space from which the consensus solution is computed. In the case of the average consensus procedure, the solution is the tree minimizing the total sum-of-squared distances to all of the m input trees in the profile P. In other words, the average consensus is computed as the closest tree to the centroid of the input profile. The computation of multiple consensus trees thus proceeds with a partition of the profile P into q subprofiles (with $q < m$). Separate consensus trees $\mathbf{T}_{c_1}, \ldots, \mathbf{T}_{c_t}, \ldots, \mathbf{T}_{c_q}$ are then derived independently from each subprofile P_t.

Given a profile P of m weighted trees, the simplest way to partition this profile is to compute $m(m\text{-}1)/2$ distances among all pairwise combination of trees. A clustering algorithm (hierarchical or not) is applied to that tree-to-tree distance matrix $\mathbf{R}$ to determine the partitions. If sum-of-squared distances $\Delta(\mathbf{T}_k, \mathbf{T}_l)$ (eq. 1) are computed between pairs of path-length matrices $\mathbf{D}_k$ and $\mathbf{D}_l$ representing the corresponding trees (Hartigan, 1967), a minimum variance method can be used to define the optimal partition of the m trees of P into q clusters. For a predetermined number of clusters, the K-means algorithm (MacQueen, 1967) can be applied to obtain such a partition of trees. A consensus can then be computed from each subprofile of trees using the same average procedure as described above (see eq. 2). However, when the number of consensus trees is not predetermined, Ward's (1963) agglomerative clustering algorithm can be applied to obtain a hierarchical separation of the profile P into nested subprofiles of trees. Cutting the resulting dendrogram at any given level thus defines a partition of input trees into q groups, for q ranging from 1 (trivial cluster) to m (all clusters are singletons).

3 Determining the optimal number of consensus trees

Let us define Q^* as the sum of Q_t, where Q_t is the sum-of-squared distances from each consensus tree to its corresponding subprofile P_t. As the number of consensus trees q increases, Q^* will decrease to reach zero when $q = m$. A stopping rule must then be defined to determine the optimal number q of consensus trees to summarize a given profile of input trees. Two such rules are proposed for comparative purposes:

$$Q_1 = \frac{Q^*}{m - q} \tag{5}$$

and

$$Q_2 = \frac{\sqrt{Q^*}}{m - q} \tag{6}$$

where q is the number of subprofiles and m is the total number of input trees (with $q < m$) in the profile P. Simulations have shown that both of these functions have one minimum, for $q < (m - 1)$. Obviously, other stopping rules could be used to determine the number of consensus, as many ways to determine the number of subprofiles of P (see Milligan and Cooper, 1985).

It can easily be shown that the sum of the squared distances from the trees in each subprofile P_t to their respective centroids is equal to the sum (computed over q groups) of the mean squared within-group distances for every P_t (see Legendre and Legendre, 1998). In the specific case of average

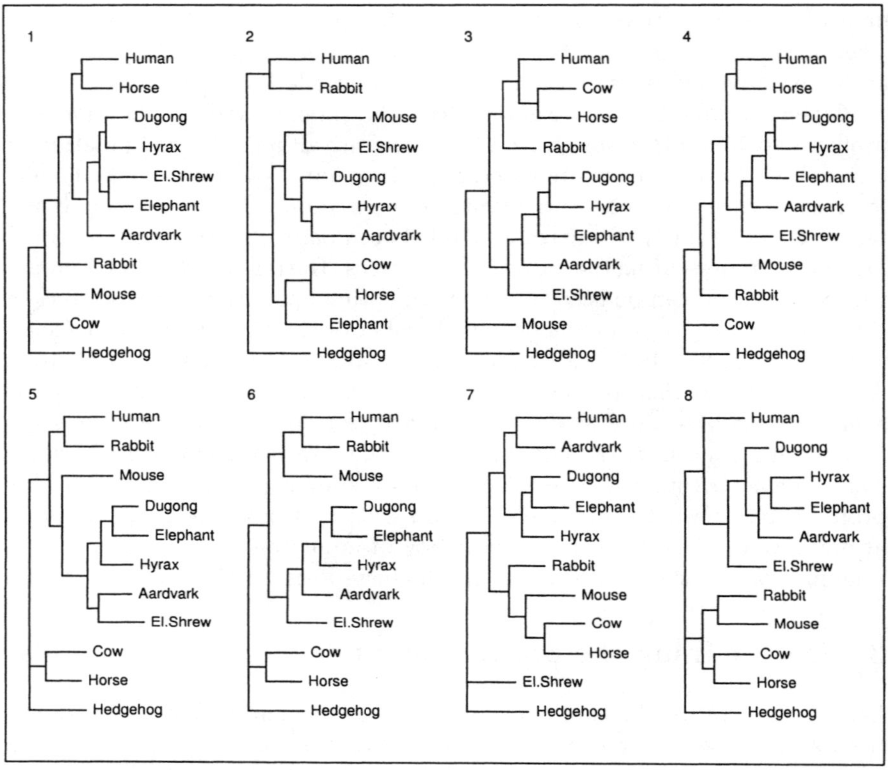

Fig. 1. A profile of eight phylogenies of mammals derived from different molecular sequences (data from Springer et al., 1999).

consensus trees, it is the distance to the centroid of each subprofile that is minimized (eq. 4), such that the optimal number of consensus can be approximated directly from the matrix **R** of pairwise distance among trees, without computing actual consensus trees.

4 Application to phylogenetic trees

To illustrate the procedure for determining the optimal number of consensus trees to represent a profile P, we have applied the approach described in this paper to eight phylogenies of mammal species, derived from independent molecular sequences (Springer et al., 1999). Figure 1 presents the input profile P of trees and the single consensus computed from all the trees in P is presented in Figure 2A. To obtain multiple consensus trees, the profile of trees was sequentially partitioned on the basis of a hierarchical clustering of trees, using Ward's (1967) algorithm applied to a matrix **R** of sum-of-squared

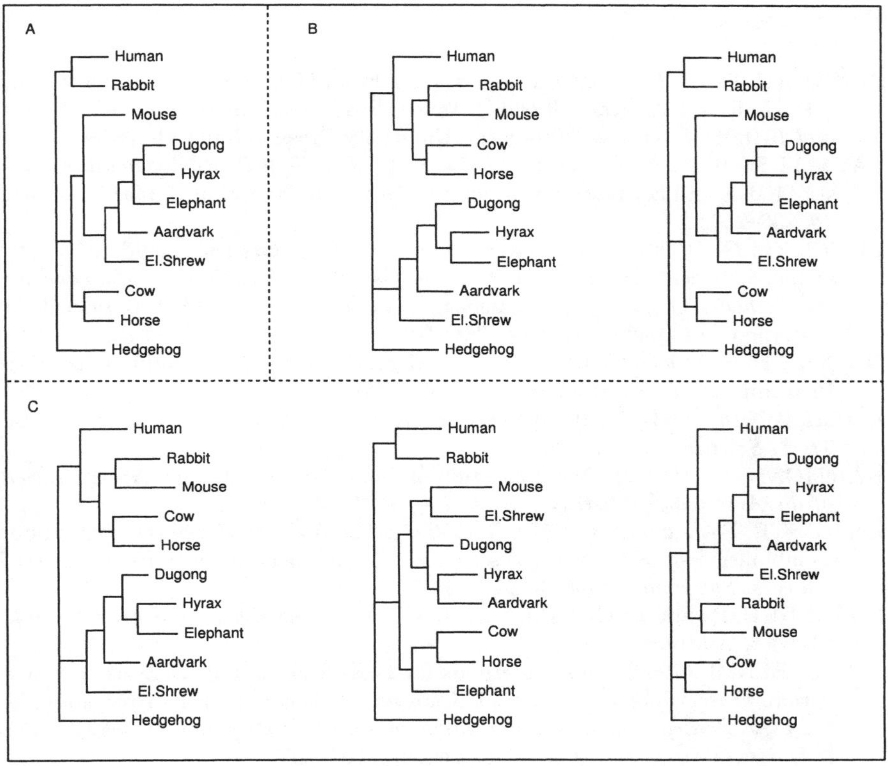

Fig. 2. Single and multiple consensus trees derived from the profile of phylogenies in Figure 1, with q=1 (A), q=2 (B), and q=3 (C). The corresponding values of the stopping-rule Q_2 are 0.051, 0.044, and 0.040 respectively.

distances (eq. 1) between all pairs of trees. The bipartition of P in two subprofiles separated the trees {6, 8} from {1, 2, 3, 4, 5, 7}. The consensus trees corresponding to this partition are presented in Figure 2B. The next partition separated the profile P into three subprofiles: {6, 8}, {2} and {1, 3, 4, 5, 7}. The corresponding consensus trees are depicted in Figure 2C. Using the Q_2 stopping rule, it was determined that these three consensus defined the optimal representation of the tree profile P (the same number of trees was obtained using Q_1). This multiple consensus solution illustrates that at least one of input phylogenies (tree 2 in Figure 1) was so different from the other trees that it must be considered separately in the consensus output.

Acknowledgments

We thank Mark Springer for the data set used in this study. This work was supported by NSERC grant no. OGP0155251 to F.-J. Lapointe.

References

BUNEMAN, P. (1971): The Recovery of Trees From Measures of Dissimilarity. In: F.R. Hudson, D.G. Kendall and P. Tautu (Eds.): *Mathematics in Archeological and Historical Sciences.* Edinburgh University Press, Edinburgh, 387–395.

CAVALLI-SFORZA, L. L., and EDWARDS, A. W. F. (1967): Phylogenetic Analysis: Models and Estimation Procedures. *American Journal of Human Genetics, 19*, 233–257.

CUCUMEL, G. (1990): Construction d'une hiérarchie consensus l'aide d'une ultramétrique centrale. In: *Recueil des Textes des Présentations du Colloque sur les Méthodes et Domaines d'Application de la Statistique 1990.* Bureau de la Statistique du Québec, Québec, 306–312.

DE SOETE, G. (1983): A Least Squares Algorithm for Fitting Additive Trees to Proximity Data. *Psychometrika, 48*, 621–626.

FITCH, W. M., and MARGOLIASH, E. (1967): Construction of Phylogenetic Trees. *Science, 155*, 279–284.

HARTIGAN, J.A. (1967): Representation of Similarity Matrices by Trees. *Journal of the American Statistical Association, 62*, 1140–1158.

LAPOINTE, F.-J. and CUCUMEL, G. (1997): The Average Consensus Procedure: Combination of Weighted Trees Containing Identical or Overlapping Sets of Objects. *Systematic Biology, 46*, 306–312.

LEGENDRE, P., and LEGENDRE, L. (1998): *Numerical Ecology, Second Edition.* Elsevier, Amsterdam.

MACQUEEN, J. (1967) : Some Methods for Classification and Analysis of Multivariate Observations. In: L. M. LeCam and J. Neyman (Eds.)*Proceedings of the Fifth Berkeley Symposium on Mathematical Statistics and Probability, Vol. 1.* University of Califo rnia Press, Berkeley, 281–297.

MAKARENKOV, V., and LECLERC, B. (1999): The Fitting of a Tree Metric According to a Weighted Least-squares Criterion. *Journal of Classification, 16*, 3–26.

MILLIGAN, G. W., and COOPER, M. C. (1985): An Examination of Procedures for Determining the Number of Clusters in a Data Set. *Psychometrika, 50*, 159–179.

SPRINGER, M. S., AMRINE, H. M., BURK, A., and STANHOPE, M. J. (1999): Additional Support for Afrotheria and Paenungalata, the Performance of Mitochondrial versus Nuclear Genes, and the Impact of Data Partitions with Heterogeneous Base Composition. *Systematic Biology, 48*, 65–75.

WARD, J. H. (1963): Hierarchical Grouping to Optimize an Objective Function. *Journal of the American Statistical Association, 58*, 236–244.

A Family of Average Consensus Methods for Weighted Trees

Claudine Levasseur and François-Joseph Lapointe

Département de sciences biologiques, Université de Montréal,
C.P. 6128, Succ. Centre-Ville, Montréal (Québec), Canada, H3C 3J7
(e-mail: Claudine.Levasseur@Umontreal.CA)
(e-mail: Francois-Joseph.Lapointe@Umontreal.CA)

Abstract. Consensus techniques represent useful tools in phylogenetic analysis, particularly for combining trees derived from different data sets. In the present paper, a family of average consensus methods for weighted trees is presented; the mean and median procedures are compared and applied to combine phylogenetic trees while taking into account their branch lengths. We also provide some recommendations about the use of these consensus techniques in relation to the more classical methods based on topological relationships alone.

1 Introduction

Consensus techniques are used in phylogenetic analysis to summarize the trees obtained from independent data sets, or combine multiples trees obtained from the same data (Bull et al., 1993). In spite of their popularity, consensus methods have been much criticized in the past (Barrett et al., 1991; de Queiroz, 1993). Because classical methods like the strict (Sokal and Rolhf, 1981) and majority rule (Margush and McMorris, 1981) consensus are based on topological relationships alone, unresolved trees are often produced by such techniques, a property considered as undesirable by some (see Kluge and Wolf, 1993). However, consensus methods for weighted trees that take into account branch lengths (Lapointe, 1998) can do better than most standard procedures, as they are more likely to produce fully resolved trees. In this paper, we describe two consensus methods for weighted trees based on different optimization criteria. These mean and median consensus procedures are used and compared to the more conservative strict consensus in an application involving phylogenetic trees.

2 A family of average consensus methods

Let $S = \{1, \ldots, i, \ldots, j, \ldots, n\}$ be a set of n objects and $P = \{\mathbf{T}_1, \ldots, \mathbf{T}_k, \ldots, \mathbf{T}_m\}$ a profile of m weighted trees defined on S. An average consensus method is defined as a function that takes as input the profile P and returns a consensus weighted tree $\mathbf{T}_c$ that is in some sense closest to P. Since there exists a one-to-one correspondence between any weighted tree $\mathbf{T}$ and its associated path-length matrix $\mathbf{D}$ (Hartigan, 1967; Buneman, 1971), it is equivalent to deal with the trees $\mathbf{T}_k$ of P or their corresponding matrices $\mathbf{D}_k$ to compute the consensus solution $\mathbf{T}_c$. Average consensus trees are thus obtained by applying a median or mean consensus function to the set $M = \{\mathbf{D}_1, \ldots, \mathbf{D}_k, \ldots, \mathbf{D}_m\}$ of path-length matrices associated with the trees of P.

2.1 The median consensus for weighted trees

The median procedure was introduced originally by Margush and McMorris (1981) to compute a consensus tree minimizing the sum of symmetric differences to the trees in P. This method applies to unweighted trees only. The same approach can be generalized, however, to combine weighted trees in a similar fashion. As such, the median consensus tree $\mathbf{T}_c$ is defined as the solution that minimizes the following average consensus function:

$$\sum_{k=1}^{m} \Delta(T_c, T_k) = \sum_{k=1}^{m}\sum_{i=1}^{n}\sum_{j=1}^{n} |d_c(i,j) - d_k(i,j)| \tag{1}$$

where the $d_c(i,j)$ are the path-length distances in the matrix $\mathbf{D}_c$ associated with the consensus tree $\mathbf{T}_c$ and the $d_k(i,j)$ are the path-length distances in the matrices $\mathbf{D}_k$ of M associated with the trees $\mathbf{T}_k$ of P.

Practically, the median consensus tree $\mathbf{T}_c$ is obtained in two simple steps. First, a median distance matrix $\mathbf{D}_{med}$ is computed from the path-length matrices of M. If the number of trees m is odd, the distances in $\mathbf{D}_{med}$ are given by the $(m+1)/2$ ordered $d(i,j)$ values in the m path-length distance matrices of M, for every pair of objects i and j. If m is even, the distances in $\mathbf{D}_{med}$ are computed as the mean of the $m/2$ and $(m/2)+1$ ordered $d(i,j)$ values in the m path-length distance matrices of M, for every pair of objects i and j. The closest path-length distance matrix from this median matrix $\mathbf{D}_{med}$ is then obtained by minimizing the following loss function using a minimum-absolute-deviation algorithm (e.g., Smith, 2001):

$$\sum_{i=1}^{n}\sum_{j=1}^{n} |d_c(i,j) - d_{med}(i,j)| \tag{2}$$

where the $d_c(i,j)$ are the fitted path-length distances in the matrix $\mathbf{D}_c$ associated with the consensus tree $\mathbf{T}_c$ and the $d_{med}(i,j)$ are the distances in the median matrix $\mathbf{D}_{med}$. The resulting tree $\mathbf{T}_c$ is the median consensus solution.

2.2 The mean consensus for weighted trees

The average consensus procedure was originally defined by Lapointe and Cucumel (1997) as a consensus function to combine weighted trees. In order to differentiate the use of the more general average consensus that also includes the median function, we will use the more specific mean consensus when referring to Lapointe and Cucumel's *average consensus* method.

The mean consensus tree $\mathbf{T}_c$ is thus defined as the solution that minimizes the following average consensus function:

$$\sum_{k=1}^{m} \Delta(T_c, T_k) = \sum_{k=1}^{m} \sum_{i=1}^{n} \sum_{j=1}^{n} [d_c(i,j) - d_k(i,j)]^2 \tag{3}$$

where the $d_c(i,j)$ are the path-length distances in the matrix $\mathbf{D}_c$ associated with the consensus tree $\mathbf{T}_c$ and the $d_k(i,j)$ are the path-length distances in the matrices $\mathbf{D}_k$ of M associated with the trees $\mathbf{T}_k$ of P.

Again, the mean consensus tree $\mathbf{T}_c$ can be computed in two simple steps. First, a mean distance matrix $\overline{\mathbf{D}}$ must be computed from the path-length distance matrices of M. The closest path-length distance matrix from this mean matrix $\overline{\mathbf{D}}$ is then obtained by minimizing the following loss function using a least-squares algorithm (e.g., Cavalli-Sforza and Edwards, 1967):

$$\sum_{i=1}^{n} \sum_{j=1}^{n} [d_c(i,j) - \bar{d}(i,j)]^2 \tag{4}$$

where the $d_c(i,j)$ are the fitted path-length distances in the matrix $\mathbf{D}_c$ associated with the consensus tree $\mathbf{T}_c$ and the $\bar{d}(i, j)$ are the distances in the mean matrix $\overline{\mathbf{D}}$. The resulting tree $\mathbf{T}_c$ is the mean consensus solution.

3 Application of the average consensus

To illustrate the use of the median and the mean consensus, we applied these procedures to combine three phylogenies of frog species derived from morphological data, mating calls and molecular sequences (data from Cannatella et al. 1998). The least-squares trees (Cavalli-Sforza and Edwards, 1967) computed from the different data sets are presented in Figure 1. It shows that phylogenies estimated from morphology (Fig. 1a) or calls (Fig. 1b) are topologically identical, whereas the tree computed from molecular data (Fig. 1c) is different. The conservative strict consensus tree (Fig. 1d) of those three phylogenies is not well resolved and contains a single species pair found in all of the input trees. On the other hand, the median (Fig. 1e) and mean (Fig. 1f) consensus trees that take into account branch lengths are much more resolved. In this particular case, the median consensus tree is the same as the phylogeny derived from the calls data (Fig. 1b). The mean consensus tree, however, differs from all input trees and contains a combination of species groups not found in the original phylogenies.

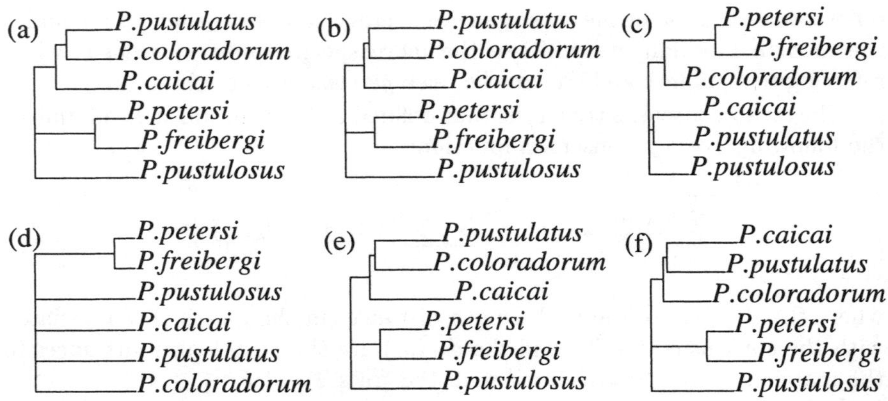

Fig. 1. Tress derived from the three different data sets and their different consensus trees: (a) morphological data, (b) mating calls, (c) molecular data, (d) strict consensus, (e) median consensus, (f) mean consensus.

4 Discussion

The median and mean consensus procedures are designed to combine weighted trees while taking into account their branch lengths. These average consensus methods are based on different optimization criteria, however, and they may not always produce identical results. For one, the mean consensus function minimizes the L_2 norm and is very likely to be affected by extreme values. On the other hand, the median function minimizes the L_1 norm; this consensus method should be more appropriate in cases involving outliers. Our application illustrates the differences between the two techniques. In this particular example, the molecular tree (Fig. 1c) clearly influenced the mean consensus since it represented a very different phylogeny compared to the other two trees (Fig. 1a,b). Both methods will produce comparable solutions when the tree profile P encompasses a homogeneous distribution of phylogenies, or when all trees are very similar (or identical in topology). Regardless of the average consensus function selected, these methods will usually produce fully resolved trees, contrary to consensus techniques that ignore branch lengths. Such average consensus techniques thus offer interesting alternatives to the classical consensus methods based on topological relationships alone.

Acknowledgments

We would like to thank Buck McMorris and Fred Roberts for their helpful suggestions about our work, and Guy Cucumel and Olivier Gauthier for their assistance with the LaTeX format. This work was made possible by a NSERC scholarship to C. Levasseur and by a NSERC grant no. OGP0155251 to F.-J. Lapointe.

References

BARRETT, M., DONOGHUE, M. J., and SOBER, E. (1991): Against Consensus. *Systematic Zoology, 40*, 486–493.

BULL, J. J., HUELSENBECK, J. P., CUNNINGHAM, C. W., SWOFFORD, D. L., and WADDELL, P. J. (1993): Partitioning and Combining Data in Phylogenetic Analysis. *Systematic Biology, 42*, 384–397.

BUNEMAN, P. (1971): The Recovery of Trees From Measures of Dissimilarity. In: F.R. Hudson, D.G. Kendall and P. Tautu (Eds.): *Mathematics in Archeological and Historical Sciences.* Edinburgh University Press, Edinburgh, 387–395.

CANNATELLA, D. C., HILLIS, D. M., CHIPPINDALE, P. T., WEIGT, L., RAND, A. S., and RYAN, M. J. (1998): Phylogeny of Frogs of the *Physalaemus pustulosus* Species Group, with an Examination of Data Incongruence. *Systematic Biology, 47*, 311–335.

CAVALLI-SFORZA, L. L., and EDWARDS, A. W. F. (1967): Phylogenetic Analysis: Models and Estimation Procedures. *Evolution, 32*, 550–570

DE QUEIROZ, A. (1993): For Consensus (Sometimes). *Systematic Biology, 42*, 368–372

HARTIGAN, J.A. (1967): Representation of Similarity Matrices by Trees. *Journal of the American Statistical Association, 62*, 1140–1158.

KLUGE, A.G., and WOLF, A.J. (1993): Cladistics: Whats in a Word? *Cladistics, 9*, 183–199.

LAPOINTE, F.-J. (1998): For Consensus (with branch lengths). In: A. Rizzi, M. Vichi, and H.-H. Bock (Eds.): *Advances in Data Science and Classification.* Springer-Verlag, Berlin, 73–80.

LAPOINTE, F.-J. and CUCUMEL, G. (1997): The Average Consensus Procedure: Combination of Weighted Trees Containing Identical or Overlapping Sets of Objects. *Systematic Biology, 46*, 306–312.

MARGUSH, T., and MCMORRIS, F. R. (1981): Consensus n-trees. *Bulletin of Mathematical Biology, 43*, 239–244.

SMITH, T. J. (2001): Constructing Ultrametric and Additive Trees Based on the L_1 Norm. *Journal of Classification, 18*, 185–207.

SOKAL, R. R., and ROHLF, F. J. (1981): Taxonomic Congruence in the Leptopodomorpha re-examined. *Systematic Zoology, 30*, 309–325.

Comparison of Four Methods for Inferring Additive Trees from Incomplete Dissimilarity Matrices

Vladimir Makarenkov

Département d'informatique, Université du Québec à Montréal, C.P. 8888, Succ. Centre-Ville, Montréal (Québec), Canada, H3C 3P8.
Institute of Control Sciences, 65 Profsoyuznaya, Moscow 117806, Russia.
(e-mail: makarenkov.vladimir@uqam.ca)

Abstract. The problem of inference of an additive tree from an incomplete dissimilarity matrix is known to be very delicate. As a solution to this problem, it has been suggested either to estimate the missing entries of a given partial dissimilarity matrix prior to tree reconstruction (De Soete, 1984 and Landry et al., 1997) or directly reconstruct an additive tree from incomplete data (Makarenkov and Leclerc, 1999 and Guénoche and Leclerc, 2001). In this paper, I propose a new method, that is based on the least-squares approximation, for inferring additive trees from partial dissimilarity matrices. The capacity of the new method to recover a true tree structure will be compared to those of three well-known techniques for tree reconstruction from partial data. The new method will be proven to work better than widely used Ultrametric and Additive reconstruction techniques, as well as the recently proposed Triangle method on incomplete dissimilarity matrices of different sizes and under different noise conditions.

1 Introduction

Incomplete dissimilarity data can arise in a variety of practical situations. For example, this is often the case in molecular biology, and more precisely in phylogenetics, where an additive or a phylogenetic tree represents an intuitive model of species evolution. The presence of missing data in a distance or dissimilarity matrix among species or taxa can be due to the lack of biological material, imprecision of employed experimental methods, or to a combination of unpredictable factors. Unfortunately, the vast majority of the widely used additive tree fitting techniques, as for example the Neighbor-Joining (Saitou and Nei, 1987), Fitch (Felsenstein, 1997), or BioNJ (Gascuel, 1997) algorithms, cannot be launched unless a complete dissimilarity matrix is available. To solve this challenging problem, some methods have been recently proposed.

There exist in the literature two types of methods, using either *indirect* or *direct* estimation of missing values, for inferring additive trees from incomplete dissimilarity matrices. The first type of methods, or indirect estimation, relies on the assessing missing cells prior to phylogenetic reconstruction using the properties of path-length matrices representing trees. An additive tree can then be inferred from a complete dissimilarity matrix by means of any available tree-fitting algorithm. The second type of methods handling missing values, or direct estimation, consists of reconstructing a tree directly from an incomplete dissimilarity matrix by using a particular tree-building procedure. As far as the direct estimation techniques are concerned, I have to mention the work by De Soete (1984) and Landry et al. (1996), who showed how to infer additive trees from partial data using either the *ultrametric inequality*:

$$d(i,j) \leq \max\{d(i,k); d(j,k)\}, \text{for any } i, j \text{and } k, \tag{1}$$

or the *four-point condition* (Buneman 1971):

$$d(i,j) + d(k,l) \leq \max\{d(i,k) + d(j,l); d(i,l) + d(j,k)\}, \text{for any } i, j, k \text{and } l \tag{2}$$

Using the properties of the ultrametric inequality and the four-point condition, one can fill out incomplete matrices; the missing cells can actually be estimated through the combinations of the available ones. As to the direct reconstruction, two tree-building algorithms allowing for missing cells in dissimilarity matrices have been recently proposed by different authors; the Triangle method of Guénoche and Leclerc (2001), see also Guénoche and Grandcolas (1999), relies on a constructive approach, whereas the MW procedure of Makarenkov and Leclerc (1999) is based on a least-squares approximation.

This paper aims first at introducing a new original method for direct reconstruction of additive trees from partial matrices. The second goal consists of proving the efficiency of the proposed method by comparing it to the Ultrametric and Additive indirect procedures, as well as to the Triangle direct reconstruction method. In order to compare the new method to the three above-mentioned existing approaches, Monte Carlo simulations were conducted with dissimilarity matrices of different sizes and with different percentages of missing cells. The performances of the four methods were assessed in terms of both metric and topological recovery. The conducted simulations clearly showed that the new method regularly provided better estimates of the path-length distances between tree leaves, as well as a better recovery of the correct tree topology than the three other competing strategies.

2 Brief description of the new method

The new method for reconstructing trees from partial matrices introduced in this article was inspired by the Method of Weights (MW) proposed in Makarenkov and Leclerc (1999). The latter method used a stepwise addition procedure to infer an additive tree from a complete dissimilarity matrix. The approximation procedure used in the MW was based on a weighted least-squares model. The new method, called *MW-modified*, is an extension of the

MW approach to partial matrices. The first attempt to use the MW method for treatment of partial matrices was made in Levasseur et al.(2000), where the MW procedure was compared to the Triangle method. However, this first attempt to employ the least squares for tree reconstruction from partial matrices showed that the direct MW procedure had to be adjusted to the treatment of missing data.

Let **D** be a given dissimilarity matrix on the set X of n taxa. Let us suppose that some entries of **D** are missing. The least-squares criterion consists in minimising the following function:

$$Q = \sum_{i<j} (d(i,j) - \delta(i,j))^2, \tag{3}$$

where $\delta(i,j)$ is an obtained estimate for an existing entry $d(i,j)$ of **D**. The function δ is a tree metric, which is associated with an additive tree; δ verifies the four-point condition.

The following approach was adopted in the MW-modified procedure to build an additive tree from a partial dissimilarity matrix **D**:

Step 1. The taxa i and j are chosen, such that $d(i,j)$ is a present entry of **D**. The tree T_2 will comprise the only edge ij of length $d(i,j)$.

Step p, (p < n). Let T_p be an additive tree with p leaves constructed at the previous steps. The leaves of T_p are associated with p taxa from X. Among the $n-p$ taxa from X that are not represented by the leaves of T_p, we have to find one to be placed in the growing tree. We propose to place in T_p, the taxon $p+1$ such that it provides the maximum number of existing dissimilarity entries of type $d(p+1,l)$, where l is a taxon of X already placed in T_p. The exact location of the new leaf $p+1$ in T_{p+1} and the lengths of the three new edges appearing after addition of a new leaf will be determined with respect to the MW procedure (see Makarenkov and Leclerc 1999).

The time complexity of such a new procedure is $O(n^3)$ for a dissimilarity matrix **D** of size (nxn). In order to improve the quality of fit in the simulation study described below, I carried out this procedure k times for each dissimilarity matrix, where k was a number of possible existing pairs of taxa i and j to be selected at the first step. Such an exhaustive strategy increases the algorithmic time complexity up to $O(n^5)$, but often enables a substantial improvement in fit.

3 Simulation design

I carried out a series of Monte Carlo simulations to compare the performances of the four competing methods for tree reconstruction from incomplete dissimilarities. Each data set was obtained as follows: first, an unrooted binary tree topology with n leaves and $2n-3$ edges was randomly generated. For each such tree topology, the length of each edge was then selected at random from a uniform distribution on the real interval [0,1], leading to an additive tree T. The corresponding tree metric tt was computed and normalized to have a unit variance. Four normally distributed random noises with mean zero and, respectively, variances $\sigma^2 = 0.0$ (noise-free condition), 0.1, 0.25, and 0.5, were added to the values of the normalized tree metric tt to obtain variants of the dissimilarity d. Then, 0 to 50 percent of entries were removed from d to obtain a partial dissimilarity used as input for the four tree-reconstructing methods compared in this study. Thus, all considered partial dissimilarity matrices were simulated according to the "tree metric + noise - missing entries" model. For each combination of values $(n, \sigma^2, miss)$, where n (matrix size) = 8, 16, and 24, and mis (percentage of missing values) = 0, 10, 20, 30, 40, and 50, I generated 100 different data sets. However, only the results obtained for $n = 16$ are illustrated in Figures 1 and 2 below.

The metric and a topological recovery provided by each of the four tree-building methods were assessed using the two following quantities:

1. The *proportion of variance accounted for* as expressed in the following formula:

$$\%Var = 100\% \times (1 - \frac{\sum_{i<j}(d(i,j) - \delta(i,j))^2}{\sum_{i<j}(d(i,j) - m(d))^2}), \qquad (4)$$

where $m(d)$ is the mean value of the initial dissimilarity d, and δ is the obtained tree metric.

2. The *Robinson and Foulds topological distance* (Robinson and Foulds 1981) between the true tree T and the solution tree corresponding to the obtained tree metric d was also considered. This important criterion of tree similarity is equal to the minimum number of elementary operations, consisting of merging or splitting vertices, necessary to transform one tree into the other.

In this Monte Carlo study, I used the additive tree generating strategy that was also employed in Makarenkov and Leclerc (1999), and Makarenkov and Legendre (2001). Another way of simulating data for the additive model was suggested by Lausen and Degens (1988).

4 Conclusion

In this paper, the performances of four different tree-building methods for incomplete dissimilarity matrices were compared. I carried out a Monte Carlo study to determine which method provides better metric and topological recoveries under different noise conditions and for different percentages of missing values. When analyzing curves illustrated in Figures 1 and 2, one can notice that the proposed MW-modified procedure generally achieves better results in terms of both metric and topological recovery than the three other methods. The new method is particularly good in case of complete absence of noise (noise = 0, in Figures 1 and 2). A number of interesting trends can

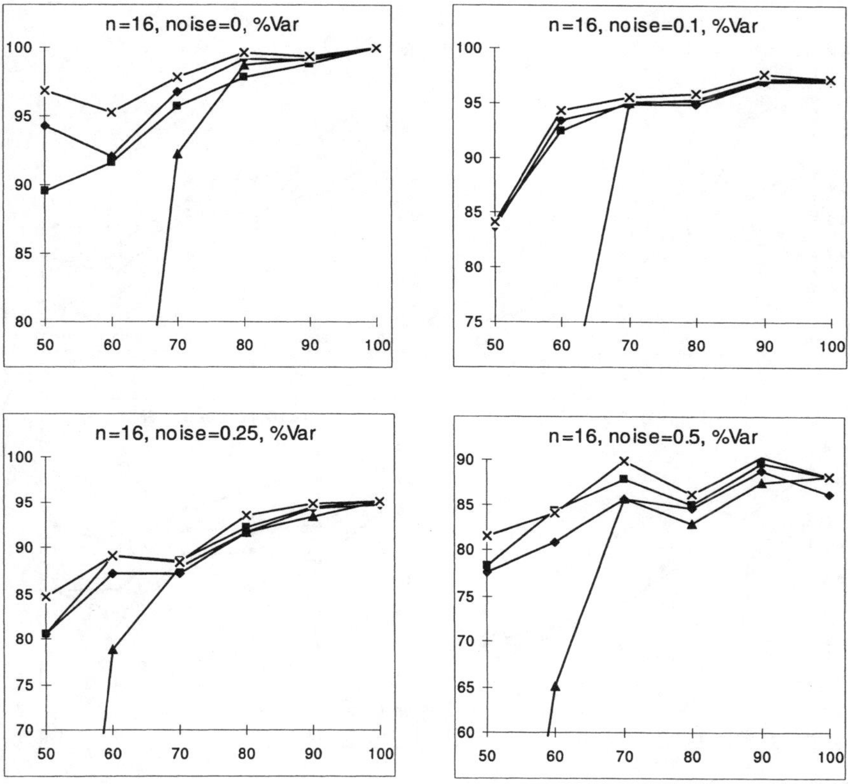

Fig. 1: Metric recovery values obtained for different percentages of missing entries and under four different noise conditions. The four competing methods [Triangle ♦, Ultrametric ■, Additive ▲, and MW-modified X] were tested on dissimilarity matrices of size (16x16). The abscissa axis represents the percentage of existing entries in a given dissimilarity matrix; the ordinate axis represents the percentage of variance of a given dissimilarity accounted for by an obtained tree metric. For each case, the mean values (over 100 simulated data sets) of the percentage of variance are given. Larger values of the percentage of variance accounted for point out a better recovery achieved by a tree reconstruction method.

be observed when examining curves behaviour in Figure 1 and 2. As to the percentage of variance: the Additive procedure becomes very unstable when more than 30 percent of entries are missing in a given dissimilarity matrix; the results provided by the Triangle method and the Ultrametric procedure are very close; the MW-modified procedure slightly outperforms both latter methods in the most of the situations. As to the normalized Robinson and

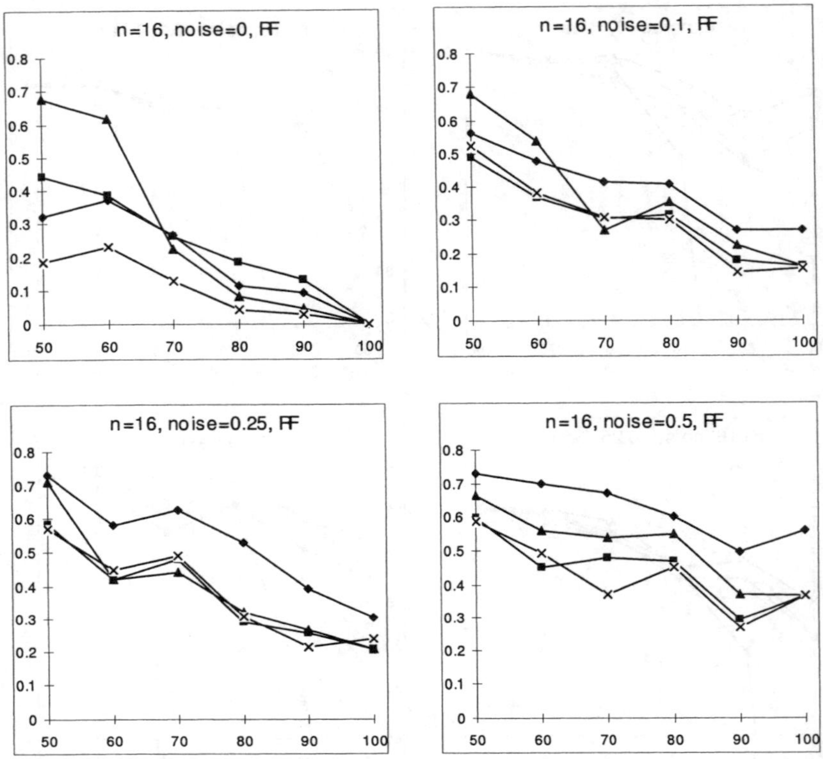

Fig. 2: Topological recovery values obtained for different percentages of missing entries and under four different noise conditions. The four competing methods [Triangle ♦, Ultrametric ■, Additive ▲, and MW-modified X] were tested on dissimilarity matrices of size (16x16). The abscissa axis represents the percentage of existing entries in a given dissimilarity matrix; the ordinate axis represents the Robinson and Foulds (RF) topological distance between a given true additive tree and an obtained tree. The computed RF distance was normalized by its largest value, which is 2n-6 for two binary additive trees with n leaves. For each case, the mean values (over 100 data sets) of the topological distance are given. Lower values of the RF distance point out a better recovery achieved by a tree reconstruction method.

Foulds topological distance: the Triangle method does not resist well to the increasing of noise; the best results are regularly obtained by the MW-modified method or the Ultrametric procedure; similarly to the percentage of variance, the Additive procedure can not cope well with big percentage of missing cells in a given dissimilarity matrix. Surprisingly, the Ultrametric procedure which does not provide good results for noise-free data, the same conclusion was also made by Landry at al. (1996) and Guénoche and Grandcolas (1999), becomes much more competitive when the noise increases. Similar trends were observed for additive trees with 8 and 24 leaves, for which the detailed results were not presented in this paper. Ultimately, I would recommend using the MW-modified method for treatment of data that are free of noise and the MW-modified method or the Ultrametric procedure for processing "noisy" data.

5 Software

The four methods compared in the framework of the present study were implemented in T-Rex (*tree and reticulogram reconstruction*) package intended for reconstructing additive trees and reticulation networks (Makarenkov, 2001). This computer application also includes a number of well-known tree fitting methods, as well as some methods for modelling reticulation networks between considered species or taxa. A tree structure obtained by means of one of these methods can be visualized using Hierarchical, Radial, or Axial drawing and then manipulated interactively. The Windows and Macintosh versions of this software were made freely available for researchers at the T-Rex web site at http://www.fas.umontreal.ca/biol/casgrain/en/labo/t-rex.

Acknowledgement

The author is grateful to Alain Guénoche and François-Joseph Lapointe for providing me with the source code of the Triangle, Ultrametric, and Additive methods.

References

BUNEMAN, P. (1971): The Recovery of Trees From Measures of Dissimilarity. In: F.R. Hudson, D.G. Kendall and P. Tautu (Eds.): *Mathematics in Archeological and Historical Sciences.* Edinburgh University Press, Edinburgh, 387–395.

DE SOETE, G. (1983): Additive-tree representations of incomplete dissimilarity data. *Quality and Quantity, 18*, 387–393.

FELSENSTEIN, J. (1997): An alternating least-squares approach to inferring phylogenies from pairwise distances. *Systematic Zoology, 46*, 101–111.

GASCUEL, O. (1997): BIONJ: an improved version of the NJ algorithm based on a simple model of sequence data. *Molecular Biology and Evolution, 14*, 685–695.

GUÉNOCHE, A. and LECLERC, B. (2001): The triangle method to build X-trees from incomplete distance matrices. *RAIRO Operations Research, 35*, 283–300.

GUÉNOCHE, A. and GRANDCOLAS, S. (1999): Approximation par arbre d'une distance partielle. *Mathématiques,Informatique et Sciences Humaines, 146*, 51–64.

LANDRY, P. A., LAPOINTE, F.-J., and KIRSCH, J. A. W. (1996): Estimating phylogenies from distance matrices: additive is superior to ultrametric estimation. *Molecular Biology and Evolution, 13*, 818–823.

LAUSEN, B. and DEGENS, P. O. (1988): Evaluation of the reconstruction of phylogenies with DNA-DNA hybridization data. In Bock, H. H. (ed.), *Classification and related methods of data analysis*, North Holland, Amsterdam, 367–374.

LEVASSEUR, C., LANDRY, P. A., and LAPOINTE, F.-J. (2000): Estimating trees from incomplete distance matrices: a comparison of two methods. In Kiers, H. A. L., Rasson, J.-P., Groenen, P. J. F. and Schader, M. (eds.), *Data analysis, Classification and Related Methods*, Springer, 149–154.

MAKARENKOV, V. and LECLERC, B. (1999): An Algorithm for the Fitting of a Tree Metric According to a Weighted Least-squares Criterion. *Journal of Classification*, *16*, 3–26.

MAKARENKOV, V. and LEGENDRE, P. (2001): Optimal Variable Weighting for Ultrametric and Additive Trees and K-means Partitioning: Methods and Software. *Journal of Classification*, *18*, 245–271.

MAKARENKOV, V. (2001): T-Rex: reconstructing and visualizing phylogenetic trees and reticulation networks, *Bioinformatics*, *17*, 664–668.

ROBINSON, D.R. and FOULDS, L.R. (1981): Comparison of phylogenetic trees. *Mathematical Biosciences*, *53*, 131–147.

SAITOU, N. and NEI, M. (1987): The neighbor-joining method: a new method for reconstructing phylogenetic trees. *Molecular Biology and Evolution*, *4*, 406–425.

Quartet Trees as a Tool to Reconstruct Large Trees from Sequences

Heiko A. Schmidt[1] and Arndt von Haeseler[2]

[1] Max-Planck-Institut für molekulare Genetik,
Ihnestr. 73, 14195 Berlin, Germany
hschmidt@molgen.mpg.de

[2] Max-Planck-Institut für evolutionäre Anthropologie,
Inselstr. 22, 04103 Leipzig, Germany
haeseler@eva.mpg.de

Abstract. We suggest a method to reconstruct phylogenetic trees from a large set of aligned sequences. Our approach is based upon a modified version of the quartet puzzling algorithm and does not require the computation of all $\binom{n}{4}$ quartets, if n sequences are aligned. We show that the phylogenetic resolution of the resulting tree depends on the number of quartets one is willing to analyze. As an biological example we study the alignment of 215 red algae sequences obtained from the European ssu rRNA Database. Finally, we discuss several modifications and extensions of the algorithm.

1 Introduction

The development of efficient sequencing techniques has led to a large amount of molecular biology data which need to be elucidated. One application especially DNA sequences have been put to is the analysis of evolutionary relationships among a collection of species. The field of molecular tree reconstruction is one of the fascinating areas of evolutionary biology and a variety of methods exist to infer phylogenies from a set of aligned sequences (Swofford et al. (1996)).

While most tree reconstruction methods aim to reconstruct the overall tree for n sequences by optimizing a global objective function, there have been attempts to reduce the problem to the computation of four-species trees (Sattath and Tversky (1977), Fitch (1981), Dress et al. (1986), Bandelt and Dress (1986)) and to build the overall tree from the collection of four-species trees. Four-species trees form the building blocks of larger trees and several criteria are available to test if a collection of four-species trees can be represented in one tree. However, for realistic data these criteria are almost never fulfilled and thus heuristics to assemble large trees have been developed.

Here we suggest an approach to reconstruct a sequence based tree from a large set of aligned sequences. Our method applies the quartet puzzling algorithm (Strimmer and von Haeseler (1996)) which uses the collection of four-species trees inferred from a sequence alignment. Since this method requires

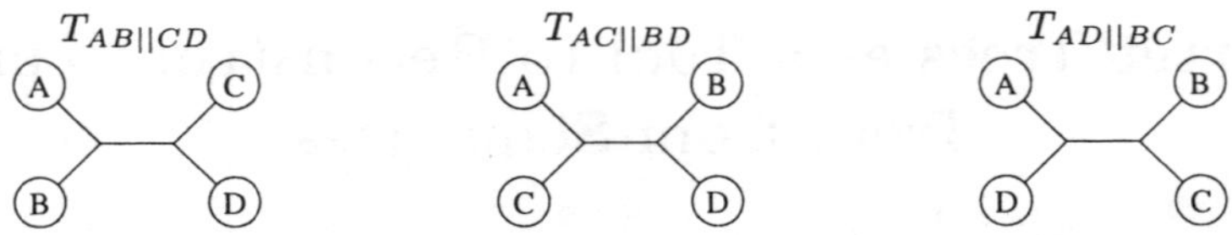

Fig. 1. The three different informative tree-topologies for the quartet $q = (ABCD)$.

the computation of gene trees for all subsets of size four, the reconstruction of species trees is of polynomial complexity. However, for datasets of 200 sequences or more the computation time required to evaluate all quartets is very large. For example the evaluation of all quartet topologies for the red algae dataset assuming rate heterogeneity would take more than 6500 hours on a SUN Enterprise 450 server. Hence we suggest a strategy that does not require the computation of all quartet trees. In order to ensure that our algorithm works, a minimal amount of pair-wise overlap, i.e., number of sequences shared by the different subsets of the alignment is required. Once the quartet trees are computed, we proceed with a modified version of the original quartet puzzling algorithm (Strimmer and von Haeseler (1996)). Finally, we compute the overall tree from a series of so-called intermediate trees, which are used to construct a majority consensus tree (Margush and McMorris (1981)). We show, that the resolution of the consensus tree critically depends on the number of quartets actually evaluated.

The outline of the paper is as follows, in section 2 we briefly summarize the relevant notation, then we explain the quartet puzzling algorithm and the adapted version for large datasets. The subsequent section gives some insights about the correlation between the resolution of the tree and the amount of computing time one is willing to spend. The final discussion section gives an outlook on further modification of the proposed method.

2 Notation

Let $\mathcal{S} = \{S_1, \ldots, S_n\}$ be the collection of n aligned orthologous sequences from different species. We denote $\left\{ {\mathcal{S} \atop 4} \right\}$ as the set, that contains all subsets of size four (quartets) of $\mathcal{S}$. The cardinality of $\left\{ {\mathcal{S} \atop 4} \right\}$ is then

$$\left| \left\{ {\mathcal{S} \atop 4} \right\} \right| = \binom{|\mathcal{S}|}{4} = \frac{n!}{(n-4)!\,4!}. \tag{1}$$

For each quartet $q = (ABCD) \in \left\{ {\mathcal{S} \atop 4} \right\}$ three informative tree-topologies, namely $T_{AB||CD}$, $T_{AC||BD}$, $T_{AD||BC}$, are possible (see Figure 1), where $A, B, C, D \in \mathcal{S}$ are the labeled leaves of the trees.

The $||$ denotes the so-called neighbor relation. The properties of $||$ and its application in tree reconstruction problems have been intensively studied and it is well known, that an n-species tree can be uniquely reconstructed from its list of neighbor relations (Bandelt and Dress (1986)).

The selection of the best quartet tree-topology is guided by the phylogenetic information provided by the sequences and the reconstruction method, i.e., maximum parsimony (Fitch (1971)), neighbor-joining (Saitou and Nei (1987)), or maximum likelihood (Felsenstein (1981)). These methods aim to optimize an objective function and the quality of the best tree can be compared to the quality of the remaining two trees. For a set of aligned sequences $\mathcal{S}$ the reconstructed quartet trees can then be used to reconstruct the overall tree $T(\mathcal{S})$ relating the n species. Suitable methods are the TREE-PUZZLE package (Schmidt et al. (in press), Strimmer and von Haeseler (1996)) or any other quartet based method (Bandelt and Dress (1986), Sattath and Tversky (1977), Fitch (1981), Dress et al. (1986)).

Especially the maximum-likelihood approach (Felsenstein (1981)) is very useful in this application, because it allows the inclusion of the specific model under which the sequences are evolving. Thus, we obtain three log-likelihoods, one for each tree topology

$$l_{AB||CD}, l_{AC||BD}, l_{AD||BC}. \tag{2}$$

This computation can be repeated for the $\binom{|\mathcal{S}|}{4}$ possible quartets. This list $\mathcal{L}(\mathcal{S})$ of quartet trees can then be used as input in the original quartet puzzling algorithm (Strimmer and von Haeseler (1996)) or the modified version that also considers the second best or even third best tree as a potential candidate (Strimmer et al. (1997)).

3 Decomposing large data sets

Central to the algorithm to construct large trees from quartets is the quartet puzzling method implemented in PUZZLE (Strimmer and von Haeseler (1996)) and in its parallelized version (Schmidt et al. (in press)).

3.1 Quartet Puzzling

Consider an arbitrary ordering of the n sequences, i.e., $A, B, C, D, E, \ldots$ and let us assume that the neighbor relation $AB||CD$ maximizes the log-likelihoods in equation 2. Then sequence E is added to the already reconstructed subtree $T_{AB||CD}$ according to the following algorithm (see Figure 2): For each quartet of the type (i, j, k, E) where $i, j, k \in \{A, B, C, D\}$, the optimal tree topology is evaluated. The resulting neighbor relations are then used to decide where to place sequence E on the four-species tree. Let us assume that $AE||BC$ is optimal, then we do not want to place sequence E on the path connecting sequences B and C in the subtree, since this would

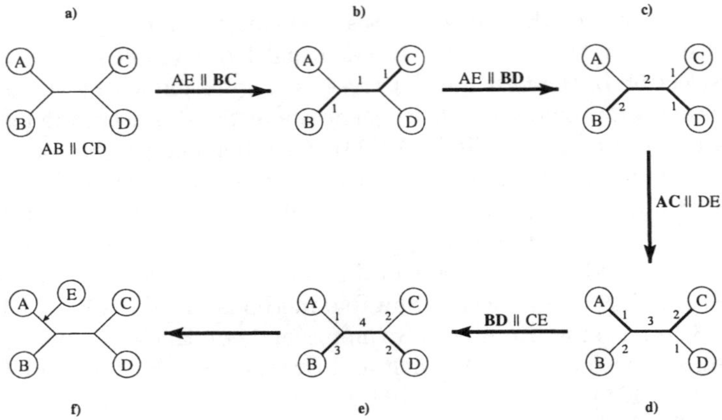

Fig. 2. Evaluation of the most probable branch to insert by penalizing. In this example $AE||BC$, $AE||BD$, $AC||DE$, and $BD||CE$ are the neighbor relations found in the previous step evaluating the quartets.

destroy the neighbor relation of BC induced by $AE||BC$, therefore a penalty value of one is added to each edge on the path connecting B and C (see Figure 2b). Then the next quartet is evaluated (see Figure 2c), until all quartets of the type (i, j, k, E) are considered. E is placed on the edge with minimum penalty (see Figure 2e and f). If more than one edge has minimal penalty, then the sequence is inserted randomly at one of them. The process of adding a single sequence to a subtree is repeated until an overall tree of n sequences is obtained. Please note, that an edge with penalty zero indicates, that all neighbor relations of the sequence to be placed and the sequences already present in the subtree are fulfilled. If this is true for all sequences, the data is perfectly tree-like. For real data, this is almost never the case. Sometimes even different input orders may result in different overall tree topologies. Therefore, the quartet puzzling step is repeated several times (Strimmer and von Haeseler (1996)) and a majority consensus tree is computed (Margush and McMorris (1981)) from the collection of so-called intermediate trees.

3.2 The modified quartet puzzling algorithm

The quartet puzzling algorithm requires the computation of $\binom{n}{4}$ quartets, which is not feasible if n is large. Here, we suggest a modified version of the algorithm by computing only quartets for a collection of subsets

$\mathcal{S}_1, \mathcal{S}_2, \mathcal{S}_3, \ldots, \mathcal{S}_k$ of $\mathcal{S}$. Moreover, we require that

$$\bigcup_{i=1}^{k} \mathcal{S}_i = \mathcal{S}, \tag{3}$$

$$|\mathcal{S}_i| \geq 4, \text{ for } i = 1, \ldots, k \tag{4}$$

$$\left| \left\{ \bigcup_{j=1}^{i} \mathcal{S}_j \right\} \cap \mathcal{S}_{i+1} \right| \geq 3, \text{ for } i = 1, \ldots, k-1. \tag{5}$$

We call equation 5 the overlap condition. It ensures that we can assemble an n-species tree from the quartet trees. Note that according to the overlap condition we need to compute only

$$\sum_{i=1}^{k} \binom{|\mathcal{S}_i|}{4} \tag{6}$$

quartets.

Before discussing some computational aspects, we will explain the algorithm for $k = 2$. If $k > 2$, then one has to apply the algorithm $k - 1$ times by setting

$$\mathcal{S}_1 = \bigcup_{i=1}^{j} \mathcal{S}_j \tag{7}$$

$$\mathcal{S}_2 = \mathcal{S}_{j+1} \tag{8}$$

for $j = 1, \ldots, k-1$. The order of the subsets is chosen randomly, fulfilling the overlap condition.

The algorithm ModPUZZLE (cf. Figure 3)

step 1: Assume that an intermediate tree $T(\mathcal{S}_1)$ is reconstructed using the quartet puzzling method.

step 2: Compute

$$\mathcal{S}_{1\cap 2} = \mathcal{S}_1 \cap \mathcal{S}_2.$$

The elements of $\mathcal{S}_{1\cap 2}$ are already in $T(\mathcal{S}_1)$.

step 3: Derive the subtree $T(\mathcal{S}_{1\cap 2})$ from $T(\mathcal{S}_1)$, that has as leaf-set the sequences in $\mathcal{S}_{1\cap 2}$ and as internal node set all nodes from $T(\mathcal{S}_1)$ that are on a path connecting two leaves from $\mathcal{S}_{1\cap 2}$. Each node of degree 2 in $T(\mathcal{S}_{1\cap 2})$ is then a root of a subtree from $T(\mathcal{S}_1)$ with leaves in $\mathcal{S}_1$ but not $\mathcal{S}_2$. Nodes of degree 2 are called root-nodes.

step 4: Pick a sequence $X \in \mathcal{S}_2 \backslash \mathcal{S}_1$ and compute the penalties of each edge in $T(\mathcal{S}_{1\cap 2})$ using the original quartet puzzling algorithm (see 3.1).

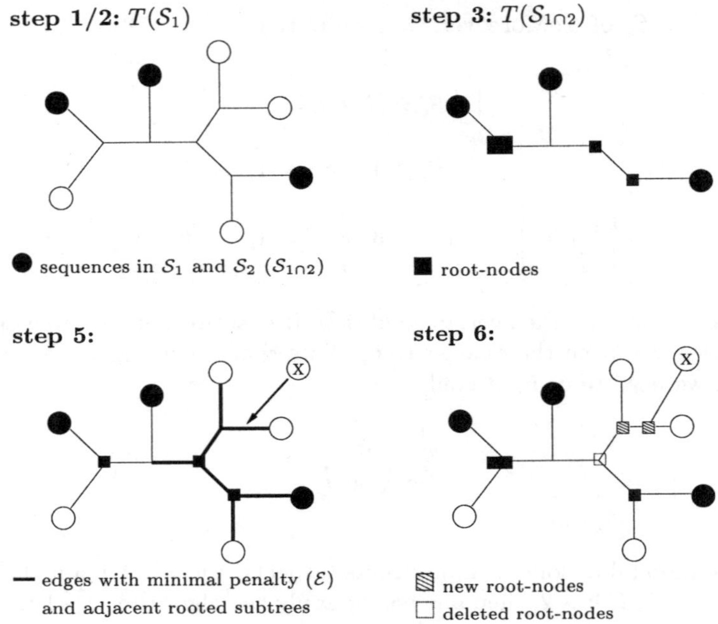

Fig. 3. The modified quartet puzzling algorithm. For details see text.

step 5: Let $\mathcal{E}$ be the collection of all edges in $T(\mathcal{S}_{1\cap 2})$ with minimal penalty, then insert X randomly either on one edge $e \in \mathcal{E}$ or on an edge from a rooted subtree of $T(\mathcal{S}_1)$ whose root node has at least two edges in $T(\mathcal{S}_{1\cap 2})$ with minimal penalty.

step 6: If X is the inserted sequence, then update $T(\mathcal{S}_{1\cap 2})$ by adding X to the leaf-set and by adding all nodes of $T(\mathcal{S}_1)$ that lie on the path between X and the root-node R of the subtree of $T(\mathcal{S}_1)$, where X was inserted, to the set of root nodes. Finally, delete R from the root-node set.

step 7: Delete X from $\mathcal{S}_2 \backslash \mathcal{S}_1$. If $\mathcal{S}_2 \backslash \mathcal{S}_1 \neq \emptyset$ goto **step 4**, otherwise goto **step 8**.

step 8: Repeat **step 1** to **step 7** of the algorithm several times for different input orders of the subsets and the sequences within the subsets.

step 9: From this collection of trees the the majority consensus tree is calculated.

4 Computational aspects

To simplify matters, we assume that we partition the data in $k \leq n$ subsets of size $m = \frac{n}{k} + o$, where o defines the overlap between two neighboring sets, i.e. $|S_i \cap S_{(i \bmod k)+1}| = o$ for $i = 1, \ldots, k$, where we require that $o \geq 3$. Although not crucial for the ModPUZZLE algorithm, for the practical application and

k subsets	m sequences	o overlap	number of quartets	% of quartets	resolved splits	% of splits
5	13	3	3,230	1.4	2	4.3
5	15	5	6,800	2.6	7	14.9
5	20	10	23,175	10.1	23	48.9
5	30	20	112,800	49.0	39	83.0
full set	50		230,300	100.0	47	100.0

Table 1. The results of the tree reconstruction using an simulated alignment of 50 DNA sequences with increasing overlap.

to add freedom to the repetitions of **steps 1** to **7** we now construct the last subset S_k overlapping the S_1.

For $k = n$ and $o = 3$ we have two compute only $n - 3$ quartets but this is not enough information to reconstruct the tree, even if the data were perfectly tree-like. Thus, one should try to increase o as much as possible, guided by the time on is willing to spent for the calculation of the

$$k\left(\binom{m-o}{4} + \binom{o}{4}\right) \tag{9}$$

possible quartets, where $o > 3$.

5 Some practical measures on the method

In phylogenetic tree reconstruction one wants to reconstruct a fully resolved tree, i.e. all inner nodes of the tree have degree 3. However, due to lack of a clear phylogenetic signal it is not always possible to achieve this goal. In those situations the tree reconstructed is not fully resolved. The amount of resolution can be measured by the number of splits, i.e. inner branches, that partition the sequences into two non-empty subsets. To analyze the influence of the size of the overlap between subsets on the resolution, we carried out a simulation study. To this end we simulated the evolution of DNA sequences on a tree with 50 leaf-vertices using the Seq-Gen package (Rambaut and Grassly (1997)).

To the resulting data the ModPUZZLE algorithm was applied by randomly splitting the 50 sequences into subsets of varying size and different overlap. Table 1 summarizes the results, which show that one should try to maximize the overlap. It is also clear, that an overlap of 20 which leads to the computation of roughly 50% of all possible quartets provides a good resolution. We were able to recover 39 from the 47 possible splits.

The ModPUZZLE algorithm was also applied the alignment of all 215 red algae ssu rRNA sequences from the European small subunit rRNA database (Van de Peer et al. (2000)) as a biological dataset. Due to the large number of quartets (86,567,815 possible quartets) we only ran tests for a limited set of values for k and m. However, this large biological dataset shows the same characteristics as the simulated one, the resolution increases with the number of quartets available for the tree reconstruction (data not shown).

The tests on both simulated and biological data show that with the minimal amount of overlap, almost no resolution of the trees can be gained. However, the resolution increases with the number of quartets used. Remarkable is the effect that the percentage of resolved splits grows faster than the percentage of quartets used. Therefore it seems to be possible to reconstruct resolved trees, even if one does not use all quartet trees. The amount of possible savings, however, depends crucially on the amount of phylogenetic information present in the alignment.

6 Discussion and possible extensions

We have presented a very simple algorithm to reconstruct phylogenetic trees from large datasets. The method is based on a modified version of the quartet puzzling algorithm (Strimmer and von Haeseler (1996)). The algorithm has the flexibility to adjust the amount of computing time one is willing to spend. If one is only interested in the coarse structure of the tree, then one needs to compute only very few quartets, thus obtaining a more or less unresolved tree. If one wants the fine details of the ramifications of the tree one needs to compute a lot more quartets by increasing the overlap between the subsets.

But, as shown in Table 1 the percentage of resolved splits seems to grow faster than the amount of quartets used. This observation leads to a strategy how to analyze large datasets, which is not fully exploited here. Instead of randomly assigning sequences to the k subsets once in some kind of linear order, one could use more decompositions to build a network of subsets. To increase the resolution of the final tree, one may also use a data guided approach. For example, a threshold graph (Barthelemy and Guenoche (1991), Huson et al. (1999)) based on the pairwise distances can be used as indicator how to group sequences. The applicability of this strategy needs to be analyzed by simulations.

Another extension of the algorithm is also possible and will be studied further. Instead of analyzing the gene tree of one set of aligned sequences, we may very well assume that each subset $\mathcal{S}_1, \mathcal{S}_2, \ldots, \mathcal{S}_k$ contains the collection of species for which a sequence alignment for gene i, $i = 1, \ldots, k$, is available. Then we can compute the big tree for the entire set of species $\mathcal{S}$, based on k different genes, without requiring that one gene sequence is known for all n species in $\mathcal{S}$. The properties of this approach need to be investigated further.

Availability

The method shown will be implemented in a future TREE-PUZZLE release, which is available from `http://www.tree-puzzle.de` (Schmidt et al. (in press)).

Acknowledgement

We thank Sonja Meyer, Roland Fleißner and Antje Krause for helpful discussions. Financial support from the Deutsche Forschungsgemeinschaft and the Max-Planck-Gesellschaft is also gratefully acknowledged.

References

BANDELT, H.-J. and DRESS, A. (1986): Reconstructing the Shape of a Tree from Observed Dissimilarity Data. *Adv. Appl. Math.*, *7*, 309–343.

BARTHELEMY, J. P. and GUENOCHE, A. (1991): *Trees and Proximity Representations.* Interscience Series in Discrete Mathematics and Optimization, Wiley, New York.

DRESS, A., VON HAESELER, A., and KRÜGER, M. (1986): Reconstructing Phylogenetic Trees Using Variants of the Four-Point Condition. *Studien zur Klassifikation*, *17*, 299–305.

FELSENSTEIN, J. (1981): Evolutionary trees from DNA sequences: A maximum likelihood approach. *J. Mol. Evol.*, *17*, 368–76.

FITCH, W. M. (1971): Toward defining the course of evolution: Minimum change for a specific tree topology. *Syst. Zool.*, *20*, 406–416.

FITCH, W. M. (1981): A Non-Sequential Method for Constructing Trees and Hierarchical Classifications. *J. Mol. Evol.*, *18*, 30–37.

HUSON, D. H., NETTLES, S. M., and WARNOW, T. J. (1999): Disk-Covering, a Fast-Converging Method for Phylogenetic Reconstruction. *J. Comp. Biol.*, *6*, 369–386.

MARGUSH, T. and MCMORRIS, F. R. (1981): Consensus n-trees. *Bull. Math. Biol.*, *43*, 239–244.

VAN DE PEER, Y., DE RIJK, P., WUYTS, J., WINKELMANS, T., and DE WACHTER, R. (2000): The European Small Subunit Ribosomal RNA Database. *Nucleic Acids Res.*, *28*, 175–176.

RAMBAUT, A. and GRASSLY, N. C. (1997): Seq-Gen: An application for the Monte Carlo simulation of DNA sequence evolution along phylogenetic trees. *Comput. Appl. Biosci.*, *13*, 235–238.

SAITOU, N. and NEI, M. (1987): The neighbor-joining method: A new method for reconstructing phylogenetic trees. *Mol. Biol. Evol.*, *4*, 406–425.

SATTATH, S. and TVERSKY, A. (1977): Additive Similarity Trees. *Psychometrika*, *42*, 319–345.

SCHMIDT, H., STRIMMER, K., VINGRON, M., and VON HAESELER, A. (in press): TREE-PUZZLE: Maximum Likelihood Phylogenetic Analysis Using Quartets and Parallel Computing. *Bioinformatics.*

STRIMMER, K., GOLDMAN, N., and VON HAESELER, A. (1997): Bayesian Probabilities and Quartet Puzzling. *Mol. Biol. Evol.*, *14*, 210–211.

STRIMMER, K. and VON HAESELER, A. (1996): Quartet Puzzling: A Quartet Maximum Likelihood Method for Reconstructing Tree Topologies. *Mol. Biol. Evol.*, *13*, 964–969.

SWOFFORD, D. L., OLSEN, G. J., WADDELL, P. J., and HILLIS, D. M. (1996): Phylogenetic Inference. In: D. M. Hillis, C. Moritz, and B. K. Mable (eds.), *Molecular Systematics*, Sinauer Associates, Sunderland, 407–514.

Regression Trees

Regression Trees for Longitudinal Data with Time-Dependent Covariates

Giuliano Galimberti and Angela Montanari

Dipartimento di Scienze statistiche, Università di Bologna,
Via Belle Arti 41, 40126 Bologna, Italy

Abstract. In this paper the problem of longitudinal data modelling in presence of time-dependent covariates is addressed. A solution based on recursive partitioning method is proposed. The key points of this solution are the definition of a suitable split function $\Phi(s, g)$, able to account for the autocorrelation structure typical of longitudinal data, the definition of splits on time-dependent covariates and the estimation procedure for the value of the step function on each element of the partition of the covariate space induced by the solution itself. The performances of the proposed strategy are studied by a simulation experiment.

1 Introduction

In many studies a number of variables are collected from the same unit repeatedly over time. The continuous improvement in the recording devices and in the computing instruments has made the problem of the statistical modelling of longitudinal data as a function of time-dependent covariates more and more actual and interesting.

This is witnessed by the publication of a long paper with discussion in one of the last issues of *JASA* (Lin and Ying (2001)) and by the increasing number of papers on the topic which have appeared in the last years (see for instance Zhang (1997), Hoover et al.(1998), Martinussen and Scheike (1999) and Chiang et al. (2001)).

In this paper a generalization of regression trees (hereafter CART), as introduced by Breiman et al. (1984), will be proposed, aimed at dealing with longitudinal data with time-dependent covariates. The ease of interpretation, the possibility of treating different types of covariates (either real valued or categorical), the ability of finding interactions and the local approximating capabilities by local subset covariate selection of regression trees have been thoroughly explored also in contexts which completely differ from the ones in which they where initially developed (see for instance the paper by Ciampi et al. (1991) on generalized regression trees, by Segal (1992) on longitudinal data with time-independent covariates and by Huang et al. (1998) and Galimberti and Montanari (2001) on survival data with time-dependent covariates) but, as far as we know, a fully satisfactory solution for the problem we are trying to solve in this paper is still lacking in the statistical literature.

2 Recursive partitioning regression

In order to better highlight how the proposed method generalizes regression trees we briefly recall the main features of recursive partitioning methodology for the study of the relationship between a scalar response variable Y and a set of covariates $(x_1, \ldots, x_p)$. Suppose that

$$y_i = f(x_{1,i}, \ldots, x_{1,i}) + e_i. \tag{1}$$

The aim is to approximate f by a step function defined as

$$\hat{f}(\mathbf{x}_i) = \sum_{m=1}^{M} a_m b_m(\mathbf{x}_i). \tag{2}$$

The functions $b_m(\mathbf{x})$ (also called basis functions) take the form

$$b_m(\mathbf{x}) = I(\mathbf{x} \in R_m), \tag{3}$$

where $I(\cdot)$ is the indicator function and $\{R_m\}_{m=1}^{M}$ (also called leaves) represent a partition of the covariate space whose elements, for real valued covariates, usually take the form of hyper-rectangular axis oriented sets. The goal of recursive partitioning is then not only to determine the coefficient values that best fit the data, but also to derive a good set of basis functions (that is, a good partition of the covariate space) by a stepwise procedure. The modelling strategy may be summarized in the following three aspects.

1) Definition of a set of questions, also called splits, regarding the covariates in order to partition the covariate space. Recursive application of these questions leads to a tree, which is binary if the questions are binary (yes/no); units for which the answer is yes are assigned to a given daughter node, those for which the answer is no are assigned to the complementary one. The set containing all the observed units at the beginning of the tree construction is called the root node; after each question a node is split in two daughter nodes, which are usually called left daughter node and right daughter node respectively. The nodes which, according to a pre-specified stopping rule, can no longer be split are called leaves.
2) Definition of a split function $\Phi(s, g)$ that, at each step, can be evaluated for any split s of any node g. The preferred split is the one which generates the purest daughter nodes. In regression trees methodology the most widely used measure of pureness is the within node sum of squares and the split function which is maximized with respect to any split s and any node g is the between daughter nodes sum of squares

$$\begin{aligned}\Phi(s,g) = (\mathbf{y}_g - \overline{\mathbf{y}}_g)'(\mathbf{y}_g - \overline{\mathbf{y}}_g) - [\,(\mathbf{y}_{gl} - \overline{\mathbf{y}}_{gl})'(\mathbf{y}_{gl} - \overline{\mathbf{y}}_{gl}) +\\ + (\mathbf{y}_{gr} - \overline{\mathbf{y}}_{gr})'(\mathbf{y}_{gr} - \overline{\mathbf{y}}_{gr})\,]\end{aligned} \tag{4}$$

where $\mathbf{y}_g$, $\mathbf{y}_{gl}$ and $\mathbf{y}_{gr}$ are vectors containing the y values for units belonging to parent node g and its two daughter nodes l and r (obtained after split s is performed) respectively and $\overline{\mathbf{y}}_g$, $\overline{\mathbf{y}}_{gl}$ and $\overline{\mathbf{y}}_{gr}$ are vectors whose elements are all equal to the related mean values. Fitting a regression tree may be interpreted as a least square fitting, within each node, of a regression model with the intercept alone or, equivalently, of a no intercept model when Y is regressed on the indicator variables which define node membership, followed by the choice of the split which produces the lowest residual sum of squares for the entire tree.

3) Definition of how to determine appropriate tree size, which in CART is accomplished by what is called pruning. A very large (possibly overfitted) tree is initially grown, so as to allow all potentially important splits. This large tree is then collapsed back, according to a given cost-complexity measure, creating a nested sequence of trees, the best of which is individuated either by cross-validation or resorting to a test sample.

3 Recursive partitioning regression for longitudinal data with time-dependent covariates

3.1 Data structure

When dealing with longitudinal data with time-dependent covariates we have to treat measurements that are taken on n units over q occasions. The number of occasions is the number of times that the measurements have been taken for each unit. In this paper we assume that all units have the same number of occasions q and the same corresponding measurement times and that neither the response variable nor the covariate vectors have missing values. For unit i $(i = 1, \ldots, n)$ at occasion j $(j = 1, \ldots, q)$, $x_{k,ij}$ and y_{ij} are respectively the measurement of the k-th covariate X_k $(k = 1, \ldots, p)$ and the observed value of the response variable Y.

The problem of interest is to model the relationship of Y to time T and the p-dimensional vector of covariates

$$y_{ij} = f\left(t_{ij}, x_{1,ij}, \ldots, x_{p,ij}\right) + e_{ij}, \tag{5}$$

where f is an unknown function and e_{ij} is the error term (with zero mean). Model (5) differs from an usual multivariate regression model as the error term e_{ij} $(j = 1, \ldots, q)$ has an autocorrelation structure, represented by a $q \times q$ matrix Σ_i, within the same unit i. In this paper we will not dwell on the specification of Σ_i, we will only assume that Σ_i may be somehow estimated and that it may be inverted (see Diggle et al. (1994) for a detailed discussion of parametric and semi-parametric modelling of (5) and Σ_i).

The aim of this paper is to suggest a solution for approximating f by a regression tree, that is by a step function defined on the covariate space (including time).

3.2 Tree construction

The presence of time-dependent covariates poses new problems as far as the definition of splits is concerned. As Segal (1992) himself stressed, "for ordered covariates the difficult lies in formulating interpretable splits that preserve ordering with respect to both time and the variable itself". He reports a strategy - which he himself deems not completely satisfactory - according to which each time-dependent covariate is regressed against time and the resulting slope and intercept are then included in the tree as time-independent covariates. Of course such an approach neglects most of the information conveyed by data, is reasonable only to the extent that the linear regression adequately describes the time-dependent covariate and affects tree interpretability.

Here we propose to handle time-dependent covariates by modifying the original meaning of split, allowing a unit to belong to more then a daughter node, that is allowing it to belong to a given daughter node only for those occasions in which the covariate values allow to answer "yes" to the split question, and to the complementary daughter node when the answer is "no" (see Huang et al. (1998) for a similar approach in the context of survival trees).

The autocorrelation structure of the error term, which characterizes longitudinal data, and our particular split definition impose to modify the split function too. The split function we propose is a modified version of the between daughter nodes sum of squares (4), allowing for correlated errors. It may be expressed as

$$\Phi(s,g) = \left[\mathbf{y} - \hat{f}_g(\mathbf{x})\right]' S^{-1} \left[\mathbf{y} - \hat{f}_g(\mathbf{x})\right] + \\ - \left[\mathbf{y} - \hat{f}_{d,s}(\mathbf{x})\right]' S^{-1} \left[\mathbf{y} - \hat{f}_{d,s}(\mathbf{x})\right] \tag{6}$$

where $\mathbf{y}$ is the $nq \times 1$ vector whose entries are the y values for each unit at each occasion, S is a block diagonal matrix whose non-zero entries are suitable estimates of Σ_i, for all i. We focus mainly on the mean structure and we assume that the covariance structure is common to all units. Of course the role of the inverse of S here is to weight the differences between the observed values and the fitted model taking into account the correlation structure.

As far as the fitted functions $\hat{f}_g(\mathbf{x})$ and $\hat{f}_{d,s}(\mathbf{x})$ (where the subscript d, s denotes the two daughter nodes obtained performing split s) is concerned, due to the error correlation, ordinary least squares fit used by CART is not a good choice as it leads to a split function which is not necessarily non-negative and therefore its maximization does not necessarily improve homogeneity.

The fit of a no intercept model where y is regressed on the indicator variables defining node membership may be best performed by resorting to generalized least squares, which have been purposely developed to deal with correlated errors. Therefore at each step $\hat{f}_g(\mathbf{x}) = B_g(\mathbf{x})\tilde{\mathbf{a}}_g$, were $B_g(\mathbf{x})$ is the matrix whose columns represent the indicator variables denoting node membership when node g is not splitted and $\tilde{\mathbf{a}}_g$ is the generalized least squares estimate of the corresponding regression coefficient vector, while $\hat{f}_{d,s}(\mathbf{x}) = B_{d,s}(\mathbf{x})\tilde{\mathbf{a}}_{d,s}$, where $B_{d,s}(\mathbf{x})$ is the matrix whose columns represent the indicator variables denoting node membership when split s is performed on node g and $\tilde{\mathbf{a}}_{d,s}$ is the vector containing the corresponding generalized least squares estimates. $B_{d,s}(\mathbf{x})$ is obtained by substituting in $B_g(\mathbf{x})$ the column that identifies parent node g with two new column that identify its daughter nodes, according to split s (the sum of these two column is then equal to the column corresponding to the parent node).

It should be noted that the use of generalized least squares also affects the zero entries of $B_g(\mathbf{x})$ and $B_{d,s}(\mathbf{x})$, that is, within each node, both the y values of unit observations belonging to the node and the y values of unit observations not belonging to it contribute in the definition of the estimate of f (with different weights), as was to be expected due to the correlation structure in the data.

At each step all the possible splits of all the terminal nodes obtained at the previous step are examined and the best of them is performed. It is worth noting that, to allow for the error correlation structure, at each step the $\hat{f}$ values in all terminal nodes are adjusted, just like in linear regression when a new covariate is added to the model.

After a large (possibly overfitted) tree is grown, the pruning procedure is essentially the same as in CART and the best tree of the nested sequence obtained by pruning may be chosen by v-fold cross-validation (in order to preserve the autocorrelation structure, when the data set is divided in v subset, all the observations on the same unit are assigned to the same subset).

4 A simulation study

In order to better highlight the potentialities of the proposed method, a simulation study was run. The instructions to generate the simulated samples and to perform the analysis were implemented in GAUSS. The variables involved in this simulation are time t, response Y and six time-dependent covariates X_1 to X_6; at each occasion j, $X_{1,j}$ is a Bernoulli with 0.5 probability of success, while covariates $X_{2,j}$ to $X_{6,j}$ take values $\{-1, 0, 1\}$ with probabilities $Pr(X_{k,j} = -1) = Pr(X_{k,j} = 0) = Pr(X_{k,j} = 1) = 1/3$, $k = 2, \ldots, 6$. The number of units in each sample n and the number of occasions q were set equal to 100 and 5 respectively. The observations for Y_j are obtained from model (5), where

$$f(t_j, \mathbf{x}_j) = f_0(t_j, \mathbf{x}_j) I(x_{1,j} = 0) + f_1(t_j, \mathbf{x}_j) I(x_{1,j} = 1), \tag{7}$$

$$f_0(t, \mathbf{x}) = -5I(t \leq 3) - 3I(t > 3) + 3x_5 + 2I(x_6 \leq -1) + I(x_6 > -1)$$

and

$$f_1(t, \mathbf{x}) = 3I(t \leq 2) + 5I(t > 2) + 3x_2 + I(x_3 \leq 0) + 3I(x_3 > 0).$$

As f does not depend on x_4, this represents a noise covariate. The error e is generated from a 5-dimensional normal distribution with covariance matrix that has 2 along the diagonal and 0.6 in the off-diagonal. 50 independent samples were generated and 50 regression trees were constructed; for each sample, Σ_i was estimated by the sample covariance matrix and the optimal tree was chosen by 10-fold cross-validation, according to the so-called rule of "one time the standard error" (see Breiman et al. (1984) for details). The main simulation results are displayed in Figure 1. The proposed method

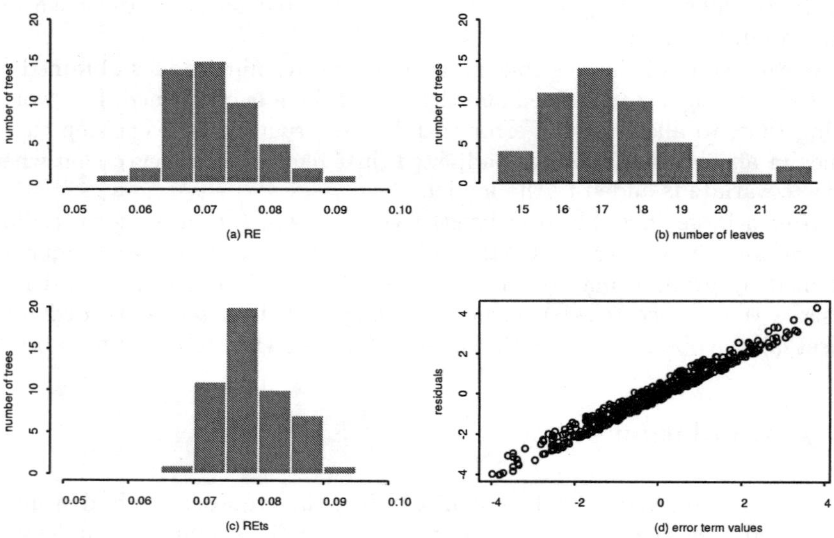

Fig. 1. Simulation results: (a) distribution of RE; (b) distribution of the number of leaves; (c) distribution of RE^{ts}; (d) error term values against averaged residuals.

performed very well for the model considered in this simulation study. Only for two samples the optimal tree contains a split on covariate x_4; this result confirms the ability of recursive partitioning methods in selecting relevant covariates. Following Breiman et al. (1984), the accuracy of each optimal tree was measured by the relative mean squared error RE, which is equal to

$$RE(\mathcal{T}) = \frac{R(\mathcal{T})}{R(t_1)}, \tag{8}$$

where $R(\mathcal{T})$ and $R(t_1)$ are respectively the weighted residual sum of squares for tree $\mathcal{T}$ and for tree t_1, which contains only one node. The value of this index decreases to zero as the accuracy of $\mathcal{T}$ increases. Figure 1.(a) shows the frequency distribution of RE calculated on the 50 optimal trees. It can be seen that RE takes very low values (for most of the samples, this index ranges between 0.065 and 0.08). However, as Figure 1.(b) shows, the optimal trees has a quite large number of leaves. This seems to be a typical feature of the considered model (7): also in simulations, not presented here, with uncorrelated errors the optimal trees have presented quite large number of leaves.

Since RE decreases as the number of leaves increases, the values of this index may represent an overoptimistic measure of the optimal trees effective accuracy. In order to have an honest measure of the accuracy, a test sample has been generated from the same model (with the same number of units and occasions). For each optimal tree, the fitted values for the test sample has been obtained and RE^{ts} has been computed on the test sample residuals, according to formula (8). The frequency distribution of RE^{ts} is shown in Figure 1.(c). As expected, RE^{ts} tends to take larger values than RE, but these differences do not seem relevant. Furthermore, there is a strong correlation between the test sample simulated error term values and the average of the corresponding test sample residuals (Figure 1.(d)). Therefore, also the use of the test sample gives support to the previous conclusions about the good performances of the proposed method.

5 Conclusions and open issues

From the results so far obtained on simulated data it seems possible to conclude that the proposed method represents an useful tool for the analysis of longitudinal data with time-dependent covariates.

However, any topic in the paper on which we have written "by now we assume" represents a field for future research. For instance we are going to study extensions of our proposal able to handle data sets with different number of occasions and different time spacing for different units and missing data either in the response variable or in the covariates or in both.

An aspect which deserves further inspection is the choice of a suitable estimate of the error covariance matrix which involves a smaller number of parameters than the sample covariance matrix. Moreover, as in Segal (1992), it might be useful to update the covariance parameters after each split, in order to take into account any possible subgroup difference in the autocorrelation structure. It may be pointed out that the proposed method is strongly dependent upon the estimate of the covariance structure. A possible way to limit the effect of this estimate is to resort to an iterative two-stage procedure. Starting with an initial estimate of the correlation structure, at the first step a regression tree is built and the residuals between the true values of y and the fitted values are computed. At the second step, a new estimate of the correlation structure is obtained on these residuals. These two steps may be iterated until a suitable stopping criteria is fulfilled (for instance, a discrepancy measure between two estimates of the correlation structure).

Finally, as one of the major disadvantages of regression trees is represented by the lack of continuity of the estimated functions, a smoother estimate of f, while retaining all properties of regression trees, could be obtained by modifying our procedure according to DART strategy (Friedman, 1996).

References

BREIMAN, L., FRIEDMAN, J. H., OLSHEN, R. A. and STONE, C. J. (1984): *Classification and Regression Trees.* Wadsworth, Belmont.

CHIANG, C., RICE, J. A. and WU, C. O. (2001): Smoothing Spline Estimation for Varying Coefficient Models With Repeatedly Measured Dependent Variables. *Journal of the American Statistical Association, 96*, 605–619.

CIAMPI, A., LOU, Z., LIN, Q. and NEGASSA, A. (1991): Recursive Partitioning and Amalgamation with the Exponential Family: Theory and Application. *Applied Stochastic Models and Data Analysis, 7*, 121–137.

DIGGLE, P. J., LIANG, K. Y. and ZEGER, S. L. (1994): *The Analysis of Longitudinal Data.* Oxford University Press, Oxford.

FRIEDMAN, J. H. (1996): *Local Learning based on Recursive Covering.* Technical Report, Department of Statistics, Stanford University.

GALIMBERTI, G. and MONTANARI, A. (2001): Recursive Partitioning Methods and Survival Analysis: a Proposal For the Study of Time-Dependent Covariates with Incomplete Information. In: C. Provasi (Eds.): *Atti del Convegno Modelli Complessi e Metodi Computazionali Intensivi per la Stima e la Previsione, Bressanone (BZ), 24-26 settembre 2001.* Cluep, Padova, 343-348.

HOOVER, D. R., RICE, J. A., WU, C. O. and YANG, L. P. (1998): Nonparametric Smoothing Estimates of Time-Varying Coefficient Models with Longitudinal Data . *Biometrika, 85*, 809–822.

HUANG, X., CHEN, S. and SOONG, S. (1998): Piecewise Exponential Survival Trees with Time-dependent Covariates. *Biometrics, 54*, 1420-1433.

LIN, D. J. and YING, Z. (2001): Semiparametric and Nonparametric Regression Model For Longitudinal Data. *Biometrika, 86*, 691–702.

MARTINUSSEN, T. and SCHEIKE, T. H. (1999): A Semiparametric Additive Regression Analysis of Longitudinal Data (with discussion). *Journal of the American Statistical Association, 96*, 103–126.

SEGAL, M. R. (1992): Tree-Structured Methods for Longitudinal Data. *Journal of the American Statistical Association, 87*, 407–418.

ZHANG, H. (1997): Multivariate Adaptive Splines for the Analysis of Longitudinal Data. *Journal of Computational and Graphical Statistics, 6*, 79–91.

Tree-based Models in Statistics: Three Decades of Research

Eugeniusz Gatnar

Katowice University of Economics, Department of Statistics, ul. Bogucicka 14, 40-226 Katowice, Poland

Abstract. The interest in tree-structured methods has been growing rapidly in statistics. In fact, all commercial statistical packages and Data Mining tools have been equipped with tree building modules. The research in this field has its roots in early 70s when early papers on recursive partitioning of the feature space (and its result which has the form of a tree) were published in statistical journals. They began intensive research in nonparametric statistical methods for classification, regression, survival analysis etc. The aim of this paper is to summarize achievements of this research and point out some still open problems.

1 Introduction

Perhaps the paper of Belson (1959) was the first paper on tree-structured methodology in statistics community, but construction of tree models began in early 60s in the social sciences. Morgan and Sonquist (1963) and Hunt et al. (1966) developed two algorithms: AID (Automated Interaction Detection) and CLS (Concept Learning System) respectively, to model the process of concept learning.

Statisticians began in early 70s when Meisel and Michalopoulos (1973) pointed out the use of tree-based methodology in classification. In the same year Breiman and Friedman started their work on trees and began to use tree models in classification. This resulted in several papers, e.g. Friedman (1973), and the famous CART book by Breiman et al. (1984).

In the 80s decision trees were beginning to be used in computer science to model the process of human learning. The induction methods based on trees play a central role in machine learning, a subfield of Artificial Intelligence, to extract knowledge from examples or build knowledge bases. Quinlan's (1986) ID3 (Iterative Dichotomizer) program was the first one used successfully to solve real world problems (chess end-games etc.)

The model building process is based on recursive partitioning of the multidimensional attribute space into disjoint regions and its consecutive steps form a structure of the tree.

Tree-based models are popular and widely used because they are simple, flexible and powerful tools for classification, regression and survival analysis. Nowadays they are integrated into Data Mining and used in the Web content and usage mining, in the analysis of complex data sets, where cases are

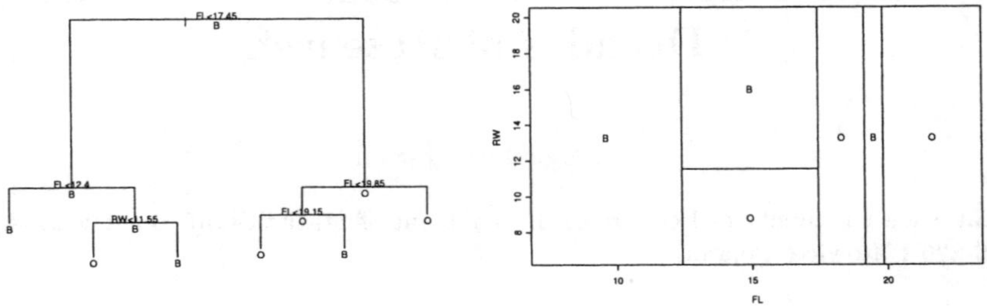

Fig. 1. Tree and associated feature space partition for crabs data.

allocated to the regions with a calculated probability (Ciampi et al. 1996), in the classification of symbolic data (Perinel and Lechevallier, 2000), etc.

In this paper we summarize achievements of this research in statistics and point out some open problems.

As an example we present a classification tree (Figure 1) for crabs[1] data collected by Campbell and Mahon(1974).

The tree-based model is extremely easy to interpret by the set of decision rules in the disjunctive normal form (DNF), e.g.

```
Color=blue if (FL<12.4) or (FL<17.45 and FL>=12.4 and RW>=11.55) or ...

Color=orange if (FL>=19.85) or (FL>=17.45 and FL<19.15) or ...
```

2 Model

We assume we have a learning sample:

$$(\mathbf{x}_i, y_i) \tag{1}$$

$(i = 1, \ldots, n)$, where $\mathbf{x}$ is the vector of M independent variables (predictors), and y is the target variable (response).

The tree corresponds to an additive model in the form of:

$$y = \sum_{k=1}^{K} a_k I(\mathbf{x} \in R_k), \tag{2}$$

where R_k are hyper-rectangular disjoint regions in the M-dimensional feature space, a_k denotes real parameters and I is an indicator function.

[1] Crabs are of two color forms, blue (B), and orange (O) and only two of their features are considered: frontal lobe (FL) and rear width (RW).

Each real-valued dimension of the region R_k is characterized by its upper and lower boundary: x_m^U and x_m^L respectively. Therefore the region induces a product of M indicator functions:

$$I(\mathbf{x} \in R_k) = \prod_{m=1}^{M} I(x_{km}^L \leq x_m \leq x_{km}^U). \tag{3}$$

If x_m is a categorical variable, the region R_k is defined as:

$$I(\mathbf{x} \in R_k) = \prod_{m=1}^{M} I(x_m \in B_m), \tag{4}$$

where B_m is a subset of the set of the variable values.

Obviously, the estimation method of parameters in (2) depends on the type of model, i.e., the nature of the response y.

2.1 Classification

If the dependent variable y is nominal and its values represent class labels (taking values from the set j=*1, ...,J*) the model (2) is a classification model.

In this case the parameter values are determined by a majorization rule:

$$a_k = \arg\max_j p(j|k) \tag{5}$$

where $p(j|k)$ is the proportion of cases in region R_k belonging to class j. In other words, a_k is the most frequent value of y in the region R_k.

2.2 Regression

If y in (2) is continuous, the model is a regression model and the parameter estimation formula depends on the way how the homogenity of a region is measured. In the simplest case, when variance is used, the best estimate is the mean of all y values in R_k:

$$a_k = \frac{1}{n(k)} \sum_{i=1}^{n(k)} y_i \tag{6}$$

where $n(k)$ is the number of objects from training set belong to region R_k.

2.3 Survival analysis

When the response variable y is censored and positive the model (2) may be refered to as a survival tree. Gordon and Olshen (1985) proposed to measure the homogenity of the region R_k by the discrepancy between two survival functions:

$$H(R_k) = \min_{c \in C} d_W(K(R_k), c) \tag{7}$$

where d_W is the Wasserstein distance between two distribution functions, $K(R_k)$ denotes the Kaplan-Meier curve in region R_k and C is a family of curves of this kind.

3 Variable selection

Several methods for selecting appropriate variablea to the model (2) have been developed. Obviously, the number of variables in the model determines the tree size.

The aim of this process is to chose thosee explanatory variables that yield the 'best' partition. The quality of partition of a region R is measured by the difference between heterogenity (in terms of y) of R and of resulted regions $R_1, R_2, ..., R_K$. In other words, it is reduction of impurity:

$$\Delta H(R, x_m) = H(R) - \sum_{k=1}^{K} H(R_k) \cdot p(k) \tag{8}$$

where H is a heterogenity measure and $p(k)$ is the proportion of objects in R_k.

If y is categorical, measures like entropy, Gini index, twoing rule etc. are involved, otherwise the variance is the most frequently used measure.

In some applications measures of association between y and x_m are used, such as χ^2 in Kaas's (1980) CHAID system.

When selecting a continuous variable, the coordinates of its split points are determined as well, and in the case of a categorical one its sets of values is divided into subsets.

Unfortunately, the selection process is biased towards variables with a large number of values at the expense of those with few ones. In other words, some heterogenity measures favour variables that divide the attribute space into many disjoint regions at one step.

To solve this problem some normalization of the variable selection measures is necessary. Moreover, Rissanen (1983) proposed to use the MDL (Mimimum Description Length) principle in order to reduce the bias.

4 Search for the best model

Since Hyafil and Rivest (1976) proved that constructing an optimal decision tree is a NP-complete problem, a (sub)optimal model is typically searched for, i.e., one which minimizes the tree size and maximizes its predictive accuracy.

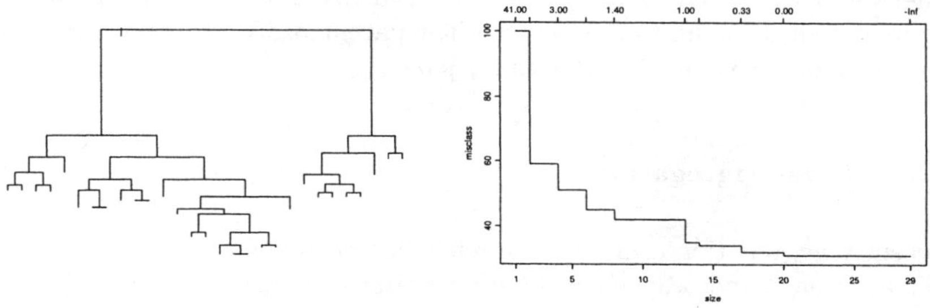

Fig. 2. Full tree for crabs data and cost-complexity parameter.

Evaluation of the model accuracy needs a test set or has to use cross-validation. Then the right sized tree is chosen from several available ones. There are three approaches to find the best model.

4.1 Pruning

First the full tree is built up (most complex model) and then it is pruned back according to some criterion. Several pruning methods have been proposed so far, e.g., pessimistic error pruning, reduced error pruning by Quinlan (1987), critical value pruning by Mingers (1987) etc.

The most frequently used method is that of Breiman et al. (1984): cost-complexity pruning based on the classification error and the model complexity:

$$E_\alpha(T) = E(T) + \alpha \cdot size(T) \tag{9}$$

where α is the cost-complexity parameter and $E(T)$ is the misclassification error (mostly determined by cross-validation) of the model T.

The parameter α governs the tradeoff between the tree size and its accuracy. Then the optimal tree is determined, e.g., by the 1SE rule proposed by Breiman et al. (1984). It is the smallest tree whose error is not more than one standard error greater than the lowest error tree in the sequence.

The tree for crabs data in Figure 1 is a tree pruned to the size 6, while the full model (presented on left side of Figure 2) is much more complex. The right side of Figure 2 is the plot of the misclassification error against size of the tree for crabs.

4.2 Shrinking

This procedure also starts with a large tree, but rather than cutting off branches, it simply smoothes the fitted values. Shrunken fitted values are determined by recursion:

$$\hat{y}(node) = \alpha \cdot \bar{y}(node) + (1 - \alpha) \cdot \hat{y}(parent) \tag{10}$$

where $\bar{y}$ is the fitted value, $\hat{y}$ the shrunken fitted value. The parameter α optimally shrinks children nodes to their parent based on the magnitude of the difference between $\bar{y}(node)$ and $\bar{y}(parent)$.

4.3 Model aggregation

Recently aggregation of multiple models has been shown to improve prediction accuracy, especially when classifying objects which are not in the training set.

There are several methods to build a 'committee' of trees. Bootstrap aggregating (or bagging) proposed by Breiman (1996) produces replicate training sets by bootstrap sampling from the learning set and a model is constructed from each of them. Then the multiple models are combined to form a composite model.

Boosting by Freund and Shapire (1997) uses all cases but each one has an associated weight. At every iteration a model is built from the weighted training cases and each case is then reweighted according to whether or not it was misclassified.

According to Breiman's (1996) decomposition of misclassification error into bias term and variance term, bagging reduces the variance significantly (but sometimes increases the bias) while boosting reduces both.

5 Predictors with missing values

Tree-based models can cope with incomplete data which is a serious problem for classical, parametric models. Several strategies have been developed to predict y for a case when a variable is missing.

The first strategy is applicable to categorical predictors. The new category 'missing' is created for that variable and then used as the other ones.

An alternative strategy is to "drop" a case down the tree as far as it can go and then use the distribution at the final node to predict y.

A third solution is to pass a case with a missing value down each branch of the tree using conditional probabilities of left and right split.

A fourth strategy, proposed by Breiman et al. (1984), uses surrogate variables. The surrogate variable is a predictor that best mimics the split done by the unobserved variable.

For each variable selected for the model (2) a list of surrogate predictors (and associated split points) is formed. If the primary splitting variable is missing the surrogate variables are used instead, in order of their importance. The order is based on the probability of splitting the attribute space in the same way as the missing predictor.

6 Summary

Like other statistical methods, the tree-structured methods have some advantages and disadvantages. We point out here the most important ones.

6.1 Advantages

Since tree-based models are nonparametric, they do neither require knowledge of the variable distribution nor the analytic form of the model. Mostly they are more accurate and easier to use for classification or simulation than the parametric ones.

The obvious great advantage is the ease of interpretation of the model (2) as a set of decision rules.

Tree-based models can handle both categorical and continuous attributes and are robust to outliers.

Unlike classical, parametric models, the model (2) is invariant to monotone transformations of predictors.

The tree-strucured methods can be used to explore very large datasets (even containing missing observations). Hence they are incorporated into all commercial statistical packages: SAS, SPSS, STATISTICA, S-PLUS etc. and Data Mining tools: Clementine, Darwin, Knowledge Seeker, Enterprise Miner, Sipina etc.

6.2 Disadvantages

The hierarchical nature of the recursive partitioning process results in a lack of stability, i.e., small change in the data values produces a quite different model. To overcome this drawback Carter and Catlett (1987) proposed 'soft' thresholds implemented in Quinlan's (1993) C4.5 system.

Nowadays averaging several trees can alleviate this problem. Recently Breiman (1999) proposed 'adaptive bagging' which is effective in reducing the bias as well as the variance. Friedman (1999) developed 'gradient boosting' of trees (implemented in the MART system) which produces highly robust models, especially appropriate for imperfect data.

Tree-based regression models demonstrate their lack of smoothness. In order to solve this problem Friedman (1991) proposed to use splines (in the MARS procedure) instead of step functions.

Another disadvantage of the model (2) is its sensivity to the coordinate rotations because of partitioning the space by hyperplanes which are parallel to axis. The use of oblique hyperplanes[2], i.e., of linear combinations of predictors can alleviate this disadvantage. Unfortunately, their interpretation is very difficult.

---

[2] Several approaches to define multivariate splits lead to more accurate models than the method of Breiman et al. (1984).

7 Open problems

Tree-structured methods are a part of nonparametric statistics, not enough explored so far. Therefore there are some open problems, three of them presented here.

The first one is concerned with non-binary splits[3]. There is no doubt that they make the trees smaller and easier to understand and also lead to more accurate models in some domains. On the other hand, multiway splits fragment the attribute space too quickly, and create regions R_k which contain only small subsets of observations.

Another drawback is concerned with bagging and boosting, successfully used to improve model stability and prediction accuracy. We point to the need to combine them into one unique theory with solid mathematical background. The work of Friedman (1999), Breiman (1999) and Hastie et al. (2001) are important steps in this direction.

A third problem is more general. Although tree-structured methods are used to analyse statistical data and to make predictions, they are often not seen as statistical methods. This is because some of their theoretical properties are still unexplored.

References

BELSON, W. A. (1959): Matching and Prediction on the Priniciple of Biological Classification. *Applied Statistics, 8*, 65–75.

BREIMAN, L., FRIEDMAN, J. OLSHEN, R. and STONE, C. (1984): Classification and Regression Trees. Wadsworth, Belmont, CA.

BREIMAN, L. (1996): Bagging Predictors. *Machine Learning, 24*, 123–140.

BREIMAN, L. (1999): Using Adaptive Bagging to Debias Regressions. *Technical Report, 547*, Statistics Department, University of California, Berkeley.

CAMPBELL, N.A. and MAHON, R.J. (1974): A Multivariate Study of Variation in Two Species of Rock Crab of Genus *Leptograpsus. Australian Journal of Zoology, 22*, 417–425.

CARTER, C. and CATLETT, J. (1987): Assessing Credit Card Applications Using Machine Learning. *IEEE Expert*, Fall Issue, 71-79.

FREUND, Y. and SCHAPIRE, R.E. (1997): A Decision-Theoretic Generalization of On-Line Learning and an Application to Boosting. *Journal of Computer and System Sciences, 55*, 119–139.

FRIEDMAN, J. H. (1999): Stochastic Gradient Boosting. *Technical Report*, Stanford University, Stanford.

FRIEDMAN, J. H. (1991): Multivariate Adaptive Regression Splines. *Annals of Statistics, 19*, 1–141.

FRIEDMAN, J. H. (1977): A Recursive Partitioning Decision Rule for Nonparametric Classification. *IEEE Transactions on Computers, 26*, 404–408.

[3] If a continuous variable is selected its value space is divided into more than two disjoint intervals. Several discretization methods are considered by Gatnar (2001).

GATNAR, E. (2001): Nonparametric Method for Discrimination and Regression. PWN Scientific Publishers, Warsaw (in Polish).

GORDON, L. and OLSHEN, R.A. (1985), Tree-structured survival analysis. *Cancer Treatment Reports*, *69*, 1065–1069.

HASTIE, T. and PREGIBON, D. (1991) Shrinking Trees. *Technical Report*, AT&T Bell Laboratories, Murray Hill, NJ.

HASTIE, T., TIBSHIRANI, R. and FRIEDMAN, J. (2001): The Elements of Statistical Learning. Springer, New York.

HAYAFIL, L. and RIVEST, R.L. (1976): Constructing optimal binary decision trees is NP-complete. *Information Processing Letters*, *5*, 15–17.

HUNT, E.B., MARIN, J., STONE, P.J. (1966): Experiments in Induction. Academic Press, New York.

KASS, G. V. (1980): An Exploratory Technique for Investigating Large Quantities of Categorical Data. *Applied Statistics*, *29*, 119–127.

MINGERS, J. (1987): Expert Systems: Rule Induction with Statistical Data. *Journal of The Operational Research Society*, *38*, 39–47.

MORGAN, J.N. and SONQUIST, J. A. (1963): Problems in the Analysis of Survey Data, and a Proposal. *Journal of the American Statistical Association*, *58*, 415–434.

PERINEL, E. and LECHEVALLIER, Y. (2000): Symbolic discrimination rule. In: H.H. Bock and E. Diday (Eds.): *Analysis of Symbolic Data. Exploratory Methods for Extracting Statistical Information From Complex Data.* Springer, Heidelberg.

QUINLAN, J.R. (1986): Induction of Decision Trees. *Machine Learning*, *1*, 81–106.

QUINLAN, J.R. (1987): Simplifying Decision Trees. *International Journal of Man-Machine Studies*, *27*, 221–234.

QUINLAN, J.R. (1993): C4.5: Programs for Machine Learning. Morgan Kaufmann, San Mateo.

RISSANEN, J. (1983): A universal prior for integers and estimation by minimum description length. *Annals of Statistics*, *11*, 416–431.

Computationally Efficient Linear Regression Trees

Luis Torgo

LIACC/FEP, University of Porto
R.Campo Alegre, 823
4150 PORTO, PORTUGAL
email: ltorgo@liacc.up.pt
URL: http://www.liacc.up.pt/~ltorgo

Abstract. This paper describes a method for obtaining regression trees using linear regression models in the leaves in a computationally efficient way that allows the use of this method on large data sets. This work is focused on deriving a set of formulae with the goal of allowing an efficient evaluation of all candidate tests that are considered during tree growth.

1 Introduction

Existing systems that build regression trees with linear models in the leaves, follow one of two paths: either they compromise the construction of the tree in order to achieve the desired computational efficiency; or they grow "correct" trees but with algorithms whose computational effort strongly impairs their applicability. In this work we describe an alternative approach that allows the construction of "correct" linear regression trees with a computational efficiency that ensures a wide range of application scenarios.

2 Existing approaches to linear regression trees

2.1 The approach of M5

M5 (Quinlan (1992)) obtains trees using a standard least squares error criterion (*e.g.* Breiman et al. (1984), Torgo(1999)). This means that at each stage of the tree growth procedure M5 chooses the split that decreases more the variability with respect to the average in the respective partitions entailed by the split. This also means that the splits are chosen assuming that the leaves will have constant models consisting of averages. One can argue that these splits are not optimal as M5 will in effect use linear models in the leaves and not averages. The consequences of this can be easily illustrated by a simple artificial example. Let us suppose that we have a dataset generated by the function, $f(X) = \text{IF } X \leq 18.5 \text{ THEN } Y = -40 + 4.2X \text{ ELSE } Y = 70 - 1.5X$. We would expect that a system using linear models in the leaves, if given a reasonable amount of training data, would easily guess it, obtaining a tree with a single split test ($X \leq 18.5$) and with the two linear models on each of the leaves generated by this test. However, due to the criterion used by M5, this system can very easily fail to identify the best model even in this simple (although artificial) example. For instance, in a random sample with 24 observations we have generated, M5 obtained the following tree:

$$\begin{aligned}&\text{IF } X \leq 15.9 \text{ THEN } Y = -6.29 + 3.89X \\ &\qquad\qquad \text{ELSE IF } X \leq 17.5 \text{ THEN } Y = 27.6 + 1.94X \\ &\qquad\qquad\qquad\qquad \text{ELSE } Y = 28 + 1.94X\end{aligned}$$

This tree with two inner nodes and three linear models in the leaves, not only completely misses the true regression function, but it also leads to larger prediction error, and it is significantly more complex than the model that would be obtained if a "correct" criterion was used to choose the best split for each node. The reason for this failure is the use of a test selection criterion that does not take into account the models that will be used in the leaves. What is then the motivation for using such criterion as M5 does? The single motivation is computational complexity. In effect, the use of averages together with the least squares criterion allows the use of fast incremental algorithms that find the best split for each candidate variable in time linear to the number of training cases (*e.g.* Torgo (1999)).

2.2 The approach of RETIS

RETIS (Karalic (1992)) is another tree-based regression system that uses linear regression models in the leaves. However, contrary to M5, RETIS chooses the tests for the tree nodes assuming that these models will be used in the leaves. This means that RETIS would find the "correct" model in the small artificial example described in the previous section.

Let us address the computational costs of the method used by RETIS. Each split node in a tree obtained by RETIS has two branches:- one with the cases satisfying the test; and other with the remaining cases. Evaluating each of these splits involves obtaining an estimate of the error on each branch. To obtain this estimate one has to obtain their respective linear models. This involves a series of standard matrix computations (*e.g.* Myers (1990)). This has to be done for all trial splits, of all variables, and for all nodes of the tree. The result is a system that hardly scales up to large problems, particularly if too many continuous variables exist[1]. This problem was not addressed by Karalic when presenting RETIS (Karalic (1992)). We claim that although the method used by RETIS can be considered more "correct" than the one used in M5, it will not scale up to large problems as well as M5 method. It is the goal of this paper to provide efficient methods of building trees similar to the ones obtained by RETIS, but without their computational drawbacks.

[1] Which are known to be the major computational bottleneck for growing tree-based models (Catlett (1991)).

3 A correct and computationally efficient approach

Recursive least squares algorithms (Goodwin and Sin (1984)) have been developed in model identification and adaptive control literature. These algorithms allow to identify the linear model parameters when the data is available sequentially instead of as a single batch. Bontempi (2000) has used these results in the context of local linear regression (*e.g.* Cleveland and Loader (1995)). This author has developed a series of recursive formulas that allow obtaining local linear models for increasingly larger bandwidths (sets of neighbouring training cases). Namely, they make possible to obtain the linear model parameters for a sample with $k+1$ cases using the model parameters obtained with k cases, without having to recalculate everything from scratch. Moreover, Bontempi has developed a similar recursive formulation for obtaining a reliable estimate of the leave-one-out cross validation error of the linear model using the PRESS statistic (Myers (1990)). The problem we face to obtain an efficient evaluation of all candidate splits of a continuous variable is similar. Let us now describe what we need to attain this goal. Our description assumes binary splits although the results generalize to the multiple branch case. We will focus on continuous variables that present the major computational challenges (Catlett (1991)).

The selection of the best split in a continuous variable for a node t, starts with the ordering of the observed values of the variable. After obtaining this ordering (say $\{v_1, v_2, ..., v_n\}$) we need to evaluate each possible test that leads to a different splitting of the cases through the left and right branches. Let us imagine that we are at a stage where we have finished evaluating the split $X_a \leq v_k$. This means that we have just built two linear models: one with the cases that satisfy this test (let this model parameters be $\widehat{\beta}_L(k)$); and other with the ones that do not satisfy this test ($\widehat{\beta}_R(k)$). If we want to evaluate a new split $X_a \leq v_{k+1}$, where $v_{k+1} > v_k$, we will have some cases changing from the right to the left branch. This means that we need to obtain two new models: $\widehat{\beta}_L(k+1)$ and $\widehat{\beta}_L(k+1)$. The model $\widehat{\beta}_L(k+1)$ is obtained with the cases used to obtain the model $\widehat{\beta}_L(k)$ plus some other cases, namely the cases defined by the set $J = \{x_i \in D_t : X_{i,a} > v_k \wedge X_{i,a} \leq v_{k+1}\}$. On the other hand we will need to obtain the model $\widehat{\beta}_R(k+1)$, which is obtained with the cases used to calculate $\widehat{\beta}_R(k)$ less the set J of observations. The goal in the remaining of this section is to obtain a set of formulas, which will allow us to obtain $\widehat{\beta}_L(k+1)$ and $\widehat{\beta}_R(k+1)$ based on $\widehat{\beta}_L(k)$ and $\widehat{\beta}_R(k)$.

Let us analyze the case of obtaining a new model resulting from removing a set J of observations from the set of cases used to obtain a previous model. Let this model be named $\widehat{\beta}_{-J}$. Within the regression tree context this would correspond to obtaining the model for the right branch of a new split where less cases fall in this branch (*i.e.* $\widehat{\beta}_R(k+1)$).

Using the standard *normal equations* we know that the parameters of our goal model could be obtained by,

$$\beta_{-J} = \left(X_{-J}^T X_{-J}\right)^{-1} X_{-J}^T y_{-J} \tag{1}$$

where X_{-J} is the matrix X with J lines (*i.e.* cases) removed; and y_{-J} the respective vector of target variable values that is obtained from y in the same way.

As mentioned before our goal is to avoid doing these calculations. Namely, inverting the matrix $X_{-J}^T X_{-J}$ that would have a cost of the order $O\left(r^3\right)$, where $r = n - j$; n is the number of lines of X; and j is the cardinality of the set J (*i.e.* the number of cases being removed from X). In effect, we want to take advantage of the model parameters obtained before the J cases were removed (i.e. $\widehat{\beta}$), thus avoinding extra computations.

Making $S = \left(X^T X\right)$, it is easy to show that the following holds,

$$S_{-J} = \left(X_{-J}^T X_{-J}\right) = \left(X^T X - X_J^T X_J\right) = S - X_J^T X_J \tag{2}$$

where X_J is the matrix consisting of the j lines of the matrix X (*i.e.* a $[j \times p]$ matrix, where p is the number of predictor variables of the domain).

Using the same reasoning we can derive,

$$X_{-J}^T y_{-J} = X^T y - X_J^T y_J$$

As $S\beta = \left(X^T X\right)\beta = X^T y$ (known as the *normal equations*) we have,

$$X_{-J}^T y_{-J} = S\beta - X_J^T y_J \tag{3}$$

¿From Equations 1, 2 and 3 we derive,

$$\beta_{-J} = S_{-J}^{-1}\left(S\beta - X_J^T y_J\right)$$

¿From Equation 2 we know that $S_{-J} = S - X_J^T X_J$, and thus $S = S_{-J} + X_J^T X_J$, which means that

$$\begin{aligned} \beta_{-J} &= S_{-J}^{-1}\left[\left(S_{-J} + X_J^T X_J\right)\beta - X_J^T y_J\right] = \\ &= S_{-J}^{-1}\left(S_{-J}\beta + X_J^T X_J\beta - X_J^T y_J\right) = \\ &= \beta + S_{-J}^{-1} X_J^T X_J \beta - S_{-J}^{-1} X_J^T y_J = \\ &= \beta + S_{-J}^{-1} X_J^T \left(X_J\beta - y_J\right) \end{aligned}$$

All this can be recast into the more convenient formulation,

$$\begin{cases} S_{-J} = S - X_J^T X_J \\ \gamma_{-J} = S_{-J}^{-1} X_J^T \\ e_{-J} = X_J\beta - y_J \\ \beta_{-J} = \beta + \gamma_{-J} e_{-J} \end{cases} \tag{4}$$

Although this formulation already allows us to obtain β_{-J} from β it still involves the inversion of matrix S_{-J}. However, we should remark that S_{-J} is a $p \times p$ matrix, and as $p \ll r$ we are already far from the complexity of solving Equation 1.

Following the same reasoning equivalent formulas can be developped for the case of obtaining a model when we add a set of J observations ($\widehat{\beta}_{+J}$). This would correspond to the model for the left branch of a new split when new cases fall in this branch ($\widehat{\beta}_L\,(k+1)$ in the description given before).

The results obtained so far can still be improved by using the Sherman-Morrison formula for optimizing the inversion of S_{-J} and S_{+J}. Let us analyse the special case where the cardinality of J is 1.

The Sherman-Morrison formula states that if A is a $p \times p$ matrix, and u and v are p vectors, the following holds,

$$\left(A - uv^T\right)^{-1} = A^{-1} + \frac{A^{-1}uv^TA^{-1}}{1 - v^TA^{-1}u} \tag{5}$$

Making $A = S = X^TX$ and $u = v = x^T$, where x is the single case to remove from X, we have

$$\begin{cases} S_-^{-1} = S^{-1} + \frac{S^{-1}x^TxS^{-1}}{1-xS^{-1}x^T} \\ \gamma_- = S_-^{-1}x^T \\ e_- = x\beta - y \\ \beta_- = \beta + \gamma_- e_- \end{cases} \tag{6}$$

Using the equivalent Sherman-Morrison formula for $\left(A + uv^T\right)^{-1}$ we could obtain the formulas for adding a single case.

In this particular case where the cardinality of J is 1, the calculus of S_-^{-1} (S_+^{-1}) is strongly simplified by the use of the Sherman-Morrison formula, resulting in simple multiplications of matrices. For larger cardinality of the set J, we can either iterate the method outlined above, or we may use the Woodburry formula that can be seen as the matrix version of the Sherman-Morrison formula.

The following is an algorithm that obtains the new model parameters, β_-, after removing a single observation from the sample of data used to obtain β, based on the results of Equation 6. With very small changes one could easily obtain a similar algorithm for obtaining β_+.

Algorithm 1 *An $O\left(3p^2\right)$ algorithm for calculating β_-.*

```
den = 0          % den will hold 1 − x_j S^{-1} x_j^T
FOR l = 1 TO p
    r1[l] = 0        % r1 will hold x_j S^{-1}
    r2[l] = 0        % r2 will hold S^{-1} x_j^T
```

```
    FOR c = 1 TO p
        r1[l] = r1[l] + x[c] × S[c,l]      % x[i] is the ith element of vector x_j
        r2[l] = r2[l] + S[l,c] × x[c]      % and S[a,b] is the element of matrix
    den = den + r1[l] × x[l]               % S_j^{-1} at line a, column b
den = 1 − den
FOR l = 1 TO p
    FOR c = 1 TO p
        r3[l,c] = r1[l] × r2[c]            % r3 holds S^{-1} x^T x S^{-1}
        r4[l,c] = S[l,c] + r3[l,c]/den     % r4 holds S_-^{-1}
p = 0
FOR l = 1 TO p
    p = p + x[l] × β[l]                    % p holds xβ
    e = p − y                              % y is the target variable value of the observation
    g[l] = 0                               % g holds S_-^{-1} x^T (xβ − y)
    FOR c = 1 TO p
        g[l] = g[l] + r4[l,c] × x[c] × e
    β_-[l] = β[l] + g[l]
```

4 Conclusions

In this paper we have addressed the issue of growing regression trees with linear models in the leaves in a computationally efficient way. We have derived a set of formulae that allow a significant reduction in the cost of evaluating all candidate test splits for each tree node. We have presented an algorithm based on these formulae, that has an average complexity of the order $O\left(3p^2\right)$. This algorithm ensures that this type of trees can be obtained in an efficient manner thus allowing the use of this method in larger data sets.

References

BONTEMPI,G. (2000): *Local Learning Techniques for Modeling, Prediction and Control.* PhD thesis, Universit Libre de Bruxelles, Belgium.

BREIMAN,L., FRIEDMAN,J., OLSHEN,R. and STONE,C. (1984): *Classification and Regression Trees.* Statistics/Probability Series. Wadsworth & Brooks/Cole Advanced Books & Software.

CATLETT,J. (1991): *Megainduction: machine learning on very large databases.* PhD thesis, Basser Department of Computer Science, University of Sydney.

CLEVELAND,W.S. and LOADER,C.R. (1995): Smoothing by local regression: Principles and methods (with discussion). *Computational Statistics.*

GOODWIN,G. and SIN,K. (1984): *Adaptive Filtering Prediction and Control.* Prentice-Hall.

KARALIC,A. (1992): Employing linear regression in regression tree leaves. In *Proceedings of ECAI-92.* Wiley & Sons.

MYERS,R. (1990): *Classical and modern Regression with Applications, 2nd edition.* Duxbury Press.

QUINLAN,J.R. (1992): Learning with continuous classes. In Adams & Sterling, editor, *Proceedings of AI'92*, pages 343–348. World Scientific.

TORGO,L. (1999): *Inductive Learning of Tree-based Regression Models.* PhD thesis, Faculty of Sciences, University of Porto.

Neural Networks and Genetic Algorithms

A Clustering Based Procedure for Learning the Hidden Unit Parameters in Elliptical Basis Function Networks

Marilena Pillati and Daniela G. Calò

Dipartimento di Scienze statistiche, Università di Bologna,
Via Belle Arti, 41, 40126 Bologna, Italy
(e-mail: pillati@stat.unibo.it - calo@stat.unibo.it)

Abstract. Radial basis function (RBF) neural networks can be used for a wide range of applications as they can be regarded as universal approximators and their training is faster than that of multilayer perceptrons. A more general version of these neural networks are referred to as elliptical basis function (EBF) networks. In this paper a robust method for EBF parameter estimation is proposed, based on hyperellipsoidal clustering and on multivariate sign and rank concepts. A simulation study on a classification problem has shown that this method represents a valid learning scheme, particularly in presence of outlying data.

1 Introduction

Artificial neural networks are a particularly rich class of non linear models, often used in regression, classification and time series prediction. In this context, radial basis function (RBF) networks provide an attractive alternative to the well known multilayer perceptrons due to their faster learning capability.

RBF networks are single-hidden-layer feedforward neural networks, in which the activation functions of the hidden units are radially symmetric, often Gaussians, each operating a non linear transformation of the p-dimensional input space and giving a localised response (see Bishop, 1995 and Ripley, 1996 for more details). In a Gaussian RBF network, the mapping $g_j(\cdot)$ performed by the j-th output unit can be represented as a linear combination of k basis functions ϕ_h (h=1,...,k), each represented by a hidden unit, as follows:

$$g_j(\mathbf{x}) = \alpha_{j0} + \sum_{h=1}^{k} \alpha_{jh}\phi_h(\mathbf{x}) \tag{1}$$

where

$$\phi_h(\mathbf{x}) = \exp\left(-\frac{1}{2\sigma_h^2}||\mathbf{x} - \mu_h||^2\right) \tag{2}$$

In (1) and (2), $\mathbf{x}$ is an input vector, the h-th basis function is defined by the centre vector μ_h and by the width $\sigma_h > 0$, α_{jh} is the weight of the connection from the h-th hidden unit to the j-th output one, and $||\cdot||$ denotes the Euclidean norm. The constant term α_{j0} is added to compensate for the difference between the average value over the data set of the basis function activations and the corresponding average value of the targets.

Replacing the Euclidean distance in the standard RBF networks by the Mahalanobis distance gives rise to more general neural network models, often referred to as elliptical basis function (EBF) networks. In these models full covariance matrices are incorporated into the basis functions. In a Gaussian EBF network, the activation function of the h-th hidden unit is defined as:

$$\phi_h(\mathbf{x}) = \exp\left(-\frac{1}{2\sigma_h^2}(\mathbf{x}-\mu_h)' \Sigma_h^{-1} (\mathbf{x}-\mu_h)\right) \qquad (3)$$

where Σ_h is the covariance matrix and $\sigma_h > 0$ controls the spread of the basis function and hence the degree of smoothness of the function realised by the network, as in (2).

The output unit in (1) simply implements a multiple linear regression in the space of hidden units. For this reason, the network parameters are usually estimated following a two-stage procedure: first, the hidden unit parameters are learned, then the weights of the connections to the output units are estimated according to the least squares optimality criterion.

The first stage of the estimation process is of crucial concern, since it strongly influences the network performance.

In this paper a robust clustering based procedure to estimate the EBF parameters is proposed. In Section 2, after a brief summary of some of the most relevant contributions in RBF networks, we illustrate the need for hyperellipsoidal clustering in EBF parameter estimation.

In Section 3, a robust version of the elliptical k-means clustering algorithm of Sung and Poggio (1998) is proposed. It involves the spatial median and the spatial Kendall's τ covariance matrix as an estimate of multivariate location and scatter, respectively.

In Section 4, we evaluate the benefits of employing this learning procedure through a preliminary simulation study in a two-class problem.

2 Clustering based estimation of RBF and EBF parameters

Several strategies for selecting the basis function parameters, so as to form a representation of the input data distribution, have been proposed.

Regarding the problem of basis function centre estimation, one solution is to locate the centres at a collection of training elements, selected by random sampling or via subset selection or orthogonal least squares. As an alternative, the use of clustering techniques to find a set of centres which more accurately reflects the distribution of the data in the input space has been proposed (see Bishop, 1995, for a review). The theoretical foundations of clustering based estimation methods may be found in Poggio and Girosi (1990), who showed that a gradient descent approach used to update the basis function centres actually "... makes the centres move toward the majority of the data, to find the position of the clusters". Therefore, basis functions with localised responses should be placed in regions of data concentration.

With regard to RBF networks, one of the most important and broadly accepted contributions is due to Moody and Darken (1989). They propose to cluster the input vectors via the k-means clustering algorithm and to locate the basis functions at the cluster centroids, each basis function being representative of a cluster of training data. Moreover, they suggest several nearest neighbour heuristics to determine the RBF widths.

As many other partitioning clustering algorithms, the k-means one is suitable for detecting hyperspherical-shaped clusters, since it uses the Euclidean metric to evaluate the distance between a point and a cluster centre. As a consequence, this approach is appropriate for RBF networks, which are weighted sums of hyperspherical basis functions. However, the use of the Euclidean distance assumes that all input features are equally important. Moreover, the RBF units in (2) achieve high accuracy in the representation of the training set when the components of the training vectors in each cluster are not correlated. If these assumptions do not hold, a large number of radial basis functions is needed to detect regions where these assumptions locally hold. The use of full covariance matrices in the Gaussian basis functions, as in (3), could improve the accuracy of the network, the number of basis functions being smaller (Musavi *et al.*, 1992). They allow for differences in the scales of, and intercorrelations among, the variables.

As k-means clustering is appropriate for RBF networks, an hyperellipsoidal clustering method (see, for example, Mao and Jain, 1996; Cerioli, 2001) should be employed to estimate the EBF parameters. It guarantees that the flexibility of EBF hidden units can be exploited in modeling the input data distribution.

In this paper, the elliptical clustering algorithm proposed by Sung and Poggio (1998) has been considered. This algorithm represents a modified version of the k-means algorithm which uses an adaptively changing *normalised Mahalanobis distance* instead of the Euclidean metric to partition the data sample. The normalised Mahalanobis distance between an input vector $\mathbf{x}_i$ and the centroid $\mathbf{m}_c$ of cluster c is defined as:

$$D_M(\mathbf{x}_i, \mathbf{m}_c) = \frac{1}{2}\left(p \ln 2\pi + \ln |S_c| + (\mathbf{x}_i - \mathbf{m}_c)' S_c^{-1} (\mathbf{x}_i - \mathbf{m}_c)\right) \tag{4}$$

where S_c is the covariance matrix of cluster c. It is the negative natural logarithm of the best fit multidimensional Gaussian distribution for cluster c. It is preferred to the standard Mahalanobis distance because of stability reasons, *i.e.* in order to avoid unusually large clusters, which could eventually overwhelm the smaller ones. The algorithm returns the k cluster centroids and the corresponding covariance matrices, which can be used to the estimate the EBF parameters.

3 A robust hyperellipsoidal clustering solution

The accuracy of the basis function parameter estimates based on clustering methods deteriorates when outlying observations are present, as well as the one of other estimate solutions (*e.g.* Musavi *et al.*, 1992).

As all variance minimisation techniques, the k-means clustering algorithm is vulnerable to outliers. More precisely, it tends to isolate large outliers. This may lead to locate some basis functions in low density regions of the input space, whereas basis functions with localised response should be placed in regions of data concentration. Moreover, even if outliers are not isolated by the clustering algorithm, they still may affect the centre estimates, moving some centroids away from the bulk of the corresponding clusters. As a result, if outliers are present the k-means algorithm leads to a set of centres which fails to reflect how the training set is scattered in the input space.

In order to cope with this problem, a possible approach is to detect and to remove outliers before applying the clustering procedure (Lay and Hwang, 1999). As an alternative, the effectiveness of robust clustering methods to locate the RBF centres has been explored in Pillati and Calò (2001).

Elliptical basis function networks can be highly sensitive to outliers as well. In fact, the presence of outlying observations may also strongly affect the estimates of the covariance matrices.

Since in an RBF hidden unit the covariance matrix Σ is set equal to the identity matrix, it suffices to robustly estimate the basis function centre to solve the problem, the width parameter σ being only related to the distance to nearby centres. On the contrary, in an EBF hidden unit the robust centre estimation can only prevent that an inaccurate estimate of the mean vector affects the sample covariance matrix, but cannot avoid extreme values to perturb it significantly. Therefore, robust estimates of both the centre and the covariance matrix are needed.

Many robust techniques for estimating multivariate location and scatter have been proposed. In this context, some multivariate sign and rank concepts have been recently introduced and their use in the estimation of both the location parameter and the covariance matrix has been investigated (Visuri *et al.*, 2000). In this framework, it is possible to define several well known multivariate medians.

We focus our attention on the so called spatial median. Given a set of n points in R^p, $\mathbf{x}_1, \mathbf{x}_2, ..., \mathbf{x}_n$ its spatial median is the vector $\hat{\theta}$ that minimises the sum of the Euclidean distances to the points, that is

$$\hat{\theta} = \arg\min \sum_{i=1}^{n} ||\mathbf{x}_i - \theta|| \tag{5}$$

in direct analogy to the univariate sample median. When contrasted with the centroid, for which the sum of the squared Euclidean distances is minimum, the spatial median reveals its robustness properties, which have been extensively investigated, and provides a solution to the problem of robustly estimating the RBF centres (Pillati and Calò, 2001).

One of its natural companions in scatter estimation is the spatial Kendall's τ covariance matrix. It can be defined as

$$TCM = ave \left\{ \mathrm{S}\left(\mathbf{x}_i - \mathbf{x}_j\right) \mathrm{S}\left(\mathbf{x}_i - \mathbf{x}_j\right)' \right\} \tag{6}$$

where S denotes the so called spatial sign function, that is $\mathrm{S}(\mathbf{x}) = \mathbf{x}/\left\|\mathbf{x}\right\|$. Other scatter estimators are defined as well, such as the covariance matrix of the n observed spatial signs with respect to the spatial median $\hat{\theta}$ (see Visuri *et al.*, 2000 for more details).

The eigenvectors of the spatial TCM are convergent estimates of the eigenvectors of the true covariance matrix. Moreover, its squared condition number estimates the theoretical one. Hence it provides only shape and orientation estimates of the data cloud. Therefore, the covariance matrix estimation must follow the strategy described below.

1) Find a matrix U, whose columns are the eigenvectors of the spatial TCM.
2) Estimate the eigenvalues using any univariate robust scale estimate.
3) Denoted with Λ the diagonal matrix containing these estimates, the covariance matrix estimate is $\hat{\Sigma} = U \Lambda U'$.

We replace the centroid and the sample covariance matrix in the clustering procedure proposed by Sung and Poggio (1998) by the spatial median and the covariance matrix $\hat{\Sigma}$, respectively.

Instead of the spatial TCM, the spatial sign covariance matrix or the spatial rank based one can be employed in the strategy as well.

4 Simulation results and concluding remarks

The robustness of the proposed solution against outlying observations has been tested in the context of classification problems. Its performances have been compared to those of the k-means based RBF network and to those of the EBF network based on Sung and Poggio's clustering algorithm. The estimation of the width parameters followed a nearest neighbour based method, as suggested in Moody and Darken (1989). The procedures have been implemented in a GAUSS code.

	UNCONTAMINATED SETTING		CONTAMINATED SETTING	
k	RBF	*robust RBF*	RBF	*robust RBF*
2	0.474 (0.022)	0.486 (0.025)	0.504 (0.043)	0.514 (0.028)
3	0.435 (0.039)	0.424 (0.030)	0.470 (0.039)	0.462 (0.036)
4	0.255 (0.037)	0.249 (0.059)	0.251 (0.023)	0.243 (0.037)
5	0.230 (0.015)	0.215 (0.015)	0.243 (0.017)	0.230 (0.014)
6	0.208 (0.016)	0.199 (0.016)	0.234 (0.021)	0.217 (0.017)
7	0.195 (0.012)	0.189 (0.013)	0.227 (0.019)	0.198 (0.018)
8	0.180 (0.013)	0.175 (0.012)	0.212 (0.023)	0.178 (0.015)

Table 1. Mean error rates and standard errors (in brackets) of RBF networks with different basis centre location methods. Columns 3 and 4 refer to corrupted training sets.

Here we report the results of a simulation experiment dealing with a 2-class problem in 2 dimensions. The classes have equal probability. Class 1 is drawn from a uniform mixture of two Gaussian distributions with common diagonal covariance matrix, with mean vectors (2,1) and (8,7) respectively and variances equal to (9,1). Class 2 is generated from a uniform mixture of two Gaussians with the same diagonal covariance matrix, with mean vectors (8,2) and (2,6) respectively and variances equal to (1,9). A contaminated setting was obtained by drawing 10% of the data points from the uniform distribution on the square $[-10, 20]^2$.

Both in the uncontaminated setting and in the contamined one, 100 training sets of 400 observations each were generated. The results are illustrated in Table 1 and Table 2, where the mean and the standard deviation of the 100 misclassification error rates, evaluated on a test set of 2000 observations, are given for each classifier.

In Table 1 the error rates of RBF networks are shown. In spite of the cluster structure of the population, the accuracy of the classifier based on the k-means algorithm improves as the number k of hidden units increases. This is consistent with the elongated cluster shape, which requires a number of hyperspherical basis functions larger than the number of data clusters. In presence of outlying observations even more basis functions are needed to get similar performances. The results show that in this context the classifier based on the robust centre estimation procedure described in Pillati and Calò (2001) is preferable. Outliers do not affect the learning of the width of the Gaussian basis functions, since it depends only on the distance between the RBF centre estimates.

Table 2 lists the misclassification error rates of the EBF networks based on the clustering algorithm due to Sung and Poggio (columns 1 and 3) and the ones based on the robust clustering algorithm we propose (columns 2 and 4).

	UNCONTAMINATED	SETTING	CONTAMINATED	SETTING
k	EBF	*robust* EBF	EBF	*robust* EBF
2	0.486 (0.043)	0.494 (0.040)	0.487 (0.046)	0.487 (0.039)
3	0.430 (0.029)	0.420 (0.026)	0.426 (0.049)	0.429 (0.037)
4	0.175 (0.033)	0.180 (0.048)	0.253 (0.019)	0.207 (0.031)
5	0.172 (0.018)	0.167 (0.018)	0.226 (0.037)	0.189 (0.027)
6	0.171 (0.016)	0.167 (0.018)	0.206 (0.029)	0.179 (0.023)
7	0.169 (0.017)	0.168 (0.015)	0.199 (0.028)	0.178 (0.023)
8	0.166 (0.015)	0.169 (0.018)	0.190 (0.024)	0.174 (0.019)

Table 2. Mean error rates and standard errors (in brackets) of EBF networks with different basis centre location methods and covariance matrix estimates. Columns 3 and 4 refer to corrupted training sets.

The number of independent adjustable parameters of RBF and EBF networks with equal number of basis functions is different. This issue must be taken into account when comparing their performances: for instance, the estimated error rate of an EBF classifier with 4 hidden units must be compared with that of an RBF network with 6 hidden units. In fact, one of the design issues is the consideration of the trade-off involved between the use of smaller number of basis functions with many adjustible parameters versus a larger number of less flexible functions.

The results of the two tables reveal that, on the uncorrupted training sets, the EBF network based on the clustering algorithm due to Sung and Poggio achieves, with only 4 hidden units, the accuracy of the RBF network based on 8 basis functions, which depends on a larger number of parameters. However, when outlying data are present, RBF networks outperform it, because they are sensitive to outliers only in the location parameter estimation. With respect to the different solutions analysed in the EBF context, Table 2 shows that in the uncontaminated setting the classifiers based on the different elliptical clustering algorithms yield similar error rates for various k values. When the training set is corrupted, the classifier based on the robust clustering algorithm we propose is affected by outliers to a lesser extent, and allows to define a more parsimonious network.

References

BISHOP, M.C. (1995): *Neural Networks for Pattern Recognition*, Clarendon Press, Oxford.

CERIOLI, A. (2001): Elliptical Clusters and the K-means Algorithm. In: *Book of Short Papers of Cladag 2001, Meeting of the Classification and Data Analysis Group of the Italian Statistical Society.*

LAY, S.R. and HWANG, J.N. (1993): Robust construction of Radial Basis Function Networks for classification, *IEEE International Conference on Neural Networks*, 1859-1864.

MAO, J. and JAIN, A. K., (1996): A Self-Organizing Network for Hyperellipsoidal Clustering (HEC), *IEEE Transactions on Neural Networks, 7*, 16-29.

MOODY, J. and DARKEN, C.J. (1989): Fast Learning in Network of Locally-Tuned Processing Units, *Neural Computation, 1*, 281-294.

MUSAVI, M., AHMED, W., CHAN, K., FARIS, K. and HUMMELS, D. (1992): On the training of radial basis function classifiers, *Neural Networks, 5*, 595-603.

PILLATI, M. and CALÒ, D.G. (2001): A Robust Clustering Procedure for Centre Location in RBF Networks. In C. Provasi (Ed.): *Modelli complessi e metodi computazionali intensivi per la stima e la previsione*, Cleup Editrice, Padova.

POGGIO, T. and GIROSI, F. (1990): Networks for approximation and learning, *Proceeding of the IEEE, 78*, 1481-1497.

RIPLEY, B.D. (1996): *Pattern Recognition and Neural Networks*, Cambridge University Press, Cambridge.

SUNG, K. and POGGIO, T. (1998): Example-Based Learning for View-Based Human Face Detection, *IEEE Transactions on Pattern Analysis and Machine Intelligence, 20*, 39-51.

VISURI, S., KOIVUNEN, V. and OJA, H. (2000): Sign and Rank Covariance Matrices, *Journal of Statistical Planning and Inference, 19*, 557-575.

Multi-layer Perceptron on Interval Data

Fabrice Rossi[1] and Brieuc Conan-Guez[2]

[1] LISE/CEREMADE, UMR CNRS 7534, Université Paris-IX Dauphine, Place du Maréchal de Lattre de Tassigny, 75016 Paris, France
[2] INRIA, Domaine de Voluceau, Rocquencourt, B.P. 105 78153 Le Chesnay Cedex, France

Abstract. We study in this paper several methods that allow one to use interval data as inputs for Multi-layer Perceptrons. We show that interesting results can be obtained by using together two methods: the extremal values method which is based on a complete description of intervals, and the simulation method which is based on a probabilistic understanding of intervals. Both methods can be easily implemented on top of existing neural network software.

1 Introduction

Interval-valued data are quite natural in many applications where they represent uncertainty on measurements (confidence intervals for instance), variability (minimum and maximum temperatures during a day), extremal behavior (maximal wind speed in a given area), etc. Many data analysis tools have been already extended to handle in a natural way interval data: Principal Component Analysis, K-means, etc. (see for instance Bock and Diday, (2000)).

In this paper we focus on nonlinear processing of interval-valued data thanks to Multi-layer Perceptrons (MLP). Several methods can be used to allow MLP to work with interval-valued data. In this paper, we present two kind of methods: the very simple extremal values approach and two probabilistic methods. Those methods can be implemented very easily on top of existing neural network software. We show that the naive center (or mean) based method should be replaced by the simulation-based approach which gives in general better results. We show on synthetic data that the simple extremal values method should be used together with the simulation-based method in order to provide meaningful results.

2 Interval processing methods for MLP

2.1 Framework

We consider in this paper that each studied individual is described by n intervals, i.e. $([\underline{x_1}, \overline{x_1}], \ldots, [\underline{x_n}, \overline{x_n}])$. The desired output can be an interval, a real output, or a class: our main concern is to be able to work as efficiently and simply as possible with interval-valued inputs.

Moreover, we consider that interval-valued inputs are kind of summary of underlying precise data. For instance, if we study the climate, we can describe a place by the minimum and maximum temperatures during the day. We consider that the interval gives a good description of temperature variations during the day. One requirement of our study is to be able to use the trained MLP both on new interval-valued inputs and on new real valued inputs. For instance, if we observe a temperature during the day, we want to be able to use it as an input to the MLP, even if it was trained with interval-valued inputs.

A very natural way to handle interval-valued inputs (and outputs) is to rely on interval arithmetic (Moore (1966)). The main idea of interval arithmetic is simply to define in a sound way interval product, sum, etc. It is easy to define an interval based MLP, which can be trained thanks to a modified back-propagation algorithm. Several authors have already worked on this kind of model (see for instance Beheshti et al., (1998), Šíma (1995) and Simoff (1996)). In this paper, we won't work with interval arithmetic for one main reason: it implies specific development, which means that this approach cannot be easily integrated in existing neural network software. Every thing (initialization, training, visualization, etc.) has to be modified and adapted to interval arithmetic and we consider this is not affordable for many practitioners.

2.2 Extremal values method

The simplest way to deal with interval-valued inputs is to transform each interval in a pair of real numbers, for instance the lower and upper bounds of the interval (or the middle and the length of the interval). We translate this way n interval inputs into $2n$ real value inputs (i.e., $([\underline{x_1}, \overline{x_1}], \ldots, [\underline{x_n}, \overline{x_n}])$ is simply replaced by $(\underline{x_1}, \overline{x_1}, \ldots, \underline{x_n}, \overline{x_n})$). The MLP is used exactly as a classical MLP with the augmented inputs. We call this approach the **extremal values method**.

In order to use a MLP trained with the extremal values method on real valued inputs, the simplest method is to replicate the data, i.e., input $(x_1, \ldots, x_n)$ becomes $(x_1, x_1, \ldots, x_n, x_n)$. It might be possible to use more elaborated methods, but this is outside the scope of this article.

2.3 Probabilistic methods

Another way to deal with interval-valued data is to consider them as simple probabilistic data. If a sample for the MLP is described by the interval $[a, b]$, a possible interpretation is to assume that in fact the sample can take any value between a and b, with uniform probability.

Considering intervals are only a way to express uncertainty, one can replace each interval by its middle (the mean) and train the network with the obtained values. We call this approach the **mean method**. When we want to use a trained MLP with new data, we replace again each interval by its middle value. We handle real valued input directly.

Another way to proceed is to replace each sample by a set of real valued samples. Those samples are obtained thanks to simulation, assuming that the interval $[a, b]$ corresponds to an uniform distribution in $[a, b]$. Moreover, if we work with multiple interval inputs, we assume that each variable (each input) is independent from the others. We call this approach the **simulation method**. For new real valued inputs, we use the trained MLP directly. For new interval-valued inputs, we generate simulated real valued inputs and we compute normally corresponding outputs. One simple way to define the output corresponding to the initial data is to use the interval of simulated outputs. The practical meaning of this interval is the variability on the output induced by the variability on the input. We can also define the output for interval-valued inputs as the mean output for simulated inputs. Note that even if the MLP as been trained with the mean approach, the simulation approach can still be used to compute the output corresponding to an interval-valued input.

3 Comparison of probabilistic methods

3.1 Theoretical discussion

The mean and simulation methods can use exactly the same neural architecture. Therefore, training a MLP with the simulation method takes longer time than training it with the mean method, simply because we have more examples for the first method. This is the main drawback of the simulation method compared to the mean method.

The main drawback of the mean approach is that the MLP is trained without any knowledge of the uncertainty on the samples and it will provide overconfident answers in difficult cases, as will be shown in section 3.2. In fact, the mean trained MLP will often have worse generalization results than the simulation trained one. Indeed, using simulated data rather than the mean is quite similar to noise injection techniques which were introduced by Sietsma and Dow, 1991. The main idea of such techniques is to add noise to input data during the training of a MLP. Simulation results in the pioneer article showed that generalization performances where much improved by this technique. Several theoretical analysis of noise injection techniques have been done (see for instance Bishop, (1995b) and Grandvalet et al., (1997)). They tend to prove that noise injection is an efficient way to improve generalization.

We can therefore consider that the simulation method is a data driven noise injection technique applied to the mean method. The main difference with noise injection is that in our case, data give a precise description of the noise which depends on the individual (and on the variable). Indeed, each input variable is associated to an observed interval value, whereas for noise injection methods, variations are artificially generated around observed values.

3.2 Simulation results

In difficult cases, uncertainty should decrease the quality of the result provided by the MLP. Let us consider a simple example. We assume that we have two classes. Elements of the first class are described by a real variable chosen uniformly in $[-1, 0]$. For the second class, the variable is chosen uniformly in $[0, 1]$. In order to take into account measurement error, each real value is replaced by a interval centered on the value and with a length of 0.2. We have 20 interval-valued examples from each class and we simulate 10 real valued examples for each original one in the simulation approach.

With the mean approach, the measurement error is not taken into account, therefore we have perfectly separated classes. A simple MLP (one input, one hidden neuron, one output) can be used to learn[1] to separate samples with no error. This is obviously an overconfident behavior. Indeed, when an input is close to zero, the MLP should not give a sharp answer (0 or 1), but on the contrary give an answer close to 0.5, as the measurement error implies that the value can come from either class.

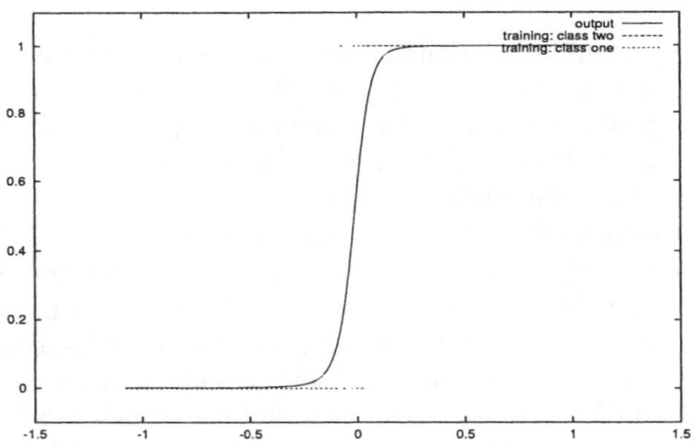

Fig. 1. Output of the MLP trained with the simulation approach

With the simulation approach, we obtain training samples from class one with strictly positive value and samples from class two with strictly negative values. When we train the MLP (similar to the one use with the mean approach), the prediction error does not reach zero (the mean square error stays around 0.02). As illustrated by figure 1, for input values close to zero,

[1] All simulations have been done with SNNS (Zell, (1995)), using Scaled Conjugate Gradient method.

the MLP does not give a sharp output, but, on the contrary, outputs varying between 0 (for class one) and 1 (for class two). In fact, as it is well known, e.g. Bishop, (1995a), the MLP approximates the posterior probability of the input to belong to the second class (which is trained with one as desired output). This behavior is better suited to imprecise inputs because it's the only correct way to show that for some inputs, we cannot obtain a class, but only a probability to belong to the classes.

It is also interesting to see what happen if we calculate the output of trained MLPs for different input intervals:

Input	mean output	interval output for mean	simulation output
$[-0.1, 0.1]$	1.00	$[0, 1]$	$[0.12, 0.95]$
$[-0.1, 0]$	0.00	$[0, 1]$	$[0.12, 0.59]$
$[-0.2, 0]$	0.00	$[0, 1]$	$[0.022, 0.59]$
$[-0.3, -0.1]$	0.00	$[0, 0]$	$[0.007, 0.12]$

It is quite clear that results provided by the simulation method are more interesting than results provided by the mean method. For the mean method, interval outputs are quite useless, even with a short interval as $[-0.1, 0]$ which cannot be classified. Moreover, the mean method classifies $[-0.1, 0.1]$ in the second class, which is not a good result. The simulation method give interesting intervals. For $[-0.1, 0.1]$, we obtain a quite broad interval which shows that it is quite difficult (and in fact meaningless) to try to classify this interval. For $[-0.1, 0]$ we obtain a wide interval, but closer to 0 than 1, therefore we can classify this interval in the first class, but the wide result shows that there is still a high probability that $[-0.1, 0]$ corresponds to a member of class two. Results obtained for other inputs show also that the simulation based approach gives more realistic results than the mean approach.

4 Comparison of simulation method and extremal values method

4.1 Theoretical comparison

The simulation method handles a n interval input with n input neurons, whereas the extremal values method needs $2n$ input neurons. Assume that we want to use a MLP with p hidden neurons and one output. Then, we have exactly $(n + 2)p + 1$ numerical parameters for the simulation based method. If we translate n intervals into $2n$ real values, we must use $(2n + 2)p + 1$ parameters. Therefore, if we have a fixed number of training patterns, we must use less hidden neurons for the extremal values method than for the simulation method, in order to obtain the same estimation quality for parameters. Obviously, this reduces the computing power of the extremal values method as illustrated in section 4.2.

Another drawback of the extremal values method is that it cannot be extended easily to arbitrary probabilistic inputs. It can obviously be applied to parametric probabilistic inputs (for instance Gaussian distributed inputs), but not to histogram inputs (as long as the number of bin is not constant).

The extremal values method has two important advantages over the simulation method. First of all, if we use roughly the same number of parameters, the training time of the simulation method is in general longer than the one of extremal values method, simply because the former uses much more examples then the latter. Moreover, in some cases, the extremal values method can classify examples than cannot be separated by the simulation method, as will be demonstrated in section 4.3.

4.2 XOR problem

As explained in previous section, extremal values method uses more numerical parameters than probabilistic methods. Let us consider the case of two dimensional interval-valued inputs. With the probabilistic methods, a MLP with two hidden neurons and one output uses 9 parameters. With the extremal values method, it uses 13 parameters (and 7 with only one neuron).

Let us consider an interval version of the XOR problem. We have four training examples, centered on $(-1,-1)$ and $(1,1)$ for the first class, and on $(-1,1)$ and $(1,-1)$ for the second class. We assume that measurement errors replace the exact values by intervals of length 0.2. It is well known that we need at least 2 hidden neurons to solve the XOR problem. The interesting point is that with the probabilistic methods, this can be done with 9 parameters, whereas we need 13 parameters for the extremal values approach.

Of course, experiments confirm this discussion (for the simulation approach, we use 10 simulated examples for each original example):

hidden neurons	method	Mean Square Error
2	extremal values	0.0
2	simulation	0.0
1	extremal values	0.17
1	simulation	0.17

4.3 Overlapping individuals

We consider a very simple two classes problem. The unique example of class one is described by the interval $[-0.5, 0.5]$, and the unique example of class two is described by the interval $[-1, 1]$. With the extremal values approach, we can use one hidden neuron to exactly classify both examples. With the probabilistic methods, the situation is far less satisfactory. Obviously, the mean method is not usable because both intervals have the same mean. For the simulation based method, we trained a MLP with two hidden neurons. Of course, we cannot correctly classify inputs that belongs to $[-0.5, 0.5]$. In fact, if we assume that simulated points are chosen uniformly in each interval and are in equal proportion, the probability that an input belongs to class one knowing that we observe a value in $[-0.5, 0.5]$ is $\frac{2}{3}$. The trained MLP agrees with this number and therefore misclassifies one third of the inputs.

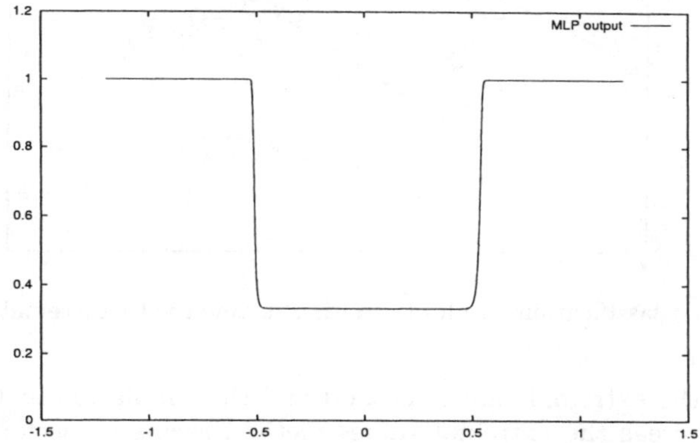

Fig. 2. Output of a MLP trained with the simulation method

Figure 2 gives that output of the trained MLP as a function of its input. As expected, the output is an approximation of the posterior probability of class two knowing the observed value. Therefore, when we submit a new input to the MLP, it gives a sound result. But if we input an interval with the simulation approach, we obtain less useful results. For instance, the output of the MLP for $[-1, 1]$ is the interval $[\frac{1}{3}, 1]$, whereas for $[-0.5, 0.5]$, we obtain approximately $[\frac{1}{3}, \frac{1}{3}]$. It is quite difficult to use this kind of result.

If we use now the extremal values MLP to classify new inputs, we also obtain unsatisfactory results which are summarized on figure 3. Any numerical input considered as a zero length interval is in fact classified into class one.

To summarize this simple numerical experiment, we have quite different results with the two proposed methods. Extremal values method gives good results on interval-valued inputs but cannot be used at all for numerical inputs (it performs exactly as a random classifier). On the contrary, simulation based method gives very useful results for numerical inputs, but cannot classify correctly interval-valued inputs.

5 Conclusion and future work

We have presented in this paper three methods that can be used to process interval-valued inputs with multi-layer perceptrons. The mean method is obviously a limited method which should be avoided as the simulation method provides in general better results (with an increased training time).

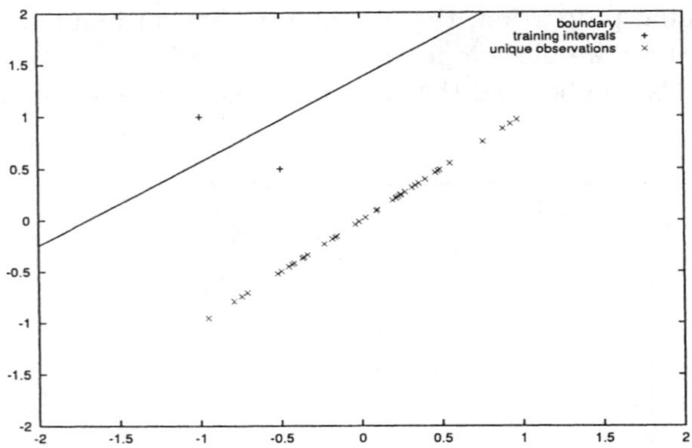

Fig. 3. Classification results for a MLP trained with extremal values

Comparing the extremal values method and the simulation method is more difficult. Whereas the extremal values method seems to perform better on interval-valued inputs, it cannot be generalized to arbitrary probabilistic inputs. Moreover, using a MLP trained with the extremal values method to classify new real valued inputs can give very incorrect results. Therefore, we recommend to use both methods together in order to add their respective qualities. We are currently implementing those methods for the ASSO (Analysis System of Symbolic Official data) European IST Project (see http://www.info.fundp.ac.be/asso/).

References

BEHESHTI, M., BERRACHED, A., DE KORVIN, A., HU, C., and SIRISAENGTAKSIN, O. (1998). On interval weighted three-layer neural networks. In *Proceedings of the 31 Annual Simulation Symposium*, pages 188–194. IEEE Computer Society Press.

BISHOP, C. (1995a). *Neural Networks for Pattern Recognition.* Oxford Press.

BISHOP, C. (1995b). Training with noise is equivalent to Tikhonov regularization. *Neural Computation*, 7(1):108–116.

BOCK, H.-H. and DIDAY, E., editors (2000). *Analysis of Symbolic Data. Exploratory methods for extracting statistical information from complex data.* Springer Verlag.

GRANDVALET, Y., CANU, S., and BOUCHERON, S. (1997). Noise injection: Theoretical prospects. *Neural Computation*, 9(5):1093–1108.

MOORE, R. (1966). *Interval Analysis.* Englewood Cliffs, New Jersey.

SIETSMA, J. and DOW, R. (1991). Creating articial neural networks that generalize. *Neural Networks*, 4(1):67–79.

SIMOFF, S. J. (1996). Handling uncertainty in neural networks: An interval approach. In *Int. Conf. on Neural Networks*, pages 606–610, Washington. IEEE.

ŠíMA, J. (1995). Neural Expert Systems. *Neural Networks*, 8(2):261–271.

ZELL, A. et al. (1995). *SNNS 4.1 user manual.* University of Stuttgart.

Part III

Applications

Textual Analysis of Customer Statements for Quality Control and Help Desk Support

Ulrich Bohnacker, Lars Dehning, Jürgen Franke, and Ingrid Renz

DaimlerChrysler Research and Technology, Information Mining, P.O.Box 2360, 89013 Ulm, Germany. Email: {ulrich.bohnacker, lars.dehning, juergen.franke, ingrid.renz}@daimlerchrysler.com

Abstract. Several business to customer applications, i.e. analysis of customer feedback and inquiries, can be improved by text mining approaches. They give new insights in the customer's needs and desires by automatically processing their messages. Previously unknown facts and relations can be detected and organizations as well as employees profit by these document and knowledge management tools. The techniques used are rather simple but robust: they are derived from basic distance calculation between feature vectors in the vector space model.

1 Introduction

The increasing use of information technology in everyone's everyday life changes the way people communicate with other people, institutions, and organizations. This development also facilitates the contact between a customer and a company with the consequence that their interaction is growing. Any process between business and customers (b2c) usually requires much effort: employees have to read, understand, process, answer the customers' messages and transform their interaction into management overviews.

Also, in near future, this message handling has to be done by human employees, but new IT-tools can be developed to support them – and additionally, to manage the messages in totally innovative ways.

For any employee, it is an easy task to really understand a customer's message. But it is a hard and in some b2c applications an unsolvable duty, to gain an overview of the customers' needs as a whole. In order to standardize any message-based process, pre-defined schemes for categorization and annotation are introduced. Such methods help to structure the customers' input and to condense and concentrate information in order to get an overview.

Although IT-tools cannot replace the employee's work, innovative approaches are able to process the customers' notes in useful but different ways. In the sections following, we present how text mining technologies are used to analyze customer feedback and inquiry messages of various domains. We start with a short overview of the techniques used.

2 Text mining: computation of relations between texts

Text mining can be defined as a special form of data mining (or knowledge discovery in databases) which is applied to large volumes of non-structured text files instead of numerical, structured data. The aim of text mining is the discovery and extraction of knowledge from text documents.

Since we investigate only written messages from customers, text mining techniques are directly applied to the customers' input.

In the following, we use techniques which are inspired from pattern recognition: Schürmann (1996) and natural language processing: Manning and Schütze (1999), Jurafsky et al. (2000) and which are also used for information retrieval tasks: Salton and McGill (1983), Sparck-Jones and Willet (1997).

Generally, we do not regard a single text on its own in isolation, but situated in a given context. This context usually consists of a whole collection where the single text is part of. Also, any knowledge discovered is grounded on this context.

According to the state of the art: Chakrabarti (2000), we think of a text as a "bag-of-words", i.e. a collection of different words, which are assembled into a "word vector", for representation. Naturally, these words can be modified by stemming algorithms or morphological analysis (which we ignore in the following short technical overview). According to their distribution in the collection, some words which are neither too frequent nor too seldom are chosen as most informative features.

TEXT
Collection: Help Desk

user is calling he is unable to access his notes mail he is getting an error msg. server not responding

WORD VECTOR	
user	1
is	3
calling	1
he	2
unable	1
to	1
access	1
his	1
notes	1
mail	1
getting	1
an	1
error	1
msg	1
server	1
not	1
responding	1

FEATURE VECTOR	
unable	0.05
access	0.17
notes	0.21
mail	0.20
getting	0.06
error	0.09
server	0.11
not	0.04
responding	0.07

Fig. 1. Exemplary text together with word and feature vector representation

For this text in Figure 1, which is part of a collection from a help desk for information technology users, only words are used as features which are not common English stop words like "is", "he", "to", etc. – defined in a stop word list by the designer of the text mining system – and which are not distributed uniformly within the collection like "user" and "calling". Also, very infrequent words of the collection – defined by system parameters – (here the abbreviation "msg") are ignored.

After the selection of most informative features, based on their frequency in the text investigated and in the whole collection, a weight for any feature is calculated (TF/IDF measure): Joachims (1998). Then, every text is represented as a normalized vector of these weights. The transformation of a text into a vector is illustrated in Figure 1. Feature selection as well as feature weighting are based solely on collection internal knowledge (i.e. frequency statistics).

Since each text is represented as a vector, common calculations on vector space models can be performed in order to acquire knowledge about the texts. For these calculations, two essential characteristics have to be taken into account: first, according to the number of features, the vectors are very large, usually consisting of tens of thousands of features, yielding a high-dimensional vector space. Second, typically a text consists only of a few dozens or hundreds of words (according to its length) resulting in very sparse vectors. These characteristics, large and sparse vectors, strongly affect subsequent computations.

The main computation on vectors is the distance calculation, a relation between two vectors. We interpret distance as the similarity between two texts: if the distance is small, the texts are similar to each other. If it is large, the texts have nothing in common. In Figure 2 (left), we give exemplary feature vectors to informally demonstrate their similarity. Furthermore, we describe the distance relation which can be computed by Euclidean distance or Cosine similarity in a 2-dimensional space, see Figure 2 (right).

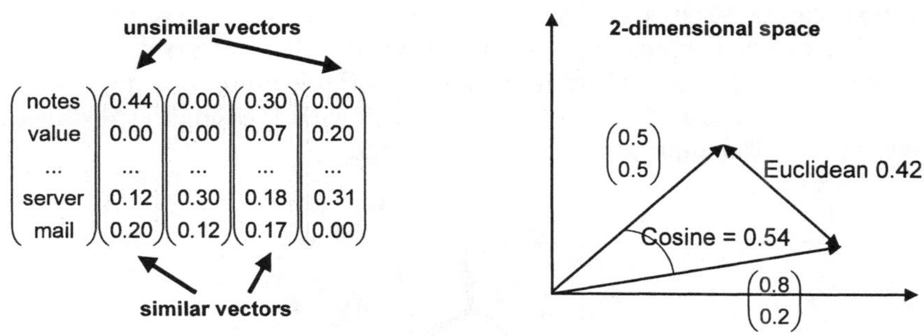

Fig. 2. Similarity relation between texts: distance calculation on vectors

Calculating the similarities between all text pairs in the collection this knowledge about all relations can be stored in an upper triangle matrix because the similarity relation is symmetric. The number of similarities is

$N * (N-1)/2$ for a collection of N texts. For further analysis of the text collection these results can be used efficiently.

When we use the Cosine similarity many of the similarities are zero (or nearly zero). The reason is that the feature vectors are – nearly – perpendicular to each other when they have – only few or – no words in common. This fact can be used to store even the similarity matrix in a sparse representation.

In the following, we use distance calculation as basic text mining technique to acquire knowledge of a text collection.

2.1 Detection of similar texts

Given a collection of texts and the distance matrix between feature vectors, which represent these texts, for any text, we can identify its similarity to all other texts. Here, similarity is contrary to distance: the smaller the distance between two vectors, the larger the similarity between the texts.

Browsing: With this representation of a text as a vector of selected features – defined by the collection itself with no human interaction – and the definition of similarity between texts, we can generate, for each text, a rank-ordered list of related – similar – texts. Due to their similarities, it is possible to offer a new browsing facility in text collections which does not search related texts by the definition of keywords given from the user, but by the collection and the starting text itself.

Categorization: If we do not consider the collection as a whole, but split it into subsets – categories – (according to a manually or automatically given label), see Figure 3, then for any of these categories a prototype representing the category can be defined. This prototype can be computed as centroid by averaging all feature vectors of one category (i.e. an artificial feature vector which does not represent an actual text). A prototype, representing a real life text, can be selected by finding that text whose feature vector is nearest to the centroid. For any new and unknown text, its category can be easily determined by calculating its distance to those prototypes representing the different categories. The category related with the smallest distance is considered as the appropriate one for this text.

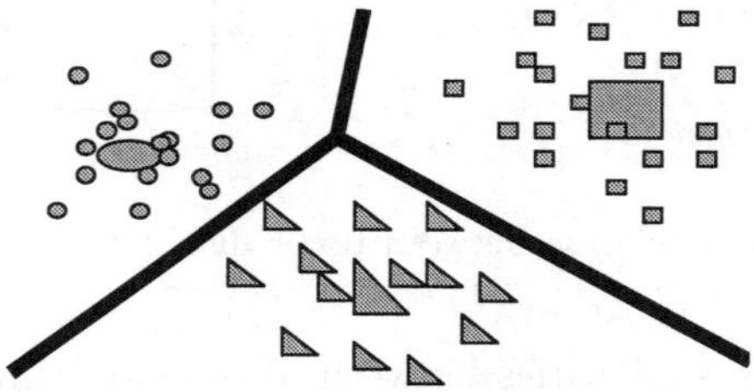

Fig. 3. Categorizing texts to computed prototypes (large symbols)

Retrieval: On the one hand, this procedure assigns each text its appropriate category. On the other hand, all the texts in the category assigned may be returned as a response of a retrieval query for related texts. This list of texts in the category can be ordered by their similarity to the prototype.

The above described techniques are all nearest neighbor approaches. These algorithms can be improved by other more elaborated techniques, like statistically motivated algorithms such as Multilayer Perceptrons, Radial Basis Functions: Schürmann (1996) or Support Vector Machines: Joachims (1998). These techniques try to separate the feature space in distinct regions by analyzing the distribution of each category. In our current applications we do not use them because they require more computational power, large and labeled data sets and the number of categories should not be too large (about 100 categories).

2.2 Detection of least similar texts: outlier detection

Having computed the distance matrix, the similarity list can be sorted in the reverse order so as to identify outliers as texts with largest distance. Different problems can be handled by this approach.

For a given subset with defined prototype(s), all texts which are least similar can be automatically detected. After manually inspecting these texts, it can be determined whether the definition of the subset was incorrect or the categorization failed, see Figure 4. On the left side of the figure, the small circles and squares represent text vectors, and the bigger ellipse is the prototype for the subset. Here, the squares are perhaps mislabeled and have to be placed manually out of the subset, or the boundary of this category should be smaller (excluding the squares).

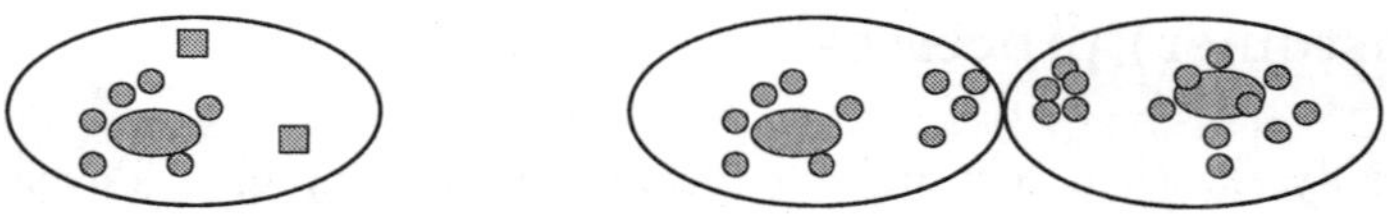

Fig. 4. Finding mis-classified texts (left) or not uniformly distributed texts (right).

Here, the distance relation is used to investigate the quality of a subset. Vectors with the largest distance to the other vectors or the prototype indicate that the corresponding texts may be in a false subset or unusual.

Sometimes just these outliers are not the problem, but the interesting information, because they show a drift to new – formerly unknown – categories, problems or topics. This situation is indicated by not uniformly distributed or badly defined categories. Especially if a group of least similar texts occurs which are very closely related to each other, see the right side of Figure 4. If the outlier group is intermediate between two subsets, they can show an unknown relation between different subsets or a new cluster of related texts which define a new category.

2.3 Detection of text groups

Different to the above mentioned applications is the detection of groups (clusters) in the text collection itself. In the above approaches, the focus was on the relation between a single text, and a whole collection or a subset of it. These subsets were defined by different tasks, sources or by labels. The relation between a single text and the text collection was applied to retrieval and classification. When the whole text collection is investigated, i.e. the relations between all texts to each other, it is a clustering task, i.e. finding unknown structures and knowledge.

This clustering task is also based on the distance calculation of feature vectors. Here, cluster algorithms: Duda (2001) are applied to split any collection into (more or less) homogeneous groups.

As a result of a cluster algorithm, a hierarchy of clusters of texts is computed and according to a threshold, groups of texts are determined. These automatically found text groups (subsets of the collection) can be afterwards defined as new categories and used like mentioned above. Furthermore, the central text of each cluster can be presented as the most typical text or as a prototype for this cluster.

Additionally the clustering result illustrates the structure of the text collection: whether there are large groups which are the topic themes in this collection or how homogeneous the collection is. Large distances between subsets indicate that they belong to very distinct topics.

3 Text mining applications in B2C (business to customer) processes

The briefly sketched text mining techniques are helpful in developing new applications for b2c processes. In the following, we describe how these innovative business applications are exploited for two different corporate services.

3.1 Analysis of customer feedback for product improvement

In the automotive industry, customer feedback is crucial for the process of product improvement. Every month, approx. 400 customers are asked about their experience with the product. Employees work with a detailed query catalogue in order to find any problem the customer had with the product. Every customer feedback is manually split into several short but homogenous statements (between 600 and 1,500 texts each month). These statements are manually labeled according to a fine-grained categorization scheme (10,396 categories which are composed of 4,597 different product parts and 284 different kind of defects), the kind of product, the production plant, and the production date. Based on this process, problem areas are found and the responsible sales department gives hints to the product development and production how to improve product quality.

The department in charge expected an automatic classification of every customer feedback – a goal which could not be reached with our technology (only 1,000 texts per month for more than 10,000 categories). But the technology is able to give a new kind of support for the whole quality process, since it enriches this process with information which was never found before and which employees could hardly find out.

The fixed categorization scheme fails when unexpected feedback is given. Even worse, it hides this feedback because more than 10,000 categories facilitate that different employees classify these statements into different classes – and no significant conclusion can be drawn anymore.

Here, the automatic detection of text groups is helpful to find clusters of unexpected feedback statements. Based on the whole collection of feedback statements, a clustering tool was developed, which proposes sensible text groups. If such a group is composed of texts differently labeled, this indicates a possible new and unexpected kind of problem area. Surely, an employee has to verify the proposed group, but not the whole collection.

Furthermore, this clustering tool gives an overview of the whole collection. The automatically clustered groups are compared to the manually labeled categories. This comparison indicates which categories correspond to clusters. The fine-grained scheme can be revised with this information.

Whereas until now mistakes can only be detected accidentally, another text mining application allows quality control of manually labeled statements. Here, it is not the whole collection but only statements of the same category which are investigated. If some statements are less similar to the other statements, this indicates that they are most probably labeled incorrectly.

These text mining tools are supplements to the familiar processes which analyze customer feedback. Their application allows a better control of these processes and new insights into the customers' needs.

3.2 Analysis of customer inquiries for user help desk

The second corporate service which uses text mining for innovative business applications is a corporate user help desk (IT-support). This help desk supports thousands of customers and processes approx. 6,000 inquiries every month.

In order to handle the customer requests, the employees use a case based reasoning (CBR) system: Aamodt and Plaza (1994). Such a CBR system compares new problems with solved ones and presents the solutions found. Basis of any CBR system are the prepared cases (combination of problem and solution).

A main task in any CBR system is to transform problem-solution pairs into cases. But for which problem is such a transformation useful and sensible? Generally, those problems which are new, interesting and numerous. So far, employees propose their problem/solutions to an authoring group which decides about new cases. Now, this process can be supported by text mining technology, too.

Given all text of the cases in the CBR database, the least similar problem/solutions texts among the new ones (respecting a frequency threshold) are automatically calculated. Besides the employees' proposals, the authoring group now uses this list of good candidates for new cases.

Here, text mining facilitates the construction of new cases, an essential part of a CBR system in any business application.

4 State of the work

In the previous section, we have described two exemplary corporate units in which text mining technology improves existing business to customer (b2c) processes. These real world applications demonstrate the benefit from our technology which is strongly influenced by pattern recognition.

For both corporate units, prototypes are transferred. For the analysis of customer feedback, a versatile tool which investigates customer feedback statements was given to the responsible sales/marketing department. At the user help desk department, our system is in use and refined for one year. Evaluation work is still in progress.

5 Future work

We will focus our future work on two main issues. First, the evaluation of our work: how do the new tools improve the existing processes? New information which results from our technology is only useful if it can be integrated into the department's work. We are monitoring the responsible departments for evaluation purposes.

The second topic is to identify further suitable business units. Tasks of customer relationship management are characterized by much interaction and communication between company and customer. In near future, we expect that this interaction is still growing and that further corporate services of this kind will benefit from our technology.

References

AAMODT, A. and PLAZA, E. (1994): Case-Based Reasoning: Foundational issues, methodological variations, and system approaches. In: *AI Communications (7:1)*, 39-59.

CHAKRABARTI, S. (2000): Data Mining for Hypertext: A Tutorial Survey. In: *SIGKDD Explorations (1:2)*, 1-11.

DUDA, R., HART, P. and STORK, D. (2001): *Pattern Classification.* Wiley, New York.

JOACHIMS, T. (1998): Text Categorization with Support Vector Machines: Learning with Many Relevant Features. In: *Proceedings of the European Conference on Machine Learning.* Springer.

JURAFSKY, D., MARTIN, J. and VAN DER LINDEN, K. (2000): *Speech and Language Processing: An introduction to Natural Language Processing, Computational Linguistics and Speech Recognition.* Prentice Hall.

MANNING, C. and SCHÜTZE, H. (1999): *Foundations of Statistical Natural Language Processing.* MIT Press, Massachusetts.

SALTON, G. and MCGILL, M. (1983): *Introduction to Modern Information Retrieval.* McGraw Hill.

SCHÜRMANN, J (1996): *Pattern Classification.* Wiley, New York.

SPARCK-JONES, K. and WILLET, P. (Eds.) (1997): *Readings in Information Retrieval.* Morgan Kaufmann.

AHP as Support for Strategy Decision Making in Banking

Czesław Domański[1] and Jarosław Kondrasiuk[2]

[1] University of Łódź, Chair of Statistical Methods
ul. Rewolucji 1905, 41, 90-214 Łódź, Poland
[2] LG Petro Bank S.A., Department of Economic Planning and Analyses
ul. Rzgowska 34/36, 93-172 Łódź, Poland

Abstract. This article is an approach to the problem of group decision making in which one of decision makers plays a decisive role. Due to our research we will focus on banking sector and possible choice concerning marketing strategy under a very limited cost budget. Our solution to the problem is an adjusted Analytic Hierarchy Process (AHP) application model.

1 Establishing the problem

Our problem is how to find an optimal marketing strategy under a very limited cost budget. In theory a Marketing Department should adjust marketing policy of the firm to appearing limitations. However, the Marketing Department also participates in the process of creation a financial plan - influencing the possible level of costs and incomes. At the same time - from a final decision maker point of view, it is crucial to find a solution profitable for the company. Due to negotiation tricks usually used in such a situation, like total negation of any possibility to decrease promotion costs (due to market competition etc.) or efforts to limit non-marketing costs, it is a must to build an objective decision support mechanism.

The approach to the problem usually will follow below described steps:

1. Building a preliminary version of a financial plan for the following period (usually a year) - enforcing making decisions concerning aims of marketing strategy and therefore proper cost budget.
2. Creating an AHP model adjusted to the complexity of the problem and decision maker/makers preferences.
3. Finding an optimal solution.
4. Sensitivity analyses (this is a final part of the AHP method but in this application we have to underline importance of this step).
5. A final decision concerning the financial plan.

2 Application of the AHP in banking marketing policy

2.1 The Analytic Hierarchy Process

The AHP is a multicriteria decision support method created by Thomas L.Saaty. This method is oriented on providing an objective way for reaching an optimal decision for both individual and group decision makers. The AHP allows selecting the best alternative from a number of alternatives evaluated with respect to several criteria. It is taken by carrying out pairwise comparison judgements that are used to develop overall priorities for ranking the alternatives. This method allows for some level of inconsistency in judgements (that is unavoidable in practice) and provides some measures for limiting that.

The AHP is a general theory of preference measurement with providing necessary information for choosing the best decision. In the AHP process there are four main stages:

1. Building a hierarchy model.
2. Identifying the preferences of decision makers.
3. Synthesis.
4. Sensitivity analyses.

The basic AHP model consists of three levels: goal, criteria level and alternatives. Depending on complexity of the problem it is possible to add as many as necessary levels of subcriteria.

The most complex problem is identification of decision maker preferences. In AHP it is done by collecting information about pairwise judgements due to a goal (for criteria), a specified criterion (for alternatives or subcriteria) or a subcriterion (for alternatives). The are a few possible scales of converting collected information into numeric form - however it is not always necessary. Having one set of information we build a matrix of ratio comparison for a given goal/criterion. It is possible to find many ways of converting the matrix $\mathbf{A}$ (matrix of ratio comparison) into the vector of priorities $\mathbf{w}$. However, the need of consistency make us to choose the eigenvalue formulation $\mathbf{A}\mathbf{w} = n\mathbf{w}$. Assuming that the priorities $\mathbf{w} = (w_1, \dots, w_n)^T$ with respect to a single criterion are known, such as the weights of stones - we can examine what we have to do to recover them. Having the matrix $\mathbf{A}$:

$$\mathbf{A} = \begin{bmatrix} \frac{w_1}{w_1} & \frac{w_1}{w_2} & \cdots & \frac{w_1}{w_n} \\ \vdots & \vdots & \ddots & \vdots \\ \frac{w_n}{w_1} & \frac{w_n}{w_2} & \cdots & \frac{w_n}{w_n} \end{bmatrix}$$

we multiply it on the right by $\mathbf{w}$

$$\mathbf{w} = \begin{bmatrix} w_1 \\ \vdots \\ w_n \end{bmatrix}$$

to obtain $n\mathbf{w}$. Elements a_{ij} of the matrix of ratio comparison represents the importance of alternative i over alternative j. In order to guarantee the judgements to be consistent relevant groups of the matrix elements have to follow the equation: $a_{ij} \cdot a_{jk} = a_{ik}$. In case, we do not have a scale at all, or do not have it convenient as in the case of some measuring devices - we can only give an estimation of w_i/w_j. It leads to the problem:

$$\mathbf{A}'\mathbf{w}' = \lambda_{max}\mathbf{w}'$$

where λ_{max} is the principal eigenvalue of $\mathbf{A}' = (a'_{ij})$ the perturbed value $\mathbf{A} = (a_{ij})$ with the reciprocal $a'_{ji} = 1/a'_{ij}$ forced. The solution is obtained by raising the matrix to sufficient large power - then summing over the rows and normalizing to obtain the priority vector $\mathbf{w}' = (w'_1, \dots, w'_n)^T$. The above mentioned process is stopped when the difference between components of the priority vector obtained at k-th power and at the $(k+1)$-st power is less than some predermined small value. The vector of priorities is the derived scale associated with the matrix of comparisons. The value zero in this scale is assigned to an element that is not comparable with the elements considered.

With the eigenvector for $n \leq 3$ normalizing the geometric means of the rows leads to an approximation to the priorities. In all the cases it is possible to get an approximation by normalizing the elements of each column of the judgement matrix these steps can lead to rank reversal (in spite of closeness of the eigenvector solution).

A simple way to obtain the exact value (or an estimate) of λ_{max} when the exact value of $\mathbf{w}'$ is available in normalized form is to add the columns of $\mathbf{A}'$ and multiply the resulting vector by the priority vector $\mathbf{w}$.

After obtaining the principal eigenvector estimate $\mathbf{w}$ we should consider the question of consistency. The problem is arisen by the fact that the original matrix $\mathbf{A}$ need not to be transitive, for example A_1 may be preferred to A_2 and A_2 to A_3 but A_3 may be preferred to A_1. The solution to this problem is the consistency index $(C.I.)$ of a matrix of comparison defined as:

$$C.I. = \frac{\lambda_{max} - n}{n-1}.$$

The consistency ratio $(C.R.)$ is obtained by comparing the $C.I.$ with the appropriate one of the following set of numbers (Table 1) each of which is an average random consistency index derived from a sample of randomly generated reciprocal matrices. The study of the problem and revision of the judgements should be completed if $\frac{C.I}{R.I.} \geq 0.1$.

Table 1. Average Random Consistency Index *R.I.* published in Saaty (1986)

N	1	2	3	4	5	6	7	8	9	10
Random Consistency Index	0.00	0.00	0.52	0.89	1.11	1.25	1.35	1.40	1.45	1.49

The above solution to the problem is consider to be classical Saaty solution (Saaty (1994) and Domański, Kondrasiuk (2000)) and is used for reaching both local and global vectors of priorities - necessary for synthesis.

Hierarchical synthesis is obtained by a process of weighting and adding down the hierarchy leading to multilinear form. There are two possible modes of the synthesis:

- the distributive mode in which the principal eigenvector is normalized to yield a unique estimate of ratio scale underlying the judgements;
- the ideal mode in which the normalized values of alternatives for each criterion are divided by the value of the highest rate alternative.

The final step is sensitivity analysis that gives an answer to a question whether the alternative chosen as the best would be changed in case of modifying criteria/subcriteria preferences.

Additionally, we present in Table 2 a scale allowing conversion the intensity of judgements into values of relative importance.

Table 2. The Fundamental Scale published in Saaty and Vargas (1994)

Intensity of Importance	**Definition**	**Explanation**
1	**Equal importance**	**Two activities contribute equally to the object**
2	Weak	
3	**Moderate importance**	**Experience and judgement slightly favour one activity over another**
4	Moderate plus	
5	**Strong importance**	**Experience and judgement strongly favour one activity over another**
6	Strong plus	
7	**Very strong or demonstrated importance**	**An activity is favoured very strongly over another; its dominance demonstrate in practice**
8	Very, very strong	
9	**Extreme importance**	**The evidence favouring one activity over another is of the highest possible order of affirmation**

2.2 Application of the AHP

After occurring a necessity of choosing a new marketing strategy we should build an AHP model adjusted both to the problem complexity and preferences of a main decision maker. Following AHP methodology we have structured the AHP hierarchy presented in Figure 1.

A new approach to the AHP methodology is applied at the level of criteria. For us a criterion is an organisation part of a bank. The pairwise preferences due to a specified criterion in this approach mean generalised preferences between possible alternatives of a department (represented by its director) or a committee (achieved by internal negotiations or even voting). The consequence of this methodological adjustment is a different meaning of pairwise comparison due to a goal - which represent our main decision maker (in our case: a president of Management Board) pairwise preferences between suggestions made by departments and committees. It is crucial for the success of the whole project to assure the president of full confidentiality of his preferences or even allow him personally to finish the model (with proper software support and internal training). Lack of a proper protection of final decision maker preferences might discourage the president from using this adjusted AHP methodology for making strategic decisions.

The final version of the model used for choosing new marketing strategy consists of four AHP criteria:

- Management Board - in case when the main decision maker is a president this criterion describes aggregated pairwise preferences of the other members of Management Board (instead of this criterion it is possible to get separate criteria for each Management Board member);
- ALCO - Assets and Liabilities Management Committee provides an opinion on bank's needs from liquidity point of view (in general: an answer to a question whether bank needs to increase a pace of growth of loans or deposits);
- Marketing Department - based of in-depth analysis of bank's marketing weaknesses and opportunities pairwise summarisation of Departments preferences;
- Planning Department - a combined efficiency, available cost budget and development targets based presentation of preferences.

In presented model we have simplified possible alternatives to:

- General Image - focusing in media on brand name promotion;
- Deposits - marketing of liability products;
- Loans - marketing of asset products;
- Services - promotion of other products like banking cards, bankassurance products etc.

Fig. 1. The tree level hierarchy used for selecting marketing strategy of the bank

Our model is adjusted to a bank with centralised system of decision making. In such an organisation the most important thing is pairwise comparison of criteria level - personally made by bank's president. After reaching an optimal solution according to the AHP methodology - sensitivity analyses provides to the final decision maker a possibility to reconsider his pairwise preferences.

We shall remember that presented model is very general - it gives only an idea of marketing policy goals. In practice, alternatives will be more specific due to specific product categories, target group of clients, media selection, timing, geographic area selection etc. Depending on all the parties involved in this decision making process, there is almost no limitation concerning level complexity or building sub-models with using AHP methodology.

3 Summary

The proposed approach to the problem of group decision making is much wider than marketing strategy. Applying the AHP method might lead to limiting internal conflicts concerning problems in which different groups (even alliances) are involved. Depending on organisation culture and tradition, the final decision maker does not have to be a president of the organisation - it might be a management board or board of directors. Furthermore, this method allows building multi-level models or a system of several models - meaning a high flexibility for applications.

Even lack of success in some cases is not a disadvantage because the AHP method enhances decision makers knowledge of the problem complexity and makes it easier to create decision making procedures. Finally, the AHP incorporates both data and judgements into one logically consistent model.

References

DOMAŃSKI, C. and KONDRASIUK, J. (in publishing): Establishing the Bank Rates with Using Statistical Methods.

DOMAŃSKI, C. and KONDRASIUK, J. (2000): Implementing of Analytic Hierarchy Process in Banking, *Acta Universitatis Lodziensis - Folia Oeconomica, 152*, WUL, Lodz.

DOMAŃSKI, C., KONDRASIUK, J., MORAWSKA, I. (1997): Zastosowanie analitycznego procesu hierarchicznego w przedsiebiorstwie. In: T. Trzaskalik (Ed.): *Zastosowania Badan Operacyjnych*, Absolwent, Lodz.

SAATY, T. L. (1986): Axiomatic Foundation of the Analytic Hierarchy Process, *Management Science, 32*, No. 7.

SAATY, T.L. (1994): *Fundamentals of Decision Making and Priority Theory with the Analytic Hierarchy Process*, Vol. VI, RWS Publications, Pittsburgh.

SAATY, T.L., Vargas, L.G. (1994) : *Decision Making in Economic, Political, Social and Technological Environments with the Analytic Hierarchy Process*, Vol. VII, RWS Publications, Pittsburgh.

Bioinformatics and Classification: The Analysis of Genome Expression Data

Berthold Lausen

Department of Medical Informatics, Biometry and Epidemiology
University of Erlangen-Nuremberg
Waldstr. 6, D-91054 Erlangen, Germany

Abstract. The availability of microarray data caused a new interest in clustering and classification methods. DNA microarrays are likely to play an important role for diagnosis and prognosis in clinical practice. Using the example of gene expression of diffuse large B-cell lymphona I introduce and review proposals for the experimental design and pattern recognition problems of gene expression experiments, the supervised learning or classification problem, the unsupervised learning or clustering problem and the potential of improving prognostic models. Moreover, I suggest boostrap methods to estimate the stability of the used hierarchical clustering. The proposal is applied to prognostic factors by micro array data for censored survival data.

1 Introduction

The recent completion of a draft sequence of the human genome resulted in huge public interest in genomics and bioinformatics. Post genome projects involve data and experiments which need biometrical considerations. For example the availability of micro array data caused a new interest in clustering and classification methods for clinical and biological applications. Gene expression analysis is already a standard tool in genomic research projects. Diagnostics based on DNA microarrays are likely to play an important role in standard clinical practice in the next years (Aitman, 2001). It was shown that gene expression of cancer tissues improve discrimination or supervised classification of known cancer types (Golub et al., 1999). Brown et al. (2000) applied support vector machines and Zhang et al. (2001) recursive partitioning. Eilers et al. (2001) discuss the problem of having more variables than observations and reviews suggestions from chemometrics as well. Alizadeh et al. (2000) demonstrate that unsupervised classification allows the identification of new cancer (sub)types and they demonstrated that gene expression has the potential to improve clinical prognosis.

In Section 2 I introduce briefly the data analytic challenges of DNA micro array data and pattern recognition problems of the gene expression experiments (Eisen, 1998). One aspect is the large number of available genetic markers on an array, e.g. 18000, and the relative small number of tissue samples, e.g. 40. Using the example of gene expression of diffuse large B-cell lymphona the unsupervised learning or clustering problem is discussed in Section 3 and the potential of improving prognostic models in Section 4.

2 Gene expression of diffuse large B-cell lymphona

Using different expression patterns of the genome it is an interesting and straightforward hypothesis to assess types of a disease on the molecular level. Alizadeh et al. (2000) analyze the overall survival of patients with diffuse large B-cell lymphoma (DLBCL) and show that it is possible to identify two distinct types of DLBCL by gene expression profiling. They use a micro array with genes that are expressed in lymphoid cells and genes with known or suspected roles in processes important in immunology and cancer. Overall they use 16384 cDNA clones and 96 normal and malignant lymphocyte tissue samples. The first step of the data analysis (cf. Eisen 1998) is to identify each spot on the micro array and to classify the spot pixels and the background pixels. The ratio of the red and green fluorescence can be estimated. I use the simple robust estimate MRAT of Eisen et al. (1998) which is defined as the median of the ratio for each spot pixel. Each spot value is centered by the median of the background pixels. I use the data available at `llmpp.nih.org/lymphoma/`. Restricting the analysis to complete data, i.e. expression spot available for each combination of marker gene and tissue sample and survival times observed, I analyse data of 7680 marker genes and 40 tissue samples.

3 Unsupervised classification of gene expression data

In the following x_{ij} denote gene expression values, e.g. $log(MRAT)$ values, where $i = 1, ..., G$, $j = 1, ..., N$. G is the number of marker genes, N the number of tissue samples analysed and $X \in R^{G \times N}$ the matrix of the standardized expression values. Eisen et al. (1999) use hierarchical cluster analysis to find groups of similar expressed genes and to display genome-wide expression patterns. Instability is a well known problem of hierarchical cluster analysis. Consequently, measures of stability should be used. Lausen & Degens (1988) suggested a bootstrap approach for distance data and Felsenstein (1988) introduced a bootstrap approach for DNA sequence data. Both suggestion can be adapted to micro array data.

Hilsenbeck et al. (1999) applied principal component analysis. Fellenberg et al. (2001) used correspondence analysis and discussed biplot representations. Alter et al. (2000) suggested a singular value decomposition (SVD) to derive a joint representation of the marker gene space and the tissue sample space. Similar to correspondence analysis such a joint representation can be seen as a first and fast solution of a seriation problem: Find an order of the marker genes and an order of the tissue samples that puts high gene expression values close to the diagonal of X and low gene expression values away from the diagonal: With $\tilde{X} = X - X1_N/N - X^t 1_G/G$ and the SVD of $\tilde{X} = \tilde{U}\tilde{D}\tilde{V}^t$; an approximate solution to the seriation problem is given by the order of the coefficients of the first vector of $\tilde{U}$ for the gene markers and by the order of the first vector of $\tilde{V}$ for the tissue samples (patients). D denotes the diagonal matrix of the singular values, which are sorted from the largest to the smallest. The figure (below) shows such a representation of the 7680×40 gene expression values. The mean expression value of 40 markers is used for one entry in the shown 192×40 matrix. Non negative means are shown white and negative black. The seriation allows to identify pairs of patient groups and gene marker groups with relatively high or low expression values. Moreover, association structure of the tissue samples and gene markers provide information regarding the possibilities to find two way structure. The SVD above can be used for a first assessment of the data and especially for a fast informative graphical representation.

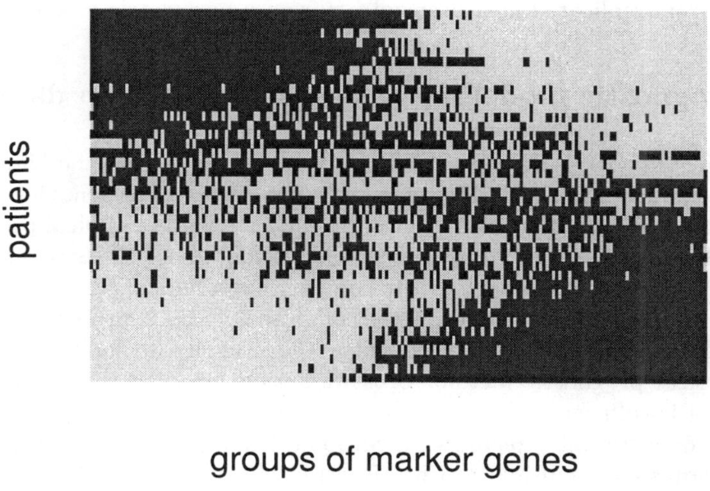

Fig. 1. Seriation analysis: Joint representation of similarities of marker gene space and tissue sample space. Mean expression value of 40 markers is used for one entry in the shown 192×40 matrix.

Hastie et al. (2000) developed a nested principal component analysis - gene shaving - as a method for identifying sets of genes with similar expression patterns, which have an interpretation in terms of functional genomics. The unsupervised version of their suggestion starts with the matrix of the standardized expression values $X \in R^{G \times N}$. Afterwards, they suggest to compute the leading principal components of the rows of X and shave of a proportion $\alpha > 0$ of marker genes having smallest inner-product with the leading principal component. Repeat the shaving until only one gene remains, which results in a sequence of nested clusters of genes $S_N \supset S_{k_1} \supset S_{k_2} \supset ... \supset S_{k_1}$. The optimal cluster of the sequence is defined by a "gap statistic" $Gap(k)$. With $SS_w = \frac{1}{N}\sum_{j=1}^{N}\left[\frac{1}{k}\sum_{i \in S_k}(x_{ij} - \bar{x}_{.j})^2\right]$, $SS_b = \frac{1}{N}\sum_{j=1}^{N}[\bar{x}_{.j} - \bar{x}]^2$, $SS_T = SS_b + SS_w$ and $R^2 = 100 SS_b/SS_T$. D_k denotes R^2 of k-th member of sequence of clusters. Using a permutation test with X^{*b} each row permutated, D_k^{*b} denotes R^2 of the permutated matrix and $\bar{D_k^*}$ denotes the mean. With the definition $Gap(k) = D_k - \bar{D_k^*}$, the "optimal cluster size " is given by $\hat{k} = argmax\ Gap(k)$. Each row of X is orthogonalized regarding the average gene in the optimal cluster $S_{\hat{k}}$ and the process above is repeated M times. Consequently, the proposal can be seen as a cluster method with M overlapping clusters. Kerr & Churchill (2001) and van der Laan & Bryan (2001) suggest to assess the results by bootstrap techniques derived from replicated measurements or from a parametric model.

4 Prognostic modelling with gene expression data

Alizadeh et al. (2000) demonstrated that gene expression has the potential to improve clinical prognosis. They show that two groups of patients have a similar prognostic value than two groups given by a standard clinical prognostic index. Hastie et al. (2001) suggest gene harvesting to use clusters of the hierarchical clustering of marker genes to compute hypothetical prognostic factors for the observed survival times. They compute average expression values for each marker gene cluster. Consequently, they suggest to look at $G(G-1)$ hypothetical prognostic factors. Crossvalidation implies that a model with one factor is sufficient. Moreover, they observe that one cluster demonstrates a high prognostic value regarding a cutpoint defined by the median, but they do not adjust for multiple testing. Bootrap evaluation of hierarchical cluster analysis allows a consideration of the stability of the estimated clusters, therefore a reduction of the multiple testing problem is given by considering stable clusters only.

Using the data of 40 subjects as introduced in section 2, I perform an overall analysis of the prognostic value of the micro array data. I compute a conservative lower bound of the exact distribution of a maximally selected logrank statistic (Hothorn & Lausen, 2001; Lausen & Schumacher, 1992 and 1996) for the overall average gene expression, and I obtain a significant ($P < 0.05$) difference between two groups of patients as well. Figure 2 shows the empirical process of the maximally selected logrank statistic and Figure 3

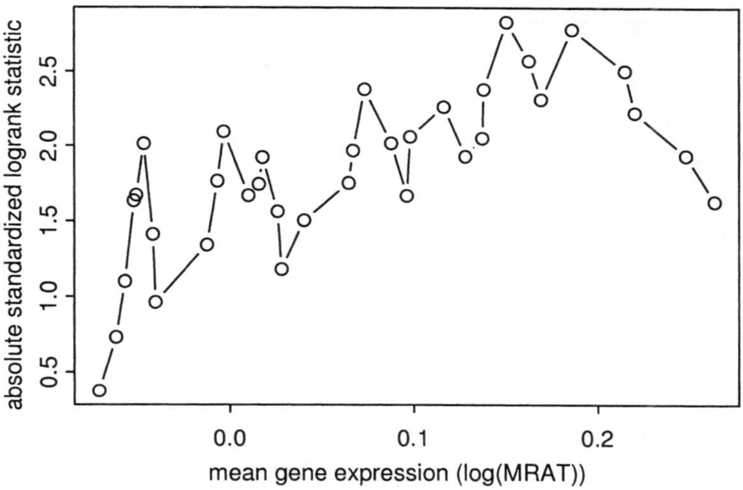

Fig. 2. Prognostic value of mean gene expression: The empirical process of the maximally selected logrank statistic. The maximum is obtained for 0.1506.

gives the Kaplan-Meier curve as graphical assessment of the prognostic value for the optimal cutpoint at 0.1506 mean gene expression.

5 Discussion

Data analysis of DNA microarray gene expression provides various challenges for unsupervised learning methods in bioinformatics. Supervised learning or classification was successfully applied to DNA microarrays in many applications in the molecular sciences. I review methods suggested for unsupervised learning or cluster analysis. Moreover, I discuss that better solutions are needed to adjust for the instability of hierarchical clustering and to adjust for the overoptimism caused by multiple tests. I suggest to use bootstrap methods for stability assessments, which were introduced for evaluation in phylogenetic inference. The prognostic modelling of DNA microarrays indicate in the example of overall survival of patients with diffuse large B-cell lymphoma (DLBCL), that there is no convincing evidence for a higher order model, i.e. for clusters of genes which outperform the mean gene expression value of all gene markers. Future randomised clinical trials or other clinical studies will

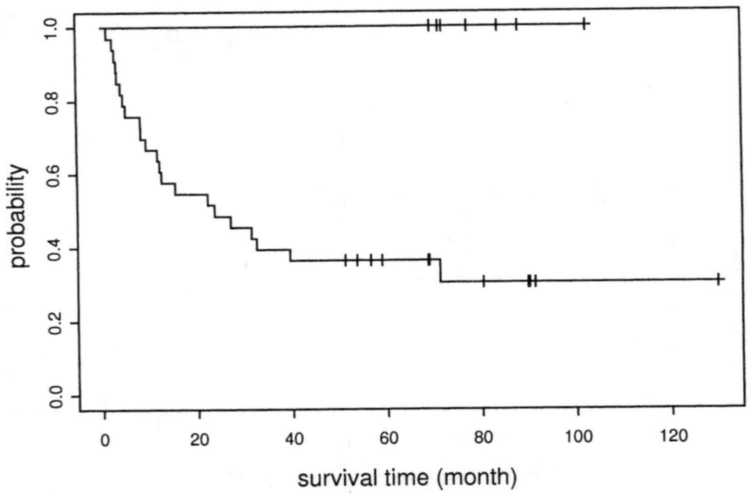

Fig. 3. Prognostic value of mean gene expression: Kaplan-Meier curve for the optimal cutpoint (Maximally selected logrank test, Hothorn and Lausen 2001, $P < 0.05$).

provide larger numbers of patients or of tissue samples with observed overall survival of patients and it will be possible to search for models with more parameters, for example for subgroups and interactions by tree models.

References

AITMAN, T.J. (2001): DNA microarrays in medical practice. *British Medical Journal, 323*, 611–615.

ALIZADEH, A., EISEN, M., DAVIS, E., et al. (2000): Distinct types of diffuse large B-cell lymphoma identified by gene expression profiling. *Nature, 403*, 503–511.

ALTER, O., BROWN, P.O., and BOTSTEIN, D. (2000): Singular value decomposition for genome-wide expression data processing and modeling. *Proceedings National Academy of Science, 97*, 10101–10106.

BROWN, M.P.S., GRUNDY, W.N., LIN, D., et al. (2000): Knowledge-based analysis of microarray gene expression data by using support vector machines. *Proceedings National Academy of Science, 97*, 262–267.

EILERS, P.H.C., BOER, J.M., OMMEN, G.J. van, and HOUWELINGEN, H.C. van (2001): *Classification of microarray data with penalized logistic regression.* Preprint, Leiden University.

EISEN, M. (1998): *ScanAlyse User Manual.* User manual, Stanford University.

EISEN, M.B., SPELLMAN, P.T., BROWN, P.O., and BOTSTEIN, D. (1998): Cluster analysis and display of genome-wide expression patterns. *Proceedings National Academy of Science, 95*, 14863–14868.

FELLENBERG, K., HAUSER, N.C., BRORS, B., NEUTZNER, A., HOHEISEL, J.D., VINGRON, M. (2001): Correspondence analysis applied to microarray data. *Proceedings National Academy of Science, 98*, 10781–10786.

FELSENSTEIN, J. (1988): Phylogenies from molecular sequences: Inference and reliability. *Annual Review of Genetics, 22*, 521–565.

GOLUB, T.R., SLONIM, D.K., TAMAYO, P., et al. (1999): Molecular classification and class prediction by gene expression monitoring. *Science, 286*, 531–537.

HASTIE, T., TIBSHIRANI, R., EISEN, M., et al. (2000): Gene shaving as a method for identifying sets of genes with similar expression patterns. *Genome Biology, 1*, research0003.1–21.

HASTIE, T., TIBSHIRANI, R., BOTSTEIN, D., and BROWN, P. (2001): Supervised harvesting of expression trees. *Genome Biology, 2*, research0003.1–12.

HEYDEBRECK, A.v., HUBER, W., POUSTKA, A., and VINGRON, M. (2001): Identifying splits with clear separation: A new class discovery method for gene expression data. *Bioinformatics, 17 Suppl*, S107–S114.

HILSENBECK, S.G., FRIEDRICHS, W.E., SCHIFF, R., et al. (1999): Statistical analysis of array expression data as applied to the problem of tamoxifen resistance. *Journal of the National Cancer Institute, 91*, 453–459.

HOTHORN, T., and LAUSEN, B. (2001): *On the exact distribution of maximally selected rank statistics.* Preprint, University of Erlangen-Nuremberg.

KERR, M.K., and CHURCHILL, G.A. (2001): *Bootstrapping cluster analysis: Assessing the reliability of conclusions from microrarray experiments. Proceedings National Academy of Science, 98*, 8961–8965.

LAUSEN, B., and DEGENS, P.O. (1988): Evaluation of the reconstruction of phylogenies with DNA-DNA hybridization data. In: Bock, H.H. (Ed.): *Classification and related methods of data analysis.* North Holland, Amsterdam, 367–374.

LAUSEN, B., SAUERBREI, W., and SCHUMACHER, M. (1994): Classification and regression trees (CART) used for the exploration of prognostic factors measured on different scales. In: Dirschedl, P., and Ostermann, R.(Eds.): *Computational statistics.* Physica Verlag, Heidelberg, 483–496.

LAUSEN, B., and SCHUMACHER, M. (1992): Maximally selected rank statistics. *Biometrics, 48*, 73–85.

LAUSEN, B., and SCHUMACHER, M. (1996): Evaluating the effect of optimized cutoff values in the assessment of prognostic factors. *Computational Statistics and Data Analysis, 21*, 307–326.

VAN DER LAAN, M.J., and BRYAN, J. (2001): Gene expression analysis with the parametric boostrap. *Biostatistics, 2*, 445–461.

YEUNG, K.Y, HAYNOR, D.R., and RUZZO, W.L. (2001): Validating clustering for gene expression data. *Bioinformatics, 17*, 309–318.

ZHANG, H., YU, C-Y., SINGER, B., and XIONG, M. (2001): Recursive partitioning for tumor classification with gene expression microarray data. *Proceedings National Academy of Science, 98*, 6730–6735.

Glaucoma Diagnosis by Indirect Classifiers

Andrea Peters, Torsten Hothorn, and Berthold Lausen

Department of Medical Informatics, Biometry and Epidemiology
University of Erlangen-Nuremberg
Waldstr. 6, D-91054 Erlangen, Germany

Abstract. Medical decision making is based on various diagnostic measurements and the evaluation of anamnestic information. For example glaucoma diagnosis is based on several direct and indirect assessments of the eye. We discuss possibilities to use a definition of glaucoma to improve supervised glaucoma classification by laser scanning image data. The learning sample consists of laser scanning image data and other diagnostic measurements.
We discuss indirect supervised classification proposed by Hand et al. (2001), which provides a framework to incorporate the full information of the learning sample as intermediate variables. We compare direct classifiers like linear discriminant analysis, classification trees and bagged classification trees with indirect classification. Comparing bagged direct and bagged indirect classifiers, we achieve a reduction of estimated misclassification error from 18.3% to 15.4%.

1 Introduction

We consider glaucoma classification as an example of medical decision making. For this neuro-degenerative disease the diagnosis is based on measurements of visual field loss, perimetry and papillometry. Visual field loss often can be observed only at a late state of the disease. Therefore methods for early diagnosis are of special interest. A common examination is the analysis of the optic nerve head morphology using a three dimensional laser scanning image. Our goal is to evaluate a classification rule which is based on measurements of the optic nerve head morphology and anamnestic variables. The indirect approach classifies patients using a definition of glaucoma based on predictions of visual field, perimetry and papillometry by Heidelberg Retina Tomograph (HRT) measurements and anamnestic data.
A methodological framework proposed by Hand et al.(2001) called *indirect classification* subdivides a given set of variables into three groups: explanatory, intermediate and response variables. We assume a priori information or a structural classification rule depending on intermediate variables. Following this framework we construct a classifier in two steps: In a first step models are fitted for the prediction of a set of intermediate variables by the explanatory variables. In a second step the data is classified by the given classification rule.

2 Glaucoma classification by laser scanning image data

Glaucoma is a slow and irreversible neuro-degenerative disease which affects the retinal nerve fibre layer, see Coleman(1999). The onset is usually not detected by the patients. There are several possibilities to examine the visual field defects and the optic nerve head (ONH) of the affected subjects. Common examinations are for example measurements of the visual field loss with the flicker test, papillometry, perimetry or a three dimensional topographical analysis of the optic nerve with the Heidelberg Retina Tomograph (HRT), see Swindale et al. (2000), Mardin et al. (1999). The HRT is a confocal scanning laser tomograph that produces a series of 32 images, each of 256 × 256 pixels, which are converted to a single topography image where each pixel represents a depth value, see Heidelberg Engineering(1997). Variables which describe the

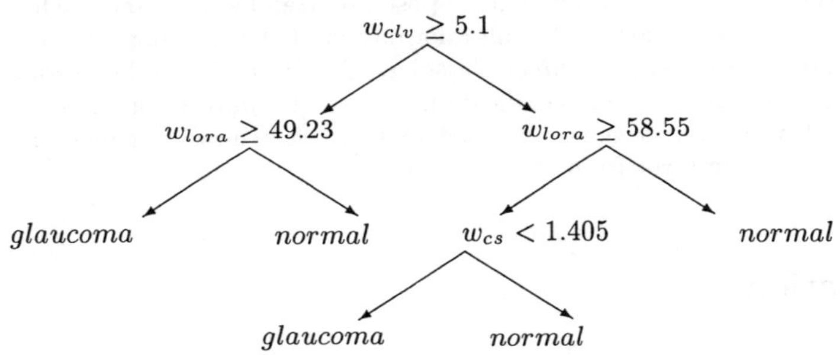

Fig. 1. Diagnosis of glaucoma: Based on the intermediate variables w_{lora} (loss of rim area), w_{cs} (contrast sensitivity), w_{clv} (corrected loss variance) a patient is classified according to the graph.

visual field defect and the proportion of retinal nerve fibres loss are used to define the disease, although a unique definition of glaucoma is controversial. For example Lee et al.(1998) observed that there are several different definitions of normal tension glaucoma, which is a special case of glaucoma. We introduce a simplified and artificial definition of glaucoma which is based on medical knowledge and depends on the following intermediate variables: Loss of rim area w_{lora}, corrected loss variance w_{clv} and contrast sensitivity w_{cs}. Loss of rim area describes the proportion of the papilla which does not consist of the rim area, i.e. which does not consist of retinal nerve fibres, and is measured using stereo optic disc photographs. Corrected loss variance describes the loss of variance of the visual field corrected by the shorttime fluctuation, it is observed by perimetry. The flicker test describes the contrast sensitivity of the eye. Based on these three intermediate variables a patient is classified as normal, if $g(\mathbf{w}) = 0$ and as glaucomatous if $g(\mathbf{w}) = 1$, where

$$g(\mathbf{w}) := \text{Inv}_{\{w_{clv} \geq 5.1\}}(w_{clv})\text{Inv}_{\{w_{lora} \geq 49.23\}}(w_{lora}) + \text{Inv}_{\{w_{clv} < 5.1\}}(w_{clv})\text{Inv}_{\{w_{lora} \geq 58.55\}}(w_{lora})\text{Inv}_{w_{cs} < 1.405}(w_{cs}), \quad (1)$$

with $\mathbf{w} = (w_{lora}, w_{cs}, w_{clv})$ and Inv(.) denotes the indicator function. The function $g(\mathbf{w})$ for glaucoma diagnosis is also displayed in Figure 1.

3 Indirect classification and bagging

In this section we discuss indirect classifiers more formally and introduce bagged indirect classification. Let $\mathcal{L} = \{(y_i, \mathbf{w}_i, \mathbf{x}_i), i = 1, \dots, N\}$ denote a learning sample of N independent observations. The responses $y_i \in \{1, \dots, J\}$ are the class labels. The intermediate variables are q-dimensional vectors $\mathbf{w}_i = (w_{i1}, \dots, w_{iq}) \in \mathbb{R}^q$. The class labels are defined by a known function $g : \mathbb{R}^q \to \{1, \dots, J\}$ which classifies an observation based on the intermediate variables only: $y_i = g(\mathbf{w}_i)$. The p explanatory variables are denoted by $\mathbf{x}_i = (x_{i1}, \dots, x_{ip}) \in \mathbb{R}^p$. The intermediate variables are related to the explanatory variables by a function $f : \mathbb{R}^p \to \mathbb{R}^q$, that is $\mathbf{w}_i = f(\mathbf{x}_i) + \varepsilon$, where ε is a random vector. The learning sample is a random sample from some distribution function F:

$$(y_1, \mathbf{w}_1, \mathbf{x}_1), \dots, (y_N, \mathbf{w}_N, \mathbf{x}_N) \overset{\text{iid}}{\sim} F. \quad (2)$$

The marginal distributions of $(y, \mathbf{x})$ and $(\mathbf{w}, \mathbf{x})$ are denoted by $F_{y,\mathbf{x}}$ and $F_{\mathbf{w},\mathbf{x}}$, respectively. Let $\mathcal{L}_{y,\mathbf{x}} = \{(y_i, \mathbf{x}_i), i = 1, \dots, N\} \sim F_{y,\mathbf{x}}$ denote the restricted learning sample without intermediate variables. Note that in the framework of indirect classification only new explanatory variables $\tilde{\mathbf{x}}$ are available for predicting future $\tilde{y}$, see Figures 2 and 3. Direct classification methods try to estimate the composition $g \circ f$. More formally, a direct classifier $C^d(\tilde{\mathbf{x}}, \mathcal{L}_{y,\mathbf{x}})$ predicts future $\tilde{y}$-values for a new observation $\tilde{\mathbf{x}}$ based on a learning sample of responses and explanatory variables. We use linear discriminant analysis

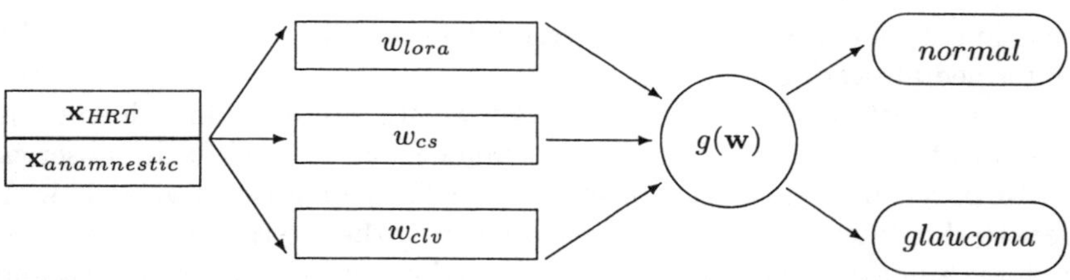

Fig. 2. Indirect classification: Models are constructed based on the explanatory variables $\mathbf{x}_{HRT}$ and $\mathbf{x}_{anamnestic}$ to predict the intermediates w_{lora}, w_{clv} and w_{cs}. Suspects are classified according to $g(\mathbf{w})$ as *glaucoma* or *normal*.

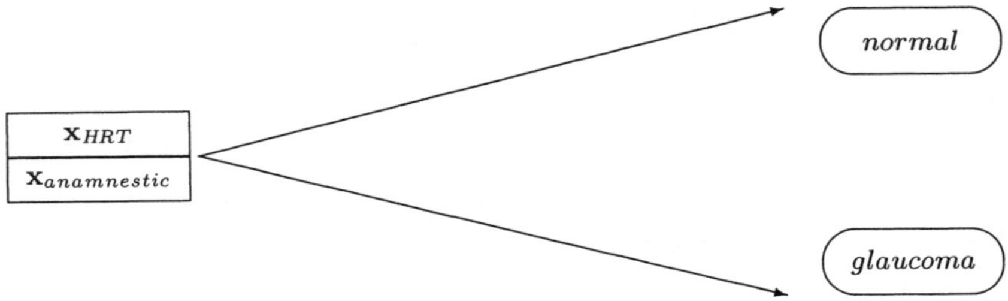

Fig. 3. Direct classification: Glaucoma classification rules are constructed based on the explanatory variables $\mathbf{x}_{HRT}$ and $\mathbf{x}_{anamnestic}$.

(LDA) and classification trees (CTREE) as classifiers C^d as well as bootstrap aggregated classification trees following Breiman(1996), Breiman(1998), denoted by "bagged-CTREE". The aggregated direct classifier C_A^d is given by

$$C_A^d(\tilde{\mathbf{x}}) = E_F C^d(\tilde{\mathbf{x}}, \mathcal{L}), \tag{3}$$

i.e. the expectation of $C^d(\tilde{\mathbf{x}}, \mathcal{L})$ with respect to the distribution of the learning sample and estimated by the bootstrap:

$$\hat{C}_A^d(\tilde{\mathbf{x}}) = E_{\hat{F}_{y,x}} C^d(\tilde{\mathbf{x}}, \mathcal{L}_{y,\mathbf{x}}^*), \tag{4}$$

where the expected value is over $\mathcal{L}_{y,\mathbf{x}}^*$, a random sample from the empirical distribution function $\hat{F}_{y,\mathbf{x}}$. $\hat{C}_A^d(\tilde{\mathbf{x}})$ is approximated by drawing a finite number of bootstrap samples, for details see Breiman(1996).

Indirect classification methods use g and estimate f instead of the composition $g \circ f$. Let $R(\tilde{\mathbf{x}}, \mathcal{L})$ predict future intermediate variables $\tilde{\mathbf{w}}$ for a new observation $\tilde{\mathbf{x}}$ based on the learning sample $\mathcal{L}$. An indirect classifier $C^{ind}(\tilde{\mathbf{x}}, \mathcal{L})$ predicts future $\tilde{y}$-values for a new observation $\tilde{\mathbf{x}}$ based on a learning sample $\mathcal{L}$ by applying g to the predicted intermediate variables:

$$C^{ind}(\tilde{\mathbf{x}}, \mathcal{L}) = g(R(\tilde{\mathbf{x}}, \mathcal{L})). \tag{5}$$

We use linear models (LM) and regression trees (RTREE) as predictors for the intermediate variables.

In both direct and indirect classification we are interested in reducing the misclassification error, that is the proportion of misclassified future observations. However, in indirect classification we optimise regression models with respect to their mean squared error for predicting the intermediate variables. Therefore, we have to deal with two loss functions: misclassification loss and squared error loss. In the following we use the framework given by Efron and Tibshirani(1997). Let Q_1 denote the misclassification loss function

$$Q_1(\hat{y}, y) = \begin{cases} 1 & \hat{y} \neq y \\ 0 & \hat{y} = y \ . \end{cases} \tag{6}$$

The expected true error rate μ_1^d for direct classification is denoted by

$$\mu_1^d = E_{F_{y,\mathbf{x}}} E_{F_{\tilde{y},\tilde{\mathbf{x}}}} Q_1(\tilde{y}, C^d(\tilde{\mathbf{x}}, \mathcal{L}_{y,\mathbf{x}})), \tag{7}$$

where the expectation is over future observations $(\tilde{y}, \tilde{\mathbf{x}})$ and learning samples $\mathcal{L}_{y,\mathbf{x}}$. For indirect classification, the expected true error rate is given by

$$\mu_1^{ind} = E_F E_{F_{\tilde{y},\tilde{\mathbf{x}}}} Q_1(\tilde{y}, C^{ind}(\tilde{\mathbf{x}}, \mathcal{L})) = E_F E_{F_{\tilde{y},\tilde{\mathbf{x}}}} Q_1(\tilde{y}, g(R(\tilde{\mathbf{x}}, \mathcal{L}))). \tag{8}$$

Here $E_{F_{\tilde{y},\tilde{\mathbf{x}}}}$ refers to expectation over future observations $(\tilde{y}, \tilde{\mathbf{x}}) \sim F_{\tilde{y},\tilde{\mathbf{x}}}$ and E_F to learning samples $\mathcal{L} \sim F$. In contrast to misclassification loss, let Q_2 denote squared error loss

$$Q_2(R(\tilde{\mathbf{x}}, \mathcal{L}), \tilde{\mathbf{w}}) = \sum_{j=1}^{q} (R(\tilde{\mathbf{x}}, \mathcal{L})_j - \tilde{w}_j)^2. \tag{9}$$

The mean squared error is defined by

$$\mu_2 = E_F E_{F_{\tilde{\mathbf{w}},\tilde{\mathbf{x}}}} Q_2(\tilde{\mathbf{w}}, R(\tilde{\mathbf{x}}, \mathcal{L})), \tag{10}$$

where again E_F is with respect to learning samples $\mathcal{L}$ and $E_{F_{\tilde{\mathbf{w}},\tilde{\mathbf{x}}}}$ refers to expectation over future explanatory and intermediate variables $(\tilde{\mathbf{w}}, \tilde{\mathbf{x}})$.

However, there is no indication that improving the prediction model R with respect to mean squared error μ_2 does improve the misclassification error μ_1 of the indirect classifier C^{ind} simultaneously. Friedman(1997) gives a discussion on the connection of mean squared error and misclassification error in the context of estimated class probabilities.

We try to improve the misclassification error μ_1^{ind} by bagging indirect classification. We define the aggregated indirect classifier as

$$C_A^{ind}(\tilde{\mathbf{x}}) = E_F C^{ind}(\tilde{\mathbf{x}}, \mathcal{L}) = E_F g(R(\tilde{\mathbf{x}}, \mathcal{L})), \tag{11}$$

where the expectation is with respect to learning samples $\mathcal{L}$. $C_A^{ind}(\tilde{\mathbf{x}})$ is estimated by the bootstrap in the usual way

$$\hat{C}_A^{ind}(\tilde{\mathbf{x}}) = E_{\hat{F}} C^{ind}(\tilde{\mathbf{x}}, \mathcal{L}^*) = E_{\hat{F}} g(R(\tilde{\mathbf{x}}, \mathcal{L}^*)), \tag{12}$$

and approximated by the following procedure:

1) Draw B bootstrap samples $\mathcal{L}^{*^{(1)}}, \ldots, \mathcal{L}^{*^{(B)}}$ from $\mathcal{L}$.
2) Construct a predictor R for each bootstrap sample $\mathcal{L}^{*^{(b)}}, b = 1, \ldots, B$.
3) A new observation $\tilde{\mathbf{x}}$ is classified by majority voting over all predictions of the response $g(R(\tilde{\mathbf{x}}, \mathcal{L}^{*^{(b)}}))$ for $b = 1, \ldots, B$.

Using regression trees we denote the procedure outlined above "BID-RTREE" (bootstrap aggregated indirect classification using RTREE). Efron and Tibshirani(1997) argue that the expected true error rates can be estimated using cross-validation. Therefore, we use 10-fold cross-validation to estimate the expected true error rates μ_1^d and μ_1^{ind}.

4 Indirect glaucoma classification

Data from a cross-sectional study including 85 glaucomatous and 85 normal eyes from the Erlangen Glaucoma Registry are used cf. Mardin et al.(1999). Only the measurements of the first examination of one eye of each patient are taken. The variables are obtained by HRT, papillometric and visual field examinations and include anamnestic information. Normal and glaucomatous subjects are matched by age and sex, to adjust for possible confounders.

The definition of glaucoma based on the intermediate variables is known (Figure 1). Furthermore it is sensible to reduce the number of necessary examinations to make less demand on the patients time and to reduce costs. Hence the aim is to evaluate classifiers that predict glaucoma on HRT and anamnestic data only. The outcome of visual field, perimetry, flicker test and papillometry examinations are predicted from the HRT measurements. The predictions are used to classify glaucoma.

direct classification		indirect classification	
LDA:	0.276	LM:	0.282
CTREE:	0.251	RTREE:	0.207
bagged-CTREE:	0.183	BID-RTREE:	0.154

Table 1. Cross-validation-estimators of the misclassification error: Direct approaches are linear discriminant analysis (LDA), classification trees (CTREE), bagged classification trees (bagged-CTREE), the indirect classification includes linear models (LM), regression tress (RTREE), bagged indirect classification using RTREE (BID-RTREE).

We assume that for future examinations only HRT and anamnestic data are available. Hence, only these sets of variables are used as explanatories $\mathbf{x}$. There are 64 HRT and 7 anamnestic explanatories. The HRT variables include several measurements of the volume and areas of the papilla in certain parts of it. The anamnestic variables include gender, age, maximum intra-occular pressure, intra-occular pressure at the examination day, familiar disposition, blood pressure and body mass index of the patients. The intermediate variables are papillometric and visual field variables w_{lora}, w_{cv} and w_{clv} as described above. Furthermore, the function $g(\mathbf{w})$ for classifying glaucoma based on the intermediate variables is known and fixed. In order to classify an new observation consisting only of the anamnestic and HRT measurements, a predictive model for the intermediate has to be fitted. Linear models and regression trees are used as predictive models R. Note that each intermediate variable is predicted by a univariate predictor, i.e. each intermediate variable is modelled separately. The bagged indirect classifier introduced in Section 3 is constructed using regression trees. The functionality of indirect and direct classification is displayed in Figure 2 and 3 for the given example of glaucoma diagnosis.

We additionally apply the direct classifiers LDA, CTREE and bagged-CTREE, which are based on a learning sample consisting of the explanatories. Bagging is performed using $B = 25$ bootstrap replications. A 10-fold-cross-validation (CV) is calculated on the given set of observations to compute an estimate of misclassification error. The results of the indirect and direct classification approaches are listed in Table 1. The CV error is 28.2% for the indirect approach using LM. The error is smaller if a regression tree is used to predict the intermediates, namely 20.7%. BID-RTREE results in a misclassification error estimate of 15.4 %. Direct classification by LDA achieves an error estimation of 27.6%. Training a CTREE on the given data set of explanatories, the misclassification error estimate is 25.1%. The error can be reduced onto 18.3% by bagging the tree. Comparing bagged direct and indirect classifiers, the reduction in estimated misclassification error from 18.3% to 15.4% indicates the gain achieved by using a priori knowledge.

5 Discussion

Indirect classification which includes a priori knowledge about the classification process is applied to glaucoma classification. We show that the estimate of the misclassification error can be reduced by incorporating the definition of glaucoma diagnosis. Bagging the classification process leads to a further reduction of the estimated error rates for our set of observations. The approach of indirect classification provides a framework which allows to model many classification problems. Lausen et al.(1994) discuss classification trees for factors measured in different scales.The combination of existing statistical models is discussed by Ciampi and Lechevallier(2000), whether it is possible to combine the output function of an artificial neural network with survival analysis. The division of variables used in the indirect classification is an approach similar to the framework of graphical modelling cf. Cox and Wermuth(1996), which can be seen as an example for an unknown relationship between intermediate and response. Another approach would be using clustering techniques in the predictive step. Torgo and da Costa(2000) proposed a method which integrates a clustering technique with regression trees. To reduce the misclassification error it could also be of interest to use the decision tree linking the intermediates and the diagnosis. Such a model can be achieved by replacing the continuous intermediate variables with binary ones which represent whether the intermediates exceed certain cutpoints or not, e.g. whether w_{clv}: $\{w_{clv} < 5.1\}$ or $\{w_{clv} \geq 5.1\}$. In summary we showed in an example of glaucoma diagnosis that it is possible to improve classification by using an indirect approach.

Acknowledgements

Financial support from Deutsche Forschungsgemeinschaft, grant SFB 539-A4/C1, is gratefully acknowledged.

References

COLEMAN, A. L. (1999): Glaucoma, *The Lancet*, *Vol. 354*, *1803–10*.

BREIMAN, L. (1996): Bagging predictors, *Machine Learning*, *Vol. 24*, 2, *123–140*.

BREIMAN, L. (1998): Arcing classifiers, *The Annals of Statistics*, *Vol. 26*, 3, *801–49*.

TORGO, L. and DA COSTA, J. P. (2000): Clustered multiple regression, in KIERS, H., RASSON, J.-P., GROENEN, P., and SCHADER, M. (Eds.): Data Analysis, Classification, and Related Methods, Springer, Heidelberg, 217–222.

FRIEDMAN, J. H. (1997): On bias, variance, 0/1-loss, and the curse-of-dimensionality, *Data Mining and Knowledge Discovery*, *Vol. 1*, *55–77*.

LEE, B. L., BATHIJA, R., and WEINREB, R. N. (1998): The definition of normal-tension glaucoma., *Journal of Glaucoma*, *Vol. 7*, 6, *366–71*.

COX, D. and WERMUTH, N. (1996): Multivariate Dependencies. Models, analysis and interpretation, Chapman & Hall, London, UK.

EFRON, B. and TIBSHIRANI, R. (1997): Improvements on cross-validation: The .632+ bootstrap method, *Journal of the American Statistical Association*, *Vol. 92*, 438, *548–560*.

HAND, D., LI, H., and ADAMS, N. (2001): Supervised classification with structured class definitions, *Computational Statistics & Data Analysis*, *Vol. 36*, *209–225*.

SWINDALE, N. V., STJEPANOVIC, G., CHIN, A., and MIKELBERG, F. S. (2000): Automated analysis of normal and glaucomatous optic nerve head topography images., *Investigative Ophthalmology and Visual Science*, *Vol. 41*, 7, *1730–42*.

MARDIN, C. Y., HORN, F. K., JONAS, J. B., and BUDDE, W. M. (1999): Preperimetric glaucoma diagnosis by confocal scanning laser tomography of the optic disc., *British Journal of Ophthalmology*, *Vol. 83*, 3, *299–304*.

HEIDELBERG ENGINEERING (1997): Heidelberg Retina Tomograph: Bedienungsanleitung Software version 2.01., Heidelberg Engineering GmbH, Heidelberg.

LAUSEN, B., SAUERBREI, W., and SCHUMACHER, M. (1994): Classification and regression trees (CART) used for the exploration of prognostic factors measured on different scales, in DIRSCHEDL, P. and OSTERMANN, R. (Eds.): Computational Statistics, Physica-Verlag, Heidelberg, 483–496.

CIAMPI, A. and LECHEVALLIER, Y. (2000): Constructing artificial neural networks for censored survival data from statistical models, in KIERS, H., RASSON, J.-P., GROENEN, P., and SCHADER, M. (Eds.): Data Analysis, Classification, and Related Methods, Springer, Heidelberg, 223–228.-

A Cluster Analysis of the Importance of Country and Sector on Company Returns

Clifford W. Sell

E-T-A
cliffsell@aol.com

Abstract. Does a company's country of incorporation or the sector of its activity have a greater influence on its equity returns? Rather than using dummy variable regressions or factor models on pre-determined characteristics, this study finds naturally occurring groups of companies based on the company returns themselves. The resulting groups are then examined in terms of their country and sector composition. The cluster analysis outcome indicates that companies group by country rather than by sector and that this effect is not diminishing.

1 Introduction

Investors, advisors, analysts and researchers use heuristics to speed up and simplify the investment decision-making process, because it is generally cumbersome and expensive to look at the entire set of investments opportunities. International investors, for example, typically aggregate international equities by country and by sector. It is interesting to know whether it is principally better to diversify across countries or to diversify across sectors, because the results affect the next steps in the investment decision. The question whether country or sector characteristics dominate equity returns is thus of fundamental importance to international investors.

Several articles address the issue of country versus industry effects on company equity returns. These studies typically use country and industry groups in dummy variable regressions or factor models. Thus, the discussion centers around the choice and definition of the regression model, as well as the breadth and length of the data series. The empirical results differ, depending on the study, not allowing for any final conclusion. One method of validating or disproving earlier findings is to apply a quantitative technique that has not been used to address this question. The cluster analysis approach used in this study analyzes the pattern of company returns over time to mathematically determine groups of companies that are alike. The companies in each group are described with respect to their country of incorporation and sector of main activity.

This study tests two hypotheses: (1) Countries have greater impact than sectors on company returns. If this hypothesis holds, we would expect companies from each country to be grouped in the same cluster. Conversely, we would expect companies from each sector to be split across different clusters.

(2) Sectors have greater impact than countries on company returns. If this hypothesis holds, we should see companies from each sector allocated to the same cluster. Conversely, we should see companies from each country divided into different clusters.

2 Literature review

The importance of country versus industry on company returns has been addressed explicitly in a number of studies. Several key papers find that the country of incorporation is of greater importance than the industry of main activity. Beckers, Grinold, Rudd, and Stefek (1992), Heston and Rouwenhorst (1994), Griffin and Karolyi (1998) find that industries explain little of the cross-sectional variation in returns, while countries explain a large portion. Beckers, Connor, and Curds (1996), Baca, Garbe, and Weiss (2000), and Cavaglia, Brightman, and Aked (2000) find an increase in sector importance over time and conclude that sector and country effects are equal for the recent past.

While the studies cited above differ in their findings and conclusions, they are typically based on some type of dummy variable regression. Their main difference lies in the choice of time period and countries covered. The goal of this study is to using an entirely different method to validate or disprove these earlier findings on countries and sectors. Cluster analysis uses data about the companies themselves in order to arrive at groups that occur naturally, rather than examining the impact of pre-defined existing company groups.

There are earlier cluster analysis studies that analyze company returns. King (1966), Meyers (1973) and Livingston (1977) use cluster analysis on equity returns to find that companies chosen from a limited range of U.S. industries will group naturally along industry lines. Farrell (1974) and Arnott (1980) find that a larger set of U.S. companies that is not constrained to a limited range of industries is grouped into four to five clusters that can be interpreted as broad sector groups. These studies generally find that cluster analysis is capable of recovering meaningful company groups, but are constrained to U.S. datasets.

3 Data and method

Any cluster analysis approach has to define "naturally occurring groups" in quantitative terms that can be mathematically optimized. This typically involves the following four steps:

Step 1: Choose a measure that characterizes the entities to be clustered. In the case of grouping companies, one such measure consists of the total split- and dividend-adjusted equity returns. This is the most interesting piece of information about a company to an investor. In this study, dollar-denominated monthly returns are taken from COMPUSTAT for 3328 to 4748 international companies for three five-year periods from 1989 to 1999. The companies are from 18 countries and distributed across all seven sectors of the Financial Times/Goldman Sachs classification scheme.

Step 2: Calculate the proximity or distance between all entities to be clustered. In most previous cluster analysis studies of financial data, the relationship between companies is expressed in terms of the Pearson correlation coefficient $\tilde{r}$. This measure is appealing because of its importance in Markowitz portfolio optimization. However, for most commercially available clustering routines, the distance cannot be negative. Taking the absolute value of $\tilde{r}$ would constrain the values of such a distance measure to be between zero and one. However, an example will illustrate the implications of such a measure: two companies that are quite alike with a correlation coefficient of 0.80 and two companies that are quite different with a correlation coefficient of –0.80 would seem to have the same relationship. Therefore, a different measure is proposed.

The distance used in this study is defined as

$$d_P(Y_1, Y_2) = \mid \tilde{r}_{1,2} - 1 \mid$$

where d_P is the Pearson distance between entities Y_1 and Y_2, and $\tilde{n}_{1,2}$ is the Pearson correlation between entities Y_1 and Y_2. Now companies with a correlation coefficient of 0.80 receive a distance measure of 0.20, while companies with a correlation coefficient of –0.80 receive a distance measure of 1.80. In other words, the measure d_P gives the companies that are more alike a lower distance value, and the companies that are more dissimilar a higher distance value. This distance measure is implemented in ClustanGraphics 4.18.

Step 3: Recover clusters from the proximity matrix. There are dozens of mathematical algorithms for finding groups, with several new algorithms appearing every month. Based on the Monte Carlo studies of clustering algorithms cited in Milligan (1996), the Ward method is chosen. This algorithm initially allocates each company to its own cluster and combines clusters such that the increase in within-cluster sum of squared distances to the cluster means are minimized at each merger.

Step 4: Evaluate the classification results. Prior to discussing the country and sector characteristics of companies in clusters, it is necessary to determine a sensible number of clusters. In this study, Beale's (1969) measure indicates two to three clusters. Next the classification results are evaluated in terms of the distribution of companies across clusters, countries, and sectors.

4 Classification results

Table 1 shows the distribution of companies across countries, sectors, and clusters for the three time periods. The majority of companies from one country typically fall into the same cluster. In all three periods, for example, over 95 percent of the Japanese companies fall into the third cluster. This indicates that the returns from Japanese companies consistently behave differently from the returns of all other companies in the sample. However, there are companies from some countries that at some point in time are almost evenly split across two of three clusters. South African companies in the 1989 to 1993 period, for example, fall into the first and the second clusters. The same is true for Canadian companies in the 1992 to 1996 period. However, these are the only two cases out of 54 possible cases in which fewer than 70 percent of the companies from a country are in the same cluster. It thus seems that a company's country of incorporation helps describe its returns behavior.

Table 1. Countries and Clusters

Country	Cluster 1			Cluster 2			Cluster 3			Total		
	89-93	92-96	95-99	89-93	92-96	95-99	89-93	92-96	95-99	89-93	92-96	95-99
USA	1131	1026	1402	23	238	3	3	8	5	1157	1272	1410
Japan	23	9	59	2	9		928	968	980	953	986	1039
United Kingdom	122	59	722	324	436	2	5	19	11	451	514	735
Canada	180	77	248	6	118			1	2	186	196	250
Germany	24	30	28	65	119	195	2	2	1	91	151	224
France	17	12	1	50	98	172				67	110	173
Australia	71	21	126	9	73		2	3	2	82	97	128
Thailand	59	8	101		88				2	59	96	103
Italy	9	9		28	83	99	1	2		38	94	99
Singapore	7	6	93	31	54					38	60	93
Netherlands	5	13	5	16	49	75				21	62	80
Chile	6	8	81	20	38	1	1		1	27	46	83
Hong Kong	3	6	72	25	47					28	53	72
Spain	7	4		19	40	56		2		26	46	56
South Africa	11	1	47	16	37			4		27	42	47
Denmark	3	2	57	25	26					28	28	57
Austria	5	3	3	24	30	44		2		29	35	47
India	20		52		25					20	25	52
Total	1703	1294	3097	683	1608	647	942	1011	1004	3328	3913	4748

The core of the three groups is stable over time: Cluster 1 contains the majority of U.S. companies, cluster 2 tends to collect most Western European companies, and cluster 3 consists almost exclusively of Japanese companies. The remaining Asian companies tend to group with European companies in the earlier period and with U.S. companies in the later period. Over time,

Cluster 2 drops almost all companies from countries that do not participate in European Monetary Union (EMU); 99 percent of the companies in cluster 2 in the final period are incorporated in EMU countries. The clusters thus seem to capture major world economic regions.

Table 2 shows the distribution of companies across sectors and clusters. Companies from all sectors spread across several clusters, and only utilities and energy companies seem to favor one cluster consistently. The other sectors have a large proportion of companies in at least two clusters. In the 1995 to 1999 period, companies in the consumer goods sector tend to concentrate in cluster 1. However, these companies do not represent the majority of companies in that cluster. In none of the three periods is one cluster dominated by a particular sector; rather, companies from one sector always comprise less than one third of any given cluster. Clusters thus do not seem to represent sectors.

Table 2. Sectors and Clusters

Sector	Cluster 1			Cluster 2			Cluster 3			Total		
	89-93	92-96	95-99	89-93	92-96	95-99	89-93	92-96	95-99	89-93	92-96	95-99
Consumer	543	449	1024	201	478	176	218	232	242	962	1159	1442
Capital	355	275	571	83	247	110	285	306	307	723	828	988
Basic	292	172	529	118	319	141	252	265	253	662	756	923
Finance	203	137	533	229	414	154	128	142	139	560	693	826
Utilities	162	140	192	21	62	37	16	18	17	199	220	246
Energy	102	85	148	4	39	10	11	12	13	117	136	171
Transportation	46	36	100	27	49	19	32	36	33	105	121	152
Total	1703	1294	3097	683	1608	647	942	1011	1004	3328	3913	4748

The results indicate that company clusters contain companies from many countries and sectors. However, companies from most countries fall into the same group while companies from most sectors fall across a number of groups. This evidence suggests that equity returns naturally group companies by country rather than by sectors.

5 Conclusion and implications

Previous research has addressed the issue of country versus sector effects on company equity returns. The results of these studies differ, depending on the choice and definition of the regression model, as well as the breadth and length of the data series, not allowing for any final conclusion. Using an entirely different method can validate or disprove these earlier findings. In this study, the patterns of company returns over time are used to mathematically determine groups of companies that are alike.

First, the split- and dividend-adjusted monthly returns for 3328 to 4748 international companies are calculated and normalized. Second, the Pearson distance between all companies is calculated for three five-year periods. Third, each company is allocated to its own cluster and the clusters are fused in the order that minimizes the sum of within-cluster distances at each step. Finally, the number of clusters is determined and the cluster composition is examined.

The constituents of the naturally occurring company groups are examined with respect to their country of incorporation and sector of main activity. The company clusters typically contain companies from many countries and sectors. However, companies from most sectors fall across a large number of groups, while companies from most countries fall into a small number of groups. This evidence suggests that companies naturally group by country rather than sectors during all three periods. These results indicate that international investors should continue to focus on country asset allocation.

References

ARNOTT, R.D. (1980): Cluster Analysis and Stock Price Comovement. *Financial Analysts Journal, 37, 6*, 56-62.

BACA, S.P., GARBE, B.L., and WEISS, R.A. (2000): The Rise of Sector Effects in Major Equity Markets, *Financial Analysts Journal, 56, 5*, 34-40.

BECKERS, S., CONNOR, G., and CURDS, R. (1996). National versus Global Influences on Equity Returns, *Financial Analysts Journal, 52, 2*, 31-39.

BECKERS, S., GRINOLD, R., RUDD, A., and STEFEK, D.(1992): The Relative Importance of Common Factors across the European Equity Markets, *Journal of Banking and Finance, 16, 1*, 75-95.

CAVAGLIA, S., BRIGHTMAN, C., and Michael AKED, M. (2000): The Increasing Importance of Industry Factors, *Financial Analysts Journal, 56, 5*, 41-54.

FARRELL, J.L. (1974): Analyzing Covariation of Returns to determine Homogeneous Stock Groupings, *Journal of Business, 48, 2*, 186-207.

GRIFFIN, J.M., and G. Andrew KAROLYI, G.A. (1998): Another Look at the Role of Industrial Structure of Markets for International Diversification Strategies, *Journal of Financial Economics, 50, 3*, 351-373.

HESTON, S.L.and ROUWENHORST, K.G. (1994): Does Industrial Structure Explain the Benefits of International Diversification?, *Journal of Financial Economics, 36, 1*, 3-27.

KING, B.F. (1966): Market and Industry Factors in Stock Price Behavior, *Journal of Business, 39, 1*, 139-199.

LIVINGSTON, M. (1977): Industry Movements of Common Stocks, *Journal of Finance, 32, 3*, 861-874.

MEYERS, S.L. (1973): A Re-Examination of Market and Industry Factors in Stock Price Behavior, *Journal of Finance, 28, 2*, 695-705.

MILLIGAN, G.W. (1996): Clustering Validation: Results and Implications for Applied Analyses.. In P. Arabie, L. Hubert, and G. De Soete (Eds.): *Clustering and Classification.* River Edge, NJ: World Scientific Press.

SELL, C.W.: Finding Homogeneous Groups of International Companies Using Different Types of Clustering Processes. In: *Studies in Classification, Data Analysis, and Knowledge Organization*, Springer-Verlag, (to be published).

Problems of Classification in Investigative Psychology

Paul J. Taylor, Craig Bennell, and Brent Snook

Department of Psychology, University of Liverpool,
Eleanor Rathbone Building , L69 7ZA Liverpool,
United Kingdom, pjtaylor@liv.ac.uk

Abstract. Problems of classification in the field of Investigative Psychology are defined and examples of each problem class are introduced. The problems addressed are behavioural differentiation, discrimination among alternatives, and prioritisation of investigative options. Contemporary solutions to these problems are presented that use smallest space analysis, receiver operating characteristic analysis, and probability functions.

1 Introduction

Investigative Psychology (IP) is a growing discipline that studies the complex interaction between offender, victim and environment with the purpose of developing models of behaviour that can be used to provide actuarial support to police inquiries. Research in this area is based on the assumption that psychologically important information about an offender may be acquired by analysing and interpreting the patterns of behaviour that emerge during their criminal activity. This premise means that many questions in IP are essentially problems of classifying the differences in offender behaviour and offender information, as well as the correspondence between these two domains. Table 1 presents a framework for conceptualising the central problems of classification in IP, and gives examples of these problems to introduce the reader to the types of data available for analyses. Problems differ in form according to whether the data are individually assigned a score from a meaningful common measurement range (Ordering), or transformed to allow partitioning into a number of defining classes (Grouping). Problems differ in reference depending on whether data is classified through an independent theory-driven interpretation of content (Unsupervised) or with respect to some measured external criterion (Supervised). These distinctions combine to form three kinds of classification problems, unsupervised differentiation of subgroups, supervised discrimination among alternatives, and a more general supervised/unsupervised prioritisation of investigative options. Boundaries among these three kinds of problems are not mutually exclusive, and experience has shown that categories identified through differentiation are often the stimulus for hypotheses about effective discriminators, and findings from both these problems are precursors to prioritisation.

		Subject of Classification	
		Unsupervised	**Supervised**
Form of Classification	**Grouping**	Type I: Differentiation	Type II: Discrimination
		Patterns of offender behaviour	Linking crimes
		Types of offence	Statement validity
		Types of interviewing	Threat credibility assessment
	Ordering	Type III: Prioritisation	
		Enquiry (resource) management	Geographical profiling
		Filtering calls to service	Risk assessment
		Targets for war against terrorism	Suspect prioritisation

Table 1. The three kinds of classification problems in Investigative Psychology.

The remainder of this paper presents examples of how current researchers are attempting to derive adequate solutions to each kind of classification problem. As is typical in IP, the solutions offered draw on disparate methods, but common to each is an emphasis on making minimal assumptions of the data. Such intrinsic analyses Shye (1998) are essential to IP because the data, collected from police investigations, are often incomplete, ambiguous, and inconsistent across cases, thus making the assumptions of many conventional methods misleading and unrealistic.

2 Behavioural differentiation

IP was originally motivated by the problem of classifying the variety of ways in which an offender interacts with a victim during an offence. Research has sought to effectively differentiate a number of subgroups of highly co-occurring offence behaviours, where the number of groups is hypothesised from theoretical explanations of criminal motivation. Any support for such grouping substantiates the proposal that offenders bring different interpersonal styles to criminal activity. Aside from theoretical interest in individual differences, an effective classification of offence behaviour is valuable to investigative research because it is the necessary first step in linking variation in crimes to differences in the people who commit them. The challenge, then, is to take information regarding the occurrence (or not) of a set of B behaviours (e.g., hitting, threatening) over N offences and rearrange the resulting dichotomous matrix B x N so that classes of behaviours can be partitioned into psychologically meaningful groups.

Although many techniques are available for analysing the structure of a B x N matrix, the currently favoured method in IP is non-metric multidimensional scaling, and in particular Smallest Space Analysis (SSA-I). The output of SSA-I represents the rank order of correlations among behavioural variables as distances between their representative points in a geometric configuration, such that the further apart two behaviours on the plot the less likely it is that they were both used by the same offender. Since the SSA-I configuration represents the interrelationships among behaviours in a single solution space and without reference to any arbitrary dimension or cluster Shye (1998), the dominant groups of criminal actions may be identified through a more relaxed criterion; coherent regions of behaviours with substantively similar meanings. The utility of this regional approach is made evident by the growing body of publications which, despite the potential high degree of variation between offenders and offences, have been remarkably consistent in finding a three-fold circular radex pattern to the interrelationships among crime scene behaviours, both in serious (e.g., homicide and arson) and mass (e.g., burglary and juvenile delinquency) crimes Canter (2000).

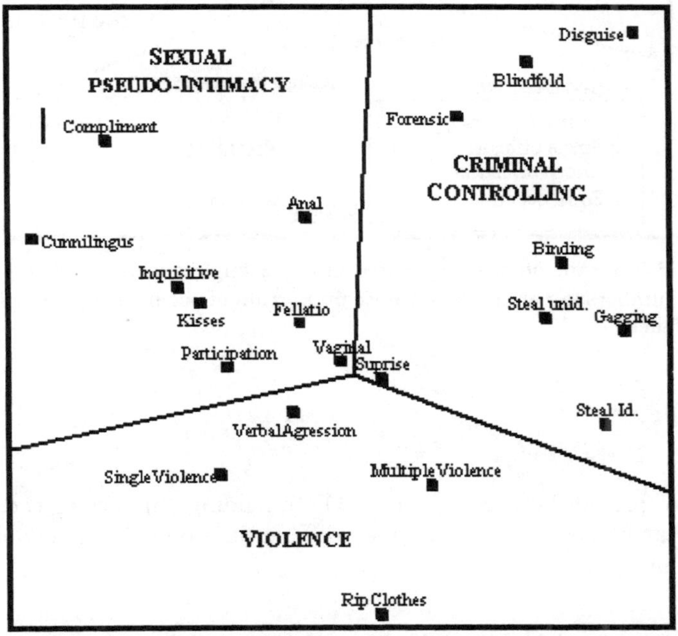

Fig. 1. SSA-I of 20 behaviours in 188 rape offences with regional interpretations. Coefficient of alienation = 0.15 in 15 iterations.

As an example of this approach to differentiating criminal behaviour, Figure 1 shows a 2-dimensional SSA-I configuration of 20 behaviours committed in 188 rape offences. The labels associated with each point correspond to one of the 20 behaviours defined in a content dictionary available from the authors. By drawing on previous research and theory to interpret the configuration, support was found for a regional classification of offenders behaviour into three themes: Sexual-Pseudo-Intimacy, Criminal-Controlling, and Violence. Evidence for these three interaction themes replicates the predominant

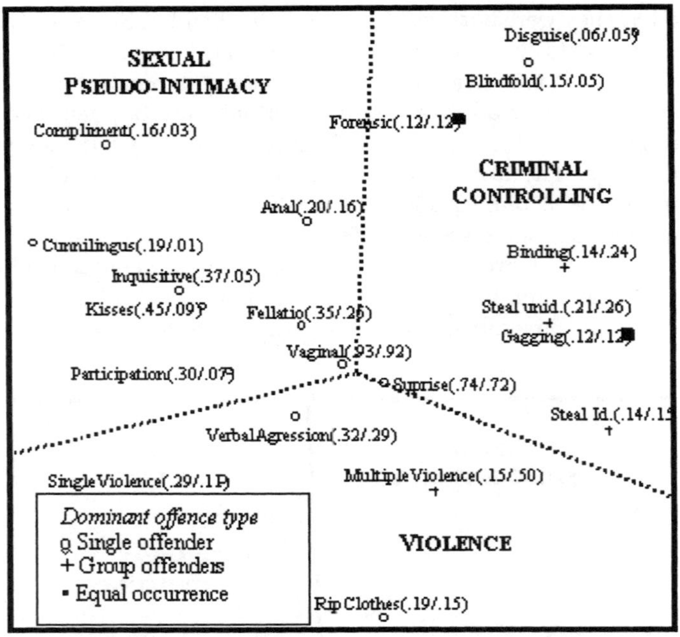

Fig. 2. SSA-I of rape offences separated into 112 single offender and 76 group rape offences. Numbers in parentheses denote proportion of occurrence for single offences / gang offences.

findings of previous studies Canter (2000), lending support to the proposal that similar behavioural themes may differentiate offender behaviour across many crimes.

The full potential of the SSA-I model emerges when the relationship between behaviours and external aspects of the data are explored. For example, of the 188 rape offences used in the SSA-I shown in Figure 1, 112 were committed by single offenders while the remaining 76 were gang rapes. Given the potential for differences between single offender and gang behaviour, Figure 2 shows the identical SSA-I plot where each behaviour is labelled according to whether it occurred relatively more often within single or group offences. As is shown in Figure 2, all of the behaviours that occurred more frequently in gang offences are situated towards the right side of the SSA-I space. This provides some initial evidence to suggest that group offences are predominantly controlled attacks on individuals that are not obviously motivated by sexual intent but often do involve extreme use of violence.

3 Discrimination

A second type of classification problem in IP requires a decision-maker to discriminate between two alternatives. Such situations exemplify a highly structured classification problem in which two possible outcomes are correct (hits and correct rejections) and occur when decisions correspond with reality, while two other outcomes are incorrect (false alarms and misses) and occur when decisions do not correspond with reality. Making a correct decision is difficult because evidence is typically ambiguous and can often arise for both alternatives Swets et. al. (2000). One goal of IP, therefore, is to identify methods that improve the decision-makers ability to identify unambiguous discriminators and set appropriate decision thresholds that define how much evidence needs to be present before particular decisions are made.

One such task involves determining whether two crimes were committed by the same offender based only on information from the crime scene Grubin et al.(2001). Previous approaches to the problem have had limited success because they do not easily allow decision-makers to evaluate the discriminatory power of various forms of behavioural evidence, or to assess how different decision thresholds affect discrimination accuracy. More recently, receiver operating characteristic (ROC) analysis has been proposed as a potential method for linking crimes Bennell (2001). Given information about a set of behavioural evidence over a number of crimes, the ROC approach begins by deriving the probability that any crime pair is linked for each item of evidence. Multiple decision thresholds are then set along this continuum of probabilities for each item, whereby any crime pair receiving a probability score above the specified threshold are classified as linked. One can then calculate the conditional probabilities of hits (pH) misses (pM), correct rejections (pCR) and false alarms (pFA) from their respective frequencies (e.g., pH=hits/(hits+misses)) and plot these probabilities on a graph as a function of the different decision thresholds Swets et al. (2000). The height of the resulting ROC curve relates to the ambiguity of the evidence used, such that the proportion of area lying beneath the curve (A) can be used as a measure of overall discrimination accuracy (perfect discrimination = 1.0, chance discrimination = 0.5). Threshold-specific accuracy measures can then be obtained by examining single points along the ROC curve.

In order to provide an empirical example of the ROC approach, behavioural information from 133 solved burglaries committed by 29 offenders was collected from a UK police force. The level of similarity between every combination of two crimes was calculated as a function of a variety of behavioural variables, including across crime distances, entry behaviours, internal behaviour, and property stolen. These 'discriminator' measures were then used in logistic regression models to calculate the probability that each crime pair was linked, and ROC curves were constructed from these probabilities using the previously described procedure. As the ROC curves in Figure 3 indicate, across crime distances are the most effective feature for classifying burglaries as linked vs. unlinked (A=0.84), though effectiveness depends largely on the threshold adopted. The practical significance of such a finding is clear when one considers how many more hits (or how many less false alarms) will be made at a particular decision threshold depending upon the evidence used for the task Swets et al. (2000). For example, a police force may decide, based on an evaluation of their available resources, that an appropriate decision threshold is one that prevents them from exceeding pFA=0.30. At pFA=0.30 an investigator would correctly identify 34 more linked crime pairs for every 100 crime pairs if across-crime distances were drawn on instead of entry behaviours (i.e., the difference between pH=0.82 and pH=0.48).

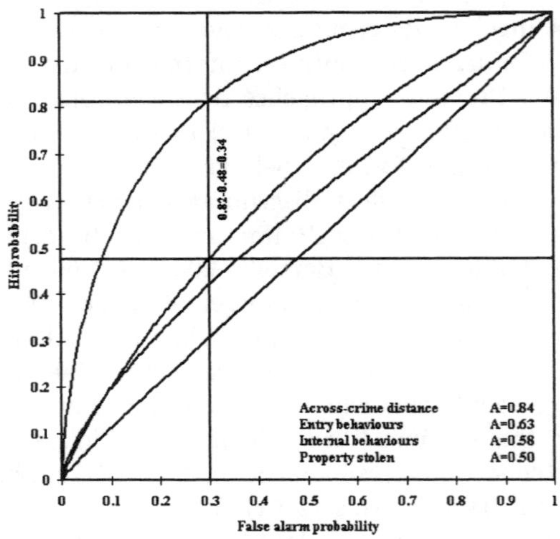

Fig. 3. Using ROC curves to classify burglaries as linked vs. unlinked.

4 Prioritisation

The third classificatory goal of IP is to identify how police investigations can make better use of resources. This goal is accomplished through some form of ordering or weighting of options, where options with a higher prioritisation are considered more likely to yield investigative success. Prioritisation is an intriguing problem that retains individual observations as the object of classification (i.e., each observation is a separate class) and instead requires modelling of the inter-relationships among these observations. This task is manageable in many investigative problems because observations must only be assigned to positions along a single measurement scale; typically the probability of detecting the offender.

One prescient example of this probabilistic modelling comes from research developing profiling systems that aim to classify geographic areas according to how likely they are to contain the offender's residence. The solution to area prioritisation begins with the assumption that crime site locations are an observed subset of an offender's overall spatial activity, and that the 'activity space' within these offences contains the offender's residence. Based on data of the x and y co-ordinates of offence locations, this activity space is classified by superimposing a grid (defined by the minimum and maximum x and y co-ordinates), and then applying a probability distance function around each of the crime locations to assign cells of the grid a probability value. Given that the accuracy of grid classification rests on the effectiveness of the prescribed function, research aimed at improving geographical profiling systems has focused on comparing the accuracy of different functions.

One popular system known as Dragnet Rossmo (2000) uses a function that assumes that the probability of an offender residing in a particular area decreases with increasing distance from an offence. Prioritisation is carried out by using one function from a family of negative exponential decay functions, defined as:

$$f(d_{ij}) = a \cdot e^{-c \cdot d_{ij}} \tag{1}$$

where $f(d_{ij})$ is the likelihood that the offender will reside at a particular location, d_{ij} is the distance from the centre of the grid cell i to an offence j, e is a coefficient chosen from a basic understanding of decay functions to indicate of the maximum likelihood of finding a home, and c is an exponent, either arbitrarily chosen or predetermined using data from solved cases, that determines the steepness of the function Levine (2000). A second alternative system called Criminal Geographic Targeting Canter et al. (2000) makes the additional assumption that there is an area around each offence where the offender is less likely to live (a buffer zone). This initial increase in probability is modelled using a spline function that starts at zero likelihood of locating the offender's home and increases to a peak likelihood (defined by user), at which point the function follows the negative exponential function given in equation 1. This additional linear part of the function is defined as:

$$f(d_{ij}) = g + b \cdot d_{ij} \tag{2}$$

where g represents the probability of an offence site also being the offender's residence (typically zero), and b is the positive constant that defines the slope of the linear function, and hence how likelihood increases with distance away from the crime site Levine (2000). The left panel of Figure 4 shows the probability curves resulting from these two approaches. In applying these functions around each crime site location, a degree of probability overlap will occur that can be summed to produce an overall probability for each grid cell. The resulting probability surface provides investigators with a method to order the classified areas within the offender's activity space in terms of the likelihood that the area contains the offender's residence.

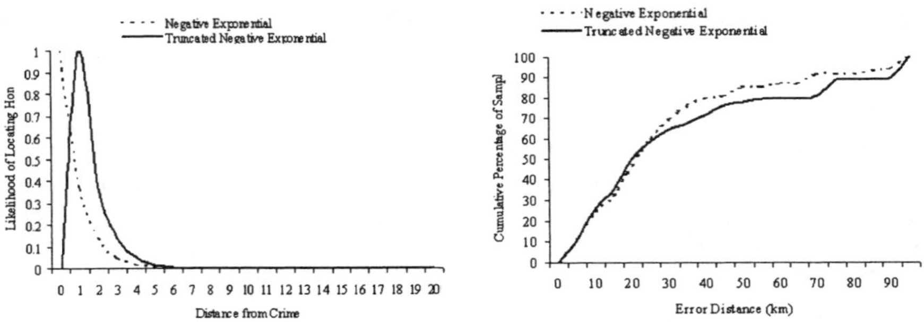

Fig. 4. Two probability distance functions and their relationship to error distance.

As a simple examination of the effectiveness of these functions, a measure known as error distance was calculated for a sample of 68 German serial murder series. Error distance is the distance between the area within the offender's activity space that is classified as most likely to contain the offender's residence and the area that actually contains the offenders residence. The right panel of Figure 4 portrays the effectiveness of the functions by plotting the percentage of sample correctly located as a function of error distance. A steeper curve indicates that home locations were, on average, closer to the point of highest probability and that, consequently, the probability distance function was more efficient. Functions were compared by examining the difference in error distance across all 68 series. This difference was non-significant (Negative exponential: median = 22 km, standard deviation = 119; Truncated negative exponential: median = 18 km, standard deviation = 122 km; Wilcoxon Z = -0.283, *ns*).

5 Conclusions

In addition to answering a variety of theoretical questions, the value of IP rests on the extent to which solutions developed from research can be integrated into effective decision support tools. Before this is done, however, it would be useful for others within the field of classification research to provide solutions to the problems identified above through other techniques. Such comparisons across methodologies will allow research to maximise model effectiveness and so improve the accuracy of inferences drawn from criminal behaviour.

Acknowledgements

PT, CB and BS are Ph.D. candidates in the Department of Psychology at the University of Liverpool. The sections of this paper dealing with discrimination and prioritisation are based in part on dissertations in progress by CB and BS respectively, while PT should be contacted for information regarding behavioural differentiation. The authors thank Louise Porter for providing rape data and Christian Setzkorn for software development and LaTeX formatting. Parts of this research were supported by Overseas Research Student Scholarships awarded to CB and BS.

References

ROSSMO, D.K. (2000). *Geographic Profiling.* Boca Raton: CRC Press.

BENNELL, C. (2001). *Using ROC analysis to link serial crimes.* Liverpool - UK. Paper presented at the Merseyside Police conference on investigative strategies for domestic burglary.

CANTER, D. (2000). Offender profiling and criminal differentiation. *Legal and Criminological Psychology,* 5:23–46.

SHYE, S. (1998). Modern facet theory: Content design and measurement in behavioural research.*European Journal of Psychological Assessment,* 14:160–171.

SWETS, J.A., DAWES, R.M., and MONAHAN J. (2000). Psychological science can improve diagnostic decisions. *Psychological Science in the Public Interest,* 1:1–26.

GRUBIN, D., KELLY, P. and BRUNSDON C. (2001). *Linking Serious Sexual Assaults through Behaviour.* London: Home Office.

LEVINE, N. (2000). *CrimeStat: A Spatial Statistics Program for the Analysis of Crime Indicdent Locations (version 1.1).* Washington, D.C.: National Institute of Justice.

CANTER, D., COFFEY, T., HUNTLEY, M., and MISSEN, C. (2000). Predicting serial killers' home base using a decision support system. *Journal of Quantitative Criminology,* 16:457–478.

List of Reviewers

Phipps Arabie
David Banks
Jean-Pierre Barthelemy
Vladimir Batagelj
Hans-Hermann Bock
Slavka Bodjanova
Paul De Boeck
Edwin Diday
Czesław Domański
Józef Dziechciarz
Anuska Ferligoj
Bernard Fichet
Eugeniusz Gatnar
Allan Gordon
John Gower
Patrick Groenen
Alain Guenoche
Peter Ihm
Krzysztof Jajuga
Henk A. L. Kiers
Carlo Lauro
Berthold Lausen
Bruno Leclerc
Wladimir Makarenkov
Sorin Moga
Fionn Murtagh
Noboru Ohsumi
Józef Pociecha
Jean-Paul Rasson
Gavin Ross
Lars Schmidt-Thieme
Andrzej Sokołowski
Javier Trejos
Rosanna Verde
Maurizio Vichi
Marek Walesiak
Claus Weihs
Klaus-Dieter Wernecke

Index

Druck: Strauss Offsetdruck, Mörlenbach
Verarbeitung: Schäffer, Grünstadt